ECOSYSTEM ECOLOGY RESEARCH TRENDS

ECOSYSTEM ECOLOGY RESEARCH TRENDS

JUNYING CHEN
AND
CHUGUANG GUÔ
EDITORS

Nova Science Publishers, Inc.
New York

For permission to use material from this book please contact us:
Telephone 631-231-7269; Fax 631-231-8175
Web Site: http://www.novapublishers.com

LIBRARY OF CONGRESS CATALOGING-IN-PUBLICATION DATA

Ecosystem ecology research trends / Junying Chen and Chuguang Guô (editors).
p. cm.
ISBN 978-1-60456-183-8 (hardcover)
1. Ecology--Research. 2. Biogeochemical cycles--Research. I. Chen, Junying. II. Guo, Chuguang.
QH541.2.E257 2007
577--dc22 2007048133

Published by Nova Science Publishers, Inc. ✢ New York

CONTENTS

Preface vii

Short Communications 1

Mineral Nutrition of Forest Plants in a Changing Environment—Some Neglected Aspects 3
P. Michopoulos

Towards the Use of Distance Sampling to Monitorize Mountain Ungulates Numbers 9
Jesús M. Pérez and Russell Alpízar-Jara

Research and Review Studies 21

Chapter 1 Soil Ecosystem: Dynamic and Sustainability 23
A. Muscolo and M. Sidari

Chapter 2 New Tools and Strategies for Biomonitoring Marine Ecosystems: Learning from Echinoderms 65
M. Sugni, A. Barbaglio, P. Tremolada and Candia Carnevali

Chapter 3 Interactions between Marine Mammals and Fisheries: Implications for Cod Recovery 107
Claude Savenkoff, Lyne Morissette, Martin Castonguay, Douglas P. Swain, Mike O. Hammill, Denis Chabot and J. Mark Hanson

Chapter 4 The Mathematics of Maintenance and Acquisition 153
Noël Bonneuil

Chapter 5 Ecological Significance of Cover Crop and No Tillage Practices for Ensuring Sustainable of Agriculture and Eco-system Service 177
Masakazu Komatsuzaki

Chapter 6	Integrated Approach for the Assessment of Biotech Soybean Pesticides Impact on Low Order Stream Ecosystems of the Pampasic Region *A.E. Ronco, P. Carriquiriborde, G.S. Natale, M.L. Martin, H. Mugni and C. Bonetto*	**209**
Chapter 7	Ecology and Prevalence of Actynomyces-Antagonists in Soils of Kyrgyzstan *T.D. Doolotkeldieva and N.E. Totubaeva*	**241**
Chapter 8	Assessing and Managing Wildfire Risk in the Wildland-Urban Interface *Tony Prato, Bob Keane and Dan Fagre*	**275**
Chapter 9	Soil Nitrogen Transformation in Different Land Uses in Indian Dry Tropical Forests *N. Tripathi and R.S. Singh*	**299**
Chapter 10	Temperature and Soil Moisture Effects on the Respiration of a Grassland Area *A.F.G. Jacobs, B.G. Heusinkveld and C.M.J. Jacobs*	**313**
Chapter 11	Crab Influences on the Export of Plant Detritus from Salt Marshes and Mangroves: A Review *Jorge L. Gutiérrez*	**327**
Index		**345**

PREFACE

Ecosystem ecology is the integrated study of biotic and abiotic components of ecosystems and their interactions within an ecosystem framework. This science examines how ecosystems work and relates this to their components such as chemicals, bedrock, soil, plants, and animals.

A major focus of ecosystem ecology is on functional processes, ecological mechanisms that maintain the structure and services produced by ecosystems. These include primary productivity (production of biomass), decomposition, and trophic interactions. Studies of ecosystem function have greatly improved human understanding of sustainable production of forage, fiber, fuel, and provision of water. Functional processes are mediated by regional-to-local level climate, disturbance, and management thus ecosystem ecology provides a powerful framework for identifying ecological mechanisms that interact with global environmental problems, especially global warming and degradation of surface water. This new book presents the latest developments in the field from around the world.

There is an increasing amount of literature on the possible effects of global climate change on forests. Most of these studies focus on tree growth and biomass production under a CO_2 enriched atmosphere. Nearly all experimental results suggest that high concentrations of CO_2 will bring about an increase on forest biomass. Few papers address the issue of soil fertility and nutrient cycling. Among these few papers nitrogen and phosphorus are the mostly examined nutrients. Base cations and micronutrients are almost totally ignored. However, forest plant nutrition equally concerns the rest of the nutrients and in this respect scientists have to deal with it. If the plant transpiration rates are reduced, as most researchers agree, then the nutrient uptake by the mechanism of mass flow will also be reduced. The nutrient translocation efficiency will probably play a major role for plant adaptation. It is also imperative to investigate if soil weathering and total deposition of base cations can make up for the increasing nutrient demands by forest species, especially in acid soils. With regard to decomposition processes, it appears that the below ground litter decomposition rates will increase, whereas the opposite will happen for the above ground litter where the C/N ratio will increase. So there will be the strange phenomenon of long residence time for nutrients in the forest floor and short residence time in the subsoil. The consequences for this discrepancy are unknown. Nitrogen saturated forest may appear to have an advantage over N limited forests.

Many management actions involving mountain ungulates require unbiased and precise estimations of their numbers. Topography, among other factors, makes difficult the use of direct methods and to assess the effective area sampled. On the other hand, social behaviour of animals increases the risk to violate the assumption that observations must be independent events. This could be avoided using the group or cluster size as a covariate and estimating density of animals through the density of clusters. In this paper the authors revise the increasing potential of Distance Sampling methodology to estimate mountain ungulate populations. The authors address different problems which compromise some of the assumptions and discuss several topics concerning the importance of assessing effective area sampled, estimation of g(0), and model selection when attempting to adapt the use of Distance Sampling to a three-dimensional scenarios, like mountainous habitats and ungulate populations.

Soil is the foundation of the entire biosphere. Due to different activities of various soil organisms, permanent transitions take place in soil between growth and decay, life and death. Thus in a biological and biochemical way soil ecosystem is the most important interface for the global interchange of matter and energy. Soil quality represents an integral value of the compositional structures and natural functions of soil in relation to soil use and environmental conditions. Among the indigenous soil components, different organisms and especially microorganisms play a key role in ecologically important biogeochemical processes. In that way soil microorganism contribute to the maintenance of the matter and energy transfer in the terrestrial environments. Thus the thin layer of soil covering the surface of the earth represents the difference between survival and extinction for most land-based life. Soil quality is a living dynamic system whose functions are mediated by a diversity of living organisms that require management and conservation. Numerous factors (environmental change; human activities; pollutions, extensive soil cultivation, salinization) can affect soil condition causing several alterations in the development and biochemical activity of soil organism. The organic contents of soil are vitally important for providing energy, substrates, and the biologically diversity necessary to sustain the numerous soil functions. The concept of soil ecosystem quality has recognized soil organic matter as an important attribute that has a vast control on many of the key soil ecosystem functions. However, soil organic matter varies among environments and management systems, generally increasing with higher mean annual precipitation, lower mean annual temperature, higher clay content with an intermediate grazing intensity, with higher crop residue inputs, and native vegetation compared to cultivated management. Chapter 1 reviews the biological properties of soil, much of the material will be familiar, the key difference is the recognition of the vital role soil microorganisms play in recycling, releasing and storing plant nutrients. Some specific issues will be focused on the role and function of organic matter in the sustainability of soil ecosystem, and on the potential of soil microbial biomass to detect management induced changes in dynamic soil quality.

Marine ecosystems are critically threatened by several anthropogenetic compounds (xenobiotics). In the last decades, among the numerous relevant xenobiotics, specific attention has been given to the Endocrine Disrupter Compounds (EDCs): these chemicals are able to alter the endocrine functions potentially causing serious health problems in humans and wildlife. In spite of the global relevance of these pollutants, most past and recent ecotoxicological research has been addressed only to a selected number of animal models, usually vertebrates, whereas invertebrates (particularly marine invertebrates) have been largely neglected. Among the scarcely considered invertebrate taxa, there are common macroinvertebrates such as echinoderms which actually are regarded as representative animals of the marine ecosystem. Chapter 2 will explore the appropriateness of echinoderm models as test species in marine ecotoxicology. Presence and role of the different echinoderm classes in marine ecosystems will be discussed in terms of their ecological relevance and susceptibility to be targets of EDCs. Moreover, echinoderms can offer a wide range of biological processes appropriate for studies on endocrine dysfunction; although basic knowledge on echinoderm endocrinology is still scarce, several works suggested that many physiological processes of these invertebrates are plausibly regulated by humoral mechanisms, often similar to those found in vertebrates. Here, the sensitivity and responsiveness of different biological processes such as reproduction and regenerative development are discussed in relation to EDCs ecotoxicological tests on two echinoderm species: the sea urchin *Paracentrotus lividus* and the crinoid (feather star) *Antedon mediterranea*. The effects of low-dose exposures to several EDCs (triphenyltin, fenarimol, *p,p'*-DDE, methyltestosterone and cyproterone acetate) were investigated by several parameters including regenerative growth, histological pattern, gamete development and gonad maturation. In most of the exposure experiments, evidences of plausible endocrine disruption were found. These experiments were also considered in terms of modelling the fate of the contaminants in the aquaria. The modelling approach in the evaluation of the exposure can help in the planning phase of ecotoxicological tests, and, later, in the comprehension of the results and in the exportability to the natural environment. Finally, problems and weak points encountered with these animal models will be critically discussed together with considerations and perspectives for future research developments.

As explained in Chapter 3, abundance of many Atlantic cod (*Gadus morhua*) and groundfish stocks in the Northwest Atlantic declined to low levels in the early 1990s, resulting in cessation of directed fishing for these stocks, thus ending one of the largest and longest running commercial groundfish fisheries in the world. The stocks of the northern (nGSL) and southern Gulf of St. Lawrence (sGSL) were closed to directed cod fishing from 1994 to 1996 for the nGSL and from 1993 to 1997 for the sGSL, followed by the opening of a small directed fishery in the two systems. In the northern Gulf of St. Lawrence, the ecosystem biomass structure shifted dramatically from one dominated by demersal fish predators (Atlantic cod *Gadus morhua*, redfish *Sebastes* spp.) and small-bodied forage species (capelin *Mallotus villosus*, mackerel *Scomber scombrus*, herring *Clupea harengus*, northern shrimp *Pandalus borealis*) to one now dominated by only small-bodied forage species. The decline of large predatory fishes has left only marine mammals as top predators during the mid-1990s, and marine mammals and Greenland halibut *Reinhardtius hippoglossoides* during the early 2000s. Large changes also occurred in the biomass structure and ecosystem functioning of the adjacent southern Gulf of St Lawrence (sGSL) but they were not as dramatic. Although predatory fishes decreased between the mid-1980s and mid-1990s, and prey consumption by seal species increased considerably, large cod remained among the most important single predators on fish in the sGSL. The changes in top-predator abundance driven by human exploitation of selected species resulted in a major perturbation of the structure and functioning of both Gulf ecosystems and represent a case of fishery-induced regime shift. Overfishing influenced community biomass structure directly through preferential removal of larger-bodied fishes and indirectly through predation release.

Species interactions are central to ecosystem considerations. In marine ecosystems, predation can be the major ecological process affecting fish populations and piscivory is often the largest source of fish removal, usually larger than fishing mortality. In both northern and southern Gulf ecosystems, predation mortality exceeded fishing mortality for most groups in recent years because fishing mortality was intentionally reduced by fisheries closures. Seals have benefited from reduced hunting (harvesting and culling/bounties) since the 1970s. Consumption of fish by marine mammals exceeded consumption by predatory fishes in the two ecosystems in the recent time periods. Since the collapse of groundfish stocks, commercial fisheries and seals have become important predators on predatory fishes – possibly slowing their recovery. In recent years, consumption by seals shifted towards species at lower trophic level (forage fishes and invertebrates), which were also the main target of fisheries. Thus, commercial fisheries and seals may have become important competitors of predatory fishes for the same resource.

Modelling dynamic systems commonly focusses on asymptotic behavior. Is the search for asymptotic patterns appropriate for modelling maintenance and acquisition? In biology, where variability and surprise prevail, moment-by-moment time, not limit sets, matters. Taking the state of a particular system at a particular moment into account requires the substitution and replacement of many of the concepts usual to asymptotic analysis: the replacement of probability by set membership; the replacement of trajectories by attainable sets; the replacement of limit sets by the *C*-viability kernel or capture domain. Rather than convergence to an attractor and dominance of an optimal or evolutionary stable strategy, the focus here is on identifying the right strategy at the right moment that allows a living system to have at least some possibility of perpetuation or collapse. Basic models of ecosystem and population genetics are revisited from this perspective. Notably, rather than proposing a mechanism and examining how it behaves under various constraints, in Chapter 4 the constraints are incorporated in the dynamic system from the start through *C* -viability multipliers. Case studies reveal that even without using the theoretical tools of *C* -viability, some experimenters have empirically invoked *C* -viability multipliers or explored *C* -viability domains.

A doubling of the global food demand projected for the next 50 years poses a huge challenge for the sustainability of both food production and global and local environments. Today's agricultural technologies may be increasing productivity to meet world food demand, but they may also be threatening agricultural ecosystems. Cover crops and no-tillage practices may be effective tools for ensuring sustainable agriculture and eco-system service; because these soil management strategies increase soil organic matter and improve soil biological diversities and activities in intensively managed summer field cropping system. Chapter 5 reviews several aspects of cover crop biology and soil eco-system ecology of no-tillage with cover crop managements in the view point of carbon and nitrogen dynamics, improving soil organic matter and soil biological community dynamics. Recent intensive research revealed that a combined with no-tillage system with cover cropping will be a new soil management strategy because they can scavenge soil residual nitrogen , improve carbon balance and their ecological functions can be utilized to establish an optimal nitrogen cycle due to higher population and diversity of soil fauna organism. Especially, the management technique of subterranean clover reseeding may have the most noticeable economical and ecological benefits, because it can eliminate the cost of cover crop introduction, substitute of fertilizer requirement, and weeds suppression. In order to maximize the potential of no-till and cover crops as a soil management tools, it is essential to obtain and utilize information that expands our ability to manage soil eco-systems and to ensure ecological services.

In the last decades Argentine farmers have massively adopted the genetically modified glyphosate resistant soybean together with no tillage managerial practices. The area cultured with soybeans increased to 15 million ha with consequent higher pesticide consumption. Although recent publications deal with some detailed aspects of pesticide effects on organisms, the overall impact of such vast agricultural intensification remains unreported. Chapter 6 reviews and integrates results from laboratory and field studies within the frame of a major project oriented towards an integrated assessment of soybean agrochemicals impacts on aquatic ecosystems from streams of the region. The main conclusions are: i) chemical analysis of water, sediments and soils indicate low impacts on nutrient levels; ii) all the biota components of the studied ecosystems (flora, insects, fish and amphibians) were affected in different extent after pesticide applications, representing an appreciable stress over the systems. Effects were sometimes low, without noticeable impact on resident populations, as observed in fishes, or being more drastic, populations tend to recover after the episodic toxicity pulses, as observed in the resident invertebrate fauna; iii) of the three most commonly used insecticides, endosulphan fumigation caused the most severe effects, followed by the terrestrial chlorpyriphos and cypermethrin applications; iv) regarding spatial extent of impacts, the herbicide glyphosate drift and runoff caused chlorophyll content reduction on adjacent riparian and, at least 250 meters downstream application, on aquatic macrophyte vegetation. Although effects were followed by later recovery, studied sites showed a rather poor ruderal composition, suggesting that reiterated applications, together with the history of land use eventually impoverished the regional community originally described for the area. Populations of non target invertebrates were the most sensitive ecosystem components, showing strongest effects, with total disappearance of ephemeroptera, odonata, amphipoda as a response to pesticide runoff, eventually at long distance from spray areas, while fish and amphibian impacts were generally circumscribed to adjacent sectors; v) *in situ* exposed experimental populations of insects, fish and amphibians showed good correspondence with observed responses of resident populations within the same watercourses; vi) finally, complementation between laboratory and field studies resulted of great significance for detecting erroneous predictions from the former and lighten observations from the latter. An integrated approach analysis seems to be a promising option towards the understanding of the overall impact of intense soybean farming on aquatic related environments. Improving managerial practices and riparian habitat restorations seems feasible goals for long term sustainable production.

The ecological adaptation and the prevalence of actynomyces of Streptomyces class in soils of Kyrgyzstan have been investigated in Chapter 7. Various types of soil were served as natural sources for isolation Streptomyces: mountain chernozems, mountain dark-chestnut, bright-brown, mountain meadow subalpine, mountain meadow alpine, mountain-meadow-steppe-alpine soils, mountain–wood dark-colored, mountain meadow chernozem like, mountain brown typical, mountain-wood black brown.

The phenotypic (taxonomic) characteristic , microscopy and 16S rRNA gene sequencing methods were used to identify the isolates of Streptomyces. Screening and bioassays methods were used for discovery the microbial secondary metabolites of novel Streptomyces isolates.

The results have shown that Streptomyces spp. were particulary abundant and contained a rich biodiversity in the high mountain- meadow-steppe-alpine soils, high mountain alpine soils, mountain chernozems , where they represented 60% of all soil microorganisms. The climatic conditions, agri-physical regime , chemical and mechanical structure of these soils have a favorable influence on the prosperity of Streptomyces groups. The major factors saving stability and developments of Streptomyces populations in these soils are high percent contents of humus , neutral or alkalescent reaction of soils and a significant accumulation of mineral elements as phosphorus, nitrogen, potassium and calcium. In these conditions, Streptomyces ssp are able to participate in decomposition of high-molecular, inaccessible organic compounds, rather than readily available.

The received data also have shown that the occurrence of antagonistic forms of Streptomyces in soils depends on a degree of cultivation of soils, their agrichemical and genetic properties. The isolates with antagonistic properties were seldom found out in soils subjected to agricultural use. The virgin soils, not used in agriculture were allowed much more cultures with antagonistic properties. The soils from non-polluted landscapes contain the greater number of Streptomyces than soils from the territories polluted with industrial wastes. A total 441 Streptomyces strains were isolated from various soil types of Kyrgyzstan, from them 22 strains were selected for the use in agriculture for biotechnological purposes, 105 strains for pharmacology.

As explained in Chapter 8, wildfires generate ecosystem benefits and socioecological costs. On the benefit side, wildfire is an essential natural process that stimulates vegetation growth, enhances biodiversity, and sustains ecosystem health. On the cost side, large, intense wildfires often cause human injury/mortality, structural damages, commercial timber losses, natural resource degradation, temporary disruption of ecosystem processes, and high suppression costs. Managing wildfire risk in the wildland-urban interface (WUI) involves balancing the benefits and costs of wildfires. Such balancing is likely to become more challenging due to the convergence of future climate change and future sprawl of residential communities into the WUI. Two conceptual frameworks are proposed for simulating how future climate and land use changes might influence wildfire risk in the WUI, and how stakeholders' can optimally respond to future changes in wildfire risk. Four stakeholders or agents are considered: forest managers and wildfire protection agencies; homeowners; land developers/homebuilders; and community/regional planners. The ex ante framework: (1) specifies alternative futures consisting of combinations of future climate change scenarios and future land use change scenarios; (2) uses biogeophysical process models to simulate how alternative futures influence wildfire risk in the WUI; and (3) employs agent-based modeling and multiple attribute evaluation to determine decision behavior that minimizes wildfire risk to stakeholders under each alternative future. Future climate change scenarios are specified by disaggregating the results of general circulation models. Future land use change scenarios are specified using a land use change model that simulates the expansion of residential communities into the WUI for alternative economic growth rates and land use policies. Potential ecosystem responses to future climate change are evaluated using a GIS-based, biogeophysical model that simulates: (1) carbon, water, and nutrient fluxes at the ecosystem scale; (2) climate change in forest stands; (3) tree growth, vegetation structure, and fuel loads in forest stands; (3) wildfire ignitions; and (4) fire behavior. The ex post framework uses adaptive management, biogeophysical simulation, agent-based modeling, and multiple attribute evaluation to identify ways stakeholders can optimally adjust their decision behavior over time in response to evolving changes in wildfire risk. Both frameworks can be used to evaluate hypotheses, such as: (1) whether convergence of future climate change and economic growth would significantly increase wildfire risk in WUIs; (2) whether limiting future urban sprawl into wildland areas would significantly reduce wildfire risk; and (c) whether lower future economic growth would significantly decrease wildfire risk.

As explained in Chapter 9, the effects of conversion of an Indian tropical forest ecosystem into alternate land uses, such as grassland and cropland on plant available N, N-mineralization, microbial biomass N (MBN) and fine root biomass in soil were studied seasonally and total plant biomass and soil particle size were studied annually for two years. The mean annual mineral N and soil microbial biomass, respectively, in three ecosystems varied significantly with season; the values being highest in summer season and lowest in rainy season, while the net nitrification rate and net N-mineralization rate values were maximum in rainy season and minimum in summer season. The MBN was positively related to root biomass and total plant biomass (i.e., above- and below-ground biomass). Interestingly, MBN and inorganic N are reciprocally and nitrification and N-mineralization are directly related to seasonal soil moisture. The soil microbial biomass, nitrification and N-mineralization are negatively related to clay content. The mean annual net nitrification and N-mineralization rates declined significantly after conversion into grassland and cropland by 43 and 54% and net N mineralization by 33 and 46%, respectively.

Above- and below-ground biomass also have had their impact on MBN, and thereby N-mineralization. Thus, in tropical forests, the conversion into different land-uses affects remarkably the nitrogen transformation process in soil. Therefore, there should be a nitrogen management scheme to prevent the nitrogen losses from the derived ecosystems.

As presented in Chapter 10, in an ecosystem the respiration is affected by many biological and physico-chemical factors, of which year round the soil temperature and soil moisture are the most important ones. During six months in 2003, the net ecosystem exchange (*NEE*) of CO_2 was measured with the eddy-covariance technique along with soil temperature and soil moisture profiles. Nighttime surface respiration measurements were analyzed and compared with model calculations using an Arrhenius relation. Not the soil temperature but the surface temperature of the grass cover, fitted best the experimental data. For dry soil wetness conditions but also for very wet soil water conditions a correction function is introduced, which appears to work well. For the soil wetness correction function, the soil water content nearest to the soil atmosphere interface appeared to work best.

While crabs occur in perhaps most salt-marshes and mangroves worldwide, they are rarely considered in studies dealing with tidal fluxes of plant detritus from these coastal wetlands to adjacent waters. However, crabs can affect the production, availability and tidal transport of plant detritus in a variety of ways. This includes controls via both assimilatory-dissimilatory mechanisms (e.g., herbivory, detritivory, seed and propagule consumption) and non assimilatory, non dissimilatory environmental modification (i.e., *physical ecosystem engineering*; e.g., sediment oxygenation via burrows, detritus excavation and burial, detritus trapping into burrows). Given the high densities and activity rates shown by many crab species, their effects on the export of plant detritus are expected to be important relative to overall marsh production. Therefore, the predictive capacity of models dealing with detritus export by salt marshes and mangroves is likely to be enhanced by considering crab influences, as presented in Chapter 11.

SHORT COMMUNICATIONS

In: Ecosystem Ecology Research Trends
Editors: J. Chen and C. Guo, pp. 3-7
ISBN: 978-1-60456-183-8

Mineral Nutrition of Forest Plants in a Changing Environment— Some Neglected Aspects

*P. Michopoulos**
Forest Research Institute of Athens, Terma Alkmanos, Athens 115 28, Greece

Abstract

There is an increasing amount of literature on the possible effects of global climate change on forests. Most of these studies focus on tree growth and biomass production under a CO_2 enriched atmosphere. Nearly all experimental results suggest that high concentrations of CO_2 will bring about an increase on forest biomass. Few papers address the issue of soil fertility and nutrient cycling. Among these few papers nitrogen and phosphorus are the mostly examined nutrients. Base cations and micronutrients are almost totally ignored. However, forest plant nutrition equally concerns the rest of the nutrients and in this respect scientists have to deal with it. If the plant transpiration rates are reduced, as most researchers agree, then the nutrient uptake by the mechanism of mass flow will also be reduced. The nutrient translocation efficiency will probably play a major role for plant adaptation. It is also imperative to investigate if soil weathering and total deposition of base cations can make up for the increasing nutrient demands by forest species, especially in acid soils. With regard to decomposition processes, it appears that the below ground litter decomposition rates will increase, whereas the opposite will happen for the above ground litter where the C/N ratio will increase. So there will be the strange phenomenon of long residence time for nutrients in the forest floor and short residence time in the subsoil. The consequences for this discrepancy are unknown. Nitrogen saturated forest may appear to have an advantage over N limited forests.

Introduction

The rising atmospheric concentration of CO_2 and its effects on forest ecosystems has caused great concern among the scientific community. There have been very good reviews on this subject (Ceulemans and Mousseau, 1994; Saxe et al. 1998; Norby et al. 1999). Part of these

* E-mail address: mipa@fria.gr

reviews refers to the forest plant nutrition, whereas a few authors have published works exclusively on forest plant nutrition in a CO_2 enriched environment (Oren et al. 2001; Kirschbaum et al. 1998; Pepper et al. 2007; Zak et al. 1993). Karnosky (2003) addressed some important issues concerning the mineral cycling in forest ecosystems, especially for N.

The concerns on mineral nutrition of forest plants has risen from the findings that total growth of trees is enhanced under elevated atmospheric CO_2 due to increased rates of photosynthesis. Accordingly, the nutrient uptake must sustain the increased biomass.

The main objective of the present paper is to examine some neglected aspects of the forest plant nutrition.

1. The Relative Significance of Nutrient Uptake and Translocation

I consider it important to present the simple relationships used in balancing nutrient inputs and outputs at the ecosystem level. The hypothesis is that if tree biomass (leaf and trunk) growth rate increases the requirements in nutrients also increases. The requirement is met by both uptake and translocation. Tree stand requirements in nutrients are estimated by adding the amounts of nutrients in live leaf biomass and the use for annual stem and branch annual increment. Tree uptake of nutrients is calculated by summing up the nutrients amounts in litterfall, leachates from foliage and annual increment of trunk and branches. It must be pointed that the below ground "litterfall" can make a substantial proportion of the total litterfall. The difference between N requirement and uptake is the amount of nutrients translocated (Cole and Rapp, 1981). The uptake and translocation efficiency varies among species and tree age. Deciduous trees take up and translocate more nutrients than conifers. Young trees depend more on uptake than translocation. As plants age, the role of translocation becomes more important. The nutrient translocation is of great adaptive significance because it enables plants to reuse these nutrients and thereby can lead to a higher nutrient retention (Aerts and Chapin, 2000). It is certain that this plant capability deserves more research in the future.

2. Will an Increased Root Biomass Take Up More Nutrients?

There is increasing evidence from controlled experiments that plants, under CO_2 enrichment, will invest photosynthates into root growth to ensure a better acquisition of nutrients in infertile soils (Matamala and Schlesinger, 2000; Pritchard et al. 2001). Also the association of tree roots with mycorrhizal symbiosis will be enhanced since available photosynthate could be provided to mycorrhizas (Saxe et al. 1998). High amounts of root biomass release high amounts of organic C into the rhizosphere. The low molecular weight exudates can solubilize P from compounds of exceedingly low solubility (Attiwill and Adams, 1993). It stands to reason that nutrient uptake will be aided by this change. There are three factors that can offset this process. Plants take up most nutrients, apart from P and K, through the process of mass flow (Marschner, 1986). The mass flow is proportional to the transpiration rate. Elevated atmospheric CO_2 concentrations will lead to a reduced stomatal conductance, and therefore transpiration rates. Based on a large number of literature data Kimball (1983) estimated that a doubling of the present CO_2 concentration would result in a reduction of the transpiration rate by 34%. The consequences in nutrient take up are not known but it will happen. Another factor is the available soil depth. This depth cannot increase proportionally to root mass. Iron and clay pan formation can narrow soil depth. More over, there are soils derived from certain parent material, which are nearly always shallow. An example is soils derived from hard limestone widely spread in the Mediterranean zone. Unless plant roots can expand horizontally, small soil depth can seriously offset the increase of root biomass. The third factor is the depletion of nutrients around the root net. This process is particularly important for P. In the highly weathered soils of the tropics the P depletion can become a serious hindrance for an increased plant biomass support.

3. Will the Litter Decomposition Rate Change?

The tree leaf composition is very likely to change in a changing climate. Nitrogen will be reallocated mainly in fine roots, which need a large amount of N compounds (Norby et al. 1992). This means that above ground litterfall will a have a C/N ratio higher than that in the past. Fresh litter with a high C/N ratio will have a slower decomposition rate than usual. On the other hand, the below ground "litterfall" with a large amount of exudates and a N reallocation will probably have a faster decomposition rate than that in the past. So there will be a build up of the forest floor just above the mineral soil and an acceleration of organic matter mineralization below the mineral soil. In these conditions, the elemental residence time in forest floors is difficult to estimate. The balancing of elemental amounts at the ecosystem level will need new formulas for calculation.

4. Base Cations and Trace Elements

The input of base cations in a forest ecosystem is carried out through deposition and weathering. Deposition makes a substantial input percentage in southern Europe (Draaijers et al. 1997). Low amounts of precipitation for prolonged periods can have a profound effect on forests grown on acid soils. On the other hand, the supply of base cations will continue through an increase of chemical weathering rates. Indeed, the high decomposition rate of the below ground "litterfall" will bring about an increase of the carbonic acid production. Accelerated weathering introduces new cations in soil solution for take up but at the same time leaching increases. Hwan OH and Richter (2004) argue that a doubling of atmospheric CO_2 may increase leaching by up to 50%. In poorly buffered soils this process will lead to rapid acidification and depletion of base cations from soil exchange sites. Chemical weathering can also be accelerated by the increased production of organic acids in the rhizosphere soil.

Trace elements are also essential for plant growth. Their acquisition by plants is carried out mainly through chelation by soluble organic compounds. Most of trace elements are easily available at low soil pH. The residence time of Fe and Cu in forest floors is much higher than those of Mn and Zn because of their higher binding strength to soil organic matter. If litter in forest floors becomes less decomposable, the trace elements take up will depend on their availability in mineral soil to a large extent. The abundance of organic acids in the rhizosphere will easily cause chelation of trace elements if they are present. Iron is ubiquitous in mineral soils so there will be no problem for take up. The availability of Cu and Zn will depend on the nature of parent material. If soil parent material cannot supply Cu and Zn through weathering, there might be reduced uptake of these nutrients. The source of Mn in mineral soils is manganese oxides. These oxides are easily depleted in acid soils through their reaction with H^+ ions (Michopoulos, 1995) and Mn^{2+} ions enter soil solution and are taken up by plants. In the long term, under a continuous acidification process, manganese oxides will be too exhausted to supply soil solution with enough Mn^{2+} ions to meet plant demands. So soil acidification caused by the elevated atmospheric CO_2 can bring about a deficiency of Mn in forest plants.

5. Some Opinions on N nutrition

If soils become further acidified, the N mineralization will shift to ammonium production although some nitrification will still continue (De Boer and Kowalchuk, 2001). Reduced nitrification rates will offset the leaching due to the nitrate production predominating in N saturated soils. Nitrogen saturated soils will also narrow the C/N ratio in the above ground litterfall and decomposition will become faster. On the other hand, trees have to compete with microorganisms and herbaceous plants for ammonium uptake. In N poor soils trees having an extended mycorrhizal net will have an advantage. Also, taking up deposition N from tree canopies will be very important.

Conclusions

In this work some less examined issues of mineral nutrition were discussed for a future scenario of elevated atmospheric CO_2. Some conclusions can be drawn, which of course are tentative.

1. The nutrient translocation efficiency will play a vital role for plant adaptation.
2. The enhanced root biomass can cause increased nutrient take up but reduced transpiration rates and shallow soil depths can offset this process.
3. Reduced and increased decomposition rates of the above ground and below ground litterfall, respectively, will have a profound effect on nutrient cycling.
4. Unbuffered acid soils will be further acidified and more prone to present deficiencies of both base and trace elements.
5. Nitrogen saturated forest ecosystems appear to be in a better position to deal with the slow down of litter decomposition rates. All plant mechanism to take up N in infertile soils will be of avail to survive.

References

Aerts, R. & Chapin, F.S. (2000). The mineral nutrition of wild plants revisited: A re-evaluation of processes and patterns. *Adv. Ecol. Res.*, **30**, 1-67.

Attiwill, P.M. & Adams, M.A. (1993). Nutrient cycling in forests. *New Phytol.*, **124**, 561-582.

Ceulemans, R. & Mousseau, M. (1994). Effects of elevated atmospheric CO_2 on woody plants. *New Phytol.*, **127**, 425-446.

Cole, D.W., & Rapp, M. (1981). Elemental cycling in forest ecosystems. In: D.E. Reichle, (Editor). *Dynamic properties of forest ecosystems* (341-409). London, Cambridge University Press.

De Boer, W., & Kowalchuk, G.A. (2001). Nitrification in acid soils: micro-organisms and mechanisms. *Soil Biol. Biochem.*, **33**, 853-866.

Draaijers, G.P.J., Van Leeuwen, E.P., De Jong, P.G.H.& Erisman, J.W.(1997). Base-cation deposition in Europe-Part II. Acid neutralization capacity and contribution to forest nutrition. *Atmos. Environ.*, **31**, 4159-4168.

Hwan OH, N. & Richter, D.D. (2004). Soil acidification induced by elevated atmospheric CO_2. *Global Change Biol.*, **10**, 1936-1946.

Karnosky, D.F. (2003). Impacts of elevated atmospheric CO2 on forest trees and forest ecosystems: Knowledge gaps. *Environ. International*, 161-169.

Kimball, B.A. (1983). Carbon dioxide and agricultural yield: An assemblage and analysis of 430 prior observations. *Agron. J.*, **75**, 779-788.

Kirschbaum, M.U.F. et al. (1998). Modelling forest-growth response to increasing CO_2 concentration in relation to various factors affecting nutrient supply. *Global Change Biol.* **4**, 23-41.

Marschner, H. (1986). Mineral nutrition of higher plants. London, Academic Press.

Matamala, R. & Schlesinger W.H. (2000). Effects of elevated atmospheric CO_2 on fine root production and activity in an intact forest ecosystem. *Global Change Biol.*, **6,** 967-979.

Micjhopoulos, P. (1995). Studies on manganese cycling in forest soils. PhD Thesis. University of Aberdeen.

Norby, R.J., Gunderson, C.A., Wullschleger, S.D., O'Neiill, E.G. & McCracken, M.K. (1992). Productivity and compensation response of yellow poplar trees on elevated CO_2. *Nature*, **357**, 322-324.

Norby, R.J., Wullschleger, S.D., Gunderson, C.A., Johnson, D.W. & Ceulemans, R. (1999). Tree responses to rising CO2 in field experiments: implications for the future forest. *Plant Cell Environ.*, **22**, 663-714.

Oren, R. et al. (2001). Soil fertility limits carbon sequestration by forest ecosystems in a CO2-enriched atmosphere. *Nature*, **411**, 469-472.

Pepper, D.A. et al. (2007). Simulated mechanisms of soil N feedback on the forest CO_2 response. *Global Change Biol.*, **13**, 1265-1281.

Pritchard, S.G., Davis, M.A., Mitchell, R.J., Prior, S.A., Boykin, D.L., Roget, H.H. & Runion, G.B. (2001). Root dynamics in an artificially constructed regenerating longleaf pine ecosystem are affected by atmospheric CO2 enrichment. *Environ. Experim. Botany*, **46**, 55-69.

Saxe, H, Ellsworth, D.S. & Heath, J. (1998). Tree and forest functioning in an enriched CO_2 atmosphere. *New Phytol.*, **139**, 395-436.

Zak, D.R., Pregitzer, K.S., Curtis, P.S., Teeri, J.A., Fogel, R. & Randlett, D.L. (1993). Elevated atmospheric CO_2 and feedback between carbon and nitrogen cycles. *Plant Soil*, **151**, 105-117.

In: Ecosystem Ecology Research Trends
Editors: J. Chen and C. Guo, pp. 9-20

ISBN: 978-1-60456-183-8

TOWARDS THE USE OF DISTANCE SAMPLING TO MONITORIZE MOUNTAIN UNGULATES NUMBERS

***Jesús M. Pérez*[1,a] *and Russell Alpízar-Jara*[2,b]**

[1]Departamento de Biología Animal, Biología Vegetal y Ecología, Universidad de Jaén, Campus Las Lagunillas, s.n., E-23071, Jaén, Spain

[2]Centro de Investigações em Matemática e Aplicaõés, Departamento de Matemática, Universidade de Évora, Rua Romão Ramalho, 59, 7000-671, Évora, Portugal

Abstract

Many management actions involving mountain ungulates require unbiased and precise estimations of their numbers. Topography, among other factors, makes difficult the use of direct methods and to assess the effective area sampled. On the other hand, social behaviour of animals increases the risk to violate the assumption that observations must be independent events. This could be avoided using the group or cluster size as a covariate and estimating density of animals through the density of clusters. In this paper we revise the increasing potential of Distance Sampling methodology to estimate mountain ungulate populations. We address different problems which compromise some of the assumptions and discuss several topics concerning the importance of assessing effective area sampled, estimation of g(0), and model selection when attempting to adapt the use of Distance Sampling to a three-dimensional scenarios, like mountainous habitats and ungulate populations.

Introduction

Line transect sampling evolved from strip sampling (plot sampling), which assumes that all objects within the strip are detected. Distance sampling methodology allows to relaxe the critical assumption of probability of detection being equal to 1 within a plot and uses the number of detected objects (animals or groups of animals, pellets, nests, sound calls, and so on) and the perpendicular (or radial) distances to a line transect (or point) (figure 1). This approach is specifically used to produce less biased estimates of density than those obtained

[a] E-mail address: jperez@ujaen.es
[b] E-mail address: alpizar@uevora.pt

when assuming perfect detectability. Statistical inference of this methodology rests on the validity of several assumptions (Burnham et al., 1980; Barry & Welsh, 2001; Buckland et al., 2001; Borchers et al., 2002): (1) objects located just on the line or point are always detected; if g(x) is the probability of detection of an object at a given distance x, then g(0) = 1, and this probability decreases as distance increases; (2) objects are detected at their initial location, prior to any movement in response to observer, and are not counted more than once; (3) distances (and angles when appropriate) are measured accurately for ungrouped data, or objects are correctly counted within the proper range or distance category if grouped data are collected; and (4) observations must be independent events. Of course, other implicit assumptions related to design and logistic considerations are needed. For instance, it is assumed that observers have enough experience to detect and recognize animals belonging to the target species. Some of these assumptions are usually unverifiable from distance data, and to know how they affect the biases and the precision of the estimator becomes difficult or even impossible (Barry & Welsh, 2001).

Photo by Jesús M. Pérez.

Figure 1. In Distance Sampling the variables of interest are the number of animals observed and their radial distance (r) (if observation is made from a point) or perpendicular distance (x) to the transect line.

Distance Sampling has become a versatile and powerful tool for estimating abundance of animal populations and can also be used as an indirect method (Marques et al.; 2001; Williams et al., 2002). Its balance between cost and effectiveness is good compared to other methods and has theoretical advantages over the strip-transect methodology or total counts (Burnham et al., 1985). Moreover, the implementation of this method does not require handling of animals from the population under study (Southwell & Weaver, 1993).

Distance Sampling has also been used in three-dimensional scenarios, such as underwater for fish (Kulbicki & Sarramégna, 1999) in which volume, not area, is sampled from lines and/or points. Regarding wild ungulates inhabiting mountainous habitats there are only few references in scientific literature all of which involve Iberian ibex, Capra pyrenaica (Escós & Alados, 1988; Palomares & Ruiz-Martínez, 1993; Pérez et al., 1994; Pérez et al., 2002 a). In this kind of habitats, to sample straight transects may become difficult or impossible, even from the air. Under these circumstances, we could choose between point transect sampling (despite of beign less efficient than line transect sampling) (Buckland et al., 2001) or curved transects (Hiby & Krishna, 2001). Moreover, when working in uneven terrain, Buckland et al. (2001) recommend to use projections onto a horizontal plane (map) of transect lines, detection distances and the study area as well, in order to facilitate the analysis of data and density estimation.

However, Southwell & Weaver (1993) pointed out the disadvantage of this method considering the high number of analytical decisions that are needed. For instance, which estimator to use, whether data must be analyzed in grouped or ungrouped form, if in grouped form how many groupings to use, whether use truncated data or not, and whether analysis must be made on sightings of individuals or clusters of animals. These authors evaluated the effect of analytical procedures, like data grouping and data truncation on accuracy and precision of density estimation. They obtained more accurate and precise estimations when data were analysed in ungrouped form and they did not observe effects of data truncation on bias or precision. Nevertheless, these decisions are also needed for all type of terrains.

Most mountain ungulates are Caprinae (Mammalia: Artiodactyla). Despite of over 70% of them having some serious conservation problems (Shackleton, 1997), they support certain degree of exploitation and are valuable big game species. Within this context accurate and precise estimates of their population numbers are required. For this purpose different direct methods have been used (Pérez et al., 2002 b), but if assumptions of such methods are met or not are issues rarely addressed or discussed in the mountain ecology literature.

Here, we revise the possibilities of applying Distance Sampling to estimate mountain ungulates numbers when monitoring their populations and certain limitations of this method which require some adaptation to work in mountainous habitats. In what follows, we discuss issues such that survey design, effective area sampled, g(0) estimation, estimator robustness, model selection, and the analyses of clustered data when applied to surveys conducted in mountainous terrain.

Survey Design

One possible modelling framework for estimating density is to use a design-based approach (randomness arises from the random selection of the transects) (Barry & Welsh, 2001). This approach can be easily implemented by using parallel transects among other design strategies (Buckland et al., 2001). Program Distance (Thomas et al., 2005) allows to incorporate geographical information and survey design into our analyses (figure 2), although this last option has already limitations. In our case, this could lead to select transects difficult to be walked and even dangerous: security of observers is an important issue when working in mountainous terrain. In our setting (e.g., Sierra Nevada mountain range), transects are fixed

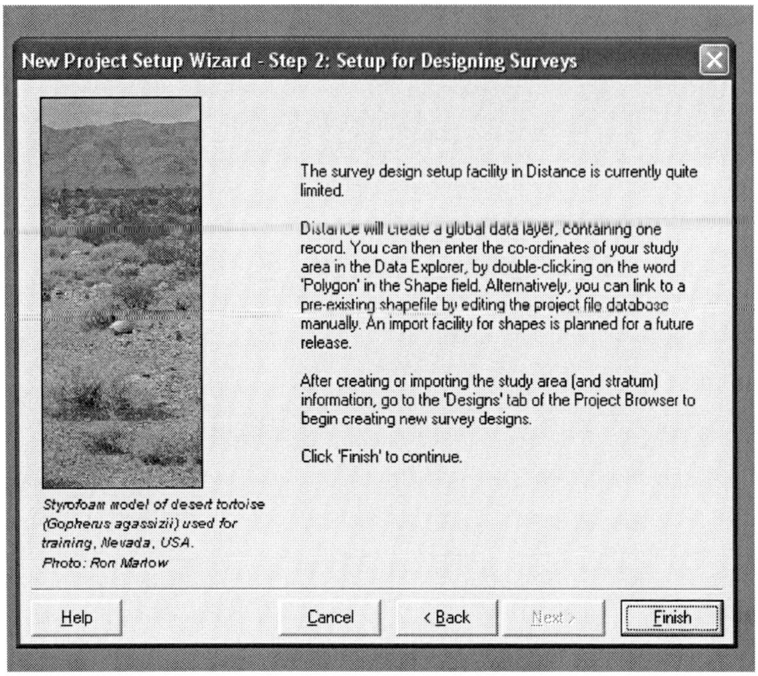

Figure 2. We can design our survey from the current version of Distance and to use geographic information. This is one of the options to be developed and improved in the future.

Photo by Jean-Paul Crampe, Park National des Pyrénées, France.

Figure 3. When populations surveyed contain marked animals the capture-recapture (mark-resighting) method can be combined with Distance sampling.

for historical and logistic limitations: they were designed including stratification (regarding height, orientation, and so on) and without prior knowledge on the ibex distribution, but which transect to sample on a given occasion is a random event is an issue to be properly addressed.

Effective Area Sampled

In a bi-dimensional scenario the covered area can be easily obtained as the product 2wL (with $L = \sum l_i$ is the total transect length and w the width of the area searched on each side of the line transect), or $k\pi w^2$ (where k is the number of covered plots searched around a point of radius w). Very often data are not included in the analysis beyond the truncation distance w (Buckland et al., 2001). Statistical inferences are commonly drawn from what is observed in the covered area to estimate the effective area sampled (Ramsey et al., 1987). Usually the estimated effective area by Distance methodology is smaller than that the real covered area (sampled area), thus resulting in an overestimation of density.

In mountainous terrain visibility can vary between different sections of a transect or between different orientations from a point because of topography and vegetation cover (although forested areas should be avoided as sampling units or parts of transects and/or points) (see figure 4c). Usually the real covered area (sampled area) is smaller than that estimated by Distance, thus resulting in an overestimation of density.

Conventional Distance sampling methodology assume that the detectability function is symmetric in relation to the transect line (Barry & Welsh, 2001). However, asymmetric distributions near a transect line are relatively frequent in mountainous areas.

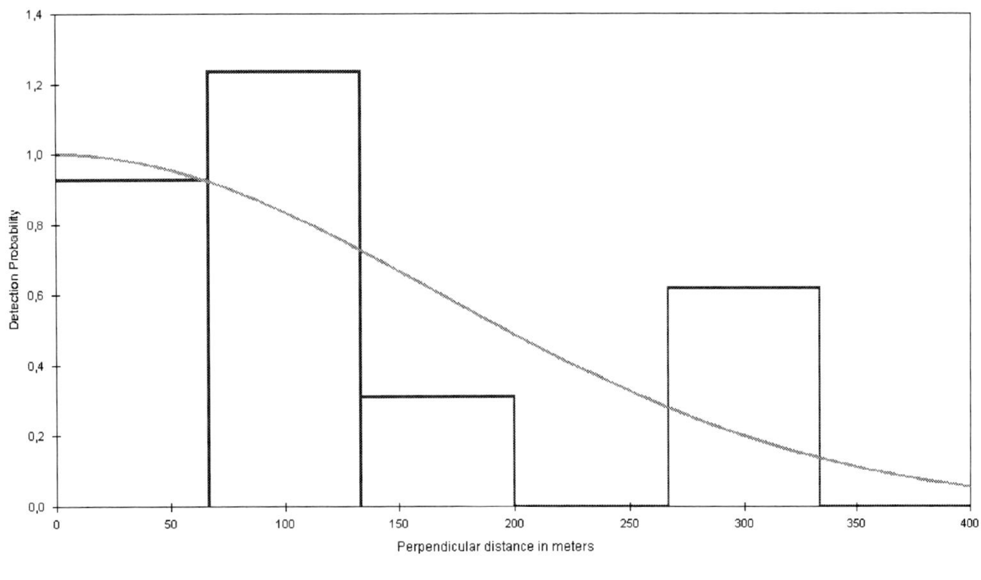

Figure 4a. Histrogram of perpendicular distances for clusters of Iberian ibex, Capra pyrenaica, from Peneda-Geres National Park–Parque Baixa Limia Serras do Xures (northern Portugal-southeastern Galicia, Spain, spring 2007). When number of observation (n) becomes small, truncation of data is more problematic.

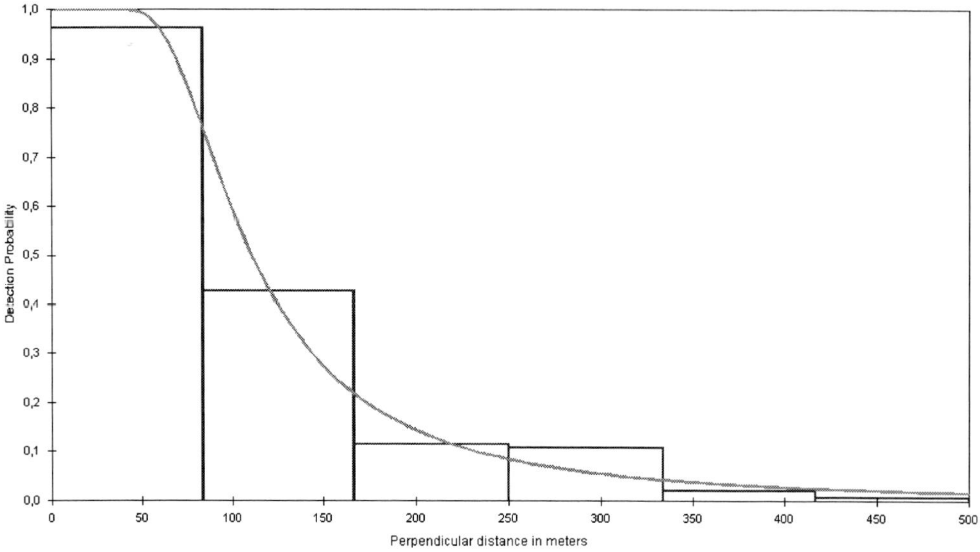

Figure 4b. Histrogram of perpendicular distances for individuals of Iberian ibex, Capra pyrenaica, from Sierra Nevada massif (southern Spain, summer 1995). Data are fitted to the hazard-rate function, which has been constrained to become strictly monotonically non-increasing.

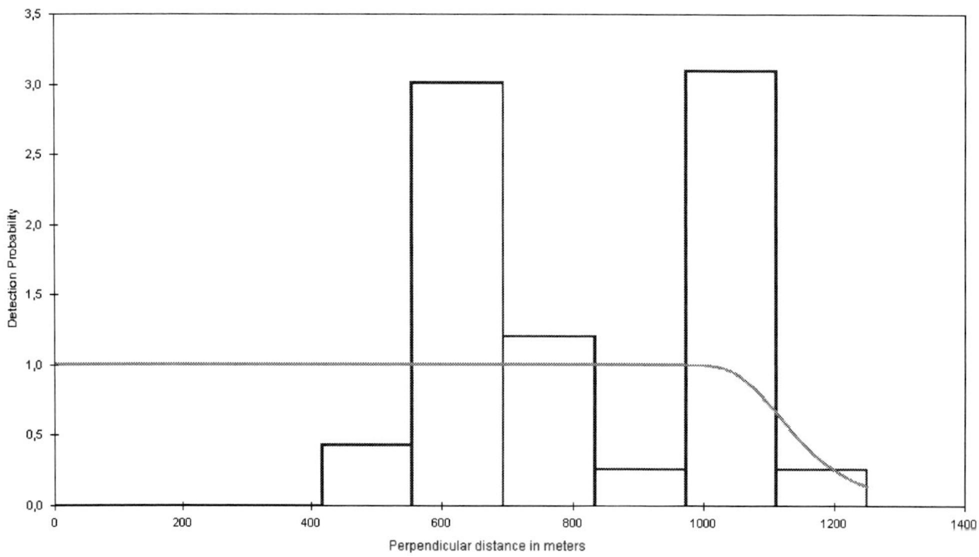

Figure 4c. Histrogram of perpendicular distances for individuals of Pyrenean chamois, Rupicapra pyrenaica, from Pyrenees National Park (southern France, fall 2002). Topography can influence probability of detection: in this case, we have no visibility in a range of 0-more than 400 m from the centre of the transect.

Estimation of g(0)

To pool distance data across transects when estimating the detection function: g, even when detectability varies by transect the estimation of abundance is still reliable. The larger sample sizes gained from pooling lead to more precise estimation of g (Fewster et al., 2005), but variance cannot be calculated empirically. We can assume then that animal distribution is Poisson.

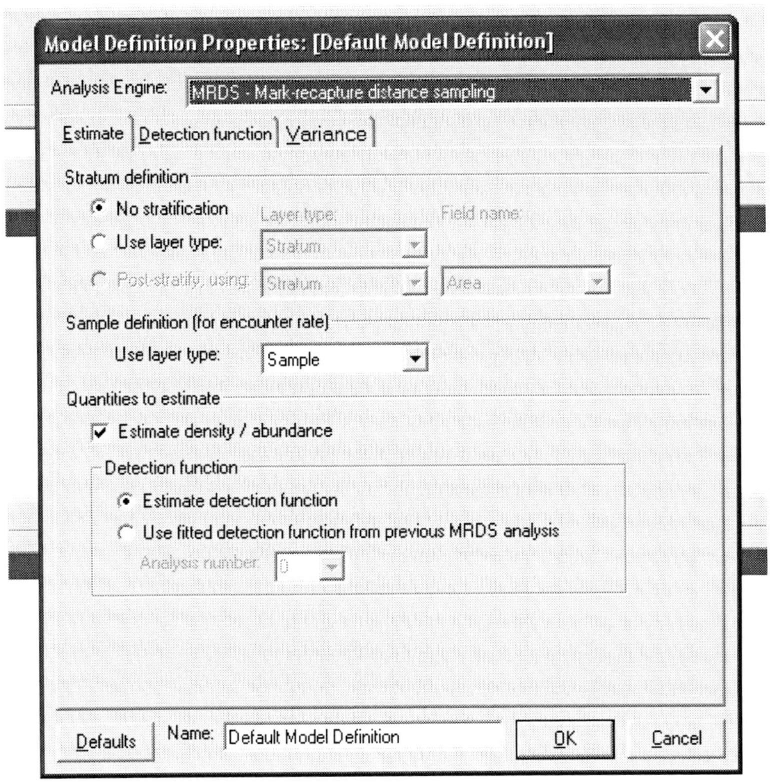

Figure 5. Current version of Distance software allows the use of mark-recapture and distance sampling to obtain robust estimations of abundance.

Alpízar-Jara & Pollock (1996, 1999) developed a combined line transect capture-recapture model which allows to estimate the probability of detecting an animal or group on the transect line (i.e., g(0)). Estimators of population size arising from such approach are more precise than those derived from using only one method. A estimator of g(0) could be given by

$$\hat{g}(0) = \frac{\hat{N}_{LT}}{\hat{N}_{LP}}$$

where $\hat{N}_{LT}$ is the line transect Distance estimator and $\hat{N}_{LP}$ is the Lincoln-Petersen estimator. Current version of Distance software may work under extensions of the combined models (Figure 5) allowing for density estimation when g(0) is less than 1 and incorporation of covariates other than distance (Borchers, 1996). In some surveys this fact occurs as the animals can react to the presence of the observers. When this happen, independent observer line transect surveys have also been proposed (Alpízar-Jara & Pollock, 1996; Chen, 1999; 2000). Two observers (or teams) detect animals independently and a third person coordinates the survey identifying which sightings have been common to both observers and which have been detected only by one of them. Then, animals or groups sighted by both observers may be considered as recaptures.

Characteristics of a Robust Estimator and Model Selection

Robust estimation of population density from a data set is based on several criteria: (i) model robustness to fit the shape of our detection function; (ii) pooling robustness, that is, the model is robust to variation in detection probability for any given distance; (iii) shape criterion, the model for the detection function should have a shoulder; and (iv) estimator efficiency in terms of precision or small variance (Burnham et al., 1979, 1980; Buckland et al., 2001).

Model selection can be carried out in Distance sampling methodology by choosing among a set of possible models (namely, half-normal, uniform, hazard-rate and negative exponential as key functions with different series expansions) and through the use of several selection criteria, such as AIC, AIC_C, BIC. However, to our knowledge, Program Distance yet lacks the possibility to do model averaging and to account for model uncertainty (see Johnson & Omland, 2004). Despite the vast evidence of data analyses in the published literature, some authors (see for instance Buckland et al. 2001 and Franzetti & Focardi, 2006) recommend avoiding the use of the negative exponential model as it does not estimate f(0) in an efficient way, and it is argued that spiked detection functions reflect poor field sampling protocols.

Clustered Populations and the Use of Cluster Size as a Covariate

Mountain ungulates usually show gregarious behaviour and sexual segregation most of the year. They are often aggregated and sighted in groups or clusters. In the case of clustered data, the sample size or the number of detected objects (n) is the number of clusters instead of the total number of observed animals, and perpendicular distances are often measured from the transect line to the geometric centre of the group. One problem associated to clustered populations is that a minimum number of sightings are recommended for robust estimation of the detection function (Mercey & Jayaraman, 1999; Buckland et al., 2001). If a group of size s_i is substituted by s_i individuals, then the assumption that each observation is an independent event is violated (Buckland et al., 2001). When we detect clusters we assume that distance is measured from the transect line to the geometric centre of the group.

Distance sampling methodology allows estimation of the density of clusters (D_s), mean cluster size, and finally the overall density of animals. An estimate of the mean cluster size is given by

$$\hat{\mathrm{E}}(s) = \bar{\mathrm{s}} = \Sigma \frac{\mathrm{Si}}{n}$$

where $\hat{E}_{(s)}$ is the mean size of cluster. And assuming that cluster size is independent from distance, then a overall density estimator is given by

$$\hat{D} = \frac{n.\hat{f}(0).\mathrm{s}}{2L},$$

where $\hat{f}(0)$ is the probability value of the density function of perpendicular distances evaluated at distance 0.

Cluster size and distance are often dependent each other, since larger groups are easily detected than small ones, giving the same perpendicular sighting distance (Chen, 1996). This fact can led to an overestimation of $E_{(s)}$ as few small groups will be observed. Therefore, $\hat{D} = \hat{D}\mathrm{s}.\bar{\mathrm{s}}$ also become overestimated. Another problem arises when at large distances some animals remain undetected and, then, the cluster size become infraestimated (Buckland et al., 2001).

Different authors recommend the use of the cluster size as a covariate and this can be carried out by the current version of Distance software. The estimated variance for mean cluster size is given by

$$\mathrm{var}(\bar{\mathrm{s}}) = \frac{\sum_{i=1}^{n}(\mathrm{s}i - \bar{\mathrm{s}})^2}{n(n-1)}$$

and under the independence assumption it could be added to the global variance of density estimator:

$$\mathrm{var}(\hat{D}) = \hat{D}^2.\left\{\left[cv(n)\right]^2 + \left[cv\left\{\hat{f}(0)\right\}\right]^2\right\}.$$

For cluster analysis the variance of $\hat{D}$ can be calculated according to the equation

$$\mathrm{var}(\hat{D}) = (C\hat{D})^2 x\,\mathrm{var}(M\hat{C}S)^2 x\,\mathrm{var}(C\hat{D}) - \mathrm{var}(C\hat{D})x\,\mathrm{var}(M\hat{C}S)$$

(Leatherwood et al., 1978).

Southwell & Weaver (1993) obtained significantly lower precision when working with clusters for all estimators they used.

One problem associated to working with clustered populations is that a minimum number of sightings are recommended for robust estimation of the detection function and, therefore, of population density in an area (Mercey & Jayaraman, 1999; Buckland et al., 2001).

Conclusion

Since its origin, Distance Sampling methodology and software are being continuously evolved, improved and tested with many species and in diverse habitats. Nevertheless, we must be careful when applying these techniques to wild ungulate populations inhabiting mountainous areas. A basic equipment to complete this kind of surveys may include GPS, telemeter and compass in order to locate animals and/or groups detected in a map with accuracy. Then, the use of a GIS software may be helpful to translate all information to a bi-dimensional view and recalculate transects length and perpendicular distances. Working under "clustered populations" option is recommended despite lacking precision in density estimates. Within the context of management programs of mountain ungulates populations, intensive capture and marking of animals is recommended not only to obtain information about the use of time and space by animals and other demographic and behavioural data, but also to obtain robust estimates of density through a prior estimation of g(0).

Running point transect may become an interesting alternative to the conventional line transect sampling, but much work is still needed to assess if under these circumstances assumptions required by Distance methodology are met.

Different analysis techniques (e.g., truncation, data pooling) allow us to gain precision of density estimates but we consider that it is necessary to apply Distance methodology to populations with known size or to artificial populations in mountainous habitats in order to assess properly the accuracy of the estimates.

References

Alpízar-Jara, R. & Pollock, K. H. (1996). A combination line transect and capture-recapture sampling model for multiple observers in aerial surveys. *Environ Ecol Stat*, **3,** 311-327.

Alpízar-Jara, R. & Pollock, K. H. (1999). Combining line transect and capture-recapture for mark-resighting studies. In Garner et al (eds.): *Marine mammal survey and assessment methods*. Rotterdam: Balkema, 99-114.

Barry, S. C. & Welsh, A. H. (2001). Distance sampling methodology. *J. R. Statist. Soc. B*, **63**, 31-53.

Beavers, S. C. & Ramsey, F. L. (1998). Detectability analysis in transect surveys. *J Wildl Manage*, **62**, 948-957.

Borchers, D. L. (1996). *Line transect estimation with uncertain detection on the trackline*. Ph. D. Thesis, Cape Town: University of Cape Town.

Borchers, D. L., Buckland, S. T. & Zucchini. (2002). *Estimating animal abundance. Closed populations*, London: Springer-Verlag.

Buckland, S. T., Anderson, D. R., Burnham, K. P., Laake, J.L., Borchers, D.L. & Thomas, L. (2001). Introduction to Distance Sampling. *Estimating abundance of biological populations*. Oxford: Oxford University Press.

Buckland, S. T., Anderson, D. R., Burnham, K. P., Laake, J. L., Borchers, D. L. & Thomas, L. (2004). Advanced Distance Sampling. *Estimating abundance of biological populations*. Oxford: Oxford University Press.

Burnham, K. P. & Anderson, D. R. (2002). *Model selection and multimodal inference: a practical-theoretic approach.* 2nd ed. New York: Springer-Verlag.

Burnham, K. P., Anderson, D. R. & Laake, J. L. (1979). Robust estimation from line transect data. *J Wildl Manage*, **43**, 992-996.

Burnham, K. P., Anderson, D. R. & Laake, J. L. (1980). Estimation of density from line transect sampling of biological populations. *Wildl Monog*, **72**, 1-202.

Burnham, K. P., Anderson, D. R. & Laake, J. L. (1985). Efficiency and bias in strip and line transect sampling. *J Wildl Manage*, **49**, 1012-1018.

Chen, S. X. (1996). Studying school size effects in line transect sampling using the kernel method. *Biometrics*, **52**, 1283-1294.

Chen, S. X. (1998). Measurement errors in line transect surveys. *Biometrics*, **54**, 899-908.

Chen, S. X. (1999). Estimation in independent observers line transect surveys for clustered populations. *Biometrics*, **55**, 754-759.

Chen, S. X. (2000). Animal abundance estimation in independent observer line transect surveys. *Environ Ecol Stat*, **7**, 285-299.

Chen, S. X. & Cowling, A. (2001). Measurement errors in line transect surveys where detectability varies with distance and size. *Biometrics*, **57**, 732-742.

Drummer, T. D. & McDonald, L. L. (1987). Size bias in line transect sampling. *Biometrics*, **43**, 13-21.

Escós, J. & Alados, C. L. (1988). Estimating mountain ungulate density in Sierras de Cazorla y Segura. *Mammalia*, **52**, 425-428.

Fewster, R. M., Laake, J. L., & Buckland, S. T. (2005). Line transect sampling in small and large regions. *Biometrics*, **61**, 856-861.

Franzetti, B. & Focardi, S. (2006). *La stima di popolazione di ungulati mediante distance sampling e termocamera a infrarossi*. Bologna: Istituto Nazionale per la Fauna Selvatica.

Hiby, L. & Krishna, M. B. (2001). Line transect sampling from a curving path. *Biometrics*, **57**, 727-731.

Johnson, J. B. & Omland, K. S. (2004). Model selection in ecology and evolution. *Trends Ecol Evol,* **19**, 101-108.

Kulbicki, M. & Sarramégna, S. (1999). Comparison of density estimates derived from strip transects and distance sampling for underwater visual censuses : a case study of Chaetodontidae and Pomacanthidae. *Aquatic Living Res.*, **12**, 315-325.

Leatherwood, S., Gilbert, J. R. & Chapman, D. G. (1978). An evaluation of some thechniques for aerial censuses of bottlenosed dolphins. *J Wildl Manage*, **42**, 239-250.

Marques, F. C.; Buckland, S. T. ; Goffin, C.; Dixon, C. E.; Borchers, D. L.; Mayle, B. A. & Peace, A. J. (2001). Estimating deer abundance from line transect surveys of dung: sika deer in southern Scotland. *J Appl Ecol*, **38**: 349-363.

Melville, G. J. & Welsh, A. H. (2001). Line transect sampling in small regions. *Biometrics*, **57**, 1130-1137.

Mercey, K. A. & Jayaraman, K. (1999). Predicting the variation in detection function in line transect sampling through random parameter model. *Environ Ecol Stat*, **6,** 341-350.

Palomares, F. & Ruiz-Martínez, I. (1993). Status and conservation perspectives for the Spanish ibex population (Capra pyrenaica Schinz 1838) of Sierra Mágina Natural Park, Spain. *Z Jagdwiss*, **39**, 87-94 [in German].

Pérez, J. M.; Granados, J. E. & Soriguer, R. C. (1994). Population dynamics of the Spanish ibex Capra pyrenaica in Sierra Nevada Natural Park (southern Spain). *Acta Theriol*, **39**, 289-294.

Pérez, J. M.; Granados, J. E.; Soriguer, R. C.; Fandos, P.; Márquez, F. J. & Crampe, J. P. (2002 a). Distribution, status and conservation problems of the Spanish ibex, Capra pyrenaica (Mammalia, Artiodactyla). *Mammal Rev*, **32**, 26-39.

Pérez, J. M.; Serrano, E.; Alpízar-Jara, R.; Granados, J. E. & Soriguer, R. C. (2002 b). The potential of distance sampling methods to estimate abundance of mountain ungulates: review of usefulness and limitations. *Pirineos*, **157**, 15-23.

Ramsey, F. L.; Wildmam, V. & Engbring, J. (1987). Covariate adjustements to effective area in variable-area wildlife surveys. *Biometrics*, **43**, 1-11.

Shackleton, D. M. (1997). Wild sheep and goats and their relatives. Status survey and conservation action plan for Caprinae, Cambridge: IUCN.

Southwell, C. & Weaver, K. (1993). Evaluation of analytical procedures for density estimation from line-transect data: data grouping, data truncation and the unit of analysis. *Wildl Res*, **20**, 433-444.

Thomas, L.; Laake, J. L.; Strindberg, S; Marques, F. F. C.; Buckland, S. T.; Borchers, D. L.; Anderson, D. R.; Burnham, K. P.; Hedley, S. L.; Pollard, J. H.; Bishop, J. R. B.; & Marques, T. A. (2005). Distance 5.0 Release 1. Research Unit for Wildlife Population Assessment, University of St. Andrews, U. K., http://www.ruwpa.st-and.ac.uk/distance

Toïgo, C.; Gaillard, J. M. & Michallet, J. (1996). La taille des groupes: un bioindicateur de l'effectif des populations de bouquetin des Alpes (Capra ibex ibex)? *Mammalia*, **60**, 463-472.

Williams, B. K.; Nichols, J. D. & Conroy, M. J. (2002). *Analysis and management of animal populations: modeling, estimation and decision making.* San Diego, Academic Press.

RESEARCH AND REVIEW STUDIES

In: Ecosystem Ecology Research Trends
Editors: J. Chen and C. Guo, pp. 23-63

ISBN: 978-1-60456-183-8

Chapter 1

SOIL ECOSYSTEM: DYNAMIC AND SUSTAINABILITY

A. Muscolo and M. Sidari
Dipartimento di Gestione dei Sistemi Agrari e Forestali
Università degli Studi "Mediterranea"
Feo di Vito, 89060 Reggio Calabria
Italy

Abstract

Soil is the foundation of the entire biosphere. Due to different activities of various soil organisms, permanent transitions take place in soil between growth and decay, life and death. Thus in a biological and biochemical way soil ecosystem is the most important interface for the global interchange of matter and energy. Soil quality represents an integral value of the compositional structures and natural functions of soil in relation to soil use and environmental conditions. Among the indigenous soil components, different organisms and especially microorganisms play a key role in ecologically important biogeochemical processes. In that way soil microorganism contribute to the maintenance of the matter and energy transfer in the terrestrial environments. Thus the thin layer of soil covering the surface of the earth represents the difference between survival and extinction for most land-based life. Soil quality is a living dynamic system whose functions are mediated by a diversity of living organisms that require management and conservation. Numerous factors (environmental change; human activities; pollutions, extensive soil cultivation, salinization) can affect soil condition causing several alterations in the development and biochemical activity of soil organism. The organic contents of soil are vitally important for providing energy, substrates, and the biologically diversity necessary to sustain the numerous soil functions. The concept of soil ecosystem quality has recognized soil organic matter as an important attribute that has a vast control on many of the key soil ecosystem functions. However, soil organic matter varies among environments and management systems, generally increasing with higher mean annual precipitation, lower mean annual temperature, higher clay content with an intermediate grazing intensity, with higher crop residue inputs, and native vegetation compared to cultivated management. This chapter reviews the biological properties of soil, much of the material will be familiar, the key difference is the recognition of the vital role soil microorganisms play in recycling, releasing and storing plant nutrients. Some specific issues will be focused on the role and function of organic matter in the sustainability of soil ecosystem, and on the potential of soil microbial biomass to detect management induced changes in dynamic soil quality.

Introduction

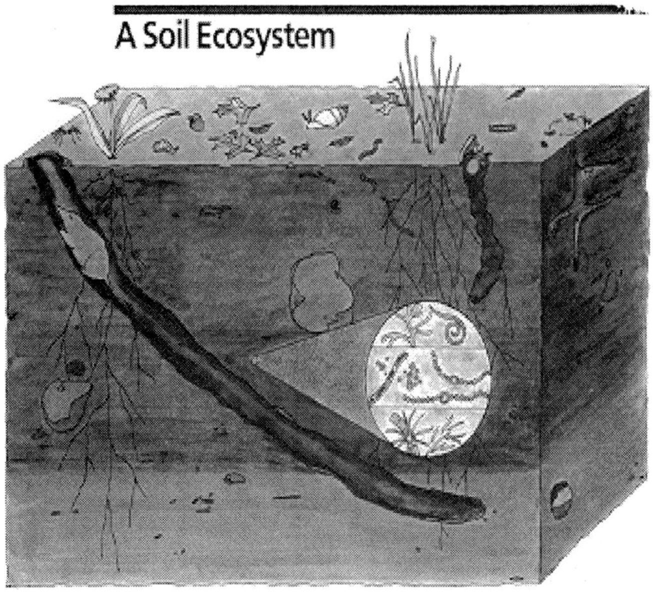

The term ecosystem was coined in 1935 by the Oxford ecologist Arthur Tansley to encompass the interactions among biotic and abiotic components of the environment at a given site. It was defined in its presently accepted form by Eugene Odum as follows: "Any unit that includes all of the organisms (i.e, the community) in a given area interacting with the physical environment so that a flow of energy leads to clearly defined trophic structure, biotic diversity, and material cycles (i.e., exchange of materials between living and non-living parts) within the system". Tansley's concept had been expressed earlier in 1913 by the Oxford geographer A. J. Herbertson, who suggested the term "macroorganism" for such a combined biotic and abiotic entity. He was, however, too far in advance of his time and the idea was not taken up by ecologists. On the other hand Tansley's concept—elaborated in terms of the transfer of energy and matter across ecosystem boundaries–was utilized within the next few years by Evelyn Hutchinson, Raymond Lindeman, and the Odum brothers, Eugene and Howard. The boundaries of an ecosystem can be somewhat arbitrary, reflecting the interest of a particular ecologist in studying a certain portion of the landscape. However, such a choice may often represent a recognizable landscape unit such as a woodlot, a wetland, a stream or lake, or, in the most logical case, a watershed within a sealed geological basin, whose exchanges with the atmosphere and outputs via stream flow can be measured quite precisely. Inputs and outputs imply an open system, which is true of all but the planetary or global ecosystem, open to energy flow but effectively closed in terms of materials except in the case of large-scale asteroid impact. Ecosystems exhibit a great deal of structure, as may be seen in the vertical partitioning of a forest into tree, shrub, herb, and moss layers, underlain by a series of distinctive soil horizons. Horizontal structure is often visible as a mosaic of patches, as in forests with gaps where trees have died and herbs and shrubs now flourish, or in bogs with hummocks and hollows supporting different kinds of plants. Often the horizontal structure is distinctly zoned, for instance around the shallow margin of a lake; and sometimes it is beautifully patterned, as in the vast peatlands of North America that reflect a very complicated hydrology. Ecosystems exhibit an interesting functional organization in their processing of energy and matter. Green plants, the primary producers of organic matter, are consumed by herbivores, which in turn are eaten by carnivores that may in turn be the prey of other carnivores. Moreover, all these animals may have parasites as another set of consumers. Such sequences of producers and successive consumers constitute a food chain, which is always part of a complicated, inter-linked food web along which energy and materials pass. At each step along the food chain some of the energy is egested or passed through the organisms as faeces. Much more is used for metabolic processes and—in the case of animals—for seeking food or escaping predators; such energy is released as heat. As a consequence only a small fraction (often of the order of 10%) of the energy captured at a given step in the food chain is passed along to the next step. There are two main types of food chains. One is made up of plant producers and animal consumers of living organisms, which constitute a grazing food chain. The other consists of organisms that break down and metabolize dead organic matter, such as earthworms, fungi, and bacteria. These constitute the detritus food chain. Humans rely chiefly on grazing food chains based on grasslands, whereas in a forest it is usual for more than 90% of the energy trapped by photosynthesis to pass along the detritus food chain. Whereas energy flows one way through ecosystems and is dispersed finally to the atmosphere as heat, materials are partially and often largely recycled. For example, nitrogen in rain and snow may be taken up from the soil by roots, built into leaf protein that falls with the leaves in autumn, there to be broken down by soil microbes to ammonia and nitrate and taken up once again by roots. A given molecule of nitrogen may go through this nutrient cycle again and again before finally leaving the system in stream outflow. Other nutrients, and toxins such as lead and mercury, follow the same pathway, each with a different residence time in the forest ecosystem. Mature ecosystems exhibit a substantial degree of stability, or dynamic equilibrium, as the endpoint of what is often a rather orderly succession of species determined by the nature of the habitat. Sometimes this successional process is a result of the differing life spans of the colonizing species, at other times it comes about because the colonizing species alter the habitat in ways that are more favorable to their competitors, as in an acid moss bog that succeeds a circumneutral sedge fen that has in its turn colonized a pond as a floating mat. Equilibrium may sometimes be represented on a large scale by a relatively stable mosaic of small-scale patches in various stages of succession, for instance in fire-dominated pine forests. On the millennial time scale, of course, ecosystems are not stable, changing very gradually owing to immigration and emigration of species and to evolutionary changes in the species themselves. The structure, function, and development of ecosystems are controlled by a series of partially independent environmental factors: climate, soil parent material, topography, the plants and animals available to colonize a given site, and disturbances such as fire and windthrow. Each factor is, of course, divisible into a variety of components, as in the case of temperature and precipitation under the general heading of climate.

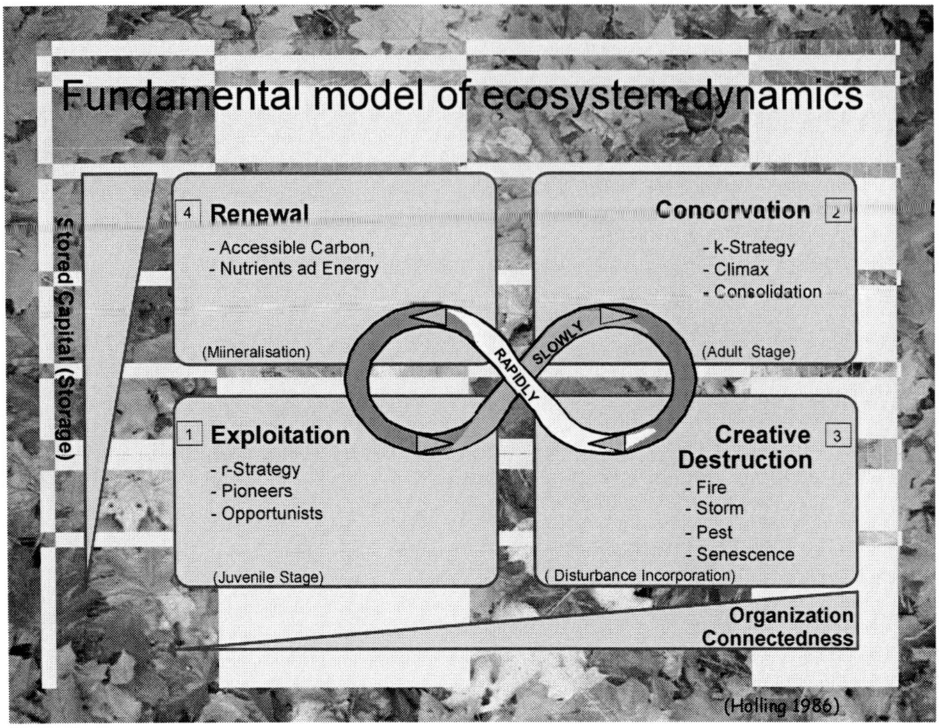

Soil Quality and Sustainability: A Conceptual Definition

Interest in evaluating the quality and health of our soil resources has been stimulated by increasing awareness that soil is a critically important component of earth's biosphere, functioning in the maintaining of local regional and global environmental quality [1]. Soil is also the basis of agricultural and of natural plant communities. Thus in a biological and biochemical way soil represents an essential natural resource that provides important ecosystem functions: a medium for plant growth, regulation and portioning of water flow in the environment [2].

Soils form the basis for all forest growth and play a key role in the hydrological, carbon and nutrient cycles of forest ecosystems. Soil quality has been defined by the Soil Science Society of America Ad Hoc Committee on soil quality (S-581) as the capacity of a specific kind of soil to function, within a natural or managed ecosystems boundaries, to sustain plant and animal productivity, maintain or enhance water and air quality, and support human health and habitation [3]. In discussion of the ambiguity of environmental terms and the need to standardize their meanings, Johnson et al. [4] defined soil quality as "a measure of the condition of soil to the requirements of one or more biological species and or to any human purpose". The term soil health is preferred by some [5, 6], because it portrays soil as a living, dynamic system whose functions are mediated by a diversity of living organisms that require management and conservation. Soil health, biodiversity, and soil resilience are several limited in extreme environments and are more sensitive to anthropogenic disturbance [7]. The terms soil quality (favoured by scientists) and soil health (favoured by farmers) tend to be used interchangeably, especially in the general press. Characterization of soil quality by scientists focuses on analytical/quantitative properties of soil with a separately defined quantitative link to the functions of soil quality. Characterization of soil health by farmers focuses on descriptive/qualitative properties of soil with a direct value judgement (unhealthy to healthy) integrated into the options for a given property; in addition, interwoven into the properties of soil per se are value-based descriptive properties of plant, water, air, and animal/human systems considered by farmers to be an integral part of soil health characterization.

The terms soil quality and soil health will be used synonymously throughout this chapter. However use of the term soil quality or soil health will generally be used in a broader sense to indicate the capacity of soil to function as a vital living system to sustain biological productivity, promote environmental quality and maintain plant and animal health. In this sense soil quality is synonymous with sustainability.

The quality of soil includes an inherent component determined by soil's physical and chemical properties within the constraints set by climate and ecosystem. In addition soil quality includes a component affected by management and land-use decisions.

Over the past two decades as the importance of the soil as an often key component of natural and man-influenced ecosystem has been recognized, there has been an increased demand to establish criteria to determine soil quality and to develop indices which may be used to rank and compare the quality of soils at different locations or at the same location through time. More recently these indices of soil quality have often been linked to attempts to evaluate the sustainability of land use activities as part of national and international programmes on sustainability.

Criteria for indicators of soil quality and health relate mainly to their utility in defining ecosystem processes and integrating physical, chemical and biological properties, their sensitivity to management and climatic variations and their accessibility and utility to agricultural specialists, producers conservationists and policy makers [8]. Public interest in soil quality is increasing throughout the world as humankind recognizes the fragility of earth's soil, water, and air resources and the need to protect them to sustain civilization. To understand soil quality however, one must first be aware of the complexity of soil and its intrinsic value. Soil is a living system that represents a finite resource vital to life on earth. It forms the thin skin of unconsolidated mineral and organic matter on the earth's surface. It develops slowly from various parent materials and is modified by time, climate, macro- and micro-organisms, vegetation, and topography. Soils are complex mixtures of minerals, organic compounds, and living organisms that interact continuously in response to natural and imposed biological, chemical, and physical forces. Vital functions that soils perform within ecosystems include:

i. sustaining biological activity, diversity, and productivity;
ii. regulating and partitioning water and solute flow;
iii. filtering, buffering, degrading, immobilizing, and detoxifying organic and inorganic materials, including industrial and municipal by-products and atmospheric depositions;
iv. storing and cycling nutrients and other elements within the earth's biosphere; and
v. providing support for socioeconomic structures and protection for archaeological treasures associated with human habitation.

Conceptually, the intrinsic quality or health of a soil can be viewed simply as "its capacity to function". By encompassing productivity, environmental quality, and health as major functions of soil, this definition requires that values be placed on specific soil functions as they relate to the overall sustainability of alternate land-use decisions. Although unstated, the definition also presumes that soil quality can be expressed by an unique set of characteristics for every kind of soil. It recognizes the diversity among soils, and that a soil that has excellent quality for one function or product can have very poor quality for another. The functions encompassed by soil quality integrate chemical, physical, and biological properties and processes occurring within every soil and are responsive to human use and management decisions. Soil quality in its broadest sense is enhanced by land-use decisions that weigh the multiple functions of soil, and is impaired by land-use decisions that focus on single functions. Soil quality can be degraded by using inappropriate tillage and cropping practices; through excessive livestock grazing or poor timber harvesting practices; or by misapplication of animal manures, irrigation water, fertilizers, pesticides, and municipal or industrial by-products. To enhance soil quality, everyone must recognize that the soil resource affects the health, functioning, and total productivity of all ecosystems. We must become more aware of potential side effects of soil management and land-use decisions. By addressing the integrative concept of soil quality, SSSA members can provide the leadership for better long-range, landscape-scale planning and land-use, thus helping to protect finite and living soil resources and fragile landscapes. To evaluate soil quality for various land-uses, assessments will be required over time and across space. For the evaluations to be meaningful, research must develop soil quality indicators that are sensitive, reliable, reproducible, and capable of detecting changes in soil physical, chemical, and biological properties, processes, and their interactions. The indicators must be qualitative or semi-quantitative for immediate use by farmers or land managers, and quantitative for the people responsible for developing long-range, process-based strategies that will improve land-use; enhance plant, animal, and human health; and sustain biological diversity within natural and fabricated ecosystems. Numerous specific soil measurements are being evaluated, but to help identify the attributes that will be most useful for soil quality evaluations, research and education are needed to:

i. identify the vital functions of soil for which assessments need to be made;
ii. determine the appropriate methods, measurements, and criteria for interpreting those assessments,
iii. develop and evaluate accessible databases as repositories for high quality, reliable soils information; and
iv. design improved land use practices that will ensure the economic viability, environmental safety, social acceptability and value that is appropriate for each soil resource.

Progress has been made in developing measures which have a wide degree of acceptability in respect to productivity. It is therefore of paramount importance that consideration of actual and potential functions of the soil are considered in soil quality evaluations. It is often difficult clearly separate soil functions into chemical physical and biological processes, because of the dynamic interactive nature of these processes. This interconnection is especially prominent between chemical and biological indicators of soil quality [9, 10, 11]. Consequently, there is seldom a one-to-one relationship between function and indicator; more likely, a given function (e.g. sustain biological productivity) is supported by a number of soil attributes, while any given soil property or process may be relevant to several soil attributes and /or soil functions simultaneously [12, 13]. A good example of the latter is soil organic matter, which plays a role in almost every soil function [12, 14, 15]. Also many soil chemical properties directly influence microbiological processes, and these processes together with soil physical-chemical properties determine the capacity of soils hold, supply, and cycle nutrients, and the movement and availability of water. A summary of the soil chemical properties cited in recent literature pertaining to soil quality in agricultural, grassland and forest soil ecosystem is provided in Table 1. Biological attributes may be very dynamic and exceptionally sensitive to changes in soil conditions, which is why there is often a preference for biological attributes for short-time evaluations. Attributes that might be measured include the population of micro, meso and macroorganisms, respiration rate or other indicators of microbial activity, and more detailed characterization of soil organic matter, widely promoted as a key indicator of soil quality.

Role and Function of Organic Matter in Soil Ecosystem

Soil Organic Matter (SOM), is a heterogeneous mixture of organic compounds of plant, animal and microbial origin in various stage of decomposition [16, 17]. At any given time, it consists of a range of materials from the intact original tissues of plants and animals to the substantially decomposed mixture of materials known as humus (Figure 1).

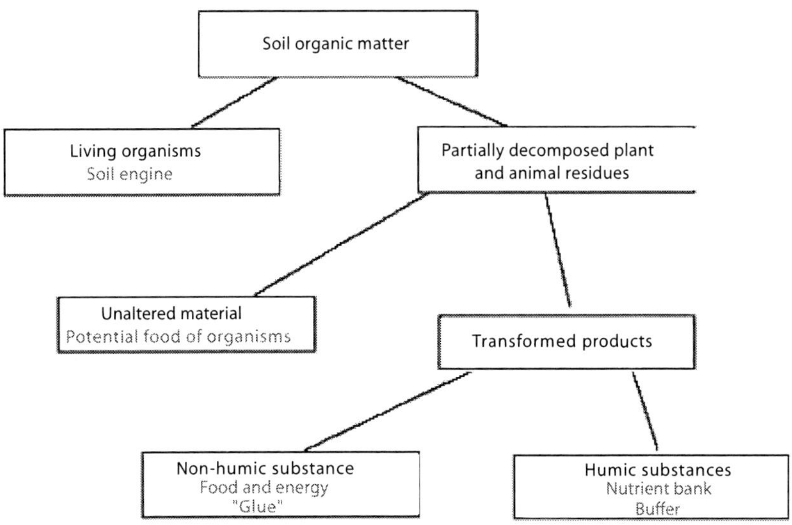

Figure 1. Components of soil organic matter and their functions.

Soil organic matter forms a large and dynamic reserve of C, which is the major part of the global C cycle. Global estimates of the C content of SOM ranges from 1100 to 3000 Pg [18,

19], understanding the mechanisms and factors that govern SOM release and retention is central to terrestrial ecosystem research.

Carbon and organic matter processing are central to the structure and function of biological systems. The terrestrial carbon cycle is a complex set of interactions between gaseous, aqueous and solid phases involving both organic and inorganic forms. In forest ecosystems, photosynthetic assimilation, the conversion of gaseous inorganic carbon to structural organic carbon is generally the principal carbon input. During decomposition, this organic carbon may be converted to soil organic carbon, aqueous organic and inorganic carbon and gaseous carbon dioxide (CO_2). As the major pool of terrestrial carbon resides in soil [18], understanding the factors that govern organic matter release and retention in soils is central to both ecosystem and environmental studies. In addition to its role in the C cycle SOM exerts an important influence on soil fertility, soil development and numerous soil chemical properties. The organic contents of soil are vitally important in providing energy, substrates and the biological diversity necessary to sustain numerous soil functions. The concept of soil quality has recognized soil organic matter as an important attribute that has a great deal of control on many of the key soil functions [9]. A direct effect of this biological activity is seen in the macrostructure of soils, through the formation of soil pores as a consequence of faunal activity and root and fungal growth. Larger, but less direct effects of biological activity are the resultants of changes in the organic compounds of SOM that result from biological breakdown and the concomitant mineralization and immobilization of nutrients [20]. The relative importance of these different functions varies with soil type, climate, and farming system [21]. Soil organic matter has a less direct, but nonetheless important effect on nutrient supply through its influence on cation exchange capacity and on the capacity to adsorb anions; and these functions have additional important implications for the impact of toxic and biocidal agrochemicals [22]. The amount and dynamics of soil C differ with soil type particularly mineralogy, with climate and with management. The capacity of soil to protect organic matter against microbial decomposition and microbial biomass against cell death or predation seems to depend on the soil clay content [23]. Under identical annual organic input, a slower organic matter turnover, a larger microbial biomass and more organic matter are expected in soil with high clay content compared to soil with a low clay content within the same climate area. Muscolo et al. (data not shown) found a different amount of soil organic matter and in particular of humic and fulvic acids in forest soils within *Pinus laricio* stands under same climate but with different clay content, the best humification process and the greater amount of SOM were detected in soils with higher clay content. In classical concepts, stable clay-organic complexes are assumed to be responsible for an increased formation of stabilized organic matter in clay reach soils. Clay may also influence the turnover of organic substrate in short term. During a 35-d incubation experiment, clay and soil surface area played a major role in controlling the decomposition of added ^{14}C –labelled glucose through stabilization and protection of the microbial biomass [24]. Many authors found a positive relationship between SOM content and soil clay content. Under natural conditions [25], and under low input conditions [23], higher clay contents led to higher SOM contents. The same was shown in long-term experiments, where clay substrate was added to sandy soils while keeping the fertiliser input identical. Muller and Hoper [26] found specific microbial turnover decreased and SMB increased with increasing clay content. This support the hypothesis of a clay dependent capacity of soils to protect microbial biomass. Because SOM is an heterogeneous mixtures of organic compounds with a wide ranges of structures and functional groups [16, 17], typically SOM is weakly acidic due to abundant carboxyl and phenolic functional groups. The dissociation of SOM and its proton-binding properties are of major importance in the chemistry and ecology of soil.

SOM decomposition is closely related to humification, mineralization, DOM dynamic, and SOM stabilization. During the first phase of decomposition, rapid mineralization of labile components is the dominant process. In the second phase of decomposition, mineralization still occurs but is slowed down to the accumulation of refractory molecules. Besides temperature and water regime, the amount and chemical composition of the primary and secondary resources control SOM dynamic and, especially, humification [27].

Dissolved organic matter (DOM) is the labile SOM fraction important substrate for microbial growth in soil, but its production is also partly mediated by microbes [28]. Therefore, DOM may be the key constituent for microbial degradation of organic matter in soils. DOM plays an important role in the cycling of carbon and nutrients, especially N, P, S, because the nutrients in the soil solution are more available, both to soil microbes and plants, than thus occurring in the soil organic matter [29]. Relating forest productivity to levels of total SOM ignores the separate contributions that different SOM components have in the soil. This has prompted researchers to shift from investigating the relationship between total SOM and productivity, to devising means to fractionate SOM into biologically meaningful pools. Presently, effort are directed at isolating the labile from the recalcitrant SOM components. Labile SOM fractions are theorized to be correlated to nutrient cycling dynamics [30]. Because prior linkages have been made between nutrient cycling and productivity [31], a link between labile SOM and nutrient cycling dynamics would be useful in predicting potential productivity changes. This in turn, could guide the development and use of land-management practices.

Arguably the most important component of SOM is humus, the well-decomposed heterogeneous organic compounds formed in soil as by-products of microbial metabolism on dead cell materials. Humus refers to the microbial resistant forms of SOM that remain after major portions of added plant and animal residues have decomposed. Humus composition and breakdown rate affect: the soil structure and porosity; the water infiltration rate and moisture holding capacity of soils; the diversity and biological activity of soil organisms. The variety and extent of these reactions and interactions indicate the highly reactive nature of humic substances (HS), thus highlighting the overall complexity of the plant-soil system

The HS were found up to now to exhibit a range of different effects on plant metabolism [32, 33, 34], depending on their origin, molecular size, chemical characteristics and concentration, and are consequently the subject of intense investigation in various fields of agriculture, such as soil chemistry, fertility, and plant physiology. According to a new view of their chemistry, HS are collections of heterogeneous, relatively low molecular-mass components forming dynamic associations stabilized by hydrophobic interactions and hydrogen bonds [35, 36, 37, 38]. Humic substances have an abundance of oxygen-containing functional groups (carboxyl, phenolic, alcoholic) which dominate their chemical properties. These groups and their relative chemistry are mainly responsible for most of the functions ascribed to HS. The relationship among HS structure, functional groups and their activity have been previously presented [33].

The molecular heterogeneity that is characteristic of HS serves a vital role in the ecological system [39]. Humic substances constitute the only natural organic material that can survive in bulk and still possess the requisite chemical reactivity to perform the various functions for sustaining soil quality and promoting plant growth. By virtue of its molecular heterogeneity, this medium is highly bio-refractory, but it still possesses the reactive functional groups needed to perform ecologically and environmentally vital tasks.

Low molecular weight (LMW) organic acid and phenolic acids, although representing only a small fraction of the SOM, exert a disproportionate influence on soil processes since they are among the most reactive forms of C in the soil [40]. Phenolic acids in soil represent an important group of compounds with ecological functions, influencing rates of nutrient turnover in soils [41].

Phenolic compounds of low molecular weight occur widely in soils [42]; they arise mainly from decomposing plant materials, although some of the compounds may be synthesized by soil micro-organisms [42, 43]. Phenolic compounds in soils may play a dominant role controlling many aspects of plant-soil interactions, interfering directly with ion transport, protein synthesis, hormone activity and energy metabolism [44, 45, 46,47]. Moreover one of the most properties of phenolic compounds is their capacity to form recalcitrant complexes with protein and thus to alter the pool and form of nutrients [48] influencing supplies of available N. Direct toxicity of some of these molecules on tree seed germination [49, 50], tree seedlings [51], and microorganism growth [52] is also well documented. Biologically active concentrations of phenolics in soils can vary throughout the year, depending on various factors such as microbial activity, adsorption on colloids, humidity, etc. [52]. Soil can be considered the "black box" of the forest ecosystem and the effect of microbial activity on the phenolic pool should not be underestimated.

In short phenolic compounds are always present in forestry soil, their amount and composition is clearly linked to seasonal changes, their effects are species-specific and concentration dependent. These characteristics may influence the density and the composition of individual plant communities inducing, in part, the cycle of natural regeneration of forest tree species.

Despite extensive research, our understanding of the basic chemistry and microbiology of humus and SOM in general, is incomplete [53]. The major constraints are the complexity of the physiochemical processes of decomposition and humification.

For these reasons there is an increasing interest in measuring the size of different C pools. These measurements have employed a wide range of techniques [54]. Early measurement of very broad, non specific chemical groups, such as humic and fulvic acids and humins, on the basis of solubility, developed into separation on the basis of more functional chemical groups, for example with ^{13}C NMR [55] and increasingly, on the basis of relative bioavailability. Measurements related to bioavailability include detection of the small but active microbial pool by fumigation techniques [56] to measurement of labile C with high potential for biological oxidation. High potential for biological oxidation has been related to surrogate measurements on the basis of physical fractionation such as the light fraction of C by density fractionation [57], and on the basis of chemical oxidation [58]. The move from very broad chemical groups, to more functional chemical groups and to pools that are more closely related to soil biological activity, increases the potential to better understand soil C dynamics. These various measurements of labile soil C need to be used more widely and, most importantly, they must be related to important aspects of soil chemical, physical, and biological fertility, to productivity, and to the broader aspects of system sustainability.

Soil fertility and agricultural systems sustainability depend upon soil organic matter. Management that causes a decline in soil quality reduces these functional abilities, appropriate soil management techniques can be expected to restore ecosystem functions once degraded

The reduction of soil disturbance through zero-tillage, the use of cover crops and the preservation of crop residues on the soil surface result in increased activity of the soil and in the accumulation of organic matter, mainly in the topsoil (Figure 2).

An argument often heard in the discussion on conservation agriculture is that it is only feasible in the humid and subhumid tropics and that the generation of sufficient biomass in semi-arid regions is the limiting factor to start implementing conservation agriculture.

However, recent research has shown that even in semi-arid areas of Morocco the application of the principles of conservation agriculture bears its fruits [1]. Mrabet [59] reports higher yields through better water use and improved soil quality; the latter caused by an increase in soil organic C and N and a slight pH decline in the seed zone [60, 61].

World soils are important reservoirs of active C and play a major role in the global carbon cycle. As such, soil can be either a source or sink for atmospheric CO_2 depending on land use and the management of soil and vegetation [62] (Figure 3). The conversion of native ecosystems (e.g. forests, grasslands and wetlands) to agricultural uses, and the continuous harvesting of plant materials, has led to significant losses of plant biomass and C [63], thereby increasing the CO_2 level in the atmosphere.

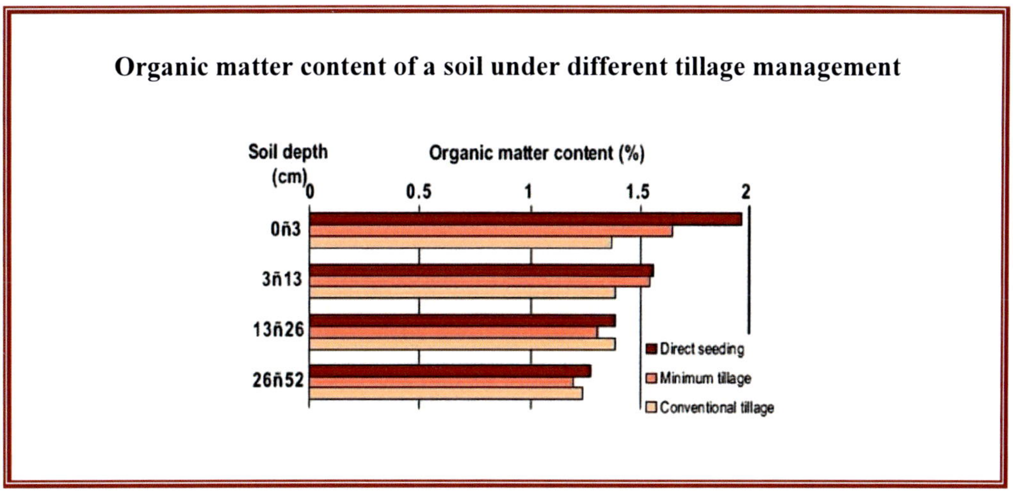

Figure 2. Organic matter content of a soil under different tillage management (Source: FAO [65]).

In particular, the practice of burning agricultural fields before cultivation has a disastrous effect on soil organic carbon content. Figure 4 shows the reduction in soil organic carbon in agricultural fields after 100 years of burning crop residues and weeds compared with an area that was not burned or ploughed during the same period. The topsoil layer (0-5 cm) represented the greatest carbon loss (36 percent) compared with the area that was not burned. Soil N stock in the same layer was reduced by 16 percent. The carbon stock was reduced not only through burning, but because of the whole land-use management, especially a drastic reduction in diversity of species as monocropping was practised [64].

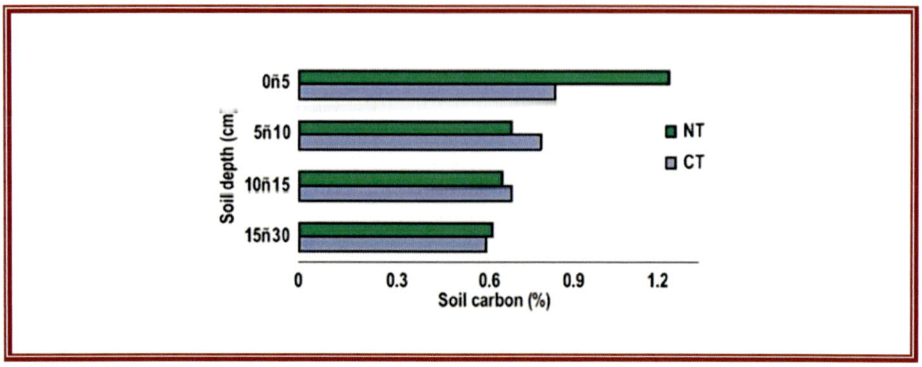

Source: Prior [66].
Note: Conventional (CT) and conservation (NT) agriculture, after two complete cropping cycles (4 years).

Figure 3. Soil carbon concentration at various soil depths affected by management system.

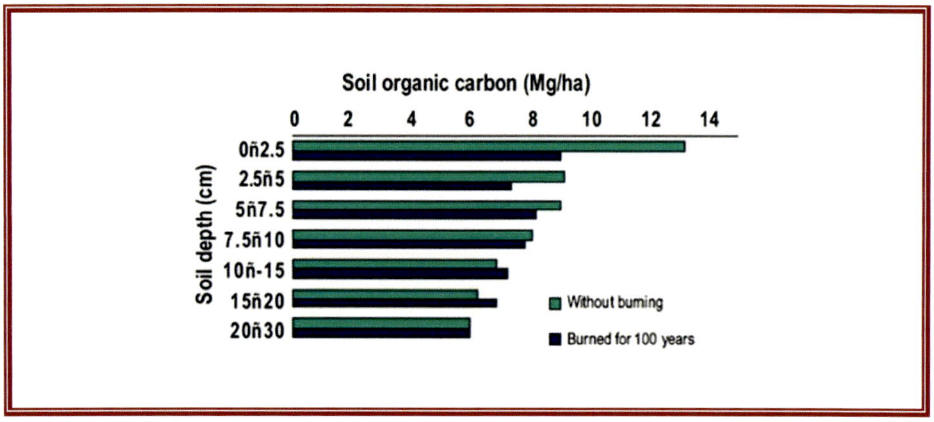

(**Source**: Spagnollo [67]).

Figure 4. Effect of crop residue burning once every two years on soil carbon stock.

Table 1 lists general practices that determine whether soil will be a sink or a source of atmospheric CO_2, putting in evidence that the soil can play a part in mitigating CO_2 levels [68]. This removal process is achieved naturally, and quite effectively, through photosynthesis. Living plants take CO_2 from the air in the presence of sunlight and water, convert it into seeds, leaves, stems and roots. Part of the CO_2 is retained or "sequestered", or stored as C in the soil when decomposed. In particular, systems based on high crop-residue addition and no tillage tend to accumulate more C in the soil than is lost to the atmosphere. Carbon sequestration in managed soils occurs when there is a net removal of atmospheric CO_2 because C inputs (crop residues, litter, etc.) exceed C outputs (harvested materials, soil respiration, C emissions from fuel and the manufacture of fertilizers, etc.) [69]. Management practices that increase soil C comply with a number of principles of sustainable agriculture: reduced tillage, erosion control, diversified cropping system, balanced fertilization, etc. In the early years of no-tillage systems, the organic matter content of the soil is increased through the decomposition of roots and the contribution of vegetative residues on the surface. This organic material decomposes slowly, and thus the liberation of C to the atmosphere also occurs slowly. In the total balance, net fixation or sequestration of C takes place; the soil is a net sink of C.

Table 1. Land use and land management determining whether soil will be a sink or source of atmospheric CO_2

Soil as a source of CO_2	Soil as a sink of CO_2
Soil properties: coarse textured soil, excessive drainage, high susceptibility to erosion	Soil properties: clayey soil, poorly drained ecosystems, depositional sites, including footslopes
Land use: seasonal crops, simple ecosystem, shallow roots and low root-shoot ratio	Land use: perennial crops, diverse ecosystem, deep roots and high root-shoot ratio
Soil management: intensive tillage based on plough, negative nutrient balance, residue removal and/or burning, continuous cropping, loss of soil and water by runoff and erosion	Soil management: no tillage, positive nutrient balance, mulch farming cover crops in rotation, cycle, soil and water conservation

(**Source**: adapted from Lal, [62]).

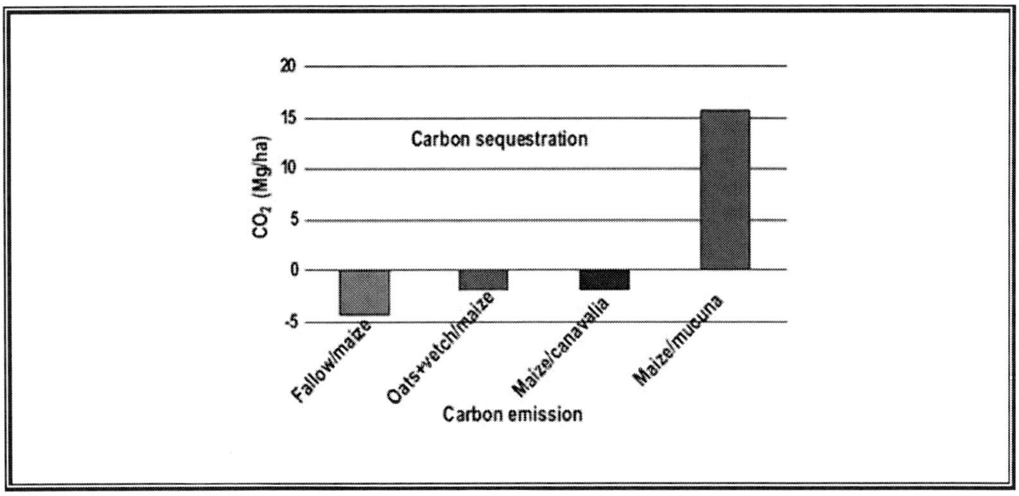

Figure 5. Estimation of emission and sequestration of CO_2 (total over 8 years) under different maize production systems with cover crops under direct sowing compared with natural vegetation in southern Brazil (**Source**: Amado [73]).

Figure 5 illustrates the fact that some cropping systems can act as a sink for CO_2. In this example, the carbon stock in soils under natural vegetation is used as a reference (steady state: DC = 0). In eight years, the fallow/maize system liberated 4.3 tonnes of CO_2 per hectare. The maize/mucuna system showed a positive balance of almost 20 tonnes of CO_2 per hectare compared with fallow/maize. Compared with soils under natural vegetation, this means a capture of atmospheric CO_2 of more than 15 tonnes/ha in eight years. Lovato [70] found an increase of 2 tonnes/ha/year over 13 years in a rotation system of oats - common vetch/maize - cowpea. These figures confirm the potential of conservation agriculture for carbon sequestration. However, the "simple" change from soil tillage to zero tillage is not enough. According to Lovato [71], a minimum addition of 4.2 tonnes/ha/year of carbon in vegetative residues in cropping systems and 4.5 tonnes/ha/year in mixed systems of pastures and crops [72] is necessary for maintaining soil organic matter at stable levels. This means that below these values CO_2 emission will or can take place and, thus, that for conservation agriculture systems to become successful in promoting carbon sequestration, it is necessary to include crops and pastures in the rotation that add large quantities of biomass. Even more C can be stored by adding leguminous cover crops to the rotation cycle. This is shown in Figure 6, based on two long-term experiments in Rio Grande do Sul, Brazil. Besides addition of C to the soil, legumes add a substantial quantity of N to the soil, which results in increased biomass production of succeeding crops.

Assuming an average carbon accumulation of 0.5 tonne/ha/year, an area like southern Brazil (Rio Grande do Sul, Santa Catarina and Paraná) under conservation agriculture would have the potential to sequester 5 million tonnes of C annually, which corresponds to 18 million tonnes of atmospheric CO_2. To compare, Brazil as emitted an estimated 84 million tonnes of CO_2 in 2000 [74]. Recent studies have shown that soil temperature is one of the main climate factors that influence CO_2 emission. High soil temperatures accelerate soil respiration and thus increase CO_2 emission [75]. This has implications for the landscape and land use in a certain area. On convex slopes and hilltops, emission is greater than in foothills, where temperatures are normally lower. The foregoing indicates that not only for soil and water conservation is it important to protect the soil with vegetation (reduction in soil temperature), but that it is advisable to cover the soil also with a view to reducing greenhouse gas emissions.

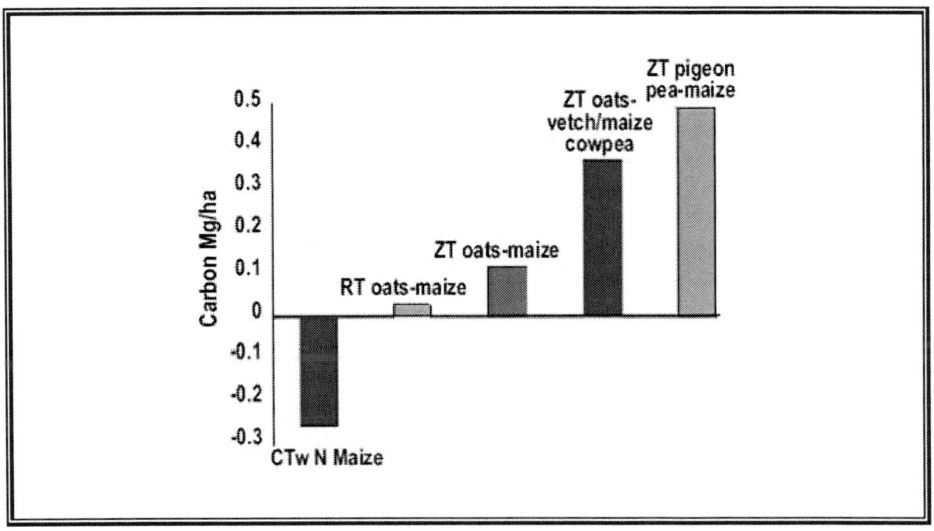

Figure 6. Potential for carbon sequestration in long-term experiment in southern Brazil (Source: Adapted from Amado [64]. Notes: CT = conventional tillage, RT = reduced tillage, ZT = zero tillage, CTwN = conventional tillage without N fertilizer. CT oats-maize (=0).

The maintenance of soil organic matter levels and the optimization of nutrient cycling are essential to the sustained productivity of agricultural systems. Both are related closely to the bioturbating activities of macrofauna and the microbially-driven mobilization and immobilization processes, which the activities of large invertebrates also encourage. Maintaining soil organic matter content requires a balance between addition and decomposition rates. As changes in agricultural practices can engender marked changes in both the pool size and turnover rate of soil organic matter, it is important to analyse their nature and impacts. Crop production worldwide has generally resulted in a decline in soil organic matter levels and, consequently, in a decline of soil fertility. Converting grasslands and forestlands to arable agriculture results in the loss of about 30 percent of the organic C originally present in the soil profile. On reasonably fertile soils with reliable water supply, yields in long-term arable agricultural systems have been maintained at very high levels by applying substantial amounts of fertilizer and other soil amendments. In low-input agricultural systems, yields generally decline rapidly as nutrient and soils organic matter levels decline. However, restoration is possible through the use of fallow lands, mixed crop-livestock and agroforestry systems, and crop rotations. Traditional mould-board plough and disc-tillage cropping systems tend to cause rapid decomposition of soil organic matter, leave the soil susceptible to wind and water erosion, and create plough pans below the cultivation depth. By contrast, reduced- or zero-tillage systems leave more biological surface residues, provide environments for enhanced soil activity, and maintain more intact and interconnected large pores and more soil aggregates, which are better able to withstand raindrop impact. Water can infiltrate more readily and rapidly into the soil with reduced tillage and this helps protect the soil from erosion. In addition, organic matter decomposes less rapidly under reduced-tillage systems. No-tillage systems have proved especially useful for maintaining and increasing soil organic matter. Crop rotation is the basis for the sustainability of direct sowing systems. A production system that includes cover crops, legumes for N fixation, crop rotation and no tillage can be adapted regionally and, therefore, contribute to the sustainability of soil management in the region. Where rainfall intensities are very high, or biomass management options are limited by a water shortage, maintenance of soil surface cover by crop canopies or crop residues during periods of high erosion risk is essential. With improved land management, at least part of the organic matter lost can be restored. The increase in soil organic matter in the absence of tillage can transform agricultural soils into carbon sinks. The relatively low levels of active organic matter fractions in zero-tillage systems have highlighted the extreme dependence of such systems on the maintenance of high levels of surface protection by crop residues. Residue accumulation, including cover crops and crop residues, increases the levels of some soil nutrients and soil organic C. The active fraction of organic matter plays a very important role in aggregate stability and rainfall infiltration. Building up active C levels in the soil in rainfed cropping systems may have a greater impact in reducing surface crusting and improving rainfall infiltration capacity than would simply changing to zero-tillage systems. Management practices designed to maximize C inputs and to maintain a high proportion of active C should be seen as essential steps towards more sustainable cropping systems. Cover crops, intercropping and crop rotations can also help to promote biodiversity both below the soil surface and aboveground. This diversity is important to maintaining a well-functioning and stable ecological system. Where many different types of organisms coexist there are: fewer problems with diseases, insects and nematodes; more competition among species; and more possibility for many types of predators to thrive. In such a situation, no single pest organism is able to reach a population of sufficient size to affect crop yield seriously.

Microorganisms in Soil Ecosystem

Minerals, organic components and microorganisms are among major solid components of soils. These components are not separate entities but rather a unified system constantly in association with each other in the environment [76]. Minerals, organic matter, and microorganisms have significant higher order interactions in terrestrial ecosystems. These interactions exert strong control over soil physical, chemical, and biological processes. Fundamental understanding of mineral–organic matter–microorganism interactions is essential for understanding, restoring, enhancing, and sustaining ecosystem integrity on a global scale. The study of the interactions of these soil components has to be considered from the molecular level to field/landscape systems and is essential to stimulate further research to uncover the dynamics and mechanisms of soil processes and the impacts on biogeochemistry and biodiversity in the terrestrial ecosystem. Huang et al. [77] showed that organic substances and microorganisms substantially affect the formation and transformation of short-range order (SRO) sesquioxides. These oxides exert great influences on biogeochemical processes in the terrestrial ecosystem. Soil mineral surfaces play a vital role in the catalysis of abiotic formation of humic substances. Soil minerals, especially noncrystalline minerals, have the ability to chemically stabilize SOM. Organic substances also act as binding agents to promote and stabilize aggregation. Therefore, noncrystalline minerals retard the microbial degradation of SOM, which in turn inhibits the transformation of noncrystalline mineral phases to more crystalline minerals. Hence, there are distinct interactive mechanisms between soil minerals, organic matter, and microbes, which should have a direct influence on the stability and cycling of C, N, P, and S and global change. The soil microbial biomass is a significant component of terrestrial ecosystems. Its activity contributes to the regulation of soil carbon (C) sequestration, C mineralization, nutrient cycling and ecosystem productivity [78,79]. Soil microbial biomass is an important parameter linking the plants to soil, and it comprises about 2–3% of total organic carbon in the soil and has been recognized as an important source of nutrients to plants because of its fast turnover [56]. Microorganisms are the main source of enzymes in soils [80], and thus the composition of the soil microbial communities strongly affects the potential of a soil for enzyme-mediated substrate catalysis [81]. Microorganisms are responsible for innumerable processes that occur in soils. Soil enzymes (intracellular and extracellular) are the mediators and catalysts of biochemical processes important in soil functioning such as nutrient mineralization and cycling, decomposition and formation of soil organic matter, and decomposition of xenobiotics (i.e., pesticides). Specifically, the assessment of the activities of hydrolases can provide information on the status of key reactions that participate in rate limiting steps of the decomposition of organic matter and transformation of nutrients in soils. Thus, knowledge of several soil enzyme activities can provide information on the soil degradation potential [82]. The decomposition of organic matter is mediated by extracellular enzymes released into the environment by microbial communities by secretion or cellular lysis. The decomposition process can be represented as a successional loop [83] in which strong interactions are established among substrate, microdecomposers, and extracellular enzymes. Substrate quality affects the microbial community, inducing the release of specific extracellular enzymes that degrade and modify the substrate. In turn, substrate break down products drive microbial succession. In this view, three major types of relationships can be recognized: enzyme activity and litter composition, enzyme activity and mass loss rate, and enzyme activity and microbial community composition. Differences in the initial litter composition, notably in the relative contents of lignin, cellulose, hemicellulose and pectin, lead to differences in enzyme activity through enzyme substrate-interactions [84]. Litter quality changes during decomposition and as a consequence changes the degree of enzyme expression. Activities of enzymes involved in the soluble saccharide degradation peak early in the decomposition process, then decline markedly; the activities of the main cellulases generally peak about midway through decomposition. On the contrary, enzyme activity involved in the breakdown of more recalcitrant compounds tends to be higher in fast-decomposing litter [83, 85, 86]. Climate conditions also influence the interaction between substrate quality and microbial community. Fioretto et al. [87] studied the patterns of microbial respiration and enzyme activity (cellulase, xylanase, laccase and peroxidase) in the leaf litter of two evergreen sclerophyll species, *Myrtus communis* and *Quercus ilex*, during decay in a low shrubland and in a coppice wood in a Natural Reserve in the Mediterranean area, suggesting a succession of functionally similar microbial communities, with changes more linked to climatic conditions than litter

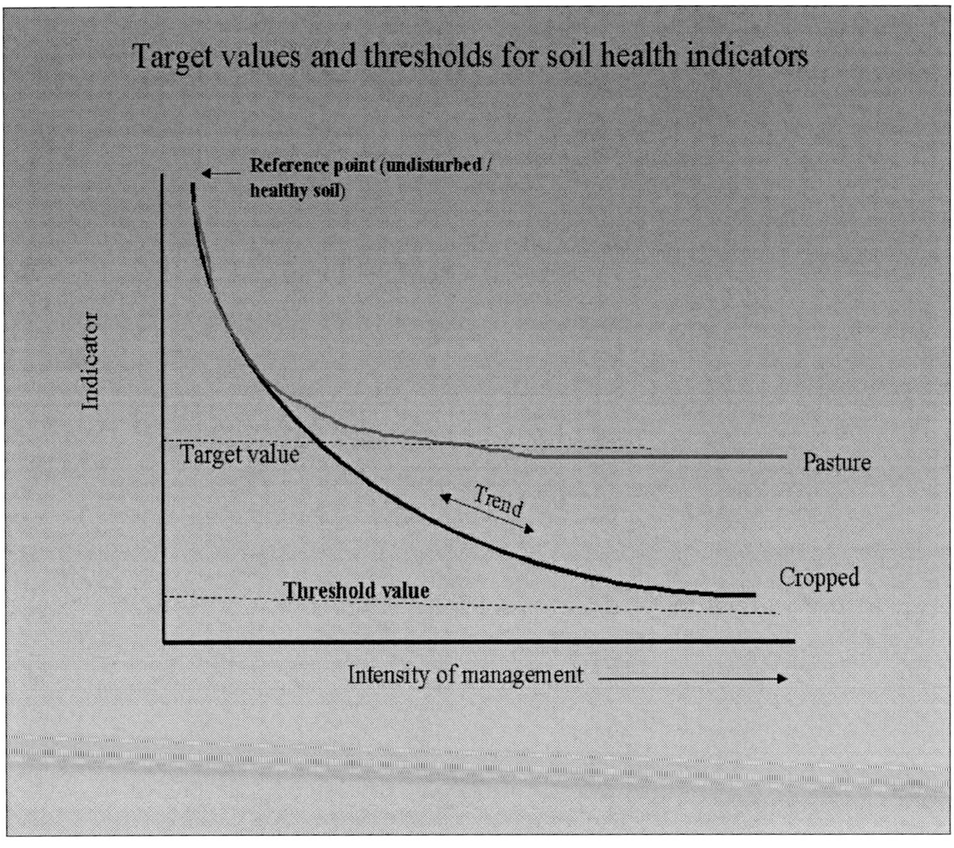

Figure 7. Soil health indicators.

quality and decomposition rate. Soil enzyme activities are the direct expression of the soil community to metabolic requirements and available nutrients. While the taxonomic diversity of soil organisms is important, the capacity of soil microbial communities to maintain functional diversity of those critical soil processes through disturbance, stress or succession could ultimately be more important to ecosystem productivity and stability. Understanding and maintaining biodiversity has become an increasingly important field of research, as well as a resource management goal. In soil microbial communities, maintaining critical functions may ultimately be more important than maintaining taxonomic diversity. Soil biota communities are one of the most diverse groups of earth's biota. Soil organisms regulate a number of processes in terrestrial ecosystems that are not only critical for productivity but are also essential for maintenance of ecosystem health. The loss of any biodiversity from the natural ecosystem levels should be regarded as detrimental. Therefore, the potential for adoption of target and threshold levels of biodiversity needs exploration. Determination of direction of trends would be important, and some indicators may respond more rapidly than others (e.g. microbial or macrofaunal diversity, soil enzymes) (Figure 7).

The first and most obvious role of biodiversity is to ensure the multiplicity of functions, which can be described to soil organisms (Figure 8).

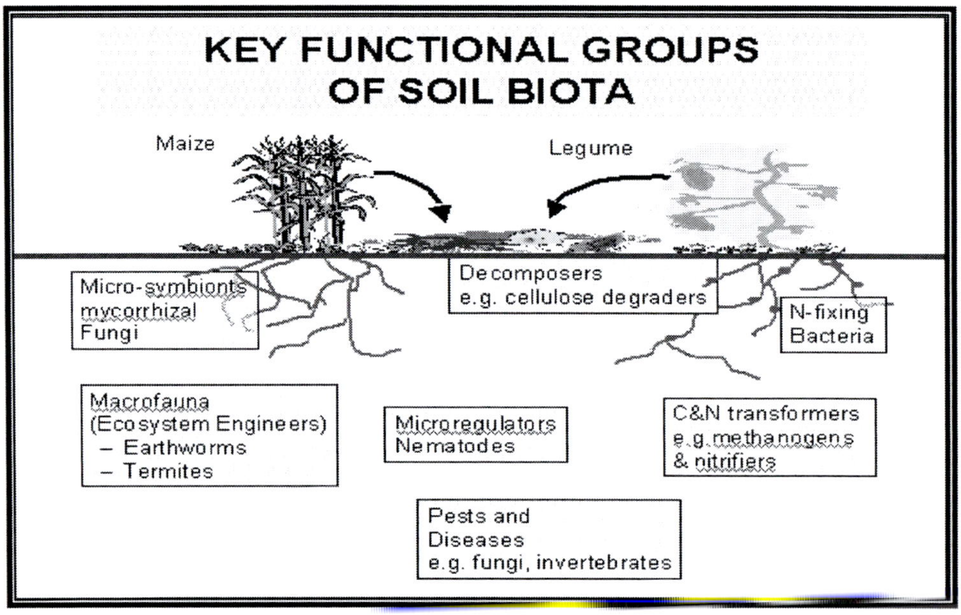

Figure 8. Groups of soil biota.

A secondary, but important, role of biodiversity is further to ensure these functions are maintained in the face of perturbations. Understanding the relationship between biodiversity and more complex functions will undoubtedly require the combined study of taxonomically distant groups of organisms which can perform specific functions, and thus belong to the same guild. Soil microbial communities represent the largest source of biodiversity on earth. Given the extreme high species diversity in soil, it is estimated that microbial communities contain such high levels of redundancy as to make small changes in soil microbial diversity insignificant. Rather, shifts among groups or species within the microbial community are considered to be of much more relevance for the functioning of terrestrial ecosystems. Shifts that might be relevant for sustainable land use include those in the relative abundance of bacteria and fungi and within groups with specific functions such nitrifying bacteria. These shifts could affect vital functions of the soil ecosystems such as the nutrient retention and antagonism against plant diseases. A greater degree of biodiversity between, or within, a given species or functional group will logically increase the inherent variability in tolerance or resistance to stress or disturbance. Genetic variability within and between species confers the potential for resistance to perturbations, whether short- or long-term. Implicit in these arguments is the assumption that a multiplicity of organisms can perform a particular function, and that the replication of the ability to perform a particular function means that a degree of functional redundancy exists. Whether organisms are ever truly redundant is a matter of debate. Though redundancy in a single function may be common among many soil biota, the suite of functions attributable to any one species is unlikely to be redundant. Furthermore, functionally similar organisms have different environmental tolerances, physiological requirements and microhabitat preferences. As such they are likely to play quite different roles in the soil system. Any novel technology that involves manipulating and managing natural resources and more specifically soil biodiversity and biotic products *in-situ* requires risk analysis. There has been considerable interest recently in evaluation of risks associated with adoption of GM crop varieties, but apart from these cases there has rarely been serious attention paid to impacts on soil ecosystem functioning. One essential microbial function in soils is the processing and recovery of key nutrients from detrital inputs and accumulated soil organic matter. This often requires the activity of extracellular enzymes to process complex organic compounds into assimilable subunits (sugars, amino acids, NH_4^+, PO_4^{-3}). Soil enzyme functional diversity can be analyzed and interpreted in a variety of ways, depending on the specific research questions. Functional diversity *between* nutrient resources could be based on specific enzyme activities against major C (cellulose), N (protein) and P constituents. Functional diversity *within* a nutrient group can be estimated by measuring cellulase and/or phenoloxidase for carbon, protease and amidase for nitrogen or phosphomono- and diesterases for phosphorus. Greater resolution within group functional diversity could be gained by focusing within a given enzyme activity; e.g., proteolytic activities separated by inhibitor class. At the simplest level, soil enzyme diversity has frequently been evaluated as differences in activity. The field of soil enzymology, including numerous methods and applications, has been extensively reviewed [88, 89]. Soil enzyme data have been the foundation for the development of conceptual models that provide a more comprehensive understanding of those key processes linking microbial populations and nutrient dynamics [90, 91]. Knowing the sources of specific soil enzyme activities would greatly enhance our understanding of which group(s) of organisms are directly accessing a given nutrient resource, thus providing greater insight into the pathways by which energy and nutrients flow through the soil food web. Developing approaches to assess soil enzyme functional diversity will increase our understanding of the linkages between resource availability, microbial community structure and function, and ecosystem processes. Influence of environmental factors on microbial population and microbial biomass plays an important role in nutrients cycling in an ecosystem. Soil physico-chemical characteristics also has a great impact on microbial biomass and microbial activity and can be used to measure soil quality [92] but it may take years for these parameters to make significant changes in soil. Soil biological and biochemical changes are very sensitive to small changes in soil conditions (degradation, erosion) and thereby gives more accurate and immediate information in soil quality because soil microbial activity has a direct influence in ecosystem stability and fertility [93]. Insam et al. [94] also proposed that the ratio of microbial biomass to total organic carbon in a soil might serve as a quantitative indicator of carbon dynamics in the soil.

Changes in size of the microbial biomass pool may also indicate changes in the soil organic matter pool that are not otherwise easily detectable. Many factors such as temperature [95, 96], moisture content [97], clay content [98] and pH [99] are known to affect microbial biomass in soil. Topography plays a critical role in modifying both microclimate and hydrological conditions within a landscape [100]. In particular, the influence of topography on the movement of water and the consequent redistribution of materials carried with water can influence or control the type and intensity of soil processes within a landscape [101, 102]. Microbial response to topographic control on moisture and nutrient redistribution within a forest landscape may alter nutrient cycling processes and, consequently, influence nutrient losses [103].

Information on soil microbial biomass in different forest ecosystems have been reported by several workers [104, 105, 106, 107, 108, 109] but information on seasonal changes in the microbial biomass in an annual cycle in forest ecosystem is limited [110, 111].

The results of numerous studies indicate that the structure of the soil microbial community varies with the composition of the forest plant community [112, 113]. There are several possible explanations for such a pattern. For example, interspecies variation in the quantity and quality of root exudates can result in the development of distinct rhizosphere microbial communities [114]. Litter inputs from aboveground can affect soil microbial community structure by altering carbon availability, nitrogen availability [115, 116], soil organic matter composition [117] and soil pH [118]. Given that soil temperature and moisture conditions can affect microbial community structure [119, 120, 121], differences among plant species in canopy morphology and phenology may also influence soil microbial communities via altered thermal, moisture and light regimes at soil surface [117, 118].

Grayston and Prescott [113], comparing forest floor microbial communities in pure plots of four tree species (*Thuja plicata*, *Tsuga heterophylla*, *Pseudotsuga menziesii*, and *Picea sitchensis*), with analyses based on CLPP and PLFA profiles, showing distinct differences in forest floors under different tree species and among sites of differing fertility. Multivariate analyses of CLPP and PLFA data indicated that forest floor layer had the greatest effect on microbial community structure and potential function, followed by site, while tree species had the least effect. The patterns were consistent among the various measures employed, and are in keeping with previous reports of the effects of these tree species and of site fertility on microbial communities. The similarity in trends among measures of nutrient availability and microbial communities based on CLPP and PLFA profiles is further evidence that these profiles provide an indication of the activity of microbes relevant to the cycling of nutrients in forest ecosystems.

Leckie et al.[112] to address the link between soil microbial community composition and soil processes, investigated the microbial communities in forest floors of two forest types that differ substantially in nitrogen availability. Cedar-hemlock (CH) and hemlock-amabilis fir

(HA) forests are both common on northern Vancouver Island, B.C., occurring adjacently across the landscape. CH forest floors have low nitrogen availability and HA high nitrogen availability. Total microbial biomass was assessed using chloroform fumigation-extraction and community composition was assessed using several cultivation-independent approaches: denaturing gradient gel electrophoresis (DGGE) of the bacterial communities, ribosomal intergenic spacer analysis (RISA) of the bacterial and fungal communities, and phospholipid fatty acid (PLFA) profiles of the whole microbial community. No differences were detected in the bacterial communities of each forest type using DGGE and RISA, but differences in the fungal communities were detected using RISA. PLFA analysis detected subtle differences in overall composition of the microbial community between the forest types, as well as in particular groups of organisms. Fungal PLFAs were more abundant in the nitrogen-poor CH forests. Bacteria were proportionally more abundant in HA forests than CH in the lower humus layer, and Gram-positive bacteria were proportionally more abundant in HA forests irrespective of layer. Bacterial and fungal communities were distinct in the F, upper humus, and lower humus layers of the forest floor and total biomass decreased in deeper layers. These results indicate that there are distinct patterns in forest floor microbial community composition at the landscape scale, which may be important for understanding nutrient availability to forest vegetation. Devi and Yadava [122] studied the soil microbial biomass C, N and P in two stands of mixed-oak forest ecosystem of Manipur, North-east India to assess the influence of abiotic variables and difference in forest stand on the seasonal variation in soil microbial biomass. They demonstrated that the soil microbial biomass exhibits strong seasonality and is highly influenced by the abiotic variables. However, soil moisture has a strong influence on the microbial biomass. The microbial C:N ratio indicates that soil fertility is influenced by the species composition of the forest stands. High microbial N during the rainy season may be considered as a nutrient conservation strategy. Further, the proportion of microbial C and N to soil C and N indicates that C and N are immobilized more during rainy season where as P is immobilized more in winter season.

Environmental conditions associated with different seasons affect the microbial biomass and activity either directly by inducing microbial response to soil changes or indirectly by influencing plant metabolism [110]. Seasonal changes in soil microbial biomass and activity have been reported in many ecosystems and have been ascribed to a combination of factors including substrate availability, root growth, microbial competition and community structure, repression of enzyme activity and the accumulation of recalcitrant and toxic compounds [123]. Seasonal changes in soil microbial biomass C and associated nutrients are important in controlling overall nutrient cycling and availability [124]. The level of inorganic P in soils depends on a combination of factors including plant uptake, adsorption/desorption and dissolution/precipitation of inorganic P, the mineralization of organic P and microbial immobilization, and fertilizer P addition [125, 126]. Seasonal changes in organic P pools are considered to be controlled by biological processes in soil which in turn are influenced by a combination of soil moisture and temperature [127, 128]. Chen et al. [129] determined how P availability and dynamics, and microbial activity in adjacent soils under grassland and forest varied seasonally in response to changes in environmental conditions, showing that levels of soil TOC, total N, total P and organic fractions of P under grassland were consistently higher, but inorganic P fractions were lower compared with forest over all seasons. These findings are consistent with those from closely related studies which showed that grassland afforestation enhances the mineralization of organic matter and associated organic P. On the other hand, the microbial biomass C and P and enzyme activities (pectin methyl esterase, PME and phosphodiesterase, PDE) in the soil under grassland were consistently higher than under forest, while soil respiration under grassland was much higher in summer and slightly lower in winter compared with forest. These findings are also consistent with results from previous studies carried out in New Zealand [130, 131]. Phosphorus recycling was mainly driven by plant P demand and sustained by mainly root litter inputs in the grassland ecosystem and leaf litter input in the forest ecosystem. Seasonal changes in environmental conditions (rainfall, soil moisture and temperature) influenced the biological and biochemical processes involved in P cycling. In general, labile organic P (e.g. BPo, MBP) was mineralized by the increasing microbial activity to meet the increasing P demand by plant growth in spring and summer, while the accumulation of organic P (labile Po) was increased by organic inputs and slow growth of plants and low microbial activity in late autumn and winter. Labile Po is an important component in P recycling in grassland and forest soils, while the microbial biomass also plays a pivotal role in P cycling. Microbial biomass varies with seasonal patterns of soil temperature, moisture, and availability of substrate [132, 133]. However, studies on the seasonal effects on microbial biomass in surface and subsurface soils have often produced inconsistent patterns. Marschner et al. [134] observed higher microbial biomass of surface soil in the wet season for the polyculture agroforestry systems in Central Amazonia. A relatively high microbial biomass C and N in summer was coupled with warm weather, high soil moisture, and high soil C and N. Possibly, the warm summer weather supported a high microbial population in the region with a mild temperature. On the other hand, many studies have indicated that soil microbial biomass tends to be highest in spring and fall and lowest in summer and winter [133, 135, 136, 137]. Muscolo and Sidari (data not shown) found a higher amount of microbial biomass in forest soil of Aspromonte Southern Italy, in spring compared to the other seasons, and consequently a better humification/mineralization processes and a major nutrient availability along soil profile. The data were in agreement with those reported by Garcia et al. [138]. Saynes et al. [139] determined the effects of forest age on microbial biomass C and N, the pools of total soil C and N and potential N transformations and evaluated the effects of rainfall seasonality on the soil and microbial nutrient dynamics in primary and secondary tropical seasonal forests. In this paper they measured the seasonal soil microbial biomass C and N (during the dry season, and at the beginning and in the middle of the rainy season), the standing pools of total soil C and N and the potential N transformations in sites of replicate chronosequence of forest succession in the center of Mexico. The results evidenced that microbial biomass C and N pools in the different age forests were affected by the start of the rainy season. Interestingly, the results of this study show a striking contrast in the effect of rainfall seasonality on the primary and late-successional forests in comparison with forests in early- and mid-successional stages of secondary succession. The start of the rainy season leads to a high increase in both the microbial biomass C pool and its ratio C:N in the soils of the early- and mid-successional forests, where N pools and N cycling are the highest; whereas in late-successional and primary forests the onset of the rain leads to a decrease in both parameters measured. Differences in response of the soil microbial biomass to the start of the rainy season in the forests under study had important consequences for nutrient cycling in these seasonal ecosystems. The response of microbial biomass to seasonal changes is important in regulating microbial turnover, which may in turn influence nutrient availability and ultimately plant nutrition [140]. Forest management practices, such as timber harvesting and mechanical site preparation have the potential to cause detrimental levels of soil and site disturbances, particularly soil compaction and organic matter removal [141, 142, 143, 144]. Many studies have investigated the impact of compaction on soil physical properties and tree growth [145, 146, 147] however, relatively few studies have focused on the effects of compaction on microbial properties and processes in forest soils [148, 149, 150]. Microbial biomass is one of the most labile fractions of soil organic matter and plays a pivotal role in mediating C and N cycling in the soil and can be a sensitive indicator of changes caused by forest management practices [151, 152]. By changing the proportions of macro- and microporosity, soil compaction could cause oxygen deficiency and protect organic material and microbial biomass against degradation by microorganisms and microfauna, and could concomitantly inhibit microbial activities and reduce N mineralization rates [153, 154]. Tan et al. [34], in the boreal white and black spruce long-term soil productivity experiment (Dawson Creek, BC), established that compaction changed soil biophysical conditions, such as reduced soil aeration porosity, and lowered growing season soil temperature and moisture content. Soil compaction decreased MBN but increased microbial C:N ratio in the mineral soil, and reduced nitrification rates in both the forest floor and mineral soils. Forest floor removal tended to decrease microbial biomass, most likely as a result of reduced substrate availability for microbial metabolism. Forest floor removal increased net N mineralization and nitrification rates in the mineral soil in certain sampling dates possibly as a result of the increased soil temperature. The hypothesis that forest floor removal would result in lower N transformation rates was not supported, suggesting that forest floor removal did not have a negative effect on N transformation rates in the short-term; however, N export in the removed organic matter was relatively high. Microorganisms have a fundamental role in the biogeochemical cycles of the elements and in the formation of soil structure [155], such that it is widely accepted that a high level of microbial activity is necessary for the maintenance of an adequate quality of soil. It has been shown that different soil management regimes affect the structure and activity of the soil microbial community [10], via changes in the quantity and quality of the plant remains that enter the soil [156].

Soil microorganisms and the processes that they govern are essential for long-term sustainability of agricultural systems [157] and are a major component in soil formation and nutrient cycling. Changes in soil chemical and physical conditions will influence microbial activity and population structure. On a primary basis, the natural environment provides long-term and seasonal fluctuations in temperature, moisture, and plant growth to which microorganisms must respond. Agriculture further alters the soil environment from its natural state, either amplifying or reducing seasonal cycles and influencing how the soil responds to climatic factors. Changes in soil physical and chemical properties that occur as a result, and subsequent changes in microbial community composition, will in turn influence soil processes [158], nutrient cycling, and overall soil quality. Nutrient fluxes through microbial biomass are of at least one order of magnitude faster than in the remaining organic matter [159] leading to the suggestion that microbial biomass could be used as an important indicator of changes in soil health and soil quality [160, 161, 162] produced by agricultural management practices.

Concurrent with changes in soil microbial biomass are shifts in the structure of the soil microbial community [123]. Tillage leads to the development of soil microbial communities dominated by aerobic microorganisms with high metabolic rates, typically bacteria, whereas under conservation practices, plant residues left at or near the soil surface encourage fungal growth and the temporary immobilization of nutrients [163]. An increase in the proportion of microbial biomass attributable to fungi was found in reduced tillage (RT) [164] and no tillage [165] systems. Microbial biomass measurements give little information about qualitative community-level changes, are limited in their ability to describe a particular microbial ecosystem, and give no indication of activity [166].

Spedding et al. [167] studied the long-term impact of tillage and residue management on soil microorganisms over the growing season in a sandy loam to loamy sand soil of south-western Quebec, growing maize (*Zea mays* L.) monoculture, demonstrating that diverse factors such as resource availability and crop growth stage, and soil physical and chemical conditions may influence soil microorganisms. These factors should be considered when using soil microorganisms as an indicator of soil quality. All communities and ecosystems are subjected to a range of extrinsic factors that can be broadly classified as stress factors (i.e. those that remain constantly limiting for organism growth and do not change markedly over time) and disturbance factors (i.e. those that involve rapid changes and often involve destruction of organism biomass). The interactive effects between stress and disturbance have been well explored for plant communities, but little is known about how these interactions influence soil communities. Williamson and Wardle [168] set up a 14-month glasshouse experiment, which consisted of varying levels of a disturbance (plant defoliation) and a stress (gradient of water limitation to plants), in a full factorial design. Response variables consisted of plant biomass and productivity, soil microbial biomass and activity (measured using multiple substrate utilization profiling), and soil microbial community structure (measured using phospholipids fatty acid analysis). Although the microbial community responded to both the stress and disturbance factors, these responses did not closely track plant responses to the same factors, pointing to a degree of decoupling between plant and microbial communities. Further, although stress and disturbance factors both limited some measures of microbial biomass and activity, the microbial community responded to these two factors in quite different ways. There were no interactive effects of stress and disturbance on any microbial response variables, with the exception of one variable that involved total microbial community substrate utilization. Although the ability of microbes to exploit water resources that are less available to plants may buffer the soil microbial community from stress, the general lack of interactions of stress and disturbance on the microbial community contrasts with what has been shown for plant communities and proposed for fungal communities. The results suggest that while stress and disturbance are both strong drivers of the microbial community, the effects of these factors largely operate independently from one another. The stability and function of a soil ecosystem depends on the cycling of nutrients by the soil microbial community [169]. They are responsible for nitrogen fixation, nutrient cycling, immobilization of essential nutrients and production of phytohormones [170, 171, 172]. Forest ecosystems depend on the soil microbial community to convert nutrients from plant biomass and soil organic matter to mineral pools primarily used by plants [169]. Correspondingly, management strategies and restoration approaches significantly affected the structure and function of soil microbial communities [173, 174].

In intensively managed forest plantations, clear-cutting and intensive site preparation may have negative effects on fundamental soil processes that determine the sustainability of these systems. Management of logging residues and soil disturbance during site preparation alter soil functions associated with forest sustainability, such as structure, aeration, water retention capacity, and can also disrupt certain biological processes involved in nutrient availability [175, 176]. Because of the large quantity of nutrients removed, whole-tree harvesting can lead to decreased soil available reserves of limiting nutrients [177, 178]. In addition to the nutrient export with biomass, nutrient losses by leaching can increase significantly after clear-cutting. Nutrient losses may be especially severe in tropical and subtropical regions, where fast-growing forest species grow on strongly acidic soils with low nutrient capital [179, 180]. Soil micro-organisms affect growth and vitality of trees by release or uptake of plant nutrients during decomposition of organic matter. One of the methods used to assess the rate and progress of soil reclamation is the monitoring of soil microbial characteristics and specifically, microbial biomass and organic matter mineralization. These parameters can provide an indication of changes in total soil organic matter long before changes in total soil C and N can be reliably detected.

Clear-cutting alters C cycling through removal of aboveground tree biomass for wood products. This induces changes in the forest ecosystem that extend to a wide range of biogeochemical processes [181, 182, 183, 184]. For example, the input of litter to the forest floor decreases for several years following clear-cutting [185]. The removal of the forest canopy also causes decreased interception of atmospheric moisture, reduced transpiration, and increased soil temperatures due to increased solar radiation input to the soil surface. These conditions will likely stimulate microbial activity and therefore increase the rate of OM decomposition [183, 186, 187, 188, 189, 190, 191]. Changes in decomposition rates and patterns after clear-cutting are likely to influence the amount, structure, and chemistry of OM in forest ecosystems.

Forest harvesting in ecosystem induces complex ecological, biogeochemical, and hydrological changes. Results from this study suggest that following clear-cutting there is a dynamic redistribution and restructuring of SOM pools and fractions in these soils. In addition, accelerated decomposition in the forest floor resulted in the preferential decomposition of *O*-alkyl C compounds and the loss of humic substances from the soil, probably through increased leaching as DOC and increased respiration losses. Changes in the composition of the humic acid fraction after harvesting indicate that humic acid plays an important role in the response of soils to clear-cutting disturbance. The magnitude of changes in the structure and composition of SOM after clear-cutting were relatively small. The absence of large changes may be attributed to large C pool sizes and longer turnover times of SOM. More labile fractions such as humic and fulvic acids, and more dynamic pools of OM such as soil solution and stream water DOC may be more sensitive indicators of disturbance effects in the short and medium terms.

Together, these results indicate that clear-cutting resulted in the loss of humic substances from the forest floor and upper mineral horizons, which was replaced by less decomposed OM in the post-clear-cut soils under the regrowing forest

Many management practices have a detrimental effect on soil microbial biomass which in turn affects soil biochemical properties and especially soil fertility. The creation of gaps in forests is an opportunity for the system to change dynamics and ecological processes [192]. It is often assumed that the increased levels of the light and nutrients [193, 194, 195, 196, 197] which occur as gaps are created act as important factors controlling natural regeneration. It is difficult to asses the role that the gaps have on regeneration processes, especially when information on the interaction between gap characteristic and soil chemistry is not well understood.

The creation of artificial gaps in forests offers the opportunity to observe in experimental way changes in soil system which are difficult to study because they occur naturally following sporadic and unpredictable treefalls.

Muscolo et al. [198] studied, over two years, change in soil chemistry in small (185 m^2) and medium (410 m^2) gaps, created in mature silver fir (*Abies alba* Mill) stands (mean height 30 m) in the Calabria Apennines (Southern Italy).

Within small gaps a greater amount of organic matter, compared to under canopy cover sites and medium gaps, was observed. On the contrary, in medium gaps a lower amount of organic matter content was detected. Moreover a different trend of the organic matter between small and medium gaps was observed. In fact, the decrease of organic matter content, associated to a lower amount of humic acid, compared to under canopy cover sites and small gaps, suggested that the organic substrate undergoes a mineralization rather than a humification process. An opposite trend is instead observed in small gaps. In fact, the increase of organic matter content, associated to an increase of humic acid amount, evidenced a better humification process compared to medium gaps and forest sites, suggesting small gaps as the best silvicultural practices.

Muscolo et.al [199] examined the impacts of small (380 m2), medium (855 m2) and large (1520 m2) gaps on microclimate, soil properties and microbial biomass, a ‘sink’ and ‘source’ of plants nutrients, in *Pinus nigra* (*Pinus laricio* Poiret) stands in Calabria Apennine Southern Italy. PAR transmittance significantly increased with increasing size of the gaps, consequently an effect of gaps size was found on microclimate: soil moisture and temperature were different among the gaps of different size showing the highest values of temperature in the large gaps and the highest values of moisture in the small gaps. Within small gaps the greatest amount of organic matter, humic matter and microbial biomass, and the highest C/N value were observed. The highest values of phosphatases, urease and FDA activities were detected in small gaps. An impact of gap size on C, N and P cycles was significant in small gaps in terms of higher availability of these nutrients and greater amount of humic matter, suggesting that small gaps may be important from an ecosystem perspective representing the appropriate management procedures for an adequate conservation of ecological functions, capable to preserve soil properties and favour *Pinus laricio* natural regeneration.

To date, understanding of the impacts of soil mineral–organic matter–microbe interactions on the transformation and degradation of anthropogenic organic compounds in soils is still relatively limited. The knowledge on detailed surface structures of the enzyme–mineral colloid and enzyme–humus complexes at the molecular level and the mechanisms of their catalysis in the transformation of organic contaminants need to be advanced also considering the possibility of using new enzymes and new mineral composite supports for catalytic degradation of a wide range of industrial and agricultural organic contaminants in soils.

Biomineralization is an important pathway of mineral formation in natural environments. Microorganisms and organics can profoundly affect the transformation of metals, metalloids and minerals in the terrestrial ecosystem. More attention should be paid to environmental impact of biomineralization on the fate of metals and metalloids. Furthermore, little is known on the influence of physico-chemical and microbiological interactions on the transformation, dynamics and toxicity of metals and metalloids in soil and related environments. The role of these interactions in remediation of metal- and metalloid-contaminated soils is an emerging issue in environmental protection.

To sum up, more research should be conducted to uncover the effects of physical–chemical–biological interactions on biogeochemical reactions, which, in turn, govern ecosystem biodiversity, sustainability and productivity. Use of advanced analytical instrumentation, i.e., synchrotron-based X-ray and infrared spectroscopy and atomic force microscopy and the like in this area of scientific pursuits should shed new light on the mystery of soil mineral–organic matter–microorganism interactions. Fundamental understanding of these interactions at the molecular level and the impacts on the terrestrial ecosystem would facilitate our development of innovative management strategies to regulate the behaviour of the terrestrial ecosystem on a global scale. Future research on this extremely important and exciting area of science should be stimulated to restore as well as sustain the ecosystem integrity.

In short, compositional changes in microbial communities are accompanied by changes in their capabilities to degrade macromolecular carbon compounds; the degradation of macromolecular carbon compounds may be related to community composition because the enzymatic capacity for the initial steps of degradation occurs in a comparatively limited number of microbial population [200] calls processes based on physiological capacities of limited distribution "narrow" processes. Several groups have reported the effect of a change in microbial community composition on process rates [158, 201]. Oxidative enzymes such as phenol oxidase and peroxidase, are produced by a relatively restricted group of microorganisms [200], but are not limited to fungi [202]. Cellulases are probably produced by more organisms than ligninases, but their relationship to shifts in community composition in natural environments is still poorly understood. Macromolecular carbon degradation is a synergistic result of many populations interacting in consortia. In addition, fungal interactions with bacteria can have profound effects on ligninocellulase activities [203]. Soil biodiversity confers an insurance against ecosystem malfunctioning under stress or disturbance Thus a broad approach that considers the components and activities of the entire soil community may be important in soil ecosystem, where frequent soil disturbance, carbon and nutrient alterations, and changes in the water and light regime can alter the composition of the microbial community and in turn the soil dynamic and sustainability.

Conclusion

It is well known that soil biota provide key ecosystem services that are responsible for naturally renewable soil fertility, however, there are areas where further research/clarifications on soil biodiversity and biological management is needed to enhance understanding on the function of soil biodiversity and ecosystem resilience in relation to land use and sustainable agricultural. On the basis of several case studies, a few gaps and opportunities for further research and development were emphasized.

1. To carry on targeted research in order to further clarify, for specific production systems, the most effective methods of organic matter management and their impact on soil life and plant-pest control.
2. To study and compile comprehensive data on soil ecosystem with a view to identifying opportunities and promoting concerted efforts for the biological restoration of soil health and fertility.
3. To carry on analyses of changes in soil biodiversity community composition and their association to soil ecosystem function and soil health and productivity, priority should be given to studies or activities that provide data on the impact of biological management of soil ecosystems, with emphasis on the environmental benefits of soil biodiversity and its management.

References

[1] Glanz, J.T. (1995). Saving our soils solutions for sustaining Earth's vital resource. Johnsons Books, Boulder, CO, USA.

[2] Larson, W.E., Pierce, F.J. 1991. Conservation and enhancement of soil quality. In: *Evaluation for sustainable land management in the developing world.* Vol.2. IBSRAM Proc. 12 (2) Int. Board for Soil Research and Management, Bankok, Thailand, pp. 175-203.

[3] Karlen, D.L., Mausbach, M..J., Doran, J.W., Cline, R.G., Harris R.F., Schuman, G.E. (1997). Soil qualità: a concept, definition, and framework for evaluation. *Soil Science Society of American Journal,* **61**, 4-10.

[4] Johnson, D.L., Ambrose, S.H., Bassett, T.J., Bowen, M.L., Crummey, D.E., Isaacson, J.S., johnson, D.N., Lamp, P., Saul, M., Winter-Nelson, A.E. (1997). Meanings of environmental terms. *Journal of Environmental Quality,* **26**, 581-589.

[5] Doran, J.W., Sarrantonio, M., Liebig, M. 1996. Soil health and sustainability. In: Sparks, D.L. (Ed.), Advances in Agronomy, Academic press, san Diego, pp.1-54

[6] Doran, J.W., Safley, M. (1997). Defining and assessing soil health and sustainable productivity. In: Pankhurst, C., Doube, B.M., Gupta, V.V.S.R. (Eds.) *Biological Indicators of Soil Health* CAB International, Wallingford, Oxon, U.K. pp. 1-28.

[7] Freckman, D.W., Virginia, R.A. (1997). Low-diversity Antarctic soil nematode communities: distribution and response to disturbance. *Ecology,* **78**, 363-369.

[8] Doran, J.W., Parkin, T.B. (1996). Quantitative indicators of soil quality: a minimum data set. In:Doran, J.W., Jones, A.J. (Eds). Methods for Assessing Soil Quality. *Soil Science Society of America,* Special Publication 49 Madison, WI, pp. 25-37.

[9] Doran, J.W., Parkin, T.B. (1994). Defining and assessing soil quality. In: Doran, J.W., Coleman D.C., Bezdicek, D.F., Stewart, B.A. (Eds.), Defining Soil Quality for a Sustainable Environment, SSSA Special Publication No 35. *Soil Science Society of America,* Madison, pp. 3-21.

[10] Reganold, J.P., Palmer, A.S. (1995).. Significance of gravimetric versus measurements of soil quality under biodynamic, conventional, and continuous grass management. *Journal of Soil and Water Conservation,* **50**, 53-80.

[11] Powers, R.F., Tiarks, A.E., Boyle, J.R. (1998). Assessing soil quality: practicable standards for sustainable forest productivity in the United States. In: Adams, M.B., Ramakrishna, K., Davidson, E.A. (Eds). The Contibution of soil science to the development and implementation of criterial and indicators of sustainable Forest Management. *Soil Science Society of America, Special Publication* No. **53**, pp.53-80.

[12] Harris, R.F., Karlen, D.L., Mulla, D.J. (1996). A conceptual framework for assessment and management of soil quality and health. In: Doran, J.W., Jones, A.J. (Eds). *Methods for Assessing Soil Quality*. Soil Science Society of America, Special Publication 49 Madison, WI, pp. 61-82.

[13] Burger, J.A., Kelting, D.L. (1999). Using soil quality indicators to assess forest stand management. *Forest Ecology and Management*, **122**, 155-156.

[14] Henderson, G.S. (1995). Soil organic matter: a link between forest management and productivity. In: Powl, P., Ladd, J.N. (Eds.), *Soil Biochemistry*, (pp. 415-471). New York: Marcell Dekker.

[15] Nambiar, E.K.S. (1997). Sustained productivity of forests as a continuing challenge to soil science. *Soil Science Society of American Journal,* **60,** 1629–1642.

[16] Rice J.A.and MacCarthy, P. (1991). Statistical evaluation of the elemental composition of humic substances, *Organic Geochemistry,* **17**, 635–648.

[17] Stevenson, F.J. (1994). Humus Chemistry: Genesis, Composition, Reactions, Second ed. ((pp 496) Wiley, New York.

[18] Schlesinger, W.H. (1997). *Biogeochemistry: An analysis of Global Change*. San Diego: Academic Press.

[19] Jobbery, E. G., Jackson, R B. (2000). The vertical distribution soil organic carbon and its relation to climate and vegetation. *Journal* of *Ecological Applications*, 423-436.

[20] Zech, W., Senesi, N., Guggemberger, G., Kaiser, L., Lehmann, J., Miano, T. (1997). Factors controlling humification and mineralization of soil organic matter in the tropics. *Geoderma,* **79**,117-161.

[21] Tiessen, H., Shang, C. (1998). Organic matter turnover in tropical land use systems. In: Bergstrom L, Kirchmann H (Eds) *Carbon and Nutrient Dynamics in Natural and Agricultural Tropical Ecosystems*. (pp 1-14) Cab International, Wallingford, United Kingdom.

[22] Woomer, P.L., Martin, A., Albrecht, A., Resck, D.V.S., Scharpenseel, H.W. (1994).The importance and management of soil organic matter in the tropics.In: Woomer, P.L., Swift, M.J. (Eds.). *The Biological Management of tropical Soil fertility*, (pp. 47-80) Wiley & Sons, New York.

[23] Körschens M. (1998). "Sustainable land use: long-term RSOM behaviour as observed in long-term field experiments", *Mitteilungen der Deutschen Bodenkundlichen Gesellschaft*, **87**, 93-105.

[24] Saggar, S., Parshotam, A., Hedley, C., Salt, G. (1999). "^{14}C-labelled glucose turnover in New Zealand soils", *Soil Biology and Biochemistry*, **31**, 2025-2037.

[25] Nichols, J..D. (1984). Relation of organic carbon to soil properties and climate in the southern Great Plains. *Soil Science society of America journal,* **48**,1382-1384.

[26] Müller, T., Höper, H. (2004). Soil organic matter turnover as a function of the soil clay content: consequences for model applications. *Soil Biology & Biochemistry,***36**,877-888.

[27] Swift, M.J., Heal, O.W., Anderson, J.M. (1979). Decomposition in Terrestrial Ecosystems. Blackwell Scientific Publication, Oxford, United Kingdom.

[28] Qualls, R.G. and Haines, B.L. (1992). Biodegradability of dissolved organic matter in forest through fall, soil solution and stream water. *Soil Science Society of America Journal,* **56**, 578-586.

[29] Herbert, B.E., Bertsch, P.M. (1995). Characterization of dissolved and colloidal organic matter in soil solutions: A review. In: Kelly JM, McFee WW (Eds) *Carbon Forms and Function in Forest Soils*. Soil Science Society of America, Madison, WI, U.S.A., pp 63-88.

[30] Wander, M.M. (2004). Soil organic matter fractions and their relevance to soil function. In: Magdoff. F., Weil, R.R. (Eds.), *Soil Organic Matter in Sustainable Agriculture*, (pp. 67-102). Boca Raton: CRC Press LLC.

[31] Cole, D.W., and M. Rapp. 1981. Elemental cycling in forest ecosystems. Pages 341–409 *in* D. E. Reichle, editor. Dynamic properties of forest ecosystems. Cambridge University Press, Cambridge, UK.

[32] Nardi, S., Concheri, G., Dell'Agnola, G., (1996).Biological activity of humic substances. Piccolo, A. (Ed.) *Humic Substances in Terrestrial Ecosystems*, Elsevier, Amsterdam, pp. 361-406.

[33] Nardi, S., Pizzeghello, D., Muscolo, A., Vianello, A. (2002). Physiological effects of humic substances on higher plants. *Soil biology & Biochemistry,* **34**, 1527-1536.

[34] Tan, X., Chang, S.X., Kabzems, R. (2005). Effects of soil compaction and forest floor removal on soil microbial properties and N transformations in a boreal forest long-term soil productivity study. *Forest Ecology and Management*, **217**, 158-170.

[35] Piccolo, A. (1996). Humus and soil conservation. In: Piccolo, A. (Ed). *Humic Substances in Terrestrial Ecosystem,*.Elsevier, Amsterdam, pp. 225-264 .

[36] Piccolo, A. (2001). The supramolecular structure of humic substances. *Soil Science,* **166**, 810-832.

[37] Sutton, R., Sposito, G. (2005). Molecular structure in soil humic substances: The new view. *Environmental Science & Technology,* **39**, 9009.

[38] Conte, p., Spaccini, R., Piccolo, A. (2006). Advanced CPMAS-^{13}C NMR techniques for molecular characterization of size-separated fractions from a soil humic acid. *Analytical and Bioanalytical Chemistry,***386**, 382-390.

[39] MacCarthy, P. (2001). The principles of humic substances. *Soil Science,* **166**, 738-751.

[40] Fox T. R. (1995). The influence of low-molecular weight organic acids on properties and processes in forest soils. In: McFee WW, Kelly JM (Eds). *Carbon forms and functions in forest soils.* Soil Science Society of America, Madison, WI, U.S.A., pp 43-62 .

[41] Inderjit, Mallik, AU (1997). Effects of *Ledum groenlandicum* amendments on soil characteristics and black spruce seedling growth. *Plant Ecology,* **133**, 29–36.

[42] Flaig, W. (1971) Organic compounds in soil. *Soil Science,* **111**, 19-33.

[43] Kuiters, A. T. and Sarink, H.M. (1986). Leaching of phenolic compound from leaf and needle litter of several deciduous and coniferous trees. *Soil Biology & Biochemistry,* **18**, 475-480.

[44] Leather G R and Einhelling F A 1988 Bioassay of naturally occurring allelochemicals for toxicity. *Journal of Chemical Ecoogy,* **14,** 1821-1828.

[45] Muscolo, A, and Sidari, M. (1998). Biological activity of humic substances and phenolic compounds extracted from two different forest soils of Aspromonte (Southern Italy). *Fresenius Environmental Bulletin,* **7**, 695-700.

[46] Muscolo, A, Panuccio, M.R., Sidari, M. (2001). Respiratory enzyme activities during germination of *Pinus laricio* seeds treated with phenols extracted from different forest soils. *Plant Growth Regululation,* **35**, 31-35.

[47] Muscolo, A., Panuccio, M.R., Sidari, M. (2002). Glyoxilate cycle in germination of *Pinus laricio* seeds: effects of phenolic compounds extracted from different forest soils. *Plant Growth Regulation* **37**, 1-5.

[48] Hättenschwiler, S. and Vitousek, P. M. (2000). The role of polyphenols in terrestrial ecosystem nutrient cycling. *Trends in Ecology Evolution,* **15**, 238-243.

[49] Odén, P.R, Brandberg, P., Andersson, R., Gref, R., Zackrisson, O., Nilsson, M.C. (1992). Isolation and characterization of a germination inhibition from leaves of *Empetrum hermophroditum* Hagerup. *Scandinavian Journal of Forest Research* **7**, 497-502.

[50] Muscolo, A., Sidari, M. (2006). Seasonal fluctuation in soil phenolics of a coniferous forest: effects on seed germination of different coniferous species. *Plant and Soil* **284**, 305-318.

[51] Muscolo A, Sidari M, Panuccio, M.R, De Santis, C., Finocchiaro, A. (2005). Early effects of phenolic compounds, extracted from two forest litters, on ammonium uptake and assimilation in *Pinus laricio* and *Pinus pinaster*. *Plant and Soil,* **269**, 309-320.

[52] Blum, U. (1998). Effects of microbial utilization of phenolic acids and their utilization of phenolic acid breakdown products on allelopathic interaction. *Journal of Chemical Ecology,* **24**, 685-708.

[53] Oades, J.M. (1995). Organic matter: Chemical and physical fractions. In: Lefroy, R.D.B., Blair, G.J., Craswell, E.T.(Eds.), *Soil Organic Matter Management for Sustainable agricolture* (pp. 135-139). Canberra: ACIAR.

[54] Lefroy, R.D.B., Blair, G.J., Strong, W.M. (1995). Chemical fractionation of soil organic matteer and measurement of the breakdown rate of residue. In: Lefroy, R.D.B., Blair, G.J., Craswell, E.T.(Eds.), *Soil Organic Matter Management for Sustainable agricolture* (pp. 149-158). Canberra: ACIAR.

[55] Oades, J.M., Waters, A.G., Vassallo, A.M., Wilson, M.A., Jones, G.P. (1988). Influence of management on the composition of organic matter in a red brown earth as shown by ^{13}C NMR. *Australian Journal of Soil Research,* **26**, 289-299.

[56] Jenkinson, D.S., Ladd, J.N. (1981). Microbial biomass in soil: Measurement and turnover. In: E.A. Jenkinson, D.S., Powison, D.S. (1976). The effects of biocidal treatments on metabolism in soil. I. Fumigation with chloroform. *Soil Biology & Biochemistry*, **8**, 167-177.

[57] Christensen, B.T. (1992). Physical fractionation of soil organic matter in primari particle size and density separates. *Advance in Soil Science,* **20**, 1-90.

[58] Blair G.L., Lefroy, R.D.B., Lisle, L. (1995). Soil carbon fractions based on their degree of oxidation and the development of a carbon management index. *Australian Journal of Agricultural Research*, **46**, 1459-1466.

[59] Mrabet, R. (2000). Differential response of wheat to tillage management systems under continuous cropping in a semiarid area of Morocco. *Field Crops Research,* **66**,165-174.

[60] Bessam, F., Mrabet, R. (2003). Long-term changes in particulate organic matter under no-tillage systems in a semiarid soil of Morocco. *Proceedings 16th ISTRO Conference*, pp. 144-149, Brisbane, Australia.

[61] Mrabet, R., Saber, N., El-Brahli, A., Lahlou, S., Bessam, F.(2001). Total, particulate organic matter and structural stability of a Calcixeroll soil under different wheat rotations and tillage systems in a semiarid area of Morocco, *Soil Tillage Research,* **57**, 225–235.

[62] Lal, R. 2005. No-till farming and environment quality. Annals of Simpósio sobre Plantio direto e Meio ambiente; Seqüestro de carbono e qualidade da agua, pp. 29-37. Foz do Iguaçu, Brasil.

[63] Davidson, E.A., Ackerman, I.L. (1993). Changes in soil carbon inventories following cultivation of previously untilled soil. *Biogeochemistry,* **20**, 161-193.

[64] Amado, T.J.C., Lovato, T., Conceição, P.C., Spagnollo, E., Campos, B., Costa, C. (2005). O potencial de sequestro de carbono em sistemas de produção de grãos sob plantio direito na região sul do Brasil. *Annals of Simpósio sobre Plantio direto e Meio ambiente; Seqüestro de carbono e qualidade da agua*, pp. 63-71. Foz do Iguaçu, Brasil.

[65] FAO. 2002. Farm management and economics. In: *Training modules on Conservation Agriculture*. FAO Land and Water Digital Media Series, Volume 22, Rome.

[66] Prior, S.A., Runion, G.B., Torbert, H.A., Rogers, H.H., Reeves, D.W. (2003). Effects of elevated atmospheric CO_2 on biomass production and carbon sequestration: conventional and conservation cropping systems. In: *Proceedings 16th ISTRO Conference*, (pp. 943-948). Brisbane, Australia.

[67] Spagnollo, E. (2004). Dinâmica da matérica orgânica em agroecosistemas submetidos a queima e manejo dos resíduos culturais. UFSM, Santa Maria, RS, Brazil. 112 pp. (PhD thesis)

[68] Paustian, K. 2002. Organic matter and global cycle. In. *Encyclopedia of soil science,*(pp. 895-898). New York: Marcel Dekker Inc.

[69] Izaurralde, R.C., Cerri, C.C. (2002). Organic matter management. In: *Encyclopedia of soil science* (pp. 910-916). New York: Marcel Dekker Inc.

[70] Lovato, T. 2001. Dinâmica do carbono e nitrogênio do solo afetada por preparos do solo, sistemas de culturas e adubo nitrogenado. Porto Alegre UFRGS. 133 pp. (PhD thesis)

[71] Lovato, T., Mielnickzuk, J., Bayer, C., Vezzani, F. (2004). Adição de C e N e sua relação com os estoques no solo e com o rendimento do milho em sistemas de manejo. *Rev. Bras. Ci. Sol.*, **28**: 175-187.

[72] Nicoloso, R.S., Lovato, T., Lanzanova, M.E. (2005). Potencial de seqüestro ou emissão de carbono em áreas de integração lavoura-pecuária no sistema plantio direito. *Annals of Simpósio sobre Plantio direto e Meio ambiente; Seqüestro de carbono e qualidade da agua*, pp. 219-222. Foz do Iguaçu, Brasil.

[73] Amado, T.J.C., Bayer, C., Eltz, F.L.F., De Brum, A.C.R. (2001). Potencial de culturas de cobertura em acumular carbono e nitrogênio no solo no sistema plantio direto e a conseqüente melhoria da qualidade ambiental. *Revista Brasileira Ciencia de Solos*, **25**, 189-197.

[74] Carbon Dioxide Information Analysis Center (2003). In: Marland, G.,. Boden, T.A., Andres, R.J. (Eds). *National CO_2 emissions from fossil-fuel burning, cement manufacture, and gas flaring:* (pp.1751-2000). (available at http://cdiac.esd.ornl.gov).

[75] Brito, L.F., La Scala Jr, N., Merques Jr, J., Pereira, G.T. (2005). Variabilidade temporal da emissão de CO_2 do solo e sua relação com a temperatura do solo em diferentes posições na paisgem em área cultivada com cana-de açúcar. *Annals of Simpósio sobre Plantio direto e Meio ambiente; Seqüestro de carbono e qualidade da agua*, pp. 210-212. Foz do Iguaçu, Brasil.

[76] Huang , P.M., Berthelin, J. Bollag, J.-M., McGill, W.B., Page, A.L. (1995). *Environmental Impact of Soil Component Interactions. Vol. 1. Natural and Anthropogenic Organics, Vol. 2. Metals , Other Inorganics,and Microbial Activities.* Baca Raton, USA:CRC Press/Lewis Publishers.

[77] Huang, P.M., Wang M.K., Chiu, C.Y. (2005). Soil mineral–organic matter–microbe interactions: Impacts on biogeochemical processes and biodiversity in soils. *Pedobiologia*, **49**, 609-635.

[78] Blagodatskaya, E.V., Anderson, T.H. (1998). Interactive effects of pH and substrate quality on the fungal-to-bacterial ratio and *Q* CO_2 of microbial communities in forest soils. *Soil Biology & Biochemistry*, **30**, 1269-1274.

[79] Blagodatskaya, E.V., Anderson, T.H. (1999). Adaptive responses of soil microbial communities under experimental acid stress in controlled laboratory studies. *Applied Soil Ecology*, **11**, 207-216.

[80] Tabatabai, M.A. (1994). Soil enzymes. In: R.W. Weaver, J.S. Angle & P.S. Bottomley (Eds.), *Methods of Soil Analysis: Microbiological and Biochemical Properties. Part 2. SSSA Book Ser.* **5,** (pp. 775-833), Madison, WI: SSSA.

[81] Kandeler, E., Kampichler, C., Horak, O. (1996). Influence of heavy metals on the functional diversity of soil microbial communities. *Biology and Fertility of Soils*, **23**, 299-306.

[82] Trásar-Cepeda, C., Leiros, M.C., Gil-Sotres, F. (2000). Biochemical properties of acid soils under climax vegetation (Atlantic oakwood) in an area of the European temperate-humid zone (Galicia, NW Spain): specific parameters. *Soil Biology & Biochemistry*, **32**, 747-755.

[83] Sinsabaugh, R.L., Carreiro, M.M., Repert, D.A. (2002). Allocation of extracellular enzymatic activity in relation to litter composition, N deposition, and mass loss. *Biogeochemistry,* **60**, 1-24.

[84] Sinsabaugh, R.L., Antibus, R.K., Linkins, A.E. (1991). An enzymic approach to the analysis of microbial activity during plant litter decomposition. *Agriculture, Agrosystems and Environment*, **34**, 43-54.

[85] Linkins, A.E., Sinsabaugh, R.L., McClaugherty, C.A., Melillo, J.M. (1990). Cellulase activity on decomposing leaf litter in microcosms, *Plant and Soil,* **123**, 15–25.

[86] Sinsabaugh, R.L., Antibus, R.K., Linkins, A.E., McClaugherty, C.A., Rayburn, G., Weiland, T. (1992). Wood decomposition over a first-order watershed:mass loss as a function of lignocellulase activity. *Soil Biology & Biochemistry*, **24**, 743-749.

[87] Fioretto, A., Papa, S., Pellegrino, A. Fuggi A. (2007). Decomposition dynamics of *Myrtus communis* and *Quercus ilex* leaf litter. Mass loss, microbial activity and quality change. *Applied Soil Ecology*, **36**, 32-40.

[88] Burns, R.G. (1978). Enzyme activity in soil. Some theoretical and practical considerations. In: Burns R.G, (Ed.), *Soil enzymes,* (pp 295-339), New York: Academic Press.

[89] Burns, R. G., Dick, R.P. (2002). *Enzymes in the Environment: Activity, Ecology, and Applications*. New York: Marcel Dekker, Inc.

[90] Sinsabaugh, R.L., Moorhead, D.L. (1994). Resource allocation to extracellular enzyme production: a model for nitrogen and phosphorus control of litter decomposition. *Soil Biology & Biochemistry,* **26**, 1305–1311.

[91] Schimel, J.P., Weintraub, M.N. 2003. The implications of exoenzyme activity on microbial carbon and nitrogen limitation in soil: a theoretical model. *Soil Biology & Biochemistry*, **35**, 1-15.

[92] Parr, J.F., Papendick, R.I. (1997). Soil quality, relationship and strategies for sustainable dryland farming system, *Annals of Arid Zones*, **36**, 181-191.

[93] Smith, L.J., Papendick, R.I. (1993). Soil organic matter dynamics and crop residue management. In: B. Metting (Ed.), *Soil Microbial Ecology*, (pp. 65-94), New York, NY: Marcel Dekker.

[94] Insam, H., Parkinson, D., Domsch, K.H. (1989). Influence of macroclimate on soil microbial biomass, *Soil Biology & Biochemistry*, **21**, 211-221.

[95] Wardle, D.A. (1992). A comparative assessment of factors which influence microbial biomass carbon and nitrogen in soil. *Biological Review of the Cambridge Philosophicals Society*, **67**, 321-358.

[96] Nicolardot, B., Fauvet, G., Cheneby, D. (1994). Carbon and nitrogen cycling through soil microbial biomass at various temperatures. *Soil Biology & Biochemistry,* **26**, 253-261.

[97] Gestel, M.V., Merckx, R., Vlassak, K. (1993). Microbial biomass and activity in soils with fluctuating water contents. *Geoderma*, **56**, 617-626.

[98] Kaiser, E.A., Muller, T., Roergensen, R.G., Insam, H., Heinemeyer, O. (1992). Evaluation of methods to estimate the soil microbial biomass and the relationship with soil texture and organic matter. *Soil Biology & Biochemistry*, **24**, 675-683.

[99] Carter, M.R. (1986). Microbial biomass and mineralizable nitrogen in Solonetzic soils: influence of gypsum and lime amendments. *Soil Biology & Biochemistry*, **18**, 531-537.

[100] Rowe, J.S. (1984). Forest Land Classification: limitations of the use of vegetation. Proceedings of the symposium on forest land classification (pp 132-147) March 18-20; Madison, WI.

[101] Huggett, R.J. (1975). Soil landscape systems: a model of soil genesis. *Geoderma*, **13**, 1-22.

[102] Pennock, D.J., van Kessel, C., Farrell, R.E., Sutherland, R.A. (1992). Landscape-scale variations in denitrification, *Soil Science Society of American Journal,* **56**, 770-776.

[103] Chen, J.S., Chiu, C.Y. (2000). Effect of topography on the composition of soil organic substances in a perhumid sub-tropical montane forest ecosystem in Taiwan, *Geoderma*, **96**, 19-30.

[104] Srivastava S.C. and Singh, J.S. (1991). Microbial C N and P in dry tropical forest soils. Effects of alternate land uses and nutrient flux, *Soil Biology & Biochemistry,* **24,** 117–124.

[105] Billore, S.K., Ohsawa, M., Numata, M., Okano, S. (1995). Microbial biomass Nitrogen pools in soils from a warm temperate grassland and from deciduous and evergreen forests in Chiba, Central Japan, *Biology Fertiity of Soils*, **19**, 124–128.

[106] Joergensen, R.G., Anderson, T.H., Wolters, V.(1995). Carbon and nitrogen relationships in the microbial biomass of soils in beech (*Fagus sylvatica* L.) forests, *Biol. Fertility Soils,* **19**, 141–147.

[107] Arunachalam, A., Kusum Maithani, Pandey, H.N., Tripathi, R.S. (1996). The impact of disturbance on detrital dynamics and soil microbial biomass of *Pinus kesiya* forest in North east India, *Forest Ecology and Management,* **88**, 273–282.

[108] Mendham, D.S., Connell, A.M.O., Grove, T.S (2002). Organic matter characteristics under native forest long term pasture, and recent conversion to Eucalyptus plantations in Western Australia: microbial biomass, soil respiration and permanganate oxidation, *Australian Journal of Soil Research,* **40**, 859–872.

[109] Lee, K.H. and Jose, S. (2003). Soil respiration and microbial biomass responses to soil water and source of nitrogen in a pecan (Carya illinoensis)-cotton (Gossypium hirsutum) temperate alley cropping system. *Book of Abstracts of the Eight North American Agroforestry Conference*, p. 18.

[110] Díaz-Raviña, M., Acea, M.J., Carballas, T. (1995). Seasonal changes in microbial biomass and nutrient flush in forest soil. *Biology and Fertility of Soils,* **19**, 220-226.

[111] Arunachalam A., Arunachalam, K. (2000). Influence of gap size and soil properties on microbial biomass in a subtropical humid forest of North-east India. *Plant and Soil,* **223**, 185–193.

[112] Leckie, S.E., Prescott, C.E., Grayston, S.J., Neufeld, J.D., Mohn, W.W. (2004). Characterization of humus microbial communities in adjacent forest types that differ in nitrogen availability. *Microbial Ecology*, **48**, 29-40.

[113] Grayston, S.J., Prescott, C.E. (2005). Microbial communities in forest floors under four tree species in coastal British Columbia. *Soil Biology & Biochemistry*, **37**, 1157-1167.

[114] Kuske, C.R., Ticknor, L.O., Miller, M.E., Dunbar, J.M., Davis, J.A., Barns S.M., Belnap, J. (2002). Comparison of soil bacterial communities in rhizospheres of three plant species and the interspaces in an arid grassland, *Applied Environment Microbiology,* **68**, 1854–1863.

[115] Broughton, L.C., Gross, K.L. (2000). Patterns of diversity in plant and soil microbial communities along a productivity gradient in a Michigan old field. *Oecologia*, **125**, 420-427.

[116] Allison, J.D., Miller, R.M., Jastrow, J.D., Matamala, R., Zak, R.D. (2005). Changes in soil microbial community composition in a tallgrass prarie chronosequence. *Soil Science Society of America Journal*, **69**, 1412-1421.

[117] Saetre, P., Baath, E. (2000). Spatial variation and patterns of soil microbial community structure in a mixed spruce-birch stand. *Soil Biology & Biochemistry,* **32**, 909-917.

[118] Priha, O., Grayston, S.J., Hiukka, R., Pennanen, T. & Smolander, A. (2001). Microbial community structure and characteristics of the organic matter in soils under Pinus sylvestris, Picea abies and Betula pendula at two forest sites. *Biology and Fertility of Soils,* **33**, 17-24.

[119] Zogg, G.P., Zak, D.R., Ringelberg, D.B., MacDonald, N.W., Pregitzer K.S., White, D.C. (1997). Compositional and functional shifts in microbial communities due to soil warning, *Soil Science Society of American Journal,* **61**, 475–481.

[120] Zak, D.R., Holmes, W.E., MacDonald, N.W., Pregitzer, K.S. (1999). Soil temperature, matric potential, and kinetics of microbial respiration and nitrogen mineralization, *Soil Science Society of American Journal,* **63**, 575–584.

[121] Fierer, J.P. Schimel and P.A. Holden. (2003). Influence of drying–rewetting frequency on soil bacterial community structure, *Soil Biology & Biochemistry,* **33**, 1827–1839.

[122] Devi, N.B., Yadava, P.S. (2006) .Seasonal dynamics in soil microbial biomass C, N and P in a mixed-oak forest ecosystem of Manipur, North-east India . *Applied Soil Ecology,* **31**, 220-227.

[123] Bardgett, R.D., Mawdsley, J.L., Edwards, S., Hobbs, J., Rodwell, J.S., Davies W.J. (1999). Plant species and nitrogen effects on soil biological properties of temperate upland grasslands. *Functional Ecology*, **13**, 650-660.

[124] He, Z.L., Wu, J., O´Donnell, A.G., Syers, J.K. (1997). Seasonal responses in microbial biomass carbon, phosphorus and sulphur in soils under pasture. *Biology and Fertility of Soils*, **24**, 421-428.

[125] Perrott, K.W., Sarathchandra S.U., Waller J.E. (1990). Seasonal storage and release of phosphorus and potassium by organic matter and the microbial biomass in a high-producing pastoral soil. *Australian Journal of Soil Research,* **28**, 593-608.

[126] Frossard, E., Condron, L.M., Oberson, A., Sinaj, S., Fardeau, J.C. (2000). Processes governing phosphorus availability in temperate soils. *Journal of Environmenalt Quality*, **29**, 15-23.

[127] Tate, K.R., Speir, T.W., Ross, D.J., Parfitt, R.L., Whale, K.N., Cowling, J.C. (1991). Temporal variations in some plant and soil P pools in two pasture soils of widely different P fertility status", *Plant and Soil*, **132**, 219-232.

[128] Fabre, A., Pinay, G., Ruffinoni, C. Seasonal changes in inorganic and organic phosphorus in the soil of a riparian forest. *Biogeochemistry*, **35**, 419-432.

[129] Chen, C.R., Condron, L.M., Davis, M.R., Sherlock, R.R. (2003). Seasonal changes in soil phosphorus and associated microbial properties under adjacent grassland and forest in New Zealand. *Forest Ecology and Management*, **177**, 539-557.

[130] Yeates, G.W., Saggar S., Daly B.K. (1997). Soil microbial C, N, and P, and microfaunal populations under *Pinus radiata* and grazed pasture land-use systems. *Pedobiologia*, **41**, 549-565.

[131] Chen, C.R., Condron, L.M., Davis, M.R., Sherlock, R.R. (2000). Effects of afforestation on phosphorus dynamics and biological properties in a New Zealand grassland soil. *Plant and Soil*, **220**, 151-163.

[132] Clarholm, M., Rosswall, T. (1980). Biomass and turnover of bacteria in a forest soil and a peat. *Soil Biology & Biochemistry*, **12**, 49-57.

[133] Sarathchandra, S.U., Perrot, K.W., Littler, R.A. (1989). Soil microbial biomass: influence of simulated temperature changes on size, activity and nutrient content. *Soil Biology & Biochemistry,* **21**, 987–993.

[134] Marschner, P., Marino, W., Lieberei, R. (2002). Seasonal effects on microorganisms in the rhizosphere of two tropical plants in a polyculture agroforestry system in Central Amazonia, Brazil. *Biology and Fertility of Soils*, **35**, 68-71.

[135] Bååth, E., Söderström, B. (1982). Seasonal and spatial variation in fungal biomass in a forest soil. *Soil Biology & Biochemistry*, **14**, 53-58.

[136] Luizzao, R.C.C., Bonde, T.A., Tosswall, T. (1992). Seasonal variation of soil microbial biomass – The effects of clear felling a tropical rainforest and establishment of pasture in the central Amazon. *Soil Biology & Biochemistry*, **24**, 805-813.

[137] Díaz-Raviña, M., Acea, M.J., Carballas, T. (1993). Seasonal fluctuations in microbial populations and available nutrients in forest soils, *Biology and Fertility of Soils,* **16**, 205-210.

[138] Garcia, C., Hernandez, T., Roldan, A., Martin, A. (2002). Effect of plant cover decline on chemical and microbiological parameters under Mediterranean climate. *Soil Biology & Biochemistry*, **34**, 635-642.

[139] Saynes, V., Hidalgo, C., Etchevers, J.D., Campo, J.E. (2005). Soil C and N dynamics in primary and secondary seasonally dry tropical forests in Mexico. *Applied Soil Ecology*, **29**, 282-289.

[140] Singh, J.S., Raghubanshi, A.S., Singh, R.S., Srivastava, S.C. (1989). Microbial biomass acts as a source of plant nutrients in dry tropical forest and savannah, *Nature* **338**, 499–500.

[141] Greacen E.L. and Sands, R. (1980). A review: Compact of forest soils. *Australian Journal of Soil Research*, **18**,163–189.

[142] Corns, G.W. (1988) Compaction by forestry equipment and effects on coniferous seedling growth on four soils in the Alberta foothills, *Canadian Journal of Forest Research,* **18,** 75–84.

[143] Jurgensen, M.F. Harvey, A.E., Graham, R.T., Page-Dumroese, D.S., Tonn, J.R., Larsen, M..J., Jain, T.B. (1997). Impacts of timber harvesting on soil organic matter, nitrogen, productivity, and health of inland northwest forests, *Forest Science,* **43**, 234–251.

[144] Kozlowski, T. (1999). Soil compaction and growth of woody plants, *Scandinavian Journal of Forest Research,* **14**, 596–619.

[145] Conlin, SS and van den Driessche, R. (1996). Short-term effects of soil compaction on growth of *Pinus contorta* seedlings, *Canadian Journal of Forest Research,* **26**, 727–739.

[146] Huang, J., Lacey, S.T., Ryan, P. J. (1996). Impact of forest harvesting on the hydraulic properties of surface soil, *Soil Science,* **161,** 79–86.

[147] Gomez, A., Powers, R.F., Singer, M.J., Horwath, W.R. (2002). Soil compaction effects on growth of young Ponderosa pine following litter removal in California's Sierra Nevada, *Soil Science Society of American Journal,* **66**, pp. 1334–1343.

[148] Piatek, K.B. and Allen, H.L. (1999). Nitrogen mineralization in a pine plantation 15 years after harvesting and site preparation, *Soil Science Society of American Journal,* **63**, 990–998.

[149] Li, Q., Allen, H.L., Wilson, C.A. (2003). Nitrogen mineralization dynamics following the establishment of a loblolly pine plantation, *Canadian Journal of Forest Research,* **33**,. 364–374

[150] Li, Q., Allen, H.L., Wollum, A.G. (2004) Microbial biomass and bacterial diversity in forest soils: effects of organic matter removal, compaction, and vegetation control, *Soil Biology & Biochemistry*, **36**, 571–579.

[151] Sparling, G.P. (1992), Ratio of microbial biomass carbon to soil organic carbon as a sensitive indicator of changes in soil organic matter, *Australian Journal of Soil Research*, **30**, 195–207.

[152] Bosatta, E., Agren, G.I. (1993). Theoretical analysis of microbial biomass dynamics in soils. *Soil Biology & Biochemistry,* **26**, 143–148.

[153] Kaiser, E.A., Walenzik, G., Heinemeyer, O. (1991). The influence of soil compaction in decomposition of plant residues and on microbial biomass. In: W.S. Wilson, Editor, *Advances in Soil Organic Matter Research: the Impact on Agriculture and the Environment*, Royal Society of Chemistry, Cambridge, pp. 207–216.

[154] Breland, A., Hansen, S. (1996). Nitrogen mineralization and microbial biomass as affected by soil compaction, *Soil Bioogy & Biochemistry,* **28**, 655–663.

[155] Roldán, A., García-Orenes, F., Lax, A. (1994). An incubation experiment to determine factors involving aggregation changes in an arid soil receiving urban refuse. *Soil Biology & Biochemistry*, **26**, 1699–1707.

[156] Christensen, N.L. (1996). Managing dynamic landscapes for heterogeneity and complexity. In S.T.A. Pickett and R. Ostveld (Eds.), Chapman-Hall, Inc., New York.

[157] Wardle, D.A. (1999). How soil food webs make plants grow, *Trends in Ecology Evolution*, **14**, 418–420.

[158] Schimel, J. (1995). Ecosystem consequences of microbial diversity and community structure. In F.S. Chapin, C. Koener (Eds.), *Arctic and Alpine Biodiversity: Patterns, Causes and Ecosystem Consequences* (pp. 239-254). New York: Springer Verlag.

[159] Dalal, R.C. (1998). Soil microbial biomass—what do the numbers really mean?, *Australian Journal of Experimental Agriculture,* **38**, 649–665

[160] Pankhurst, C.E. (1994). Biological Indicators of soil health and sustainable productivity. In: *Soil Resilience and Sustainable Land Use.* Greenland, D.J., Szabolcs I. (Eds) CAB International, Wallingford, U.K, pp. 331-352.

[161] Sparling, G.P. (1997). Soil microbial biomass, activity and nutrient cycling as indicators of soil health. In: C.E. Pankhurst, B.M. Doube and V.V.S.R. Gupta, Editors, *Biological Indicators of Soil Health*, (pp. 97–119) CABI Publishing, Wallingford

[162] Zelles, L. (1999). Fatty acid patterns of phospholipids and lipopolysaccharides in the characterization of microbial communities in soil: a review. *Biology and Fertility of Soils,* **29**, 111-129.

[163] Pankhurst, C.E., Pierrot, A., Hawke, B.G., Kirby, J.M. (2002). Microbiological and chemical properties of soil associated with macropores at different depths in a red-duplex soil in NSW Australia. *Plant and Soil*, **238**, 11-20.

[164] Frey, S.D.; Elliott, E.T.; Paustian, K. (1999). Bacterial and fungal abundance and biomass in conventional and no-tillage agroecosystems along two climatic gradients. *Soil Biology & Biochemistry* **31**, 573-585.

[165] Beare, M.H., Hu, S., Coleman, D.C., Hendrix, P.F. (1997). Influence of mycelial fungi on soil aggregation and agronomic matter storage in conventional and no-tillage systems after thinning. *Applied Soil Ecology*, **5**, 211-219.

[166] Dick, R. P. (1992). A review: long-term effects of agricultural systems on soil biochemical and microbial parameters. *Agricultural Ecosystems Environmental,* **40**, 25-36.

[167] Speeding, A., Hamel, C., Mehuys, G.R., Madramootoo C.A. (2004). Soil microbial dynamics in maize-growing soil under different tillage and residue management systems. *Soil Biology & Biochemistry*, **36**, 499-512.

[168] Williamson, W.M., Wardle, D.A. (2007). The soil microbial community response when plants are subjected to water stress and defoliation. *Applied Soil Ecology*, **37**, 139-149.

[169] Rogers, B.F., Tate, R.L. (2001).Temporal analysis of the soil microbial community along a toposequence in Pineland soils. Soil Biology & Biochemistry, **33**, 1389-1401.

[170] Curl, E.A. and Truelove, B. (1996). The Rhizosphere, Springer-Verlag, Berlin (1986) pp. 288.

[171] Turkington, R. Holl, F.B., Chanway, C.P., Thompson, J.D. (1998). The influence of microorganisms, particularly *Rhizobium*, on plant competition in grass-legume communities. In: A.J.

[107] Davy, M.J. Hutchings and A.R. Watkinson, (Eds), *Plant Population Ecology*, Blackwell Scientific Publications, Oxford UK, pp. 343–366.

[172] Perry,D. and Amaranthus, M. (1990). The plant-soil bootstrap: microorganisms and reclamation of degraded ecosystems. In: J. Berger, Editor, *Environmental Restoration: Science and Strategies for Restoring the Earth*, Island Press, Washington, DC pp. 94–102.

[173] Bossio, D.A., Scow, K.M. (1995). Impact of carbon and flooding on the metabolic diversity of microbial communities in soils. *Applied and Environmental Microbiology*, **61**, 4043-4050.

[174] Buyer, J.S., Drinkwater, L.E. (1997). Comparison of substrate utilization assay and fatty acid analysis of soil microbial communities. *Journal of Microbiological Methods*, **30**, 3-11.

[175] Ross D.J., Sparling, G.P., Burke, C.M. and Smith, C.T. (1995). Microbial biomass C and N, and mineralizable-N, in litter and mineral soil under *Pinus radiata* on a coastal sand: influence of stand age and harvest management. *Plant and Soil*, **175,** 167–177

[176] Smethurst P.J. and Nambiar, E.K.S., 1990. Distribution of carbon and nutrient fluxes of mineral nitrogen after clear-felling a *Pinus radiata* plantation. *Canadian Journal of Forest Research,* **20,** 1490–1497.

[177] Johnson, D.W. and Todd, D.E. (1998). Harvesting effects on long-term changes in the nutrient pool of mixed oak forest. *Soil Science Society of American, Journal,* **62,** 1725–1735.

[178] Olsonn, B.A., Lundkvist, H. and Staaf, H. (2000). Nutrient status in needles of Norway spruce and Scots pine following harvesting of logging residues. *Plant and Soil,* **23**, 161–173.

[179] Hopmans P., Stewart, H.T.L. and Flinn, D.W., 1993. Impacts of harvesting on nutrients in a eucalypt ecosystem in southeastern Australia. *Forest Ecology of Management,* **59,** 29–51.

[180] Folster H., Khanna, P.K. (1997). Dynamics of nutrient supply in plantations soils. In: Nambiar, E.K.S., Brown, A.G. (Eds.), *Management of Soil Nutrient and Water in Tropical Plantation Forests*. Australian Centre for International Agricultural Research, Canberra, pp. 339–378.

[181] Likens, G.E., Bormann, F.H., Johnson N.M., Fisher D.W., Pierce R.S. (1970). Effects of forest cutting and herbicide treatment on nutrients budget in the Hubbard Brook watershed ecosystem. *Ecology Monography*, **40**, 23-47.

[182] Boyle, J. R., Phillips, .J., Ek, A.R. (1973). Whole tree harvesting: Nutrient budget evaluation. *Journal of Forestry*, **71**, 760-762.

[183] Bormann, F.H., Likens, G.E., Siccama, T.G., Pierce, R.S., Eaton G.S. (1974). The export of nutrients and recovery of stable conditions following the forestation at Hubbard Brook. *Ecology Monography*, **44**, 255-277.

[184] Mann, L.K., Johnson, D.W., West, D.C., Cole, D.W., Hornbeck, J.W., Martin, C.W., Riekerk ,H., Smith, C.T., Swank W.T., Tritton, L.M., Van Lear, D.H. (1988). Effects of whole-tree and stem-only clear-cutting on postharvest hydrologic losses, nutrient capital, and regrowth. *Forest Science,* **34**, 412-428.

[185] Hughes, J.W., Fahey, T.J. (1994). Litterfall dynamics and ecosystem recovery during forest development. *Forest Ecology and Management,* **63**, 181-198.

[186] Clayton, J.L., Kennedy, D.A. (1985). Nutrient losses from timber harvest in the Idaho Batholith. *Soil Science Society of American Journal*, **49**, 1041-1049.

[187] Hendrickson, O.Q., L. Chatarpaul, and D. Burgess. (1989). Nutrient cycling following whole-tree and conventional harvest in northern mixed forest. *Canadian Journal of Forest Research,* **19**, 725–735.

[188] Mattson, K.G. and Swank, W.T. (1989). Soil and detrital carbon dynamics following forest cutting in the Southern Appalachians. *Biology & Fertility of Soils,* **7**, 247–253.

[189] Moore, T.R. (1989). Dynamics of dissolved organic carbon in forested and disturbed catchments, Westland, New Zealand. 1. Miamai. *Water Resource Research*, **25**, 1321-1330.

[190] Dahlgren, R.A., Driscoll, C.T. (1994). The effects of woole-tree clear cutting on soil processes at the Hubbard Brook Experimental Forest, New Hampshire, USA. *Plant and Soil*, **158**, 239-262.

[191] Johnson, C.E., Driscoll, C.T., Fahey, T.J., Siccama, T.G. (1995). Carbon dynamics following clear-cutting of a northern hardwood forest. In: W.W. McFee, J.M. Kelly (Eds.), *Carbon Forms and Functions in Forest Soils*Soil Science Society of America, Madison, WI, pp. 463-488.

[192] Wright, E.F., Coates, K.D., Bartemucci, P. (1998). Regeneration from seed of six tree species in the interior cedar-hemlock forests of British Columbia as affected by substrate and canopy gap position. *Canadian Journal of Forest Research,* **28**, 1352-1370.

[193] Hartshorn, G.S. (1978). Tree falls and tropical forest dynamics. In: Tomlinson PB, Zimmermann MH (Eds) Tropical trees as living system. Cambridge University Press, Cambridge pp 617-638.

[194] Denslow J.S. (1980) Patterns of plant species diversity during succession under different disturbance regimes. *Oecologia,* **46**, 18-21

[195] Riddoch, I., Lehto, T., Grace, J. (1991) Photosynthesis of tropical tree seedlings in relation to light and nutrient supply. *New Phytologyst,* **85**, 137-147.

[196] Coates KD, Burton PJ (1997) A gas-based approach of development of silvicultural system to address ecosystem management objectives. *Forest Ecology and Management,* **99**, 337-354

[197] Gagnon, J.L., Jokela, E.J., Moser, W.K, Huber ,D.A. (2003). Dynamics of artificial regeneration in gaps within a longleaf pine flatwoods ecosystem. *Forest Ecology and Management,* **172**, 133-144.

[198] Muscolo, A., Sidari, M., Mercurio, R. (2006). Variations in soil chemical properties and microbial biomass in artificial gaps in Silver fir stands. *European Journal of Forest Research,* **126**, 59-65.

[199] Muscolo, A., Sidari, M., Mercurio, R. (2007). Influence of gap size on organic matter decomposition, microbial biomass and ntrient cycle in Calabrian pine (*Pinus laricio*, Poiret) stands. *Forest Ecology and Management,* **242**, 412-418.

[200] Hammel, K.E. (1997). Fungal degradation of lignin. In: G. Cadish & K.E. Giller (Eds.), *Driven by nature: Plant litter Quality and Decomposition*, Wallingford, UK: CAB International, pp. 33-45.

[201] Balser, T. C., Kinzig, A., Firestone, M.K. (2001). Linking soil microbial communities and ecosystem functioning. In: A. Kinzig, A., Pacala, S., Tilman, D. (Eds.), *The Functional Consequences of Biodiversity: Empirical Progress and Theoretical Extensions* (pp. 295-293). Princeton: Princeton University Press.

[202] Falcon, M.A., Rodriguez, A., Carnicaro, A., Regalado, V., Perestelo, F., Milstein, O., Fuente, D.G.L. (1995). Isolation of microorganisms with lignin transformation potential from soil of Tenerife island. *Soil Biology & Biochemistry,* **26**, 1305–1311.

[203] Lang, E., Eller G., Zadrazil, F. (1997). Lignocellulose decomposition and production of ligninolytic enzymes during interaction of white rot fungi with soil microorganisms. *Microbial Ecology,* **34**, 1-10.

In: Ecosystem Ecology Research Trends
Editors: J. Chen and C. Guo, pp. 65-105
ISBN: 978-1-60456-183-8

Chapter 2

NEW TOOLS AND STRATEGIES FOR BIOMONITORING MARINE ECOSYSTEMS: LEARNING FROM ECHINODERMS

M. Sugni, A. Barbaglio, P. Tremolada and Candia Carnevali
Dipartimento di Biologia, Università degli Studi di Milano,
Via Celoria 26, 20133, Milano, Italy

Abstract

Marine ecosystems are critically threatened by several anthropogenetic compounds (xenobiotics). In the last decades, among the numerous relevant xenobiotics, specific attention has been given to the Endocrine Disrupter Compounds (EDCs): these chemicals are able to alter the endocrine functions potentially causing serious health problems in humans and wildlife. In spite of the global relevance of these pollutants, most past and recent ecotoxicological research has been addressed only to a selected number of animal models, usually vertebrates, whereas invertebrates (particularly marine invertebrates) have been largely neglected. Among the scarcely considered invertebrate taxa, there are common macroinvertebrates such as echinoderms which actually are regarded as representative animals of the marine ecosystem. This contribution will explore the appropriateness of echinoderm models as test species in marine ecotoxicology. Presence and role of the different echinoderm classes in marine ecosystems will be discussed in terms of their ecological relevance and susceptibility to be targets of EDCs. Moreover, echinoderms can offer a wide range of biological processes appropriate for studies on endocrine dysfunction; although basic knowledge on echinoderm endocrinology is still scarce, several works suggested that many physiological processes of these invertebrates are plausibly regulated by humoral mechanisms, often similar to those found in vertebrates. Here, the sensitivity and responsiveness of different biological processes such as reproduction and regenerative development are discussed in relation to EDCs ecotoxicological tests on two echinoderm species: the sea urchin *Paracentrotus lividus* and the crinoid (feather star) *Antedon mediterranea*. The effects of low-dose exposures to several EDCs (triphenyltin, fenarimol, *p,p'*-DDE, methyltestosterone and cyproterone acetate) were investigated by several parameters including regenerative growth, histological pattern, gamete development and gonad maturation. In most of the exposure experiments, evidences of plausible endocrine disruption were found. These experiments were also considered in terms of modelling the fate of the contaminants in the aquaria. The modelling approach in the evaluation of the exposure can help in the planning phase of ecotoxicological tests, and, later, in the comprehension of the results and in the exportability to the natural environment. Finally, problems and weak points encountered with these animal models will be critically discussed together with considerations and perspectives for future research developments.

1. Marine Environment, Pollution and the Endocrine Disrupting Compounds (EDCs)

There is increasing evidence that human and wildlife populations are daily exposed to adverse compounds affecting many physiological processes of the organism. Growing concern about the effects of environmental chemicals upon animal species has made ecotoxicology a continuously expanding and developing area with clearly important applied benefits.

Many different chemicals enter the environment through a number of routes and have various fates in terms of movement, storage and transformation. Marine environment is critically involved in these cycles since it acts as a huge container of many anthropogenetic compounds (xenobiotics): it collects and stores direct waste as well as water-soluble and sediment-bound xenobiotics also far discharged but transported to the sea *via* rivers and streams or atmosphere. Many xenobiotics have typically low water solubility, thus displaying high affinity for particulate matter suspended in the water. Once they reach the sea, these xenobiotics are rapidly partitioned into sediments where they may persist for a long time, particularly when they are characterised by slow degradation (stability to hydrolysis, microbial attack, etc…). Therefore, these compounds represent a potential threat especially for all those organisms living in contact with sediments. The increasing concern and scientific proof of the hazard of these substances has caused the implementation of regulations in many states regarding the introduction and use of new chemicals as well as the control of point pollution sources. Instead, relevance and danger of secondary sources such as marine sediments are often underestimated. The effects of toxic pollutants on marine biological diversity are undoubtedly significant at innumerable localities and might well be significant on a very wide scale as well, although this is often difficult to prove. It is well known, however, that exposure to toxicants can cause immediate death, diseases, immune system impairment, reduced reproductive success, and developmental malformations. (Colborn et al. 1996; de Fur et al. 1999).

Among the large amount of environmentally relevant xenobiotics, in the last decades many scientific and governative institutions have posed a specific attention on the emerging issue of the Endocrine Disruptor Compounds (EDCs), i.e. "exogenous substances or mixtures that alter function(s) of the endocrine system and consequently causes adverse health effects in an intact organism, or its progeny, or (sub)populations" (as established by the International Program for Chemical Safety).

EDCs are a wide class of environmental contaminants (including flame retardants, plasticisers, cosmetic ingredients, pharmaceuticals and agricultural chemicals) which are well known for their potential to cause serious health problems in humans and wildlife even at very low doses (Janssen, 1996). Most of these compounds enter in the environment through the water and are very common in the marine compartment (Maugh, 1987). Due to their interaction with the hormonal system, EDCs can significantly affect a wide range of physiological processes under its direct or indirect control. Once a pollutant with endocrine disrupting activity has entered a living organism, it can exert different types of deleterious actions at different levels of the organism, from the molecular (genes, biochemical markers, etc.) to the population level, producing effects also on cells, tissues and whole organisms (Colborn et al. 1996; de Fur et al, 1999). EDCs may dramatically affect the organisms (and the ecosystem) since they are known to display adverse effects on fundamental physiological processes such as growth, reproduction, development as well as immunity defence; in fact, these processes are strictly regulated by a combination of endocrine-neuroendocrine mechanisms and local growth regulators (Le Blanc et al. 1999).

EDCs can exert their disruptive effects through different mechanisms: they may bind to cellular receptor of hormones mimicking their biological activity (agonistic effect) or preventing their binding and so their cellular response (antagonistic effect); alternatively they could interact with transport proteins in body fluids thus altering the amounts of circulating endogenous hormones; finally, they may interfere with the metabolic processes of the hormones, affecting their synthesis or catabolism (de Fur et al. 1999).

EDCs interfering with sexual hormones are one of the main areas of concern, due to the magnitude of their potential risk: in fact, compounds displaying estrogenic/anti-estrogenic activity, as well as androgenic/anti-androgenic activity may produce severe effects on most reproductive parameters (gonad development, sexual differentiation, fecundity,...) which are subsequently important for species survival (Colborn et al. 1996; de Fur et al. 1999). Moreover it is now well established that steroids control a much wider range of physiological functions in different animals, being involved in brain, skeletal and muscle development and maintenance, immunity responses and behavioural pattern (de Fur et al. 1999). Steroids are usually present and act at extremely low concentrations in the organisms. It is then clear that substances inferring with their synthesis, metabolism or action, might be very dangerous even at minimal concentrations.

Despite the relevance of these pollutants, most past and recent ecotoxicological research on EDCs have been predominantly addressed to a selected number of animal models, usually vertebrates, and representative organisms of the different ecosystems. Instead, invertebrates (particularly marine invertebrates) have been largely neglected and only a few reports have considered these animals with due attention (for some reviews see de Fur et al. 1999; Ohelmann and Schulte-Ohelmann, 2003; Zou, 2003; Sugni et al. 2007a). However invertebrates represent more than 95% of the extant animal species in natural ecosystems and they often play structural and functional key roles. In addition, some of the most famous and dramatic cases of adverse EDC effects actually regard invertebrates: the phenomena of *imposex* and *intersex* (i.e. the partial or total superimposition of male sex organs on female specimens) in gastropod molluscs are probably one of the most well-documented examples of endocrine disruption, with also adverse population-level effects (Bryan et al. 1986). Ubiquitous decline in some crustaceans population has been reported together with an apparent increase in incidence of sexually ambiguous (gynandromorphic) individuals. Olmstead and LeBlanc (2007) showed that some pesticides, together with other environmental factors, can stimulate the occurrence of the gynandromorphic phenotype. Although in recent years interest has grown in using aquatic invertebrates as tools for environmental risk assessment, representative animal models for the marine environment are still quite scarce.

Considering the limited information on endocrine disruption in invertebrates the "real ecological threat" of EDCs is certainly underestimated, and their analysis is highly incomplete. Widening the approach to many representative invertebrate groups, and increasing the number of valuable experimental models for laboratory tests on EDCs should be strongly encouraged by ecotoxicologists. In addition, EU future policy will probably restrict and limit the experimental employment of vertebrates, and will be necessarily more addressed to the strategic employment of alternative animal models, ethically more acceptable. Thus, invertebrates actually represent a still largely unexplored resource of animal models on which future ecotoxicological studies, particularly those focussing on EDCs, should be encouraged and specifically addressed.

Among marine macroinvertebrates, lack of data and experimental employment are particularly evident in some very common taxa, such as Echinoderms, which actually are widespread and representative animals of the marine ecosystem. In the following paragraph appropriateness of echinoderm models as test species in marine ecotoxicology will be explored taking into account different aspects of their biology, ecology and physiology. Presence and role of the different echinoderm classes in marine ecosystems will be discussed in terms of their ecological relevance and susceptibility to be targets of EDCs.

2. The Echinoderms: Biology, Ecology and Physiology of the Different Classes

Echinoderms are among the most familiar marine animals, occurring in all the oceans from shallow waters of intertidal zones to the deep abyssal plains. They are common and key inhabitants of coral reefs and are often foremost members of many benthic communities, in terms of biomass (more than 90%, in some cases) and ecological importance (Chia and Harrison, 1994, Brusca and Brusca, 1990).

Echinoderms occupy a key position in the "zoological tree". They are deuterostome invertebrates, being therefore phylogenetically more related to chordates (and vertebrates) than to other invertebrate phyla. This phylogenetic closeness may suggest the existence of common basic mechanisms between the two groups, including processes of hormonal regulation (Candia Carnevali, 2005; Sugni et al. 2007a).

Among the typical features of echinoderms (pentamerous radial symmetry, mesodermal endoskeleton composed, etc...) the ceolomic water-vascular system is perhaps the most characteristic one. This last structure is a complex of fluid-filled canals whose numerous external projections (tube feet) play a variety of functions, such as locomotion, respiration and excretion (Brusca and Brusca, 1990); it therefore represents a wide exchange-surface with the marine environment. As a consequence primary uptake across external epithelia can easily occur, representing important routes of entry of many dissolved aquatic pollutants (Sugni et al. 2007a). This process is facilitated by their benthic habits: living in close contact with sediments makes echinoderms favourable targets of environmental contamination and easily susceptible of pollutant bioaccumulation.

There are about 7,000 extant species of echinoderms, all of which exclusively marine (Brusca and Brusca, 1990). Within the phylum, traditional taxonomy recognises five classes of living echinoderms: Crinoids (sea lilies and feather stars), Asteroids (starfishes), Ophiuroids (brittle stars), Echinoids (sea urchins and sand dollars) and Holoturoids (sea cucumbers).

2.1. Crinoids

Members of this class represent the only survivors of the most ancient group of echinoderms, which were extremely abundant in the past. Some thick limestone beds dating to the mid- to late-Paleozoic are entirely made up of disarticulated crinoid fragments (Brusca and Brusca, 1990).

To non-expert eyes crinoids can be easily mistaken for marine plants or algae; in fact their body is mainly constituted by long thin arms (from ten to hundreds) originating from a small central calyx. Several lateral branches depart along each arm, giving it the appearance of a soft feather (from this their common name "feather stars"). Some species possess a long stalk through which the animal is connected to the substratum, so that they resemble sea flower (in fact these are commonly called sea lilies) (Hymann, 1955). All crinoids are filter feeders and have benthic attitude, being nearly sessile: most of them, in fact, spend almost all the day attached to rocks, corals or algae, spreading their arms to filter the water flow (Brusca and Brusca, 1990). Therefore they can be considered representative elements of a specific location, providing information about that particular ecosystem, including environmental pollution conditions.

In shallow waters of tropical regions crinoids represents one of the most common species, being extremely frequent on coral reefs and considerably abundant in terms of biomass.

2.2. Asteroids

Starfishes are perhaps the most well-known representative members of the echinoderm phylum. Their star-shaped body is commonly found on most sea floors, from rocky bottoms to sandy or muddy grounds (Brusca and Brusca, 1990). Most forms have a flattened and flexible body with a varying number (often five but up to 35-40) of arms, or rays, arranged around a central disc. Their size ranges from less than 1 centimetre to nearly 1 meter, depending on the species. The oral surface (the side with the mouth) faces the substrate, whereas the "top side" (aboral surface) contains the anus. They usually avoid sunlight: most starfishes have nocturnal habits and those living in shallow water shelter in crevices or under the rocks. Although suspension feeders can be found, most asteroids feed on animal organisms, both living or dead; they can be second or third level predators in the marine food web, being therefore susceptible of biomagnification processes. Both dietary uptake of EDCs from contaminated preys and direct uptake from surrounding water contribute to starfish EDCs body residues which can be found at higher levels than in their preys. This has been clearly shown for those species mainly feeding on mussels, which are well known pollution bioindicators and easily accumulate contaminants (Shim et al. 2004). Some asteroid species can be considered real sea-sweepers, since they feed on anything edible they find on their way. They live in strict contact with the substratum, slowly moving on the grounds with their tube feet, and some species spend most of the day hidden into sand, mud or sediments. Their feeding and living habits make them among the most favourable target of EDCs. Different studies reported high contamination levels in starfishes collected from the field. Shim et al. (2004) found relevant concentrations of the organotin triphenyltin (TPT, a well-known EDC, very common in the sea), in specimens collected from costal areas of Korea; on the basis of their results they suggested starfishes as optimal target organisms for monitoring TPT in the marine environment, due to their high accumulation and low degradation capacity towards this compound. Coteur et al. (2003) found that in starfishes collected from the Norwegian fjords lead is chronically accumulated, without apparent control, whereas cadmium is subjected to a regulating mechanism which, however, is overwhelmed over long-term exposures. Cadmium has been reported to alter starfishes steroid metabolism and steroid levels (Voogt et al., 1987).

Several asteroid species reproduce asexually, i.e. without using gametes; during this phenomenon (called fission) the animal splits in two parts and each of them subsequently regenerate the missing part. In this way two genetically identical specimens (clones) are obtained. Fission is not used as only reproductive strategy, but it is exploited in particular physiological or environmental conditions instead of sexual reproduction. Regenerative processes occurring after splitting as well as arm lost are extremely sensitive developmental phenomena which are controlled by a complex cascade of endocrine-neuroendocrine signals (Candia carnevali, 2005) and therefore can be easily disturbed by xenobiotics such as EDCs (Candia Carnevali, 2005; Sugni et al. 2007a).

2.3. Ophiuroids

Brittlestars resembles starfishes in their body shape since they usually have five thin arms departing from a central roundish disk (more clearly distinguishable than in asteroids; Chia and Harrison, 1994). They generally have smaller sizes than starfishes, but some species (*Gorgonocephalus stimpsoni)* can reach about 70 centimetres in diameter. They can be found from tropical regions to poles, and from costal areas to abyssal plains: some specimens have been observed in deep-sea trench at about 8,000 meters of depth (Brusca and Brusca, 1990). They live in any marine habitat: some species creep among rocks, algae and in crevices, some other sink their body in sediments or sand, with the arm tip fluctuating in the water column. These last species are usually suspension-feeders and some of them can even absorb dissolved organic matter directly from the water (Brusca and brusca, 1990); in areas with high organic pollution they are potentially very susceptible of adverse contaminant action. Many ophiuroids have carnivorous-detritivorous habits, feeding on small organisms or dead animals found among particulate material, therefore they have a remarkable ecological role.

2.4. Holoturoids

Sea cucumber are perhaps the most unusual echinoderms, since the typical pentamerous radial symmetry is hardly recognisable. In fact, they have a cylindrical body which can resemble a crawling worm, with the oral opening and the anus located at the opposite ends; in some species the body shape is short (even 2,5 cm) and squat, whereas some other can nearly reach 1 m of length, such as *Ophiodesoma spectabilis,* very common on sandy bottoms of Hawaii. The oral region is surrounded by specialised tube feet called tentacles which can reach one third of the body length (Chia and Harrison, 1994). They have feeding function and their shape depends on the feeding habit: in filter-feeder species tentacles are long, often extremely branched, and they act as a fishing-net which actually filters any organism or particulate matter suspended in the water column (Brusca and Brusca, 1990). Therefore they can easily come into contact with those compounds partitioned into floating particles. In most of the Mediterranean sea cucumber, tentacles are leaves- or shield-shaped and are used to dig and to sweep the muddy or sandy bottom searching for organic particles to feed. In areas with highly polluted marine sediments these species can be prime target of contaminant bioaccumulation and adverse action. Similarly, in the order Apoda, most species use to feed on detritus and organic particles collected from the substratum. In the "medusa worms" (so called from the mythological woman bearing snakes on her head) tube feet are lacking and most respiratory and excretory exchanges occur throughout the very thin epidermis; the latter then become a very sensitive route of entry of many molecules, including pollutants.

Detritivorous holoturoids have a remarkable ecological role. Their continuous digging action allows a constant mixing up of the substratum which guarantees an appropriate oxygenation and recirculation of nutrients and minerals in the different bottom layers (Brusca and Brusca, 1990). Their presence is fundamental for the maintenance of the local ecosystem.

2.5. Echinoids

Echinoids have a globular or discoidal body which totally lacks arms and is covered by spines. Their endoskeleton is highly developed as a rigid box-shaped armour containing the viscera and the gonads (Chia and Harrison, 1994). Similarly to starfishes, echinoids are well-known inhabitants of the sea and are very common in most marine ecosystems. Regular echinoids (sea urchins) have a hemispherical body and usually live on hard bottoms, often hidden among rocks or in crevices (Tortonese, 1965). Some species can excavate niches in the wall by using their spines and teeth (such as the european *Paracentrotus lividus*). Some sea urchins have a specialised diet and feed on sponges, cnidarians or ascidians; most species feed on algae or marine plants which unceasingly graze all the day, but they do not despise dead animals or sessile organisms. Sea urchin foraging activity remarkably affect the composition and the dynamics of algal communities and rocky littoral pool, making them key elements in costal ecosystem maintenance (Lawrence, 1975). When their density is high grazer sea urchins are responsible of the formation of the "*barren grounds*" or "*urchin grounds*", which are areas predominantly occupied by encrusting calcareous algae; vice-versa a decline in sea urchin population or an intensive over-fishing can cause uncontrolled growth of erect algae (Nedelec, 1982). It has been demonstrated that in case of *P. lividus*, the main Mediterranean herbivorous-grazer, a density of four individuals per square meter can reduce to 30-50% of algal cover whereas eleven specimens per square meter can virtually eliminate it at all (Kitching and Ebling, 1961). In the 1950s, the fundamental role of sea urchins in kelp forest communities maintenance came to light when southern Californian kelp forests failed to recover from storm damage and increased sea urchin densities led to formation of barren grounds and inhibited kelp recruitment (Pearse and Hines, 1987; Lafferty and Kushner, 2000). The recovery of many kelp forests takes place in association with sea urchin control programs (North 1983).

Having said this, it is clear that any factor altering sea urchin population dynamics, as reproductive impairment due to long term exposure to EDCs, will thus affect the whole community and threaten the equilibrium of the marine ecosystem where they live.

Although with a minor ecological role with respect to their cousins sea urchins, sand dollars and heart urchins (discoidal echinoids) can potentially be more easy target of EDCs, since most of them live buried in soft substrates. Therefore they are chronically exposed to all those compounds stored in sediments and slowly but continuously released in the surrounding environment.

3. Endocrinology of Echinoderms: State of the Art

Despite their remarkable ecological presence and role, echinoderm species are not certainly the most exploited experimental model in ecotoxicological research and, with particular reference to the EDCs field, they have been only occasionally employed (Sugni et al. 2007a). Among the different reasons of this limited use, is the still incomplete knowledge of their basic endocrinology. Lack of information on which and how hormones control echinoderm physiology makes very difficult to explain results from experiments aimed to endocrine disruption assessment in these animals.

Among the different types of hormones and related humoral mechanisms well-known in both vertebrates and invertebrates, in this paper we will mainly focus on sexual steroids. However, it is important to recall that many echinoderm processes (including reproduction, development, etc..) are regulated by neuroendocrine factors (Le Blanc et al. 1999; Thorndyke and Candia Carnevali, 2001). In starfishes, for example, spawning can be stimulated by a specific nervous factor (Gonad Stimulating Substance, GSS, or Radial Nerve Factor, RNC), probably of peptidic nature, that is released by the radial nerve cords (Caine & Burke, 1984); follicular cells surrounding the oocyte are induced by the GSS to release another hormone, 1-methyladelanine, which, in turn, stimulates the production and activation of a maturation promoting factor (MPF) in the developing oocyte. MPF brings on the final oocyte maturation by inducing the germinal vescicle breakdown, i.e. the nuclear envelop rupture occurring just before meiotic division and second polar body emission (Smiley, 1990; Mita, 2000).

In the last decades several authors suggested a plausible involvement of steroids in different physiological processes such as reproduction, development, regeneration (Watts et al. 1994; Candia Carnevali, 2005; Sugni et al. 2007a). Overall, results emerging from past and recent research showed that:

Echinoderms have vertebrate-like steroids (including cholesterol, progesterone, androstenedione, testosterone, 17β-estradiol and estrone) and a number of poli-oxysteroids which were detected by several methods, namely bioassays, chromatography, radioimmunoassay and gas chromatography-mass spectrometry (Dielman & Schoenmakers, 1979; Voogt et al 1992; Watts et al. 1993; 1994; Shubina et al., 1998). Their presence was revealed in different echinoderm classes, mainly asteroids, and in different tissues. In starfishes they have been found in gonads, perivisceral fluids and pyloric caeca, the latter being the storage sites of nutrients necessary for the gametogenesis (Dielman and Schöenmakers, 1979; Schöenmakers and Dieleman, 1981; Voogt and Dieleman, 1984; Xu and Barker, 1990; Hines et al., 1992), whereas in crinoids they were detected from whole body homogenates (Lavado et al. 2006a). In echinoids they were detected in gonads of a couple of species (*Lythechinus pictus* and *Paracentrous lividus*; Wasson et al. 2000a; Lavado et al. 2006b; Barbaglio et al. 2007). In addition, according to recent studies, testosterone (T) and estradiol (E_2) are present in the coelomic fluids of the sea urchin *P. lividus* (Lavado et al., 2006b). The presence of these hormones in the circulatory system could be an indication of their release in the circulatory fluids and their plausible endocrine action at distance on specific targets, as it happens in vertebrates (Barbaglio et al. 2007). The involvement of coelomic fluid and its free-swimming cell population (coelomocytes) in steroid transport and/or processing is consistent with previous studies by Harrington & Ozaki (1986). According to their results, E_2 administration to cultured echinoid coelomocytes induced the synthesis of a novel transient protein in females of two different echinoid species, this effect being observable only on not-mature specimens. Employment of coelomic fluid as a mean of transport in echinoderms is consistent with their reproductive anatomy. In starfishes pyloric caeca are strictly associated to gonads and both organs are surrounded by reduced coelomic spaces (Schoenmakers et al 1981a); sea urchins lack pyloric caeca and the digestive tube (which actually shows some steroid synthesizing activity, see below) is only partially connected to gonads, the latter nearly hanging in the highly-developed coelomic cavity (Hymann, 1955). In both classes the germinal epithelium is located just after the haemal sinus (containing a coelomic-derived fluid) which in turn is surrounded by a genital coelom (Smiley, 1990). Morphologically similar cells can be observed in both the haemal sinus and the germinal epithelium of the echinoid *P. lividus*, thus suggesting possible transfer of these cell elements between the two layers (personal observations). Broertjes et al. (1982) observed interconnections between the germinal epithelium and haemal system, implying that the latter could be important for nutrient and hormone transfer. Kanatani and Shirai (1968) previously demonstrated that GSS is released from nerves in the coelomic cavity. As already suggested by Schöenmakers et al. (1981a), it is then plausible to speculate that also steroid transport is carried out *via* the coelomic/haemal fluid from their production sites to target tissues.

Echinoderms have steroid metabolic pathways comparable to those of vertebrates, which allow the synthesis of the hormones from cholesterol. Several laboratory research focussing on androgens demonstrated the ability of various echinoderms tissues (gonads, pyloric coeca, digestive tube but also arms and body wall) to synthesize and convert progesterone (P), androstenedione (A), T and their derivatives. Voogt et al. (1990) showed that the synthesis of A, the precursor of androgens and estrogens, occurs *via* both the Δ^4 and the Δ^5 pathway, as it happens in mammals. In these latter animals, the proximal fate of A and its metabolites is determined by the activity of several steroidogenic enzymes. Most of them were detected also in echinoderms, including 17β-hydroxysteroid dehydrogenase, 5α-reductase, 3β- and 3α-hydroxysteroid dehydrogenase, 11β-hydroxilase and 11β-hydroxysteroid dehydrogenase, and their activity showed a tissue-, and sometimes sex-specific pattern (Schoenmakers and Voogt 1980, 1981;Watts et al., 1993, 1994; Janer et al., 2005a,b). However, in echinoids conversion rates from A to T are 5-20 folds lower than in vertebrates (Janer et al. 2005a). In addition, Hines et al. (1994) showed that echinoid gonads can rapidly convert E2 and estrone (E1) into other metabolites, but they failed to detect P450-aromatase activity, the enzyme responsible of androgens conversion to estrogens in vertebrates. At present just a few studies suggested endogenous synthesis of estrogens from androgens (i.e. aromatase activity) in echinoderms (Den Besten, 1998; Lavado et al., 2006b, Barbaglio et al. 2007). Lavado et al. (2006b) reported the presence of aromatase activity in both digestive tube and gonads of *P.lividus*, the former showing enzyme activity up to 20-fold higher than in the latter. These data suggest that the digestive tube could be the main biosynthesis site of estrogens in echinoids (Lavado et al., 2006b). Tissue-specific differences were detected also in androgen metabolic pathways (Janer et al., 2005a): in gonad tissues of *P. lividus* T was primary transformed to dihydrotestosterone (DHT), whereas in the digestive tube it was rapidly metabolized in an inactive form. Interestingly, DHT is one of the most active androgen in humans (Wilson,

2001) and in many vertebrates (Sperry and Thomas, 1999a, b). Similarly to the sea urchins, in humans T is peripherally transported *via* circulatory fluid (the blood) to target tissues where it is transformed into DHT by 5α-reductase. Although to a minor extent, in echinoderm tissues T is also partially converted to 11-oxy androgens (including 11-ketotestosterone; Watts et al. 1993; Janer et al. 2005a) which are known to be the bioactive androgens in teleost fishes (Fostier et al. 1983).

Finally, also the activity of enzymes involved in phase II of androgen metabolism (i.e. conjugation) were detected in asteroids and echinoids (Voogt and van Rheenen, 1986; Janer et al., 2005b). Phase II metabolism enzymes usually reduce the biological activity of the steroid and decrease or increase their liposolubility, thus facilitating their removal from the organism tissues or their storage in inactive forms (Janer et al. 2005b). As in other invertebrates, such as molluscs (gastropods and bivalves), they might play a key role in controlling endogenous steroid levels (Janer et al., 2005b). Therefore, exogenous compounds interfering with these enzymes are likely to alter endocrine homeostasis with adverse consequences on many physiological processes (Janer et al. 2005b).

Active involvement of steroids in echinoderm reproductive processes has been postulated several times in the last decades (Watt al. 1994). Pilot-experiments by Donahue & Jennings (1937) showed that gonadal extracts from the echinoid *Lytechinus variegatus* could induce a potent estrogen-like response in the vagina and uterus of ovariectomized rats. Later, several studies showed that steroid levels vary according to the reproductive cycle and in a sex-specific manner. In the ovaries of the starfish *Asterias rubens,* high P levels were correlated to maturative stages in which low overall gonad growth but several oogonia mitotic events and the beginning of oocyte growth could be observed (Schoenmakers & Dielman, 1981). These data suggest that P might acts as an activator and a promoter of oogenesis. On the other hand, estrone (E1) was supposed to be a vitellogenesis inducer, since maximum levels of this hormone were registered right at the beginning of yolk incorporation processes by the developing oocytes (Schoenmakers & Dielman, 1981). In the testes of the same species P and E1 reach their maximum values when gonads start to growth, suggesting their involvement in the initial spermatogenesis phases (Voogt and Dielman, 1984). Similarly to what observed in the starfishes *Asterias vulgaris* e *Sclerasterias mollis* (Xu and Barker, 1990; Hines et al. 1992), in the echinoid *Lytechinus variegatus,* Wasson et al. (2000a) reported E2 peaks in correspondence of the initial stages of both ovary and testes reproductive cycle, i.e. when the gonad is plausibly accumulating nutritive material and oogonia/spermatogonia proliferation occurs. Recently, Barbaglio et al. (2007) showed significantly higher T levels in mature *P. lividus* males, suggesting its involvement in advanced spermatogenesis and possibly in spawning events. It seems that a drop in male T level corresponds to the onset of spermatogenesis and progressively increasing concentrations occur as the spermatogenic stages advance (Sugni et al., unpublished data). The decrease of T level at the beginning of the new reproductive cycle appears to be a male-specific signal, since a comparable trend is not observed in females, and right during these stages the most evident differences in T level can be detected between the two sexes (Barbaglio et al. 2007). In the ovaries of this species, but not in the testes, this hormone is inversely related to overall gonad growth (expressed as gonad index, GI), i.e. its level clearly decreases as the GI increases (Sugni et al. unpublished data). Sex-specific differences were reported also in the overall E2 titres, higher mean levels being measurable in females rather than in males (Barbaglio et al., 2007). Moreover, in the ovaries (but not in the testes) increased E2 level were detected in correspondence of those stages with higher activity of nutritive phagocite, i.e. the somatic cells of the echinoid gonad (Sugni et al., unpublished data), a situation similar to that described in *L. variegatus* by Wasson et al. (2000a). These results suggest that, at least in echinoid females, E2 could be involved in the regulation of phagocitosis processes and/or nutrient accumulation rather than directly in oocyte development. This hypothesis is supported by personal observations in which a higher lipid content is measured in E2-treated specimens.

A more specific involvement of steroids in echinoderm reproduction is also suggested by a number of experiments of direct hormone administration, in which steroid treatment of individuals or cultured tissues resulted in appreciable physiological effects. Male specimens of the echinoid *Pseudocentrotus depressus* fed with A- or E1-enriched food developed bigger gonads (higher GI valus) compared to controls; the same E1-treated animals displayed a 2-fold increase in advanced spermatogenic stages, possibly suggesting E1 involvement in late processes of spermatogenesis (e.g. differentiation/maturation of the sperm, spermiogenesis) (Unuma et al. 1999). Wasson et al. (2000b) showed that dietary administration of E2 and /or 5α-reduced androgens in combination with T promote oocyte growth in *L. variegates.* In asteroids *in vitro* and *in vivo* E2 or E1 treatment increases oocyte diameter, protein amount in the ovary and the number of male germ cells but not the overall gonadal growth (Barker and Xu, 1993; Schoenmakers et al. 1981b; Takahashi and Kanatani 1981; Takahashi 1982a, 1982b).

The action mechanism of steroids in echinoderms is still to be explained. Wasson et al. (2000a) suggested a receptor-mediated effect for these hormones. This hypothesis was supported by Roepke et al. (2005) results which showed abnormal development in sea urchin larvae exposed to specific agonist/antagonist of estrogen receptor. Radioreceptorassays showed the presence of T and E2 binding sites in asteroids, echinoids and crinoids (De Waal et al. 1982; Lutz et al., 2004), however no definitive evidence was obtained on the existence of steroid receptors in echinoderms. Preliminary screening of the Sea Urchin Genome, recently released by The Sea Urchin Genome Sequencing Consortium (2006), could not reveal the presence of classical estrogen or androgen receptors (ERs and ARs). Lack of nuclear ERs and ARs was observed also in the genome of the chordate *Ciona intestinalis* (Tunicates), that, however, possesses a member of the estrogen-related receptor family (ERRs, i.e. orphan receptors that do not bind to any known, naturally occurring steroid hormone (Dehal et al. 2002). Nevertheless, ERs do not appear to be a vertebrate innovation, since an estrogen receptor ortholog was recently reported in the marine gastropod *Aplysia californica* (Thornton et al. 2003). However, according to the authors, this is not a functional estrogen receptor involved in endocrine mechanisms, but is instead an evolutionary vestige lost from the arthropod lineage and not hormonally active. Coming back to echinoderms, it is then possible that in these animals steroids mediate their action through different transducers. In other words, since steroids are present in echinoderms and they are endogenously synthesised, some kind of receptor/transducer is likely to be present as well. However this would not be similar, at least in gene sequence, to the well known nuclear receptors of vertebrates. Interestingly, recent finding showed that in mammals, besides classical ER-mediated mechanisms, rapid physiological effects of estrogens may occur *via* non-ER membrane associated protein, such as G-protein coupled receptors (Hewitt et al. 2005).

4. Echinoderms as Useful Experimental Models for Studies on Endocrine Disruption

In the following paragraphs we will discuss some examples of our research on EDCs (together with previous studies) in which echinoderm species were employed as test animals. Different types of animal models, biological problems as well as experimental approaches were considered, in order to give a wide view on possible employment of echinoderms in these types of research and their usefulness. These studies considered two different aspects of echinoderm biology, reproduction and regeneration, and their sensitivity and responsiveness to EDCs. Some of these same exposure experiments will be also critically considered in terms of modelling the fate of the different contaminants and their main distribution phenomena, including bioaccumulation and biotrasformation in these animal species (Par. 5.2). This last approach can help in the planning phase of this type of ecotoxicological test, and, later, in the comprehension of the results and in the exportability to the natural ecosystems

4.1. Echinoderm Reproductive Biology and EDCs

Echinoderms are almost exclusively gonocoristic animals, although both simultaneous and protandrous hermaphrodites have been found within all the classes (Chia and Bickell, 1983). Fertilization is external: the numerous gametes (some species can produce up to some millions of eggs!) are released in the sea, where they will eventually give origin to planktonic larvae. The number of gonads and their position within the coelom are generally characteristic for each class (Hymann, 1955). In echinoids they are contained within the theca, suspended along the inner surface of the aboral emisphere, and are organised in clusters of acini all connected to a main central gonoduct (Fig. 1a); this last structure reaches the gonopores at the level of the genital plates, on the aboral apex. In regular echinoids they are usually in number of five, whereas in most irregulars they are four (Hymann, 1955). In crinoids, gonads are much more numerous and are usually located within the genital cavity in the lateral branches (pinnules) of each arm (Fig. 1b); Kanatani and Nagahama, 1983), specifically in the proximal-intermediate region. In asteroids, two pairs of gonads are contained within each arm (Hymann, 1955). Structural organization of the gonads is relatively simple. From outside to inside, crinoid gonads consist of a squamous visceral peritoneum, nerve cell processes and muscle cells, an haemal lacuna in which few cells are embedded, and, finally, a germinal epithelium, composed both of germinal and non germinal cells lining the central lumen. In sea urchins a further layer, the perihemal coelom or genital sinus, can be found between the peritoneum and the myoepithelium (Chia and Bickell, 1983).

Most research on endocrine disruption is addressed to the assessment of EDC effects on reproductive parameters of the organisms. This is mostly due to the wide spectrum of adverse consequences caused by an impaired reproduction: it is clear that in this case not only molecular, cell and tissue level effects can be observed but also the population and even the ecosystem might be markedly affected. Reproductive impairment is likely to reduce the animal fitness and, if adverse conditions persist for a prolonged time, it might plausibly cause a population decrease, thus influencing the whole ecosystem.

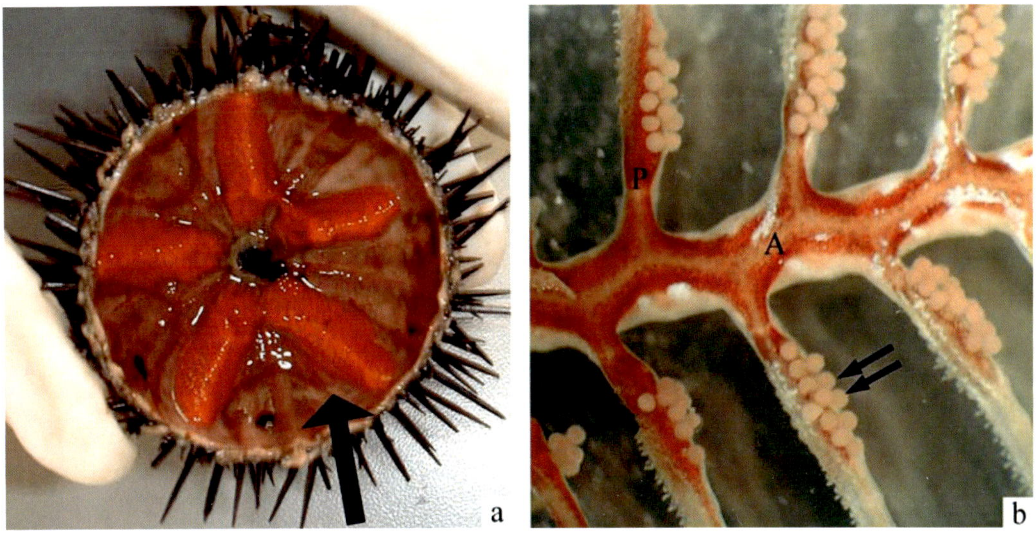

Figure 1. Echinoderm gonads. (a) Aboral half of the echinoid *Paracentrotus lividus* containing five orange gonads (arrow). (b) Proximal-intermediate region of a crinoid arm (A) with its lateral branches (pinnules, P) containing the gonads. Several spawned eggs (double arrows) are attached on the pinnule surface.

Several evidences of endocrine disruption affecting reproduction were reported in the animal kingdom. In humans and other animals, reduced sperm density, genital malformations, infertility and increased number of breast, prostate and test cancer were reported following exposure to EDCs (Sharpe & Skakkebaek, 1993; Colborn et al., 1996). In vertebrates, there are many famous example of endocrine disruption including the hermaphroditic polar bears and beluga exposed to polychlorinated biphenyls, PCBs (De Guise et al. 1994; Wiig et al. 1998), the reduced reproductive success in birds and several malformations in the reproductive organs of alligators, marine birds and amphibians (Fox, 1992; Guillette et al. 1994; Hayes et al., 2003). As previously anticipated, among the most dramatic EDC effects in invertebrates are the well-known *imposex* and *intersex* phenomena of molluscs. In the former male genital organs develop in female specimens which, however, maintain also their original reproductive system (Bryant et al., 1986); in the latter the female reproductive apparatus is progressively transformed in the male prostatic gland (Matthienssen et al., 1999). Organotin compounds used in antifouling paints, such as tributyltin (TBT), are among the most well-known imposex/intersex inducers (Smith & McVeagh, 1991; Bryan et al., 1986; Horiguchi et al. 1997).

Concerning echinoderms and the effects of EDC on their reproductive biology, data are quite limited. Previous research reported a reduced steroid synthesis and altered P and T levels in pyloric caeca and gonads of the asteroid *Asterias Rubens* following exposures to cadmium and PCBs. These changes appeared to be associated with a decreased ovary growth and with a lower quality of the offspring (den Besten et al., 1989, 1991a,b). Short-term exposure experiments on the same species confirmed that steroid metabolism as well as P and T levels in pyloric caeca are altered by cadmium and zinc (Voogt et al., 1987). More recently, Lavado et al. (2006a,b) reported altered endogenous T levels in the crinoid *Antedon mediterranea* and the echinoid *Paracentrotus lividus* following treatments with different putative EDCs: particularly, in *A. mediterranea* testosterone levels were significantly increased after exposure to triphenyltin (TPT), fenarimol and methyltestosterone (Lavado et al., 2006a), whereas in *P. lividus* TPT increases E2 level in the gonads and, similarly to the crinoid, T levels in the coelomic fluid but decreases E2 concentration in the circulatory fluid (Lavado et al. 2006b). Contrasting data were reported about EDCs effects on phase II metabolism in this same echinoid species: in fact, Janer et al. (2005b) showed that *in vitro* testosterone conjugating activity (esterification and sulfation) is inhibited by TPT, TBT and fenarimol, the formers probably acting as irreversible inhibitors, whereas *in vivo* experiments by Lavado et al. (2006) described a significant increase in both metabolic pathway. Long term exposures of *Antocidaris crassispina* specimens to phenol caused several anomalies in spermatogenic processes (such as abnormal mithocondria development) plausibly correlated to a reduced sperm motility (Au et al.2003). Finally, field-based studies showed delayed gonad development in sea urchin specimens *(Strongylocentrotus intermedius)* collected from polluted areas of the Sea of Japan (Vaschenko et al. 1993).

4.2. Crinoid Regeneration and EDCs

Up to some decades ago it was generally thought that steroids involvement in physiological processes was restricted to the reproductive functions. An increasing number of research and evidences demonstrated how this view was originally limited and incomplete. It is now well accepted that, at least in mammals, steroids play a key role in the correct development of the nervous, cardiovascular and skeletal system and are implicated in fundamental cell processes such as proliferation and apoptosis (Lorenzo, 2003). For example, their involvement in bone development and maintenance is of main importance, as demonstrated by the occurrence of osteoporosis in menopausal women (i.e. which actually display reduced estrogen levels). According to Kousteni et al. (2002) this important regulatory function occurs through the inhibition and the stimulation of osteocytes apoptosis and osteoclasts proliferation respectively, the former cell population being responsible of mineral deposition, the latter of reabsorbing processes.

Although no research demonstrated the existence of a clear relationship between steroids and regeneration, it is plausible to speculate that these molecules might influence also regenerative phenomena, since in the latter, physiological processes similar to those above described typically occur. In fact, the first regenerative phases are usually characterised by a considerable cell proliferation, followed by morphogenesis, differentiation and developmental processes. In echinoderms (which are well-known for their remarkable regenerative capacities), among the different structures to be re-built, a primary importance is addressed to the nervous system, whose contribution is fundamental for the correct progress of the regeneration itself, and to the skeletal tissue, which is the main body component in term of biomass (Candia Carnevali and Bonasoro, 2001). Although highly speculative, it could be hypothesised that even in echinoderm regeneration hormonal mechanism are involved.

Embryonic and larval stages are known to be extremely sensitive to environmental changes, as demonstrated by their wide use in many ecotoxicological researches and environmental health assessment programs (Angelini et al., 2005; Schröder et al., 2005). Regenerating organisms are very sensitive as well and can be considered optimal bio-indicators of environmental stress since they reproduce in adult organisms most of those processes (cell proliferation, differentiation, etc…) characteristic of the first developmental stages (Candia Carnevali, 2005). Differently from many embryos, regenerating structure are never protected by any type of cover (such as the egg shell) thus potentially being even more sensitive to external environmental changes.

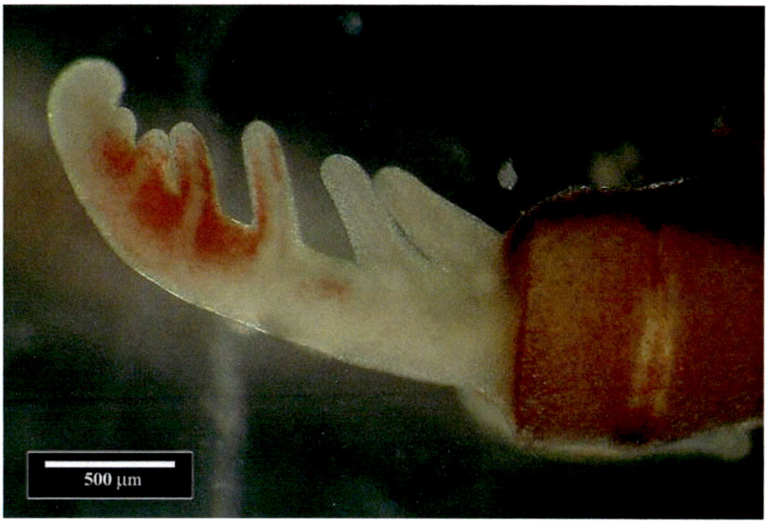

Figure 2. *A. mediterranea* regenerating arm.

Concerning echinoderms, regenerative phenomena are widely used for the replacement of both internal and external body parts, lost after traumatic injury or automutilation (autotomy; Candia Carnevali and Bonasoro, 2001). In some asteroid and ophiuroid species regeneration can be exploited for asexual reproduction, since the animal splits in two parts each of one can completely regenerate the whole individual. A similar phenomena is observed in most echinoderm larvae, where such cloning processes might be ecologically advantaging due to an increase in fecundity and in chances of settlement (Eaves and Palmer, 2003). Among echinoderms, crinoids are well known for their remarkable regenerative capabilities and particularly in the common comatulids most of the internal and external organs can be completely replaced (Candia Carnevali and Bonasoro, 2001). In nature, crinoids with regenerating arms can be frequently observed, as spontaneous autotomy can easily occur following unfavourable environmental conditions (rapid temperature or salinity changes, lack of oxygen, etc…) or mechanical stimulation (wave action or predator attacks). Crinoid arm regeneration has been extensively investigated by Candia Carnevali and Bonasoro (2001), which particularly focused on *A. mediterranea*. In this species the arm can be replaced in a very short time comparing with other echinoderms; in fact, after only 3-4 weeks an in miniature but functional regenerate can be observed (only the size differentiate it from a fully grown arm; Fig. 2). The process occurs through different phases: 1) a repair phase (0-24 hours post-amputation, p.a.), characterised by a quick wound healing and a re-epithelialization of the stump; 2) an early regenerative phase (24-72 hours p.a.), in which the first genuine regenerative processes in terms of marked cell proliferation, growth and early differentiation can be observed; these processes lead to the formation of the regenerative bud (blastema), mainly composed of presumptive stem cells; 3) advanced regenerative phase (72 hours-4 weeks p.a.), characterised by extensive phenomena of morphogenesis, cell and tissue differentiation as well as the progressive apical growth.

Studies regarding possible EDC effects on regenerative development are extremely limited and all of them regard echinoderms. These can be summarised as follows:

- Long-term exposures of the ophiuroid *Ophioderma brevispina* to organotin compounds (TBT and TPT) caused an inhibition of arm regeneration; the effect was observable as a decrease in both regenerate length and diameter (Walsh et al. 1986).
- Exposure to PCBs (Araclor 1260) at concentrations comparable to those of moderately polluted costal area induced significant variations in timing and extent of *A. mediterranea* arm regeneration. The advanced regenerative phase was accelerated and the newly formed structures displayed anomalies in morphogenetic and histogenetic pattern. In addition, the compound appeared to stimulate the cell proliferation at the level of the regenerating bud (blastema), mainly composed of presumptive stem cells (Candia Carnevali et al. 2003).
- On the contrary, exposure experiments carried out on the same crinoid species with 4-nonylphenol (4-NP, a well-known pseudoestrogenic EDC) showed an inhibitory effects on arm regeneration, as underlined by reduced regenerative growth and proliferation rates in exposed specimens (Candia Carnevali et al. 2001).
- More recently, Barbaglio et al. (2006) and Sugni et al. (2007) reported altered arm regeneration in *A. mediterranea* after exposure to selected EDCs with presumptive androgenic and antiandrogenic activity. The arm regenerative growth appeared to be differently affected by the diverse compounds; particularly, samples exposed to the putative androgenic compounds (TPT, FEN, MET) displayed an unusual accelerated growth, suggesting a stimulatory effect of these substances on regeneration. At the microscopic level, several anomalies were observed after the exposures, the most interesting regarding the skeletal development (reduced or increased) and an unusual recruitment and recycling of differentiated cells (e.g. myocytes). According to the authors and to parallel biochemical analyses (Barbaglio et al. 2006; Lavado et al. 2006a,b), most of these alteration could be related to steroid or neuroendocrine dysfunctions.

4.3. Materials and Methods

Animal models. The two selected species used in our studies belong to two representative echinoderm classes: the European echinoid (sea urchin) *Paracentrotus lividus* (Fig. 3a) and the crinoid (feather star) *Antedon mediterranea* (Fig. 3b). *P. lividus* is the most common sea urchin of the Mediterranean Sea, where it represents one of the main herbivorous species on rocky bottoms. As previously underlined (par. 2.5), its foraging activity is fundamental for the maintenance and the equilibrium of many Mediterranean benthic communities. Moreover, it is a well known edible species, their gonads being an appreciate food in many Mediterranean regions. Finally, due to its littoral habitat, it lives strictly in contact with many anthropic activities, and therefore is easily subjected to marine contamination. Although less abundant, *A. mediterranea* is also a widespread species in the Mediterranean Sea and a representative component of the benthic fauna of many marine ecosystems. Due to its microsuspension-filter feeding behaviour, persistent sediment-bound pollutants can have a direct and immediate impact on this animal; this is partially due to its capacity to accumulate persistent and hydrophobic contaminants directly by primary uptake from the surrounding water and suspended sediment, and by secondary uptake from the food (Candia Carnevali, 2005). Finally, its sedentary habits make it a reliable indicators of the environmental conditions of the studied area. *A. mediterranea* arm regeneration (Fig. 2) was used as biological test-process to assess the potential adverse effects of EDCs. Several authors suggested that this process is regulated by endocrine-neuroendocrine mechanisms (Thorndike and Candia Carnevali, 2001; Candia Carnevali, 2005; Sugni et al. 2007a), thus being potentially susceptible to EDCs.

Figure 3. Specimens of the seaurchin *Paracentrotus lividus* (a) and the crinoid *Antedon mediterranea* (b; a centimetre scale is also shown).

Selected EDCs

The selected EDCs were: two compounds suspected for their pseudo-hormonal androgenic activity, the antifouling paint component triphenyltin (TPT) tested as triphenyltin-cloride, and the pesticide fenarimol (FEN), together with the AR agonist methyltestosterone (MET) as positive control, and one compound suspected for its pseudo-hormonal antiandrogenic activity, the DDT derivative *p,p'*-DDE (DDE) together with the AR antagonist cyproterone acetate (CPA) as positive control.

Exposure experiments. The experimental approach employed laboratory tests with individual compounds, as previously described by Sugni et al. (2007). Briefly, adult *P. lividus* and *A. mediterranea* specimens, collected from the Tyrrhenian Sea (Italy), were maintained at 16±1°C in 50 l glass aquaria provided with internal circulation system. For each tested compound five different experimental groups were set up: control (only artificial sea water), solvent control and three different exposure concentrations. Exposure tests were performed in semi-static conditions (20% water renewal in 24 hr). Exposure periods of 2 and 4 weeks were selected for the crinoids; overall period of 4 weeks were employed for the sea urchins. Groups of 18-22 sea urchins and 6-16 crinoids were tested in each aquarium. In the case of regeneration experiments, groups of 25-35 specimens of *A. mediterranea* were employed in each aquarium. Experimental regeneration was induced before the start of the exposure test by amputating three arms for each individual at the autotomy plane; the animals were then put in the test-aquaria and left to regenerate for the prefixed periods (72 hr, 1 week and 2 weeks). The selected nominal concentrations were as follow: TPT 100, 225 and 500 ng/l; FEN 24,

240 and 2400 ng/l; MET 10, 100 and 1000 ng/l; p,p'DDE 100, 500 and 2500 ng/l; CPA 30, 300, 3000 ng/l. Regeneration experiments were performed with only DDE and CPA. All the compounds were dissolved in ethanol (Merck) and for each tested concentration stock solutions were prepared; for each chemical the exposure medium was obtained by adding to the aquaria an appropriate volume of the corresponding ethanol-dissolved stock solutions, so that the prefixed nominal concentrations could be reached at the start of the experiment and then maintained throughout the overall experimental period with the daily medium renewal. In the exposure experiments the ethanol concentrations were strictly the same as in the solvent-containing aquaria (solvent control). The final ethanol concentration in the aquaria was below 10μl/l. During the experiments the actual water concentrations of the tested EDCs were analytically checked, as recommended in official ecotoxicological tests (EC, 1996). Details of the chemical analyses are reported in Tremolada et al. (2006).

Reproductive stage analysis. At the end of the exposure period reproductive organs were removed from control and exposed animals. Sea urchins were divided in two halves and one out of the five gonads was collected from the internal side of aboral half; in case of crinoids, a pair of genital pinnules (which contain the gonads) was gently detached from the right side (oral side up) of each sampled specimens. Six or five reproductive stages, determined by histological analyses, were considered for *P. lividus* and *A. mediterranea* respectively, according to a slightly modified version of Unuma (2002) for the echinoid (further subdivision of the Recovery stage) and to Lavado et al. (2006) for the crinoid: *Recovery (Basal* and *Intermediate,* in case of *P. lividus)*, *Growing*, *Premature*, *Mature* and *Spent*.

Regeneration process analysis. The regeneration/exposure periods were selected as follows: 72 hr, 1 week and 2 weeks, so that they could correspond to well defined and established regenerative stages. At the end of each period regenerating arms were removed for morphometric and histological analyses.

Morphometric analyses. Fresh sea urchin eggs spontaneously released during animal sacrifice were collected and immediately photographed under a light microscope. Egg diameters were measured by employing the computer program ArcView 3.2 (GIS). The same program was used for measuring crinoid oocytes from histological sections, following Vail protocol (Vail, 1987). Regenerating arms were photographed under a stereomicroscope, and the regenerate length was measured with the same computer program previously described (ArcView 3.2 - GIS). A statistical analysis of the obtained data was performed by using one-way ANOVA test (whenever normality and homogeneity of variance were verified) or Kruskall-Wallis test (Tukey and Dunn's test for multiple comparisons, respectively). Statistical analyses were performed by the computer program GraphPad Prism 4.

Histological analyses. Standard methods for light microscopy were employed, as described by Barbaglio et al. (2006). Briefly, exposed and control samples were prefixed with 2% glutaraldehyde in 0.1 M cacodylate buffer for 4-5 hr, then, after overnight washing in the same buffer, postfixed with 1% osmium tetroxide in the same buffer. After standard dehydration in an ethanol series, the samples were embedded in Epon-Araldite 812. The semithin sections, cut with a Reichert Ultracut E were stained by conventional methods (crystal violet-basic fuchsin) and then observed in a Jenaval light microscope. Regenerates were processed also a specific immunocytochemical protocol for detecting cell proliferation (BrdU method). Cell proliferation was monitored by using *in vivo* incorporation of the substituted nucleotide, 5-bromodeoxyuridine (BrdU), then revealed by a monoclonal antibody against BrdU (Cell proliferation kit: Amersham). Control reactions were carried out by omitting the primary antibody. For use with semithin Epon-Araldite sections, the standard BrdU-immunocytochemistry protocol for paraffin sections was modified as described in detail by Candia Carnevali et al. (1995).

4.5. Effects on Reproduction

General histopathology

Hypertrophy and hypotrophy of the gonad acini could be detected in CPA and FEN samples respectively (Waterman et al., unpublished data). However, none of the compounds induced clear histophatological alterations or significantly affected the overall gonad growth, expressed as Gonad Index (GI), and a high variability was encountered in all the experimental groups. Although several authors indicated the GI as a hormonal-influenced parameter, it is probably controlled by a combination of exogenous and endogenous factors; the endocrine component could be masked by some other factors, such as food assumption, temperature, etc....so that even if some hormonal dysfunction occurs its effects on the GI are not easily observable. In natural conditions, sometimes the GI shows marked variations apparently not correlated with the reproductive cycle (Lozano et al. 1995).

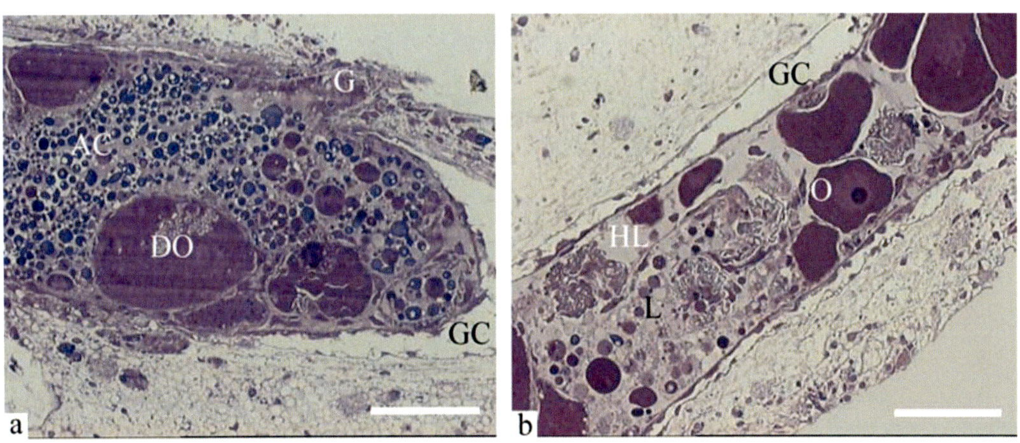

Figure 4. Longitudinal semithin section of A. mediterranea control ovary (a) and TPT-exposed ovary (b). In the latter the number of accessory cells (AC) is markedly reduced and the gonad lumen (L) is nearly deprived of cells. DO: degenerating oocyte; G: gonoduct; GC: genital coelom; HL: hemal lacuna; O: oocyte. Bars= 50 μm.

Several histological anomalies were detected in crinoids, which resulted more sensitive than sea urchins with respect to this parameter. The most evident effects were detected in organisms exposed to the putative androgenic compounds TPT and FEN. In TPT exposed *recovery* females the hemal lacuna resulted hypertrophic and the number of accessory cell was reduced (Fig. 4). According to these data, TPT could stimulate reabsorption of nutrients from their storage site in the lumen (the accessory cells phagocytes) to the hemal lacuna or/and increase phagocyte migration from the lumen to the hemal fluid. In the prosobranc snail, *Hinia reticulata,* TPT was associated to ovary atrophy rather than hypertrophy (Schulte-Oehlmann et al., 2000). In FEN exposed *premature* and *spent* females oocyte development was delayed suggesting an interference of the compound with oogenesis process. In addition FEN exposed females displayed a slightly hypotrophic hemal lacuna. A decrease in volume/size of gonad components was similarly observed in sea urchins exposed to FEN (see above), although males specimens were affected in that case (Waterman et al., unpublished data); this observations suggest a conserved mechanism of action of FEN in different echinoderm species. Finally, male crinoids exposed to TPT and FEN as well as both sexes exposed to all the other tested compound were not apparently affected by the chemicals, although a future ultrastructural analysis is foreseen to confirm these data.

Maturative stage

Some clear tendencies and sex-specific consideration on sea urchin gonad maturation could be profitably derived, although statistically significant difference between exposed and control group were not found (P>0,05). Particularly, male individuals resulted more clearly influenced than females. Low and intermediate TPT concentrations (100 and 225 ng/l) appeared to promote spermatogenesis as shown by the presence of specimens in more advanced maturative stage comparing to the solvent control. However, exposure to the highest dose induced the activation of phagocitic activity, which led all the individuals to a *recovery* (basal) stage (Fig. 5a). A similar stimulation of non-gametogenic processes was observed after exposure to high concentrations of MET and CPA, where a higher frequency of specimens in gonad resting phases (*basal* and/or *intermediate recovery*) was found (data not shown). On the other hand, DDE levels at the highest tested concentrations slightly inhibited the spermatogenesis as indicated by the presence of males in earlier maturative stage (Fig. 5b). Finally, as described by Sugni et al. (2007), TPT exposure significantly affected also female gametogenesis, inhibiting the oogenesis processes and stimulating phagocytic and cellular rearrangement processes within the ovary (at low-medium tested concentrations,).

Similarly to sea urchins, no significant differences were found in reproductive stages' relative frequency of exposed and control crinoid samples (P>0,05). Nevertheless, some suggestions could be obtained from graph observation. A common effect registered in exposed males (TPT, FEN, MT, DDE) was a slight increase of the maturity level compared to the solvent control and particularly in FEN and DDE experiments a dose-dependent relationship was detected. After 4w of exposure, a similar effect was appreciable in MT exposed samples, whereas completely disappeared in fenarimol exposed specimens. Differently from what the histopathological analysis revealed, these results confirmed also males being sensitive to the exposure. Female reproductive cycle resulted less affected by ED exposure. However, interesting results were an increase of recovery stage frequency in TPT and fenarimol exposed females (2w) and a higher maturity rate in females exposed to DDE and CPA. Fig. 5c,d shows results related to TPT exposed females and Fenarimol exposed males respectively and refer to a 2w exposure period.

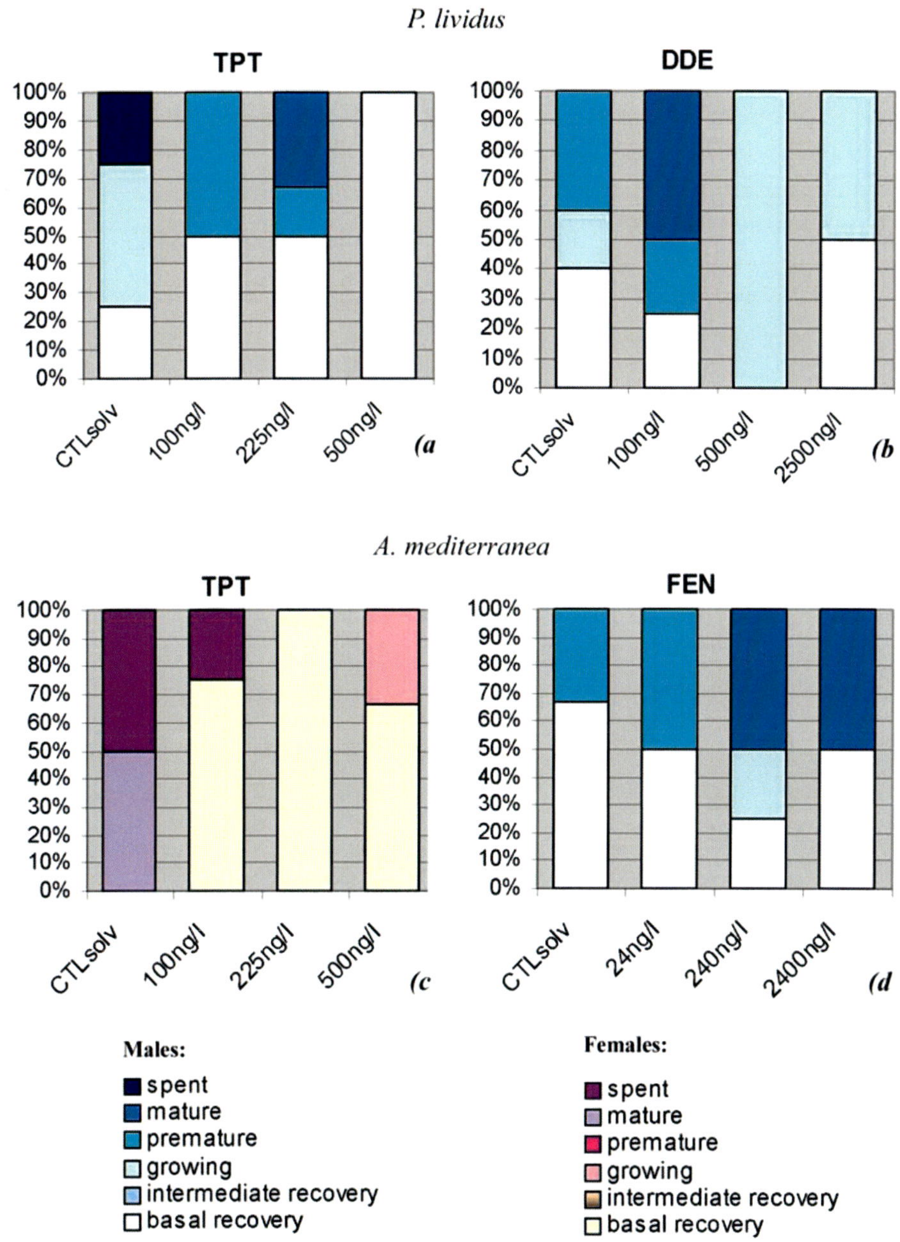

Figure 5. Reproductive stage frequency in *P. lividus* males exposed for 4 weeks to TPT (a) and DDE (b) and *A. mediterranea* specimens exposed for 2 weeks to TPT (c; females) and Fenarimol (d; males).

Overall, the maturative stage of both echinoderm species was more or less influenced by the different compounds. In both species males resulted generally more sensitive than females. Spermatogenic events leading to a complete maturation of the spermatozoon from the gonial stage occur in a shorter time than oocyte maturation. Holland and Giese (1965) showed that in sea urchin the formers take place in only twelve days, whereas oogenesis can last for several months (personal observations). An exposure period of four (echinoids) or two (in crinoids) weeks could not be sufficient to markedly influence the ovary maturative stage but enough to partially affect male gonad maturation in these echinoderm species. Considering the specific compounds, differences and similarities were found between *P. lividus* and *A. mediterranea* responses. Particularly, in both species TPT appeared to promote spermatogenic processes and inhibit oogenesis, stimulating the phagocitosis activity. This results are in agreement with what found in mollusks. In *Hinia reticulata* this compound restrained follicular development and induced ovary atrophy (Schulte-Oehlmann et al., 2000), whereas in the female abalone, *Haliotis gigantea,* it was a strong promoter of spermatogenesis within the ovary (Horiguchi et al. 2002); finally, in *Haliotis madama* a decrease in population size was registered following a displacement of male and female reproductive cycle in organisms exposed to TPT (Horiguchi et al., 2000). Considering the partial divergence of male and female maturative stage frequency registered in our exposure tests, a similar effect can be hypothesized for *P.lividus* and *A.mediterranea* populations following marine exposure to EDCs. In crinoids also fenarimol seems to behave in agreement with its putative role of androgenic compounds, promoting male maturation. DDE exerted different responses in the two experimental models, inhibiting spermatogenesis in *P. lividus* but promoting it in *A. mediterranea.* In mammals this compound is supposed to act as antiandrogenic substance, thus negatively interfering with male reproductive processes (Kelce et al., 1995,1997a,b; Kang et al., 2004). Although further studies are necessary, a similar mechanism of action could be hypothesized for the echinoid, whereas in the crinoid a different type of interaction should be considered.

Oocyte diameter

Egg diameter appeared to be a very sensitive reproductive parameter; moreover its analysis was useful in confirming the maturity level of females exposed to EDs compounds.

As described by Sugni et al. (2007), all the tested compounds could induce significant variations in sea urchin egg size. Almost all the chemicals caused a reduction of the parameter, even if the observed effect was not always strictly dose-dependent. Only with FEN an increase in egg diameter was induced. At least in case of TPT and FEN these variations could be correlated with changes in estradiol levels (Sugni et al. 2007a): an increase in the hormone concentration measured after the exposure corresponded to larger egg production whereas reduced estradiol levels were correlated to small size eggs. The existence of a similar estrogen-dependent influence on egg development is suggested by previous studies on asteroids. Schoenmakers & Dieleman (1981) measured maximum estrone levels during early vitellogenesis phases (when oocytes start to grow up) and increased egg diameter and protein amount in ovaries were reported after *in vivo* and *in vitro* treatments with both estrone and estradiol (Barker and Xu, 1993; Takahashi, 1982a,b; Takahashi and Kanatani, 1981).

Crinoid oocyte diameter appeared to be less affected by the tested EDCs rather than sea urchin eggs. This could be partially due also to the limited number of measurable oocytes. In spite of this, in some cases, statistically significant differences were registered between solvent control and exposed samples. As for histopatholgical analysis, also this crinoid parameter was more affected by androgenic compounds. TPT (Fig. 6a) and FEN (Fig. 6b) exposed females produced smaller oocytes compared to solvent control specimens ($P<0.05$ in both cases). In TPT exposed samples, the retarding effect on oogenesis is associated with an hypertrophy of the hemal sinus, thus suggesting that, as previously hypothesised, TPT could promote a massive reabsorption of nutrients from the ovarian lumen. This reabsorbed material would have not been given back to the developing oocytes leading to a decrease of oocyte mean diameter. In MT, DDE, CPA experimental groups no evident relationship was detected between the exposure and oocyte diameter.

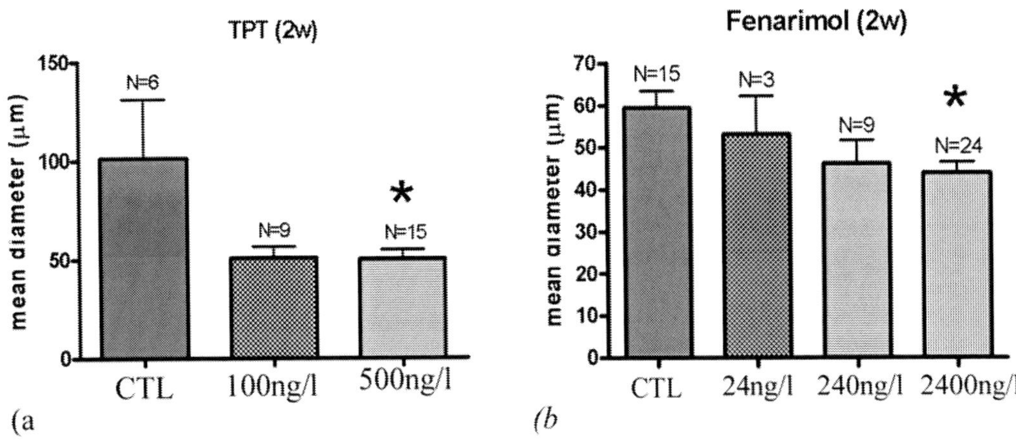

Figure 6. Oocyte diameter in TPT-Cl (a) and Fenarimol (a) experimental groups after 2 weeks of exposure. Data are expressed as mean±SEM. * P<0.05 vs solvent control (CTL).

Similarly to the previous parameter (maturative stage) TPT exposure caused comparable effects in both echinoderm species i.e. egg/oocyte size reduction, whereas different responses were observed after FEN exposure.

Overall, an important output of these exposure experiments consists in the definition of key endpoint in these echinoderm species (hystopathological parameters, reproductive stage frequency, mean oocyte diameter) to be used in future ecotoxicological analysis. Furthermore, the parallel effect of the same EDCs, registered on *A. mediterranea* and *P. lividus* hormonal levels (T and E2; Lavado et al., 2006), strongly encourages exploring echinoderm endocrinology and confirms the importance of a multidisciplinary approach in performing ecotoxicological tests.

4.6. Effects on Regeneration

After 1 week of exposure, the arm regenerate length was not affected by DDE and CPA. A similar lack of effect was observed after the same period of exposure with the putative androgenic compounds MET and FEN, whereas the organotin TPT (another supposed androgenic chemical) could induce an increase in regenerate length (Barbaglio et al. 2006; Sugni et al. 2007a). After 2 weeks of exposure the two tested compounds exerted different effects: DDE was again ineffective whereas CPA displayed significantly reduced regenerate length compared to their controls (P<0.05/0.01; Fig. 7). Interestingly, the putative androgenic compounds (TPT, FEN, MET) were reported to have a stimulatory effect on regeneration (i.e. increased regenerate lenght), although an hormetic dose-response relationship was usually observed – rather than a strictly dose-dependent relationship (Sugni et al. 2007a). As previously described, arm regenerative phases are characterised by several cell processes (proliferation, migration, etc...) whose occurrence and entity is different according to the phase itself (Candia Carnevali and Bonasoro, 2001), and which all together contribute to the macroscopic parameter of the regenerate length. Cell proliferation detected at the level of the regenerate was generally comparable in both control and DDE exposed samples after 1 and 2 weeks of exposure, whereas specimens exposed to the highest concentration at 72 hr showed a slightly reduced proliferation at the level of the stump (data not shown). Interestingly, histological sections of DDE 72 hr samples revealed reduced regenerative buds comparing with controls, possibly as a consequence of the decreased proliferation rate occurred during the first three days of exposure. Differently, CPA exposed samples showed a reduced proliferative activity at the level of the regenerate at 1 week stage (Fig. 8), but not after 72 hr. This could explain the reduced regenerative length observed in CPA exposed specimens at 2 week (Fig. 7), since a decreased contribution of cell elements in the previous period (from 1 to 2 weeks) is likely to slow down the regenerative growth and result in shorter regenerates. Although common aspects can be recognised between DDE and CPA action (they both directly or indirectly reduced the cell cycle activity), their specific mechanisms remains unknown and they probably act on different targets and at different times. Considering possible relationship with steroid dysfunctions, some interesting observation can be derived. In fact, biochemical analyses carried out on these same experimental samples showed that DDE does not affect nor testosterone level neither estradiol concentration after 2 weeks of exposure, whereas CPA act as antiandrogenic compound by reducing testosterone level at the highest concentrations (Lavado et al. 2006a).

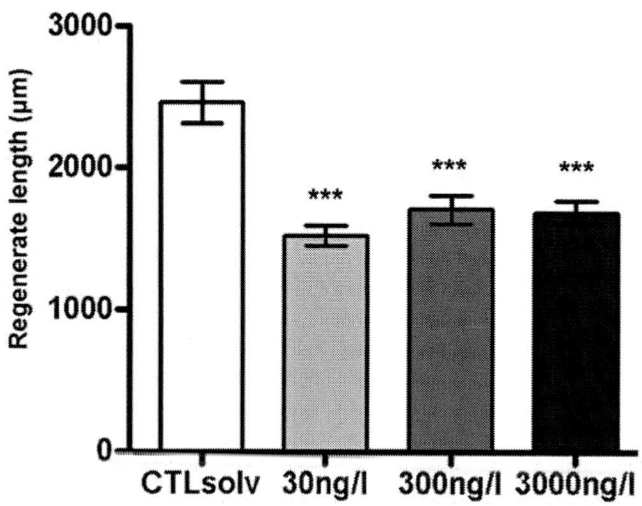

Figure 7. Regenerate length (expressed as mean±SEM) in *A. mediterranea* specimens exposed to different concentrations of CPA for 2 weeks. ***P<0.001 vs solvent control

Microscopic anatomy and histopatological analysis were actually the most useful method employed in our tests to show morphological alterations at both cellular and tissue level. These kinds of alterations are in fact regarded as the most sensitive index of ecologically relevant, individual-level responses (Hinton, 1997).

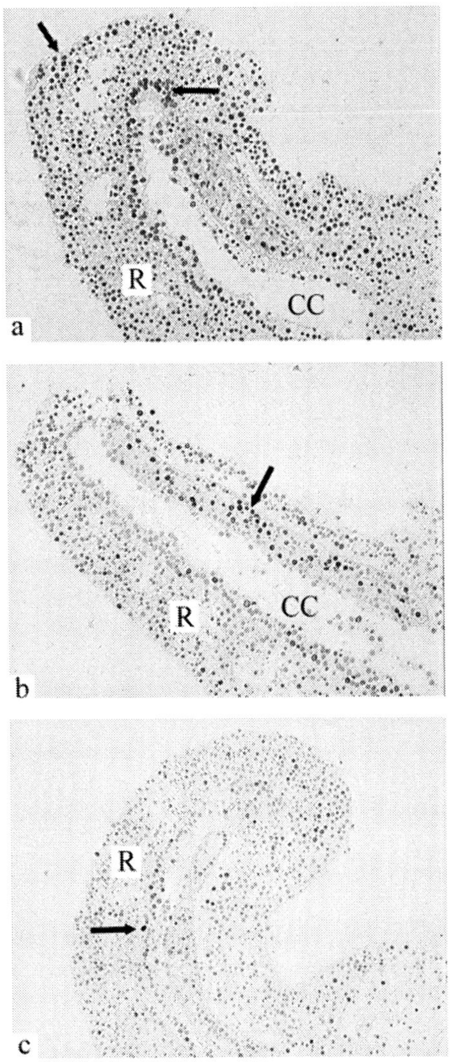

Figure 8. Proliferative activity (BrdU method) detected on sagittal semithin section of *A. mediterranea* regenerate exposed to CPA for 1 week. (a) Control. The immunolabelling (arrow) is marked and widespread throughout the tissues. The number of proliferating cells appears to be reduced in regenerates exposed to 30 ng/l (b) and 3000 ng/l CPA.

Short-term exposure (72h regenerative stage) to medium and high DDE and CPA concentrations induced an unusual activation of morphallactic processes at the level of the muscle bundles in the stump (4.9a,b): these phenomena are characterised by recruitment and recycling of differentiated cells, particularly myocytes, to be used in the regenerative processes (Candia Carnevali and Bonasoro, 2001). These marked dedifferentiation phenomena in exposed samples of both the compounds were observable as a progressive loose of the typical elongated shape and internal structural organization of the contractile apparatus. In some of these samples "detached" and dedifferentiating myocytes could be observed within the coelomic canals (Fig. 9b,c), which actually represent the easiest way to reach the regenerating portion where these cell elements will be probably recycled. An increased number of "free" dedifferentiating myocytes were also observed infiltrating in the stromal and connective tissues, their number being more evident in exposed samples of the highest concentrations. These anomalies were detectable since the earliest regenerative stages considered (72 hr, that corresponded to short-term exposure periods), but often became more evident at the advanced regenerative stages (1 and 2 weeks): the muscle bundle portion interested by these phenomena increased and also more distal bundles were affected. Although a certain physiological rate of muscular rearrangement can be observed also in normal conditions (Candia Carnevali and Bonasoro, 2001), in control samples regeneration was mostly achieved by the contribution of undifferentiated stem cells, which are the main responsible for the blastema formation (Barbaglio et al., 2006; Candia Carnevali, 2005). Overall, considering this specific parameter, the two tested compound exerted a similar effect. An increased involvement of morphallactic phenomena was previously reported also for other suspected EDCs, such as PCBs, 4-NP and TPT (Candia Carnevali et al. 2001b, 2003; Barbaglio et al. 2006; Sugni et al. 2007a). Although this response does not appear to be strictly specific for one substance (even if some differences could be detected in terms of entity, timing and cell type involvement), it has to be recalled that in echinoderms muscle maintenance is probably under endocrine-neuroendocrine influence (Sugni et al. 2007b). This is not surprising if we consider that also in mammals muscle development and maintenance is subjected to an hormonal control, particularly of steroids.

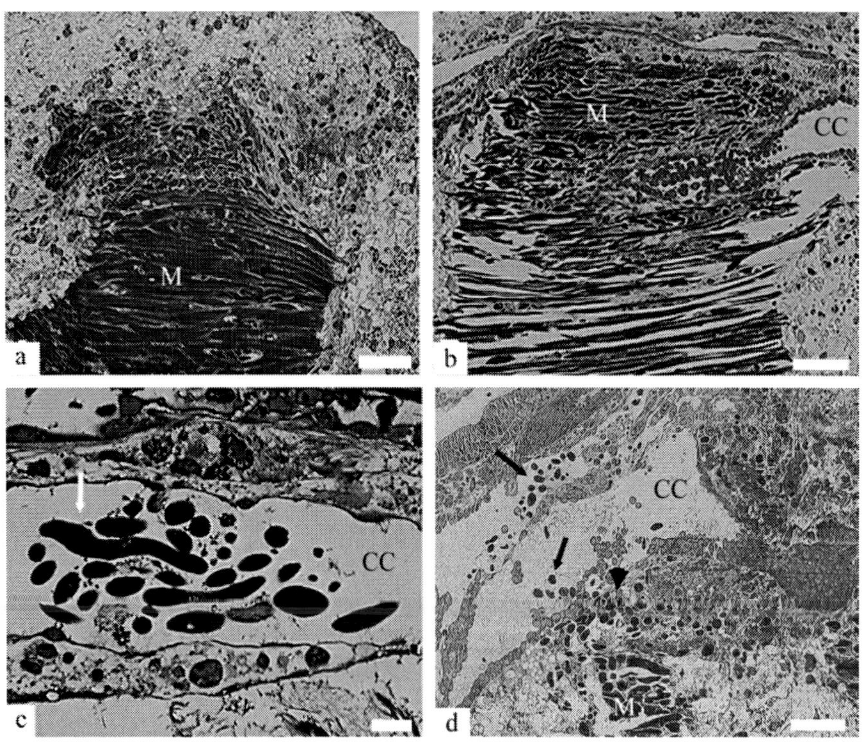

Figure 9. Sagittal semithin sections of the *A. mediterranea* stump showing morphallactic phenomena at the level of the muscles. (a) Control muscle bundle (M) displaying a normal compact structure. Bar=50 µm. (b) DDE exposed muscle with clear signs of disintegration. Bar=50 µm. "Free-swimming" and dedifferentiating myocytes (arrows) can be observed in the coelomic cavities (CC) of both DDE (c; Bar=10 µm) and CPA (d; Bar=50 µm) exposed samples.

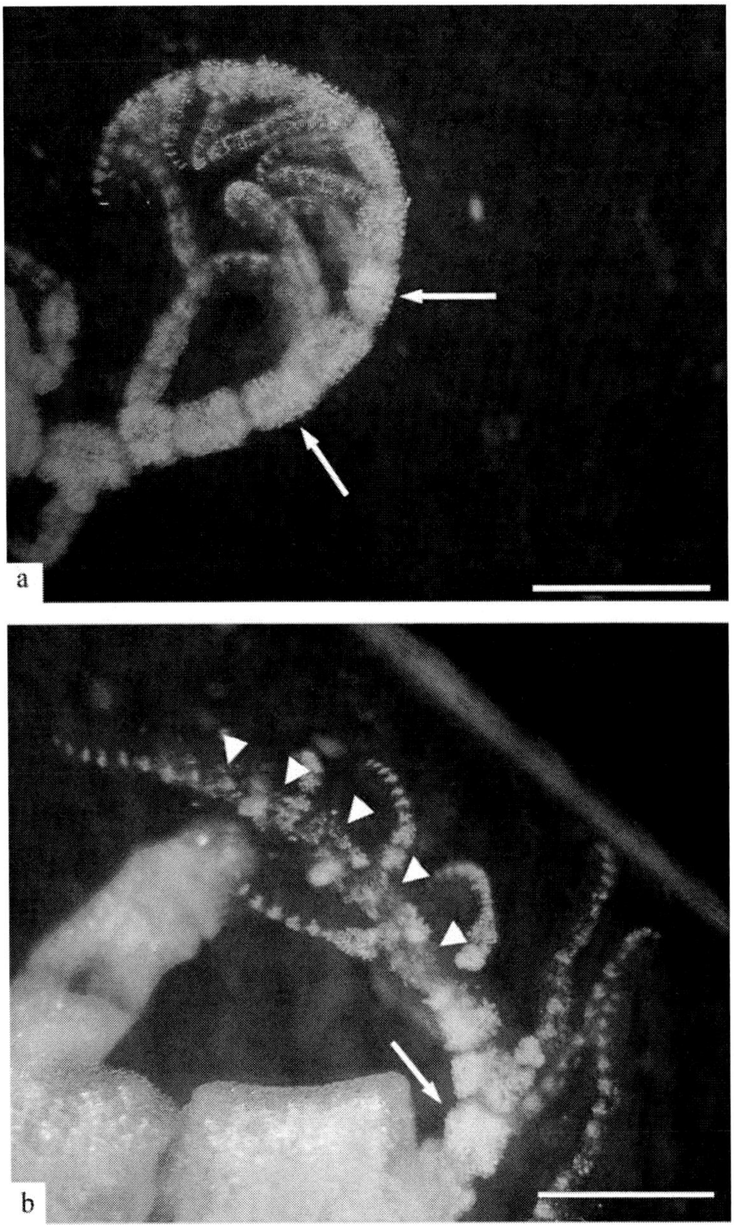

Figure 10. Steromicroscopic view of 2w samples. (a) Control regenerates displaying well developed ossicules (arrow). Bar=500 μm. (b) Regenerate exposed to 2500ng/l DDE displaying scarcely developed skeletal elements (arrowheads) in the intermediate portion. Bar=500 μm.

Considering DDE experiment, another important alteration regarded the developing skeletal elements (ossicles) of the regenerate. After both 1 and 2 weeks of exposure the regenerating samples exposed to medium-high concentrations displayed reduced ossicles in terms of both size and skeletal density and also their developmental pattern appeared to be partially affected by the compound (Fig. 10). An inhibited skeletogenesis during arm regeneration was reported also following exposure to other EDCs, particularly FEN, which is commonly considered to act with a different mechanism from DDE (Sugni et al. 2007a).

The teratogenic effects of suspected EDCs on crinoid skeletogenesis are particularly intriguing results since in vertebrates, steroids (mainly estrogens) are well known regulators of bone development and maintenance (Lorenzo, 2003; Kousteni et al., 2002). Skeletal development could be similarly influenced by hormones also in echinoderms. This fascinating hypothesis is also supported by the documented presence in crinoids of skeletal deposition processes analogous to those described for mammalians and of cellular elements at least functionally similar to osteocytes and osteoclasts (Heinzeller and Welsh, 1994). In addition, DDE is also reported to interfere with physiological activity of estrogens (Sohoni and Sumpter, 1998), the main bone maintenance regulators.

Overall, results obtained so far highlighted a differential pattern of action of DDE and CPA; although some general similarity could be observed (i.e. reduced cell proliferation), their specific effects, and probably their mechanism of action, are necessarily different. In ecotoxicology CPA is commonly considered a "positive control" (or model antiandrogen; Tilmann et al. 2001) since its mechanism of action should be well known, at least in mammals. In fact, CPA is a synthetic derivative of 17-hydroxyprogesterone and is specifically synthesized to exert antiandrogenic activity through antagonistic interaction with the androgen receptors. However this same assumption can not be automatically derived for echinoderms and for most the invertebrate models usually employed in endocrine disruption studies, whose androgen receptors were not definitively found. Even if homolog AR to those of mammals were found, differences in protein structure could be sufficient to change the interactions with CPA or prevent them at all. Therefore, in these animal models substances such as CPA cannot be considered real "positive control".

Overall, these results, together with our previous studies (Barbaglio et al. 2006; Sugni et al. 2007a) show that crinoid regenerative processes can provide original and sensitive models for ecotoxicological studies. Echinoderm regenerative processes can be regarded as an "alternative point of view" to explore in adult organisms either basic developmental processes such as growth and morphogenesis either crucial cellular processes such as proliferation, migration and differentiation; in addition they allow to apply an integrated approach, from whole organism to cellular and molecular level, with an high potential also in terms of applied biology.

5. Problems, Considerations and Methodological Suggestions

As in most (probably all!) research, different problems and/or incongruences were encountered during our researches and with this type of animal models. Here we report several considerations on what we consider of primary importance.

5.1. Chemical Fate in the Experimental System

In aquatic ecotoxicological tests, two problems are related with the actual exposure level in the aquaria. One refers to the possible discrepancy between the added amount and the resulting actual concentration, in the way that the final concentration in water is not a simply ratio of the amount given and the total water volume (nominal concentration). The second problem is related to the maintenance of the actual medium concentration to a constant level. Both problems are a key-point and a primary pre-requisite in ecotoxicological aquatic tests (EEC, 1967). Specially in long-term tests, it is not easy to maintain the water concentrations constant and close to the nominal ones, even if recommended standard precautions are adopted (ECETOC, 1996). Analytical data often show that the exposure concentrations, far to correspond to the nominal ones, tend also to slightly increase or decrease. This behaviour depends on the number of processes ongoing in the system once the chemical is added in. These processes are of different importance in function to the chemical properties, the animals and the experimental conditions. A preventive evaluation of such events became necessary for a correct experimental planning and in order to avoid inconvenient and surprises. In a standard aquarium exposure system, a number of different loss processes may occur. They involve both abiotic processes such as volatilisation, unspecific adsorption to the glass, plastic and rubber materials and chemical degradation and biotic processes such as bioconcentration and biotransformation (Fig. 11).

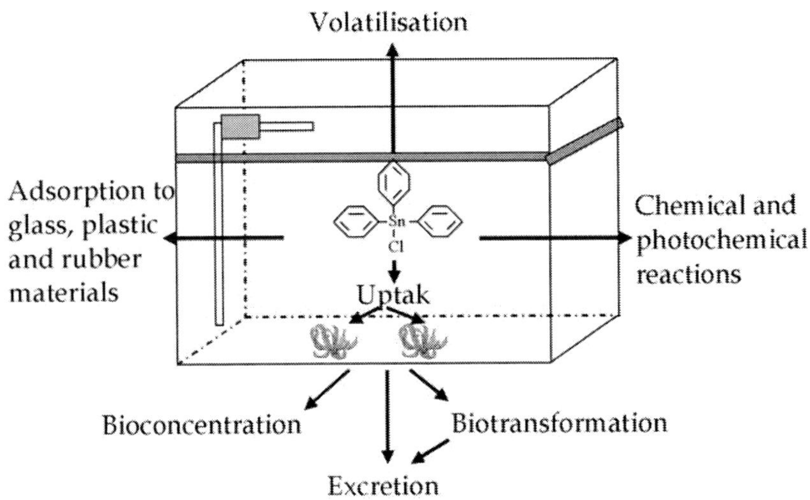

Figure 11. Scheme of the main processes occurring to the molecule once is added to the system.

All these events are controlled by several factors: a) physico-chemical properties of the compound, b) system parameters and c) physiological characteristics of the organisms. It is known that the physico-chemical properties are the main responsible for environmental partitioning and fate (Mackay, 2001). They indicate the relevance of several processes such as volatilisation and bioconcentration. The system parameters such as volumes, temperature, salinity and input and output flows are essential factors for environmental distribution as well. The organism by itself is very important too in terms of metabolism and storing capacity (e.g. lipid fraction). A preventive evaluation of these elements is essential for a preliminary evaluation of the goodness of the experimental apparatus, at least from a qualitative point of view.

5.2. Modelling Approaches in the Planning Phase of the Experiments

Modelling approaches allow to perform an evaluation of the chemical fate in the experimental system also from a quantitative point of view. Predictive models are currently used in risk assessment procedure for predicting environmental concentrations in a variety of ecosystems (EC, 1996) and can be adapted also to micro-ecosystem level (Tremolada et al., 2004). The possibility to predict the extent of the different loss processes and forecast the actual exposure concentrations, before the run of the experiment, can be a great advantage in the planning phase of an ecotoxicological test. This can lead to an accurate choice of the appropriate experimental conditions such as water/animal volume ratio or the compound renewal rate. Here two example of model application are presented: the first one very simple is derived from the level II of the Mackay Fugacity-Model, while the second one, more complex, is derived from the level IV of the same Model (Mackay, 2001). The fist one has been developed and validated with analytical data on triphenyltin compounds (TPT) and *Antedon mediterranea* as test species. Predicted and measured concentration in water and biota are presented in Fig. 12. This simple model takes into account three compartments (air, water and biota) and the main loss mechanisms, such as advection (out and into the system) and reaction (biotransformation into the animals). Details can be found in Tremolada et al. (2006). Water and biota concentrations were calculated daily as the water change frequency (daily partial water renewal of 20% of water volume). The concentrations predicted by the model in water and biota fit quite well with those actually measured for the overall exposure period (28 days). For the biota, despite the few experimental values (due to the limited number of animals available for chemical analyses) predicted concentrations *vs.* measured ones were within or near a factor 2. The increasing concentration trend in the biota and in water after the first day is determined by the daily input of new compound, throughout the water changes, greater than the chemical output (water with depleted compound concentrations by the uptake in the biota).

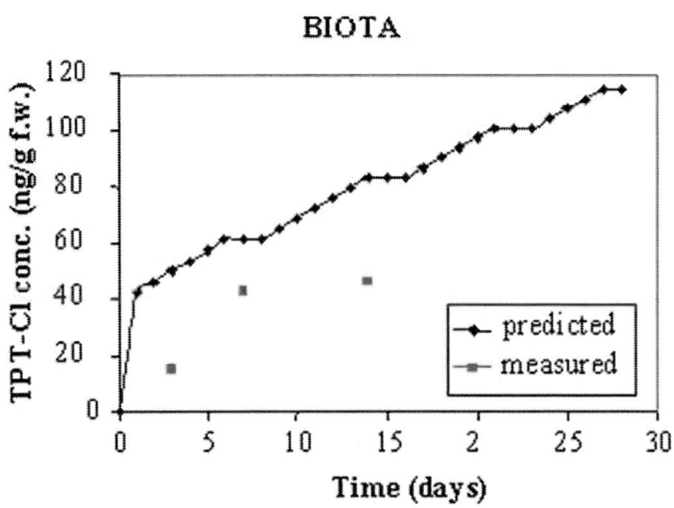

Figure 12. Predicted vs. measured TPT concentrations in water and biota during the 28 days of the exposure experiment.

The second example deals with the application of a more complex model with DDE as exposure compound in the same experimental system. In Fig.13 predicted concentrations in water and biota were reported as fugacity values in order to represent with the same unit (same scale) the two concentration units (ng/l and ng/g f.w. for the water and biota concentration, respectively). This application is able to describe in continuous (unsteady state condition) what happen in the water and biota as consequence of the chosen experimental conditions. Each day, as consequence of the partial water renewal (by which a new chemical input is done), a fugacity peak (which correspond to a concentration peak) happen in the water compartment, then the fugacities (concentrations) rapidly decrease by the chemical uptake into the animals. The uptake kinetics in the animals are sufficiently fast to cause these intense peaks for each day period. Day by day the uptake of the chemical into the biota leads the chemical concentration in the animals to slowly increase. For DDE biotransformation and excretion are of minor importance. By this model a very precise view of what could happen in the aquaria is given.

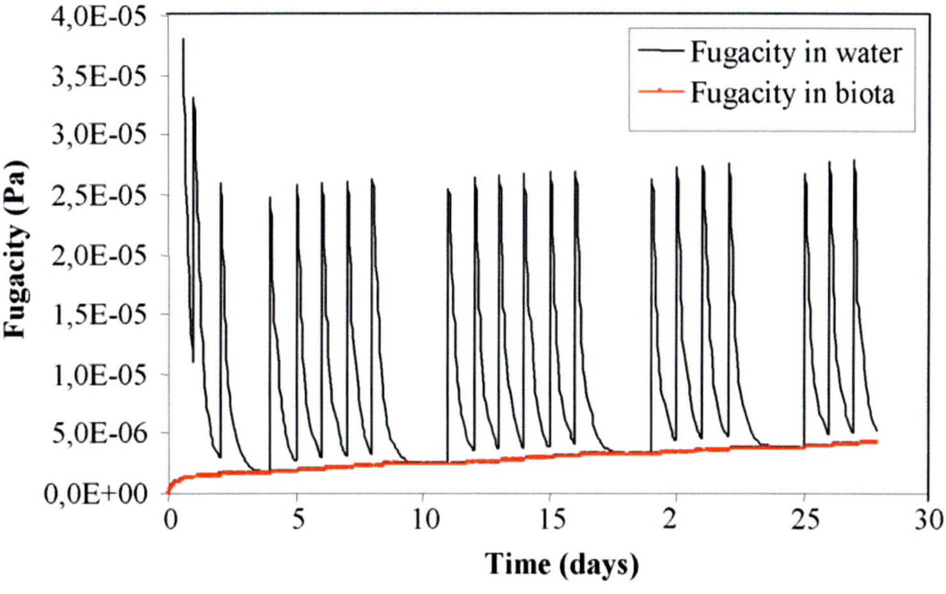

Figure 13. Predicted fugacity of DDE in water and biota for the 28 days of the exposure experiment.

However these examples are just a first attempts, they prove the potentiality of the model applications in ecotoxicological tests. They explains quite clearly the main behaviour of the compound in both water and biota during the exposure period, predicting, quite well, the resulting water concentrations and their time evolution. Transformation processes, probably, should be considered more carefully, because in this case too, they could explain in part the differences between predicted and measured concentrations (measured concentration were always lower than the predicted ones, Fig. 12). The first model application can be considered as a useful very simple tool in particular for the evaluation of the relative importance of the main loss processes in the course of the experiment. The second one, more complex, can be view as a more refined elaboration useful, in particular, for the evaluation of the actual water and biota concentrations.

5.3. The Solvent Problem

The employment of an organic solvent to facilitate the compound solubilization in the aqueous exposure medium is made necessary by the hydrophobicity of many chemicals. In our experiments it was decided to use ethanol since it was one of the solvent suggested also in the EU official methods (EU, 1967). The presence of ethanol made necessary the set up of a further tank in which no compound was tested and only ethanol was added exactly at the same concentration used in the other exposure aquaria. The results obtained from the solvent tank should represent the main comparison elements (solvent control). Nevertheless it was also important to set up a parallel "pure" control, containing only sea water: this allowed the evaluation of possible solvent effects (although they should not be present and the two controls should result as replicates). These two control tanks are both important when using animal models on which literature information and testing experiments are scarce, as in the case of echinoderms. In the present research the ethanol concentration was always below 100 μl/l, that is the accepted threshold below which this solvent should not interfere with the animal physiology (EU, 1967).

In preliminary tests slight different results, even statistically significant, were observed between the solvent control aquaria and the 'pure' water control one. These observations lead to lowering the final solvent concentration in the later exposure experiments from 25 μl/l to 10 μl/l (10 time less the suggest threshold). Although any conclusion about the "solvent problem" can be done by our experiments (it outcome our investigation items), the observation made led us to several considerations which can be of a general interest:

- Even at very low concentrations, the solvent can not be considered *a priori* without effects on the selected experimental animals. This is particularly true for those animals not commonly used in ecotoxicological research and about which few examples of similar experiments are available in literature.
- The choice of solvent type and its concentrations should be considered of primary importance when planning an exposure experiments. The ideal situation would be not to use the solvent at all, however, this is not always feasible, especially when testing scarcely water-soluble compounds.
- The established threshold levels should be considered with due caution since they can be based on experiments with other animals and for other endpoints. If preliminary tests cannot be performed, a good solution would be to use at least a 100-fold lower concentration than the proposed threshold in order to minimize any possible solvent effect.
- Finally, any observation of suspected 'solvent effect' should be also carefully considered and not neglected. The importance of informing the scientific community also about those "noisy and problematic" results should not be underestimated. Each research experiment is based and developed also on previous published works and often the most useful obtained information, for planning or understanding our own experiments, are those related to problems or mistakes encountered during similar researches.

6. General Conclusion and Future Perspective

On the whole the studies reported in this paper showed some examples of exploiting typical echinoderm species in ecotoxicological research and the potential usefulness of these animals. The present work showed that some supposed EDCs could interfere more or less markedly with two basic aspects of echinoderm biology: reproduction and regenerative processes. As previously underlined, one of the most important aims of this work was to propose new and original animal models which could contribute to ecotoxicological studies. The proposed approach mainly based on the individual-level analyses but also considering cell/tissue analyses and molecule kinetics is fundamental for the comprehension of the global results and for eventual exportability to situations occurring in natural ecosystems. According to the reported research, reproductive phenomena of echinoid and regenerative processes of crinoids can be considered reliable and alternative experimental models for studying the effects of exogenous substances interfering with basic processes such as reproduction, growth, development, differentiation and, generally, endogenous regulated mechanisms. The different sensitivity demonstrated by the two experimental models and the wide pattern of parameters considered for each of them produced a broad set of different information that, however, are complementary and all together contribute to the comprehension of the adverse effects exerted by these type of compounds on echinoderm physiology.

Future ecotoxicological research (particularly those on EDCs) should consider a much wider type of experimental animals and echinoderms can represent useful models to use. Moreover future investigations should also specifically focus on basic endocrinology of these animals: as previously underlined, a deeper knowledge of which hormonal factors control their biological processes and how these factors act is a fundamental starting point to develop reliable test models.

Employment of representative species from a wide spectrum of different taxa is crucial. The same biomarker could differently respond to EDCs in the different taxon and predicting the effects in phylogenetically distant species might be difficult and often lead to wrong evaluation. Although similar responses could be present, especially in phylogenetically close species, this should not be inferred *a priori* and without a sufficient knowledge of their basic endocrinology. Therefore, the more limited is the diversity of experimental animal models, the weaker is the power of previsions in environmental risk assessment and the ecological relevance and reality of the analysis.

References

Angelini C., Aluigi M.G., Sgro M., Trombino S., Thielecke H., Falugi C. (2005). Cell signalling during sea urchin development: a model for assessing toxicity of environmental contaminants. In: Matranga, V. (ed). *Echinodermata, Progress in Molecular and Subcellular Biology, Subseries Marine Molecular Biotechnology*, pp: 45-70. Berlin, Heidelberg, Germany: Springer-Verlag.

Au D.W.T., Yurchenko O.V., Reunov A.A. (2003). Sublethal effects of phenol on spermatogenesis in sea urchin (*Anthocidaris crassispina*). *Environ. Res.,* **93**, pp 92-98.

Barbaglio A., Mozzi D, Sugni M, Tremolada P., Bonasoro F., Lavado R., Porte C. e Candia Carnevali M.D. (2006). Effects of exposure to ED contaminants (TPT-Cl and Fenarimol) on crinoid echinoderms: comparative analysis of regenerative development and correlated steroid levels. *Mar. Biol.,* **149 (1),** 65-77.

Barbaglio A., Sugni M., Di Benedetto C., Bonasoro F., Schnell S., Lavado R., Porte C. & Candia Carnevali M.D. (2007). Gametogenesis correlated with steroid levels during the gonadal cycle of the sea-urchin *Paracentrotus lividus* (Echinodermata: Echinoidea). *Comp Biochem Physiol A Physiol* **147**, 466-474.

Barker MF and Xu RA (1993) Effects of estrogens on gametogenesis and steroid levels in the ovaries and pyloric caeca of *Sclerasterias mollis* (Echinodermata: Asteroidea). *Invert Rep Dev* **24 (1),** 53-58.

Brusca R.C. and Brusca G.J. (1990). Phylum Echinodermata. In: Brusca R.C. and Brusca G.J. (eds). *Invertebrates* (pp 801-839). Sunderland, MA: Sinauer Associates.

Bryan G.W., Gibbs P.E., Hummerstone L.G., Burt G.R. (1986). The decline of the gastropod Nocella papillus around south-west England: evedence for the effect of tributyltin from antifouling paints. *J. Mar. Biol. Ass. UK,* **66**, 611-640

Caine G.D. & Burke R.D. (1984). Immunohistochemical localization of gonad stimulating substance in the seastar *Pycnopodia helianthoides.* Proceedings of 5th International Echinoderms Conference. Rotterdam, The Netherlands: Balkema,.

Candia Carnevali M.D. (2005). Regenerative response and endocrine disrupters in Crinoid Echinoderms: an old experimental model, a new ecotoxicological test. In: Matranga V. (ed). *Echinodermata, Progress in Molecular and Subcellular Biology,* Subseries Marine Molecular Biotechnology (pp. 167-199). Berlin, Heidelberg, Germany: Springer-Verlag,.

Candia Carnevali MD, Bonasoro F, Ferreri P, Galassi S (2003). Regenerative potential and effects of exposure to pseudo-estrogenic contaminants (4-nonylphenol) in the crinoid *Antedon mediterranea.* In: Feral JP (ed). *Echinoderm research 2001* (pp: 201-207). Rotterdam, The Netherlands: Balkema.

Candia Carnevali M.D. and Bonasoro F. (2001). Microscopic overview of crinoid regeneration. *Microscopy Research and Techniques*, **55**, 403-426.

Candia Carnevali MD, Bonasoro F, Patruno M, Thorndyke MC, Galassi S. (2001). PCB exposure and regeneration in crinoids (Echinodermata). *Mar Ecol Progr Ser* **215**,155-167

Candia Carnevali, M.D., Bonasoro, F., Lucca, E. and Thorndyke, M.C. (1995). Pattern of cell proliferation in the early stages of arm regeneration in the feather star *Antedon mediterranea. J. Exp. Zool.* **272**, 464-474.

Chia F.S., Bickell L.R., 1983. Echinodermata. In: K.G. Adiyodi & R.G. Adiyodi (eds) *Reproductive Biology of Invertebrates.* Vol. 2: Spermatogenesis and Sperm Function. John Wiley & Sons: pp 545-620.

Chia F.S. and Harrison F.W. (1994). Introduction to the Echinodermata. In: Harrison F.W. and Chia F.S. (eds), Microscopic Anatomy of Invertebrates, Vol. 14, *Echinodermata* (pp:1-8). New York: Wiley-Liss.

Colborn T., Petersen Myers J., Dumanoski D. (1996). *Our Stolen Future, How we are threatening our fertility, intelligence and survival,* Boston: Little, Brown and Co

Coteur G., Pernet P., Gillan D, Joly G, Maage A., and Dubois P. (2003). Field contamination of the starfish *Asteria rubens* by metals. Part 1: short- and long- term accumulation along pollution gradient. *Env. Tox. Chem.* **22 (9),** 2136-2144

de Fur P.L., Crane M., Tattersfield L., Ingersoll C.G., Stahl R., Matthiessen P., Le Blanc G. (1999). Introduction to EDIETA Workshop. In: de Fur P.L., Crane M., Ingersoll C. and

Tattersfield L. (eds). *Endocrine disruption in invertebrates: endocrinology, testing and assessment*. Pensacola, The Netherlands: SETAC Press.

De Guise S., Lagace A., Beland P. (1994). True hermaphroditism in a St.Lawrence beluga whale (*Delphinapterus leucas*). *J. Wildl. Dis.*, **30**, 287-290.

De Waal M., Poortman J. and Voogt P.A. (1982). Steroid receptors in invertebrates. A specific 17β-oestradiol binding protein in a seastar. *Mar. Biol. Lett.*, **3**, 317-323.

den Besten, P.J. 1998. Cytocrome P450 monoxygenase system in echinoderms. *Comp Biochem Physiol C Pharmacol Toxicol Endocrinol* **121**: 139-146.

den Besten P.J., Elembaas J.M.L., Maas E.J.R., Dielman S.J., Herwig H.J.,Voogt P.A. (1991a). Effects of cadmium and polychlorinated biphenyls (Clophen A50) on steroid metabolism and cytochrome P-450 monooxygenase system in the sea star *Asterias rubens* L. *Aquat. Toxicol.*, **20**, 95-110.

den Besten P.J., van Donselaar E.G., Herwig H.J., Zandee D.I., Voogt P.A. (1991b). Effects of cadmium on gametogenesis in the sea star *Asterias rubens* L. *Aquat. Toxicol.*, **20**: 83-94.

den Besten P.J., Herwig H.J., Zandee D.I., Voogt P.A. (1989). Effects of cadmium and PCBs on reproduction of the sea star *Asterias rubens*: Aberrations in the early development. *Ecotoxicol. Environ. Safety*, **18**, 173-180.

Dieleman S.J., Schoenmakers H.N.J. (1979). Radioimmunoassays to determine the presence of progesterone and estrone in starfish *Asteria rubens*. *Gen. Comp. Endocrinol.*, **39**, 534-542.

Donahue J.K., Jennings D.E., 1937. The occurrence of estrogenic substances in the ovaries of echinoderm. *Endocrinol*, **21**, 875-877.

Eaves A.A. and Palmer A.R. 2003. Widespread cloning in echinoderm larvae. *Nature*, **425**,146

Fostier, A., Jalabert, B., Billard, R., Breton, B., Zohar, Y. (1983). The gonadal steroids. In: Hoar WS, Randall DJ and Donaldson EM (eds.), *Fish Physiology*, vol. IXA (277-372), New York: Academic press.

Fox G.A. 1992. Epidemiological and pathological evidence of contaminant-induced alterations in sexual development in free-living wildlife. In: Colborn T. & Clement C. (eds). *Chemically induced alteration in sexual and functional development: the wildlife-human connection. Advances in Modern Environmental Toxicology* (147-158), Princeton, New Jersey: Princeton Scientific Publishing Co., Inc,.

EC - European Commission (1996). *Technical Guidance Document on Risk Assessment (TGD) in Support of the Commission Directive 93/67/EEC on Risk Assessment for New Notified Substances and Commission Regulation (EC) No. 1488/94 on Risk Assessment for Existing Substances and Directive 98/8/EC of the European Parliament and the Council Concerning the Placing of Biocidal Products on the Market*, 2nd ed, Part II. Office for Official Publication of the European Commission, Luxemburg.

ECETOC - European Centre for Ecotoxicology and Toxicology of Chemicals (1996) *Aquatic Toxicity Testing of Sparingly Soluble, Volatile and Unstable Substances*, Monograph No. 26, ECETOC, Brussels, Belgium, pp.67.

EEC – European Economic Community (1967). *Annex V of the Dir 67/548/EEC on the Classification, Packaging and Labelling of Dangerous Substances* as successively amended and technically apdated. European Chemical Bureaux, http://ecb.jrc.it

Guillette L.J., Gross T.S., Masson G.R., Matter J.M., Percival H.F., Woodward A.R. (1994). Development abnormalities of the gonad and abnormal sex hormone concentrations in juvenile alligators from contaminated and control lakes in Florida. *Environ. Health Perspect.,* **102**, pp 680-688.

Harrington F.E. and Ozaki H. 1986. The effect of estrogen on protein synthesis in echinoid celomocytes. *Comp Biochem Physiol B Biochem Mol Biol B*, **84(3),** 417-421.

Hayes T.B., Hasto K., Tsui M., Hoang A., Haeffele C., Vonk A. (2003). Atrazine induced hermaphroditism at 0,1ppm in American leopard frogs (*Rana pipiens*): laboratory and field evidence. *Environ. Health Perspect*., **111 (4),** 568-575.

Heinzeller T. and Welsh U. (1994). Crinoidea. In: Herrison F. W. and Chia F.S. (eds). *Microscopic Anatomy of Invertebrates. Echinodermata,* Vol 14, 9-148. New York: Wiley-Liss.

Hewitt S.C., Deroo B.J., Korach K.S. (2005). A new mediator for an old hormone? *Science,* **307**, 1572-1573.

Hines, G.A., Watts, S.A., McClintock, J.B. (1994). Biosynthesis of estrogen derivatives in the echinoid *Lytechinus variegatus* Lamark. In: David B, Guille A, Feral JP, Roux M (eds.), *Echinoderms through time* (711-716). Rotterdam, The Netherlands: Balkema.

Hines, G.A., Watts, S.A., Sower, S.A., Walker, C.W. (1992). Sex steroid levels in the testes, ovaries and pyloric caeca during gametogenesis in the sea star *Asterias vulgaris*. *Gen. Comp. Endocrinol.* **87**, 451-460.

Hinton D.E. (1997). Histological responses signifying exposure or effect ad their application to aquatic ecotoxicology. In: Zelikoff J.T. (ed). Proceedings of the OECD Workshop on Ecotoxicology: Responses, Biomarkers and Risk Assessment. Fairhaven, New Jersey: SOS Publications.

Holland N.D., Giese A.C., (1965). An autoradiographic investigation or the gonads of the purple sea urchin (*Strongylocentrotus purpuratus*). *Biol. Bull.,* **128**, 241-258

Horiguchi T., Shiraishi H., Shimizu M. & Morita M. (1997). Imposex in sea snails, caused by organotin (Tributyl tin and Triphenyltin) pollution in Japan: a survey. *Appl. Organometal. Chem.,* **11**, pp 451-455.

Horiguchi T, Takiguchi N, Cho HS, Kojima M, Kaya M, Shiraishi H, Morita M, Hirose H, Shimizu M (2000). Ovo-testis and disturbed reproductive cycle in the giant abalone, *Haliotis madaka*: possible linkage with organotin contamination in a site of population decline. *Mar. Environ.Res.* 2000; 50(1-5): 223-229.

Horiguchi T., Kojima M., Kaja M., Matsuo T., Shiraishi H., Morita M., Adachi Y. (2002). Tributyltin and triphenyltin induce spermatogenesis in ovary of female abalone, *Haliotis gigantea. Mar. Environ. Res.,* **54**, 679-684.

Hyman L.H. (1955). Echinodermata. In: The invertebrates. Vol. IV. New York, Toronto, London: Mc Graw-Hill Book Company Inc. pp34-131.

Janer G., Le Blanc G.A., Porte C. (2005a). A comparative study on androgen metabolism in three invertebrate species. *Gen. Comp. Endocrinol.,* **143**, 211-221.

Janer G., Sternberg R.M., LeBlanc G.A., Porte C. (2005b). Testosterone conjugating activities in invertebrates: are they targets for endocrine disruptors? *Aquat. Tox.,* **71**, 273-282.

Jenssen B.M., (1996). An overview of exposure to, and effects of petroleum oil and organochlorine pollution in grey seals (*Halichoerus grypus*). *Sci. Total Envir.,* **186**, 109-119.

Kanatani H. and Nagahama Y. (1983). Echinodermata. In: K.G. and R.G. Adiyoudi (eds). *Reproductive biology of invertebrates*. Vol. 1: Oogenesis, Oviposition and Oosorption (443-495). Chichester, England: John Wiley & Sons: pp 611.

Kanatani H. and Shirai H. (1968). Does GSS or MIS Act as a Pheromone? Problems Concerning the Participation of a Pheromone in Starfish Spawning. *The Zoological Society of Japan*, **77**, 207-212.

Kang I.H., Kim H.S., Shin J.H., Kim T.S., Moon H.J, Kim I.Y., Choi K.S., Kil K.S., Park Y.I., Dong M.S., Han S.Y. (2004). Comparison of anti-androgenic activity of flutamide, vinclozolin, procymidone, linuron, and, p,p'DDE in rodent 10-day Hershberger assay. *Toxicology*, **199**, 145-159.

Kelce W.R, Lambright C.R., Gray L.E.Jr, Roberts K.P. (1997b). Vinclozolin and p,p'DDE alter androgen-dependent gene expression: in vivo confirmation of an androgen receptor-mediated mechanism. *Toxicol. Appl. Pharmacol.*, **142**(1), 192-200.

Kelce W.R, Wilson E.M. (1997a). Environmental antiandrogens: developmental effects, molecular mechanisms, and clinical implications. *J. Mol. Med.*, **75**, 198-207.

Kelce W.R., Stone C.R., Laws S.C., Gray L.E., Kemppainen J.A., Wilson E.M. (1995). Persistent DDT metabolite p,p'DDE is a potent androgen receptor antagonist. *Nature*, **375**, 581-585.

Kitching, J.A. and Ebling E.J. 1961. The ecology of Lough Ine XI, The control of algae by *Paracentrotus lividus* (Echinoidea). *J. Anim. Ecol*, **30**, 373-383

Kousteni S., Chen J.R., Bellido T., Han L., Ali A.A., O'Brien C.A., et al. (2002). Reversal of bone loss in mice by nongenotropic signalling of sex steroid. *Science*, **298**, 843-846.

Lafferty K.D. and Kushner D.J. (2000). Population regulation in the purple sea urchin *Strongilocentrotus purpuratus* at the California channel island. In: Brown DR, Mitchell KL, Chang HW (eds). *Fifth California Islands Symposium* (379–381). Santa Barbara, California: Minerals Management Service..

Lavado R., Barbaglio A., Candia Carnevali M.D., Porte C. (2006a). Steroid levels in crinoid echinoderms are altered by exposure to model endocrine disruptors. *Steroids* **71**, 489-497.

Lavado R, Sugni M, Candia Carnevali MD, and Porte C (2006b). Triphenyltin alters androgen metabolism in the sea urchin *Paracentrotus lividus*. *Aquat. Tox*, **79**, 247-256.

Lawrence J.M. 1975.On the relationships between marine plants and sea urchins. *Oeanogr. Mar. Biol*, **13**, 213-286.

Le Blanc G.A., Campbell P.M., den Besten P., Brown R.P., Changs E.S., Coats J.R. et al. (1999). *The endocrinology of invertebrates*. In: de Fur P.L., Crane M., Ingersoll C. and Tattersfield L. (eds). Proceedings of the Workshop on Endocrine disruption in invertebrates: endocrinology, testing and assessment (23-106). Noordwijkerhout, The Netherlands: SETAC Press.

Lorenzo J. (2003). A new hypothesis for how sex steroid hormones regulate bone mass. *J. Clin. Invest.*, **111**, 1641-1643.

Lozano J., Galera J., Lopéz S., Turon X., Palacin C., Morera G. 1995. Biological cycles and recruitment of Paracentrotus lividus (Echinodermata:Echinoidea) in two contrasting habitats. *Mar. Ecol. Prog. Ser.*, **122**: pp179-191.

Lutz I., Sugni M., Candia Carnevali M.D. Lavado R., Porte C., Sculte-Oehlmann U., Kloas W. 2004. First evidence for specific [3H]-testosterone and [3H]-estradiol binding sites in echinoderms. Proceedings of the CREDO Cluster Workshop on the "Ecological Relevance of Chemically Induced Endocrine Distruption in Wildlife".

Mackay D. (2001). *Multimedia Environmental Model. The Fugacity Model,* 2nd ed.. Lewis Publisher, Boca Raton, USA, pp. 261.

Maugh T.H., 1987. Chemicals: how many are there? *Science,* **199**, 162.

Matthienssen P., Reynoldson T., Billinghurst Z., Brassard D.W., Cameron P., Chandler J.T. 1999. Field assessment for endocrine disruption in invertebrates. In: de Fur P.L., Crane M., Ingersoll C., Tattersfield L. (eds) Endocrine disruption in invertebrates: endocrinology testing and assessment, SETAC, Pensacola, pp 199-270.

Mita M (2000). 1-Methyladenine: a starfish oocyte maturation-inducing substance. *Zygote, 8* (supplement), 9-11.

Nédelec H., 1982. Ethologie alimentaire de *Paracentrotus lividus* dans la baie de Galerie (Corse) et son impact sur les peuplements phytobenthiques. Thése 3e cycle, Univ. P. Et M. Curie et Univ. Aix-Marseille, I : 175

North, W. J. (1983). The sea urchin problem. In: Bascom, W. (ed.), *The effects of waste disposal on kelp communities.* Southern California Coastal Water Research Project (243-255), Long Beach, CA.

Oehlmann J. and Schulte-Oehlmann U. (2003). Endocrine disruption in invertebrates. *Pure Appl. Chem*. **75**, Nos. 11-12, 2207-2218.

Olmstead AW, LeBlanc GA. (2007). The Environmental-Endocrine Basis of Gynandromorphism (Intersex) in a Crustacean. *Int J Biol Sci;* **3**, 77-84.

Dehal P, Satou Y., Campbell R.K., Chapman J, Degnan B., et al. (2002). The Draft Genome of *Ciona intestinalis*: Insights into Chordate and Vertebrate Origins. *Science,* **298**, 2157 - 2167

Pearse J.S., Hines A.H. (1987). Long term population dynamics of sea urchin in a central California kelp forest: rare recruitment and rapid decline. *Mar. Ecol. Prog. Ser.* **39**, 275-283.

Roepke, T.A., Snyder, M.J., Cherr, G.N. (2005). Estradiol and endocrine disrupting compounds adversely affect development of sea urchin embryos at environmentally relevant concentrations. *Aquat. Toxicol* .**71**, 155-173.

Schoenmakers H.J.N., Voogt P.A. (1980). In vitro biosynthesis of steroid from progesterone by the ovaries and pyloric caeca of the starfish *Asterias rubens*. *Develop. Growth and Differ.,* **23(2),** 125-135.

Schoenmakers, H.J.N. and Voogt, P.A. (1981). In vitro biosynthesis of steroids from androstenedione by the ovaries and pyloric ceca of the starfish *Asterias rubens*. Gen. Comp. *Endocrinol.* **45**: 242-248.

Schoenmakers, H.J.N. and Dieleman, S.J. (1981). Progesterone and estrone levels in the ovaries, pyloric ceca and perivisceral fluid during the annual reproductive cycle of starfish *Asterias rubens*. *Gen. Comp. Endocrinol.* **43**, 63-70.

Schoenmakers H.J.N, Colenbrander P.H.J.M., Peute J., van Oordt P.G.W.J. (1981a). Anatomy of the ovaries of the starfishes *Asteria rubens* (Echinodermata). *Cell. Tiss. Res.* **217 (3),** 577-597.

Schoenmakers H.J.N., Van Bohemen C.H.G., Dielman S.J. (1981b). Effects of oestradiol-17β on the ovaries of the starfish Asteria rubens. *Develop., Growth and Differ.* **23(2):** pp 125-135.

Schröder H.C., Di Bella G., Janipour N., Bonaventura R., Russo R., Müller W.E.G., Matranga V. (2005). DNA Damage and Developmental Defects After Exposure to UV and Heavy Metals in Sea Urchin Cells and Embryos Compared to Other Invertebrates. In:

Matranga, V. (ed). *Echinodermata, Progress in Molecular and Subcellular Biology*, Subseries Marine Molecular Biotechnology, pp: 45-70. Berlin, Heidelberg, Germany: Springer-Verlag.

Schulte-Oehlmann U., Watermann B., Tillmann M., Scherf S., Markert B., Oehlmann J. (2000). Effects of endocrine disruptors on prosobranch snails (Mollusca: Gastropoda) in the laboratory. Part II: Tripheniltin as a xenoandrogen. *Ecotoxicology* **9,** 399-412.

Sharpe R.M., Skakkebaek N.E. (1993). Are oestrogens involved in falling sperm counts and disorders of the male reproductive tract?. *Lancet,* **341**, 1392-1395.

Shim W.J, Yim U.H., Kim N.S., Hong S.H., Oh J. R., Jeon J.K. and Okamura H. 2004. Accumulation of butyl- and phenyltin compounds in starfish and bivalves from the coastal environment of Korea. *Env. Poll.* **133** (3), 489-499.

Shubina L.K., Federov S.N., Levina E.V., Andryaschenko P.V., Kalinovsky A.I., Stonik V.A., Smirnov I.S. 1998. Comparative study on polyhydroxylated steroids from echinoderms. *Comp Biochem Physiol B Biochem Mol Biol* **119**, 505-511.

Smiley S. 1990. A Review of Echinoderm Oogenesis. *J. Electr. Micr. Tec.* **16** (2), 93-114.

Smith P.J., McVeagh M. (1991). Widespread organotin pollution in New Zeland coastal waters as indicated by imposex in dogwhelks. *Mar. Pollut. Bull.,* **22***,* 409-413.

Sohoni P. and Sumpter J.P. (1998). Several environmental oestrogens are also anti-androgens. *J. Endocrinology,* **158**, 327-339.

Sperry, T.S. and Thomas, P. (1999a). Characterization of two nuclear androgens receptors in Atlantic croacker: comparison of their biochemical properties and binding specificities. *Endocrinology* **140***,* 1602-1611.

Sperry, T.S. and Thomas, P. (1999b). Identification of two nuclear androgens receptors in Kelp bass (*Paralabrax clathratus*) and their biding affinities for xenobiotics: comparison with the Atlantic croacker (*Micropogonias undulatus*) androgen receptors. *Biol. Reprod.* **61**, 1152-1161.

Sugni M., Mozzi D., Barbaglio A., Bonasoro F., Candia Carnevali M.D. (2007a). Endocrine Disrupting Compounds and echinoderms: new ecotoxicological sentinels for the marine ecosystem. *Ecotoxicology*, **16**: 95-108

Sugni M, Wilkie IC, Burighel P, Candia Carnevali MD. (2007b) New evidence of serotonin involvement in the neurohumoral control of crinoid arm regeneration. *J Invert Biol, in press.*

Takahashi N. and Kanatani H. (1981). Effect of 17β-estradiol on growth of cultured ovarian fragments of the starfish *Asterina pectinifera. Develop., Growth and Differ.* **23**(6): pp565-569

Takahashi N. (1982a). The relation between injection of steroids and ovarian protein amounts in the starfish *Asterina pectinifera. Bull. Jpn. Soc. Sci. Fish.,* **48***:* pp 509-511.

Takahashi N. (1982b). Effect of injection of steroids on the starfish testes *Asterina pectinifera. Bull. Jpn. Soc. Sci. Fish.*, **48**, 513-515.

The Sea Urchin Genome Sequencing Consortium. (2006) The Genome of the Sea Urchin Strongylocentrotus purpuratus. *Science*, **314**, 941-952.

Thorndyke M.C. and Candia Carnevali M.D. (2001). Regeneration neurohormones and growth factors in echinoderms. *Can J Zool,* **79**, 1171-1208.

Thornton J. W., Need E., Crews D., (2003). Resurrection of the ancestral steroid receptor: ancient origin of estrogen signalling. *Science* **301**, 1714.

Tillmann M., Schulte-Oehlmann U., Duft M., Markert B., Oehlmann J. 2001.Effects of endocrine disruptors on prosobranch snails (Mollusca: Gastropoda) in the laboratory. Part III: Cyproterone Acetate and Vinclozolin as antiandrogens. *Ecotoxicology,* **10**, 373-388.

Tortonese E. 1965. Echinodermata. In: *Fauna d'Italia*. Calderoni, Italia.

Tremolada P., Bernardinelli I., Colombo M., Spreafico M. and Vighi M. (2004): Coumaphos distribution in the hive ecosystem: case study for modelling applications. *Ecotoxicology,* **13**, 589-601.

Tremolada P., Bristeau S., Mozzi D., Sugni M., Barbaglio A., Dagnac T. and Candia Carnevali M.D. (2006). A simple model to predict compound loss processes in aquatic ecotoxicological tests: calculated and measured triphenyltin levels in water and biota. *International Jurnal of Environmental Anal Chem*, **86**, 171-184.

Unuma T., Yamamoto T., Akiyama T., 1999. Effects of steroids on gonadal growth and gametogenesis in the juvanile red sea urchin *Pseudocentrotus depressus*. *Biol. Bull.,* **196**, 199-204.

Unuma T. (2002). Gonadal growth and its relationship to aquaculture in sea urchins. In: Yokota Y., Matranga V., Smolenicka Z. (eds). *The Sea Urchin: from basic biology to aquaculture.* Lisse, The Netherlands: Swets & zeitlinger B.V.

Vail L. Reproduction in five species of crinoids at Lizard Island, Great Barrier Reef. (1987) *Mar. Biol.* **95**, 431-446.

Vaschenko M.A., Zhadan P.M., Karaseva E.M., Lukyanova O.N., (1993). Disturbance of the reproductive function of the sea urchin *Strongylocentrotus intermedius* in polluted areas of Peter the Bay, Sea of Japan. Russian. *J. Mar. Biol.,* **19**, 35-40

Voogt, P.A. and Dieleman, J.S. (1984). Progesterone and oestrone levels in the gonads and pyloric caeca of the male sea star *Asterias rubens*: a comparison with the corresponding levels in the female sea star. *Comp Biochem Physiol A Physiol,* **79**, 635-639.

Voogt, P.A and Van Rheenen J.W.A. (1986). Androstenedione metabolism in the seastar Asterias rubens L. studied in homogenates and intact tissue: biosynthesis of novel fatty-acyl testosterone. *Comp Biochem Physiol B Biochem Mol Biol* **85**, 497-501.

Voogt P.A., den Besten P.J., Kusters C.M., Messing M.W.J. (1987). Effects of cadmium and zinc on steroid metabolism and steroid level in the sea star *Asterias rubens*. *Comp Biochem Physiol C Pharmacol Toxicol Endocrinol* **86(1),** 83-89.

Voogt, P.A., den Besten, P.J., Jansen, M., (1990). The Δ^5-pathway in steroid metabolism in the sea star *Asterias rubens L. Comp Biochem Physiol B Biochem Mol Biol* **97**, 555-562.

Voogt P.A., Lambert J.G.D., Grannman J.C.M., Jansen M. (1992). Confirmation of the presence of oestradiol-17β in the sea star *Asterias rubens* by GC-MS. *Comp Biochem Physiol B Biochem Mol Biol* **101**, 13-16.

Walsh G.E., McLaighlin L.L., Louie M.K., Deans C.H., Lores E.M. (1986). Inhibition of arm regeneration by *Ophioderma brevispina* (Echinodermata, Ofiuroidea) by tributyltin oxide and triphenyltin oxide. *Ecotoxicol Environ Saf,* **12**, 95-100.

Wasson K.M., Gower B.A., Hines G.A., Watts S.A., 2000a. Levels of progesterone, testosterone and estradiol, and androstenedione metabolism in the gonads of *Lytechinus variegatus* (Echinodermata: Echinoidea). *Comp Biochem Physiol C Pharmacol Toxicol Endocrinol* **126**, 153-165.

Wasson K.M., Gower B.A., Watts S.A., 2000b. Response of ovaries and testes of *Lytechinus variegatus* to dietary administration of estradiol, progesterone and testosterone. *Mar. Biol.,* **137**, 245-255.

Watts, S.A., Hines, G.A., Byrum, C.A., McClintock, J.B., Marion, K.R., Hopkins, T.S., (1993). 11-oxy androgens in echinoderms? XII International Congress of Comparative Endocrinology, Toronto, Canada.

Watts S.A., Hines G.A., Byrum C.A., McClintock J.B., Marion K.R. (1994). Tissue- and species-specific variations in androgen metabolism. In: David B., Guille A., Feral J.P., Roux M. (eds). *Echinoderms Through Time* (pp. 155-161). Rotterdam, The Netherlands: Balkema,.

Wiig O., Derocher A.E., Cronin M.M., Skaare J.U. 1998. Female pseudohermaphrodite polar bears at Svalbard. *J. Wildl. Dis*., **34**(4), 792-796.

Wilson, J.D., 2001. The role of 5α-reduction in steroid hormone physiology. *Reprod. Fertil. Dev.* **13**, 673-678.

Xu R.A. and Barker M.F. 1990. Annual changes in the steroid levels in the ovaries and the pyloric caeca of Sclerasterias mollis (Echinodemata:Asteroidea) during the reproductive cycle. *Comp Biochem Physiol A Physiol* **95**(1): pp 127-133.

Zou E. (2003). Current status of environmental endocrine disruption in selected aquatic invertebrates. *Acta Zoologica Sinica,* **49**(5), 551-565.

In: Ecosystem Ecology Research Trends
Editors: J. Chen and C. Guo, pp. 107-151
ISBN: 978-1-60456-183-8

Chapter 3

INTERACTIONS BETWEEN MARINE MAMMALS AND FISHERIES: IMPLICATIONS FOR COD RECOVERY

***Claude Savenkoff*[a,*], *Lyne Morissette*[a,b], *Martin Castonguay*[a], *Douglas P. Swain*[c], *Mike O. Hammill*[a], *Denis Chabot*[a] *and J. Mark Hanson*[c]**

[a]Pêches et Océans Canada, Institut Maurice-Lamontagne, 850 route de la Mer, Mont-Joli, Québec, G5H 3Z4, Canada
[b]Fisheries Centre, The University of British Columbia, Aquatic Ecosystems Research Laboratory (AERL), 2202 Main Mall, Vancouver, British Columbia, V6T 1Z4, Canada
[c]Fisheries and Oceans Canada, Gulf Fisheries Centre, 343 Université Ave, Moncton, New Brunswick, E1C 9B6, Canada

Abstract

Abundance of many Atlantic cod (*Gadus morhua*) and groundfish stocks in the Northwest Atlantic declined to low levels in the early 1990s, resulting in cessation of directed fishing for these stocks, thus ending one of the largest and longest running commercial groundfish fisheries in the world. The stocks of the northern (nGSL) and southern Gulf of St. Lawrence (sGSL) were closed to directed cod fishing from 1994 to 1996 for the nGSL and from 1993 to 1997 for the sGSL, followed by the opening of a small directed fishery in the two systems. In the northern Gulf of St. Lawrence, the ecosystem biomass structure shifted dramatically from one dominated by demersal fish predators (Atlantic cod *Gadus morhua*, redfish *Sebastes* spp.) and small-bodied forage species (capelin *Mallotus villosus*, mackerel *Scomber scombrus*, herring *Clupea harengus*, northern shrimp *Pandalus borealis*) to one now dominated by only small-bodied forage species. The decline of large predatory fishes has left only marine mammals as top predators during the mid-1990s, and marine mammals and Greenland halibut *Reinhardtius hippoglossoides* during the early 2000s. Large changes also occurred in the biomass structure and ecosystem functioning of the adjacent southern Gulf of

* E-mail address: SavenkoffC@dfo-mpo.gc.ca. (Corresponding author.)

St Lawrence (sGSL) but they were not as dramatic. Although predatory fishes decreased between the mid-1980s and mid-1990s, and prey consumption by seal species increased considerably, large cod remained among the most important single predators on fish in the sGSL. The changes in top-predator abundance driven by human exploitation of selected species resulted in a major perturbation of the structure and functioning of both Gulf ecosystems and represent a case of fishery-induced regime shift. Overfishing influenced community biomass structure directly through preferential removal of larger-bodied fishes and indirectly through predation release.

Species interactions are central to ecosystem considerations. In marine ecosystems, predation can be the major ecological process affecting fish populations and piscivory is often the largest source of fish removal, usually larger than fishing mortality. In both northern and southern Gulf ecosystems, predation mortality exceeded fishing mortality for most groups in recent years because fishing mortality was intentionally reduced by fisheries closures. Seals have benefited from reduced hunting (harvesting and culling/bounties) since the 1970s. Consumption of fish by marine mammals exceeded consumption by predatory fishes in the two ecosystems in the recent time periods. Since the collapse of groundfish stocks, commercial fisheries and seals have become important predators on predatory fishes – possibly slowing their recovery. In recent years, consumption by seals shifted towards species at lower trophic level (forage fishes and invertebrates), which were also the main target of fisheries. Thus, commercial fisheries and seals may have become important competitors of predatory fishes for the same resource.

Key words: food webs, inverse methods, regime shift, seal predation, cod collapse, Gulf of St. Lawrence (Canada).

Introduction

The shelf ecosystems in Atlantic Canada each have unique features, but they all experienced simultaneous collapses of their groundfish fisheries in the early 1990s (CAFSAC 1994, Swain and Castonguay 2000, Gascon 2003). In spite of moratoria on commercial fishing, few stocks have shown signs of recovery (Rice and Rivard 2003). Several reasons for this have been proposed, including a disturbing decline in the productive capacity of these populations (Shelton *et al.* 2006). An important component of this decline in productive capacity is an increase in the natural mortality of adult cod (Dutil and Lambert 2000, Shelton *et al.* 2006). Dutil and Brander (2003) showed that, during the late 1990s, northwest Atlantic cod stocks became less productive and hence less resilient, i.e., less able to resist and respond to perturbations, whether man-made or environmental. Ecosystem change, partly caused by overfishing, may also have led to an increase in natural mortality (Frank *et al.* 2005, Bundy and Fanning 2005, Savenkoff *et al.* 2007 a, b).

In the northern and southern Gulf of St. Lawrence ecosystems, exploitation influenced community structure directly through preferential removal of larger-bodied fishes. The biomass structures of both the northern and southern Gulf ecosystems shifted dramatically from one dominated by large demersal fish predators and small-bodied forage species in the 1980s to one now dominated by small-bodied forage species (Savenkoff *et al.* 2007a, b). Shifts from ecosystems dominated by demersal fish to those dominated by pelagic fish or invertebrate species have been documented in the Atlantic and the Baltic (Worm and Myers 2003, Frank *et al.* 2005) and other coastal ecosystems (Jackson *et al.* 2001).

The decline in Atlantic groundfish coincided with major increases in seal populations in Atlantic Canada, which saw total seal numbers increasing from approximately 4 million animals in the mid-1980s to about 6.3 million by the early 2000s (Hammill and Stenson 2002, 2005, and 2006). Marine mammals, because of their large size and abundance, are often considered to be important predators in marine ecosystems, with significant direct negative effects on their prey species and indirect positive effects on the prey of these species (Yodzis 2001, Molenaar 2002, Plagányi and Butterworth 2002; Morissette *et al.* 2006). Although it has been suggested that seal predation has limited recovery in Atlantic groundfish (e.g., Chouinard *et al.* 2005), their role in the non-recovery of cod remains controversial (Bundy 2001, Savenkoff *et al.* 2004a, Morissette *et al.* 2006) and their overall effect on the foodweb remains unknown (Bowen 1997, Yodzis 2001).

Here, we compare the northern and southern Gulf of St. Lawrence ecosystems in terms of changes over time in the main prey of marine mammals and in the relative effects of marine mammal predation (particularly by seals) versus fishing on the vertebrate and invertebrate communities of these two systems. Comparative studies have proven useful in understanding how ecosystems change and the impacts of perturbations (Cury *et al.* 2005, Bundy 2005, Savenkoff *et al.* 2007a, b). Our approach used modelling and a suite of ecosystem indicators to assess competition between marine mammals and commercial fisheries and possible effects of marine mammals on the recovery of depleted cod populations.

Material and Methods

Study Area

The study area in the northern Gulf of St. Lawrence (Northwest Atlantic Fishing Organization [NAFO] divisions 4RS) covers an area of 103,812 km^2 (Figure 1). The nearshore region (depths < 37 m) was not included in the models. In the southern Gulf of St. Lawrence (NAFO division 4T), our study area was defined as the region with depths between 15 and 200 m, covering an area of 64,075 km^2, which excluded the deep parts of the Laurentian Channel and the St. Lawrence Estuary (Figure 1). In both study areas, the nearshore zone was excluded from the models because it is not covered by annual summer bottom-trawl surveys and because exchanges between the infra-littoral and offshore zones are still poorly understood. Therefore, American lobster (*Homarus americanus*) and rock crab (*Cancer irroratus*), which support important fisheries (lobster alone > 20,000 t yr^{-1}) in the sGSL (Hanson and Lanteigne 2000), were not included in the models.

The Gulf of St. Lawrence is a stratified semi-enclosed sea connected to the North Atlantic Ocean through the Cabot Strait to the southeast and the Strait of Belle-Isle to the northeast. The bathymetry of the Gulf is dominated by the Laurentian Channel, which divides the Gulf into two very distinct systems: the deep northern Gulf (nGSL), characterized by a number of deep channels with depths greater than 200 m, and the southern Gulf (sGSL) represented by a shallow shelf, the Magdalen Shallows, with depths mostly less than 100 m (Koutitonsky and Bugden 1991).

Table 1. Descriptions of the northern and southern Gulf ecosystems. In parentheses, values are reported per km^2.

	Northern Gulf			Southern Gulf	
	Mid-1980	Mid-1990	Early 2000s	Mid-1980	Mid-1990
Area (km^2)	103,812	103,812	103,812	64,075	64,075
Inshore limit (m)	35	35	35	15	15
Greatest depth sampled (m)	530	530	530	200	200
Main marine mammal species	Fin whales, harp seals, minke whales, and grey seals			Harp seals, grey seals, minke whales, and white-sided dolphins	
Main predatory fishes	Redfish, cod, Greenland halibut, and American plaice			American plaice, cod, and flounders	
Main forage fishes	Capelin, herring, and mackerel			Herring, mackerel, and capelin	
Biomass of marine mammals and seabirds (t)	25,869 (0.25)	42,403 (0.41)	54,912 (0.53)	18,630 (0.29)	28,439 (0.44)
Biomass of predatory fishes (t)	2,267,003 (21.84)	403,352 (3.89)	471,258 (4.54)	946,509 (14.77)	536,929 (8.38)
Biomass of forage fishes (t)	2,551,021 (24.57)	1,922,926 (18.52)	5,098,186 (49.11)	972,423 (15.18)	908,226 (14.17)
Total catches (t yr^{-1})	189,819 (1.83)	48,857 (0.47)	65,701 (0.63)	180,148 (2.81)	127,690 (1.99)
Total catches of predatory fishes (t yr^{-1})	150,165 (1.45)	4,005 (0.04)	9,972 (0.10)	86,323 (1.35)	6,423 (0.10)
Total catches of forage fishes (t yr^{-1})	21,920 (0.21)	21,256 (0.20)	23,771 (0.23)	69,731 (1.09)	96,292 (1.50)

Table 2. Functional groups used in modelling in the northern and southern Gulf of St. Lawrence for each time period.

Group name	Main species
Cetaceans[a]	Mysticeti (baleen whales): Fin whale *Balaenoptera physalus*, minke whale *Balaenoptera acutorostrata*, humpback whale *Megaptera novaeangliae* Odontoceti (toothed whales): White sided dolphin *Lagenorhynchus acutus*, white beaked dolphin *L. albirostris*, common porpoise *Phocoena phocoena*
Harp seals	*Pagophilus groenlandica*
Hooded seals	*Cystophora cristata*
Grey seals[b]	*Halichoerus grypus*
Harbour seals[b]	*Phoca vitulina*
Seabirds	Double-crested cormorant *Phalacrocorax auritus*, ring-billed gull *Larus delawarensis*, herring gull *Larus argentatus*, great black-backed gull *Larus marinus*, common tern *Sterna hirundo*, Arctic tern *Sterna paradisaea*, black guillemot *Cepphus grylle*, northern gannet *Morus bassanus*, black-legged kittiwake *Rissa tridactyla*, common murre *Uria aalge*, razorbill *Alca torda*, Atlantic puffin *Fratercula arctica*, Leach's storm-petrel *Oceanodroma leucorhoa*
Large Atlantic cod (> 35 cm)	*Gadus morhua*
Small Atlantic cod (≤ 35 cm)	*Gadus morhua*
Large Greenland halibut (> 40 cm)[c]	*Reinhardtius hippoglossoides*
Small Greenland halibut (≤ 40 cm)[c]	*Reinhardtius hippoglossoides*
Large American plaice (> 35 cm)[d]	*Hippoglossoides platessoides*
Small American plaice (≤ 35 cm)[d]	*Hippoglossoides platessoides*
Flounders	Witch flounder *Glyptocephalus cynoglossus*, yellowtail flounder *Limanda ferruginea*, winter flounder *Pseudopleuronectes americanus*
Skates	Thorny skate *Amblyraja radiata*, smooth skate *Malacoraja senta*, winter skate *Leucoraja ocellata*
Redfish	*Sebastes mentella*, *Sebastes fasciatus*

Table 2. Continued

Group name	Main species
Large demersal feeders	Black dogfish *Centroscyllium fabricii*, white hake *Urophycis tenuis*, marlin spike *Nezumia bairdi*, Atlantic halibut *Hippoglossus hippoglossus*, wolffish *Anarhichas* spp., common lumpfish *Cyclopterus lumpus*, haddock *Melanogrammus aeglefinus*, longfin hake *Urophycis chesteri*, large eelpouts (Zoarcidae), monkfish *Lophius americanus*, grenadiers (Macrouridae)
Small demersal feeders	Sculpins (Cottidae), small eelpouts (Zoarcidae), Atlantic hagfish (*Myxine glutinosa*), fourbeard rockling *Enchelyopus cimbrius*, cunner *Tautogolabrus adspersus*, gunnels (Pholidae), lumpsuckers *Eumicrotremus* spp., snailfishes (Liparidae), blennies (Stichaeidae), juvenile large demersals
Capelin	*Mallotus villosus*
Sand lance[e]	*Ammodytes* spp.
Arctic cod[f]	*Boreogadus saida*
Large pelagic feeders	Spiny dogfish *Squalus acanthias*, pollock *Pollachius virens*, silver hake *Merluccius bilinearis*
Piscivorous small pelagic feeders	Atlantic mackerel *Scomber scombrus*, piscivorous myctophids and other mesopelagics, short-finned squid *Illex illecebrosus*, piscivorous juvenile large pelagics
Planktivorous small pelagic feeders	Atlantic herring *Clupea harengus*, Atlantic argentine *Argentina silus*, planktivorous myctophids and other mesopelagics, planktivorous juvenile large pelagics
Shrimp	Northern shrimp *Pandalus borealis*, striped shrimp *P. montagui,* Arctic argid *Argis dentata*, Greenland shrimp *Eualus macilentus*, circumpolar eualid *E. gaimardii*
Crabs[g]	Snow crab *Chionoecetes opilio*, other non-commercial species (e.g., toad crabs *Hyas* spp.)
Echinoderms	Sand dollar *Echinarachnius parma*, sea urchin *Strongylocentrotus pallidus*, brittle star *Ophiura robusta*
Molluscs	Wedgeclam *Mesodesma deauratum*, propeller clam *Cyrtodaria siliqua*
Polychaetes	*Exogone hebes*
Other benthic invertebrates	Miscellaneous crustaceans, nematodes, other meiofauna
Large zooplankton (> 5 mm)	Euphausiids, chaetognaths, hyperiid amphipods, cnidarians and ctenophores (jellyfish), mysids, tunicates >5 mm, ichthyoplankton

Table 2. Continued

Group name	Main species
Small zooplankton (< 5 mm)	Copepods (mainly *Calanus finmarchicus*, *C. hyperboreus*, and *Oithona similis*), tunicates < 5 mm, meroplankton, heterotrophic protozoa (flagellates, dinoflagellates, and ciliates)
Phytoplankton	Diatom species such as *Chaetoceros affinis*, *C.* spp., *Leptocylindrus minimus*, *Thalassiosira nordenskioeldii*, *T.* spp., *Fragilariopsis* spp., and a mixture of autotrophic and mixotrophic organisms including Cryptophytes, dinoflagellates, Prasinophytes, and Prymnesiophytes
Detritus	Sinking particulate organic matter including both large particles (consisting of animal carcasses and debris of terrigenous and coastal plants) and fine particles (mostly from planktonic organisms, including feces, moults, phytoplankton aggregates, and bacteria)

[a]: Cetaceans were separated into two groups in the northern Gulf for the early 2000s model: the mysticeti (baleen whales) and the odontoceti (toothed whales).

[b]: Grey and harbour seals were grouped for the northern Gulf model of the early 2000s due to a combination of limited data and a weak trophic role in the mid-1980s and mid-1990s ecosystems for harbour seals. Harbour seals accounted for 11-14% of this group biomass for each time period.

[c]: Aggregated as Greenland halibut for the southern Gulf models.

[d]: Aggregated as American plaice for the northern Gulf models.

[e]: Included in the planktivorous small pelagic feeders for the southern Gulf models and for the northern Gulf model of the early 2000s due to very low biomass for sand lance (e.g., northern Gulf biomass: 0.003 t km^{-2} in the early 2000s vs 2.398 t km^{-2} and 0.120 t km^{-2} for the mid-1980s and mid-1990s, respectively).

[f]: Included in the capelin group for the southern Gulf models and for the northern Gulf model of the early 2000s due to a combination of limited data and a weak trophic role.

[g]: Crabs were separated into small (≤ 45 mm carapace width, CW) and large (> 45 mm CW) crabs for the northern Gulf model of the early 2000s.

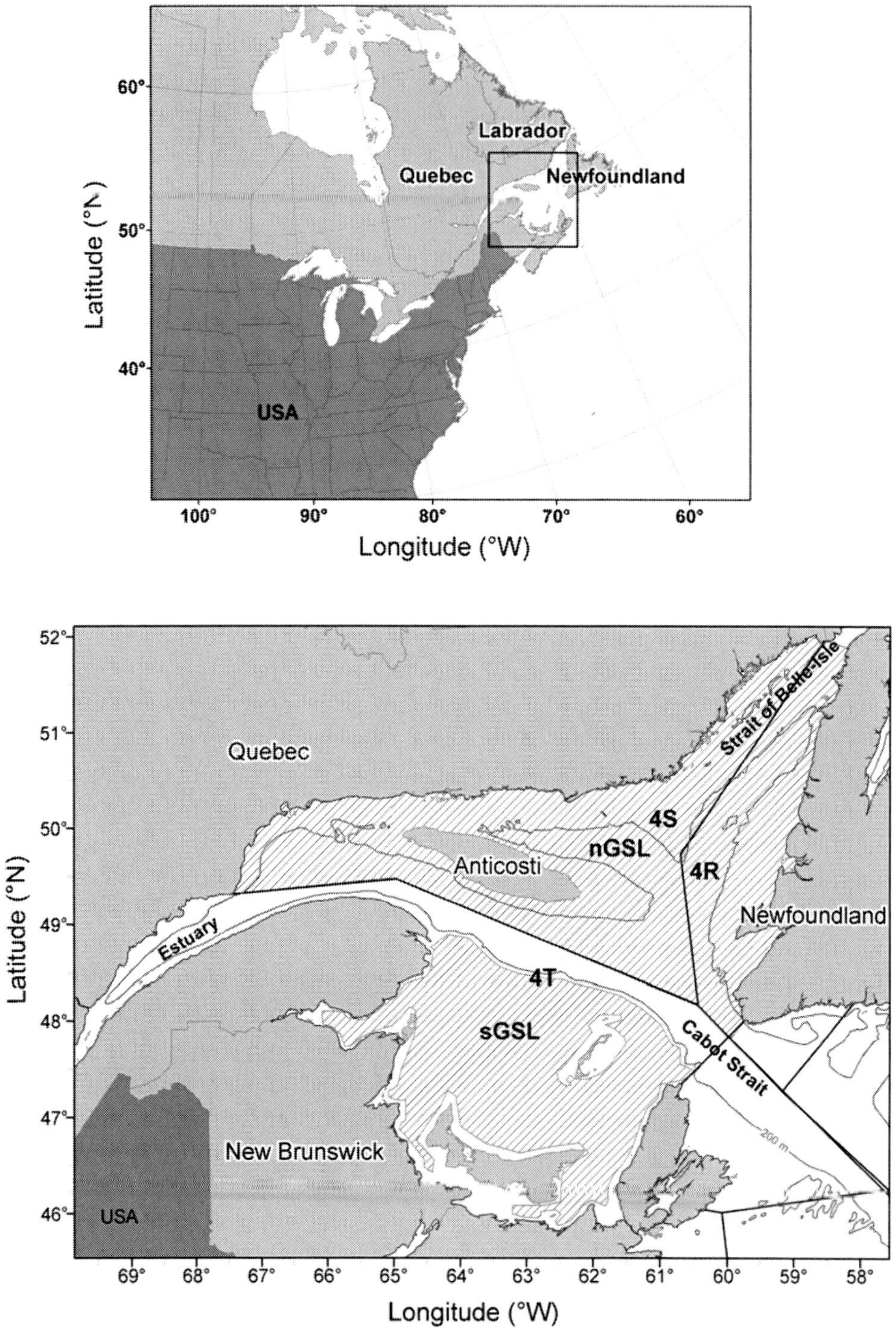

Figure 1. Study area: northern (NAFO divisions 4RS) and southern (NAFO division 4T) Gulf of St. Lawrence. The St. Lawrence Estuary and the deep parts of the Laurentian Channel were not included in this study. The 200-m isobath shows the deep channels.

The two areas support distinct faunas, with limited food-web connections between them (Table 1). The mesozooplankton community is characterized by small copepod species (mainly *Oithona similis*, *Temora longicornis*, *Pseudocalanus* spp.) in the southern Gulf, with larger calanoid copepods (*Calanus finmarchicus/glacialis* and *C. hyperboreus*) more common in the deep channels of the northern Gulf (Roy *et al.* 2000). The northern shrimp (*Pandalus borealis*) dominates the shrimp biomass in the northern Gulf where it is widely distributed at depths of 150-350 m and is the target of a commercial fishery (DFO 2000). In the southern Gulf, the shrimp community is dominated by *Argis dentata*, *Pandalus montagui*, and *Eualus macilentus* and the abundance of large shrimp is insufficient to support a commercial fishery (Hanson and Lanteigne 1999). Snow crabs (*Chionoecetes opilio*) are an important component of the benthic community on the Magdalen Shallows, where they support a large fishery, but are rare in the deep channels to the north. Atlantic herring (*Clupea harengus*) and Atlantic mackerel (*Scomber scombrus*) were the dominant pelagic fishes in the sGSL during the 1980s. Capelin (*Mallotus villosus*) was the most important forage fish in the nGSL in both the 1980s and 1990s, and increased in biomass in the sGSL as well during the 1990s (DFO 2001, Savenkoff *et al.* 2004b, c). Deep-water species (e.g., redfish [*Sebastes* spp.] and Greenland halibut [*Reinhardtius hippoglossoides*]) are important components of the fish community in the nGSL, but are rare in the sGSL (Ni 1982, DFO 2002). In contrast, mid- and shallow-water flatfish species (e.g., American plaice [*Hippoglossoides platessoides*], winter flounder [*Pseudopleuronectes americanus*], yellowtail flounder [*Limanda ferruginea*]) are important components of the fish community and fisheries in the sGSL but are rare in the deep waters of the nGSL during the summer feeding season. Atlantic cod have been an important component of both ecosystems but little mixing occurs between the cod populations from the two areas (Gascon *et al.* 1990, Swain *et al.* 2001). The only other large gadoid species in both systems is white hake (*Urophycis tenuis*), which has morphologically distinct populations in the nGSL (Laurentian Channel) and sGSL (Hurlbut and Clay 1998). Other large commercially-fished gadoids, haddock (*Melanogrammus aeglefinus*) and pollock (*Pollachius virens*) are uncommon in both ecosystems. Four species of pinnipeds occur in the Gulf of St. Lawrence: harp seals (*Pagophilus groenlandica*), grey seals (*Halichoerus grypus*), hooded seals (*Cystophora cristata*), and harbour seals (*Phoca vitulina*). Harp and grey seals are the most abundant pinnipeds in the Gulf.

Data Used in Modelling

The periods covered by this analysis were the pre-collapse (mid-1980s: 1985–1987) and post-collapse (mid-1990s: 1994–1996) of the groundfish stocks. For the northern Gulf, another period was also studied: the early 2000s (2000–2002), a few years after a limited cod fishery had resumed (the nGSL moratorium lasted from 1994 to 1996). The sGSL stock was placed under moratorium in 1993 followed by the opening of a small directed fishery in 1998. The fisheries for white hake and redfish have been closed in the Gulf of St. Lawrence since 1995 (DFO 2004, Gascon 2003).

Based on data availability (representative information on diets is critical) and the ecological and commercial significance of the species, the whole-system model of the nGSL was divided into different functional groups or compartments (32 for the mid-1980s and mid-1990s, 31 for the early 2000s) representing the main pelagic, demersal, and benthic species present (Table 2). The model structure for the early 2000s differs slightly from those of the mid-1980s and mid-1990s to take additional knowledge into account (Savenkoff *et al.* 2005; Table 2). In the sGSL, we distinguished 30 functional groups for the mid-1980s and mid-1990s (Table 2). As yet, no model has been constructed for the early 2000s in the sGSL. For the northern Gulf, details of the methods and data for the parameter estimates for each group, as well as the modelling structure, are given in Morissette *et al.* (2003) for the mid-1980s, Savenkoff *et al.* (2004b) for the mid-1990s, and Savenkoff *et al.* (2005) for the early 2000s. For the southern Gulf, this information is given in Savenkoff *et al.* (2004c). These reports are available online at www.osl.gc.ca/publications/index.jsp?lg=en.

Inverse Modelling: Ecosystem Structure and Sensitivity Analysis

Ecosystem inverse modelling is based on combining compartmental mass balance equations with data equations and eco-physiological constraints on the energy flows (Savenkoff *et al.* 2004a). The mass balance equations specify that, for each consumer group, the sum of inflows (consumption) is balanced by the sum of outflows (production, respiration, and egestion), a net change in the biomass variable (ΔBi) and a residual term (ε). We assumed that there was no change in biomass ($\Delta Bi = 0$) during each time period and that net migration was zero (migration out of, or into the study area, food intake of predators that are not part of the system, etc.) (steady-state assumption). Also, production was equal to the biomass lost to fishing, predation, and natural mortality other than predation (hereafter termed other mortality causes). Other mortality causes include other natural causes of death such as disease or could reflect unsuspected processes occurring in the ecosystem, such as misreported catches (e.g., Savenkoff et al. 2004a), unsuspected migration, or other processes not accounted for in the model.

For phytoplankton, the net (corrected for respiration) production must balance the sum of the outputs (consumption of phytoplankton and other mortality causes of phytoplankton including the egestion term). For the detritus group, the inputs (egestion and other natural causes of death for other groups) must balance the sum of the outputs (consumption of detritus, bacterial remineralization of detritus, and burial).

The models are useful in constraining observations into coherent pictures (Savenkoff *et al.* 2004a); however, since the number of flows to be solved exceeds the number of independent mass balance relations (i.e., an underdetermined system), these models do not have a unique solution. Also, the results are sensitive to choices made regarding the modelling structure, and valid solutions other than those reported here are possible. However, for each model presented in this study, the final solution is always the mean of one solution without perturbation (the "initial solution") plus 30 iterations with random perturbations of the input data (to a maximum of their standard deviations) to provide an overall view of the ecosystem and to identify robust patterns. Simulated inverse analyses have shown that the general flow structure of ecosystems can be recovered with these techniques, although the details can be inaccurate (Vézina and Pahlow 2003). Details of the model structure and method of solution for the different equations are given in Savenkoff *et al.* (2004a and 2007a) for the northern Gulf and in Savenkoff *et al.* (2007b) for the southern Gulf.

Ecopath Modelling: Basic Emergent Properties and Network Analysis Indices

The inverse approach was useful to obtain a first balanced solution by finding the solution that minimizes (objective least-square criterion) both the sum of squared flows (and thus the total sum of energy flows through the food web) and the sum of squared residual errors (i.e., minimizes the imbalances between inputs and outputs) consistent with the constraints (Vézina and Platt 1988). Each of these 31 balanced inverse solutions were then transposed into Ecopath software to estimate the basic emergent properties and network analysis indices for the studied time periods in the two ecosystems (Christensen *et al.* 2005). The indices used include the mean trophic level of the catch, the marine mammal overlap index and the niche-overlap index.

Mean Trophic Level of the Catches

The mean trophic level of fishery catches (TL_C) is a dimensionless index computed by Ecopath and is an indicator of the state of the fisheries (Christensen *et al.* 2005, Pauly and Watson 2005). The TL_C is based on Lindeman's (1942) concept of trophic levels, but can be fractional (Odum and Heald 1975). A routine assigns a trophic level (TL) of one to producers and detritus. For other species, TL is calculated as:

$$TL = 1 + \left(\sum_i p_j * TL_j \right) \qquad \text{Eq. 1}$$

where p_j is the proportion (in mass) of each prey group in the diet of its predator. TL_j is the trophic level of each prey group j.

The mean trophic level of the catches is calculated from:

$$TL_C = \sum_i \left(TL_i \left(\frac{Y_i}{\sum Y} \right) \right) \qquad \text{Eq. 2}$$

where Y_i is the total landings of species i (in tonnes), ΣY is the sum of landings for all species, and TL_i is the trophic level for species i.

Marine Mammal-Fishery Overlap Index

An overlap index between marine mammals and fisheries for each ecosystem was developed by Morissette (2007) using a modified version of the resource overlap index of Kaschner (2004). The resource overlap index presented here ($\alpha_{f,m}$) uses a weighting factor representing the relative importance of fisheries *versus* marine mammals as consumers within a given ecosystem:

$$\alpha_{f,m} = \left(\frac{2 * \sum_k (p_{m,k} * p_{f,k})}{\sum_k p_{m,k}^2 + \sum_k p_{f,k}^2} \right) * \left(\frac{Q_m}{(Q_m + C_f)} * \frac{C_f}{(Q_m + C_f)} \right) \quad \text{Eq. 3}$$

where $\alpha_{f,m}$ is the quantitative overlap between a fishery f and a marine mammal group m in the ecosystem, and the first term expresses the qualitative similarity in diet/catch composition between the marine mammal group m and fishery f sharing the resource or trophic group k, with $p_{m,k}$ and $p_{f,k}$ representing the proportion of trophic group k in the diet of marine mammals m or the catch by fishery f. This term is multiplied by the product of the proportion of total food consumption by marine mammals Q_m and the proportion of total fisheries catches C_f in the ecosystem. This index scales from 0 (no overlap) to 0.250 (identical resource). When resource use is identical between two groups, the first term of Eq. 3 is equal to 1 and each proportion of the second term is 0.5 (or 0.25 for the product). We used the same calculation to estimate an overlap index between large cod and fisheries.

Niche-Overlap Index

The niche-overlap index is a simple index of the extent of overlap in prey (or predators) between two groups (Christensen *et al.* 2005). It is modified from Pianka's (1973) niche-overlap index and estimated as:

$$O_{jk} = \left(\frac{2 * \sum_i (p_{ji} * p_{ki})}{\sum_i (p_{ji}^2 + p_{ki}^2)} \right) \quad \text{Eq. 4}$$

where p_{ji} and p_{ki} are the proportions of the resource i used by groups j and k, respectively. The index is symmetrical and assumes values between 0 and 1. A value of 0 suggests that the two groups do not share resources, 1 indicates complete overlap, and intermediate values show partial overlap in resource utilization (Christensen *et al.* 2005).

Results

Differences in Observed Biomass between Time Periods in Each Ecosystem

In the northern Gulf, biomass of marine mammals (cetaceans and seals) increased from 0.25 t km^{-2} for the mid-1980s to 0.53 t km^{-2} for the early 2000s (Figure 2 and Table 1). There was a sharp decrease in biomass of predatory fishes from 21.84 t km^{-2} for the mid-1980s to 3.89 t km^{-2} for the mid-1990s, and a slight increase in the early 2000s (4.54 t km^{-2}) (Figure 2). Predatory fishes include all fish species except capelin, sand lance (*Ammodytes* spp.), Arctic cod (*Boreogadus saida*), and small pelagic fishes (mostly herring and mackerel), which fall into the forage fish group. The largest declines were for large cod (15-fold, from 3.93 to 0.27 t km^{-2}), redfish (8-fold, from 12.15 to 1.43 t km^{-2}), large demersals (mainly white hake

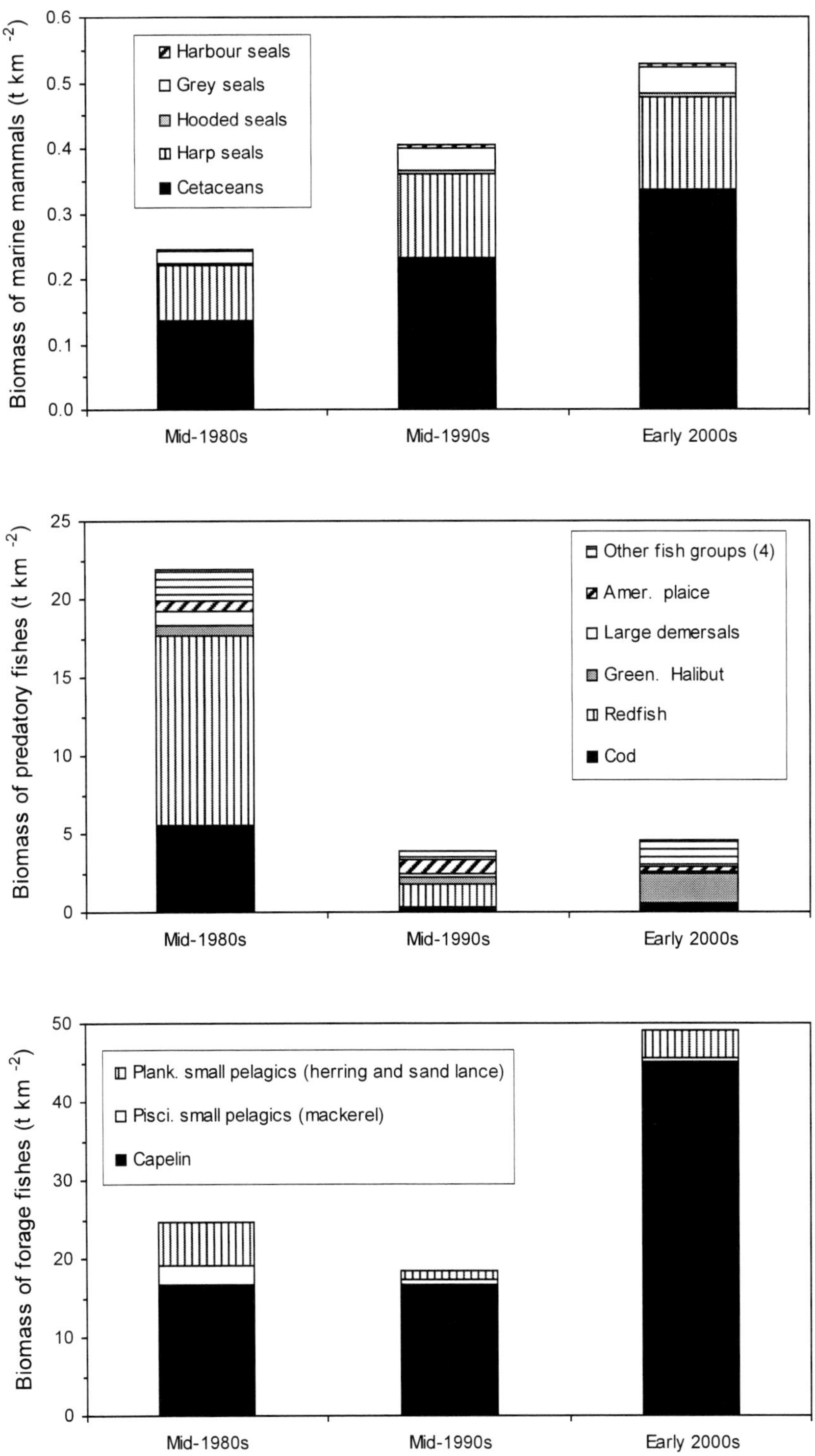

Figure 2. Distribution of observed biomass by trophic groups from the mid-1980s to the early 2000s in the northern Gulf.

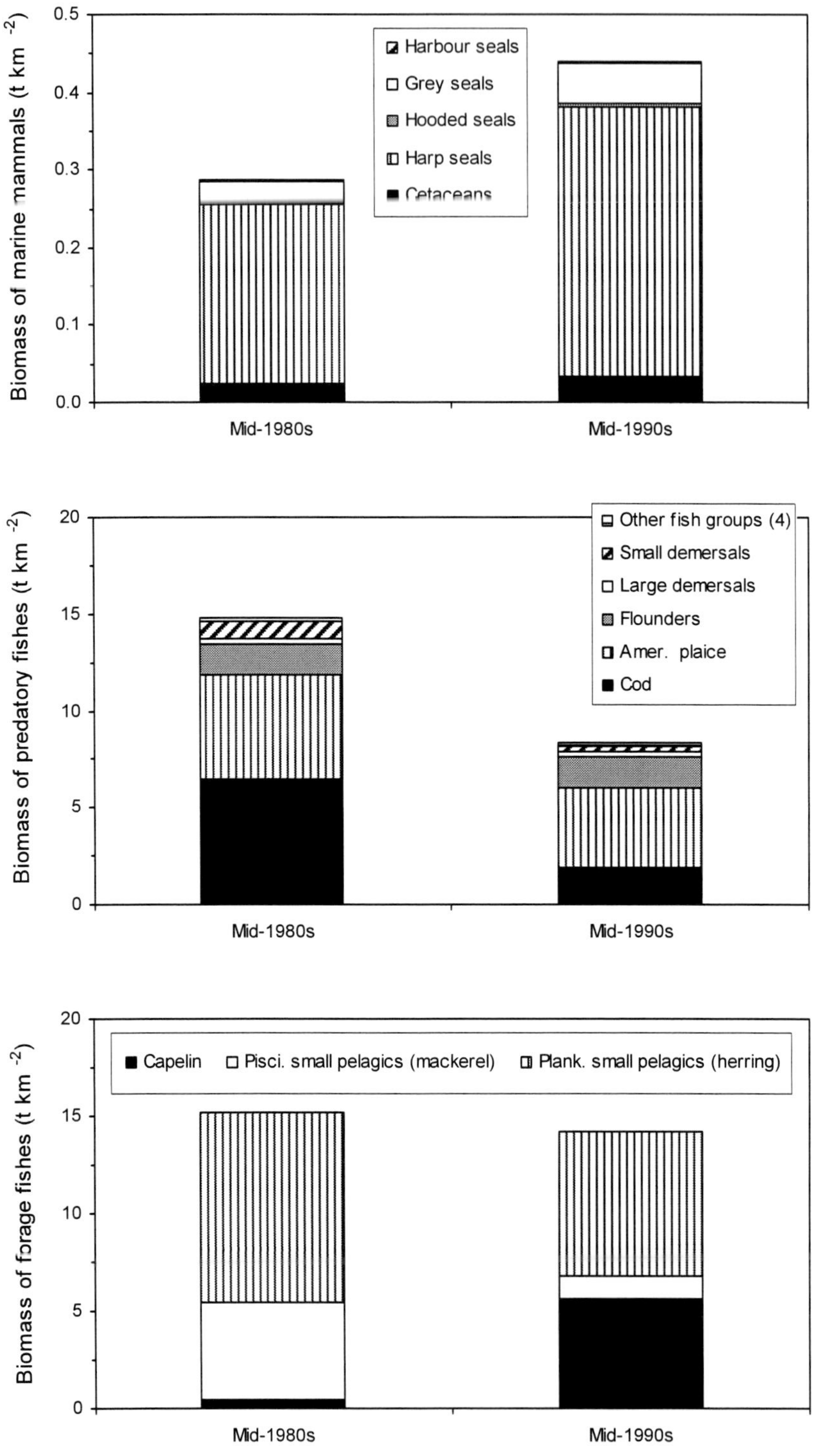

Figure 3. Distribution of observed biomass by trophic groups from the mid-1980s to the mid-1990s in the southern Gulf.

[*Urophycis tenuis*], black dogfish [*Centroscyllium fabricii*], and Atlantic halibut [*Hippoglossus hippoglossus*]; 4-fold, from 0.97 to 0.22 t km^{-2}), small cod (27-fold, from 1.64 to 0.06 t km^{-2}), and flounders (6-fold, from 0.35 to 0.06 t km^{-2}). Forage fish biomass decreased somewhat from the mid-1980s to the mid-1990s, followed by a sharp increase to the early 2000s. Among the forage fishes, the biomass of both piscivorous (mainly mackerel) and planktivorous (mainly herring and sand lance) small pelagics decreased from the mid-1980s to the mid-1990s (piscivorous: from 2.46 to 0.56 t km^{-2}; planktivorous: from 5.42 to 1.27 t km^{-2}), while capelin biomass increased over time. However, capelin biomass was only a gross approximation because catchability of capelin is unknown and may vary between years for the bottom trawl used in the survey.

In the southern Gulf, the biomass of marine mammals increased from 0.29 t km^{-2} for the mid-1980s to 0.44 t km^{-2} for the mid-1990s (Figure 3 and Table 1). The declines in the biomass of large predatory fishes were less dramatic in the sGSL (from 14.77 to 8.38 t km^{-2}). For example, large cod declined 3-fold (from 5.25 to 1.61 t km^{-2}) and large American plaice two-fold (from 0.55 to 0.26 t km^{-2}). For forage fishes, an increase in capelin biomass during the mid-1990s offset decreases in biomasses of piscivorous small pelagics (mainly Atlantic mackerel) and planktivorous small pelagics (mainly Atlantic herring) (Figure 3).

Differences in Landings between Time Periods in Each Ecosystem

There was a near 4-fold decrease in total landings in the nGSL from the mid-1980s to mid-1990s (from 1.8 to 0.5 t km^{-2} yr^{-1} or 189,819 to 48,857 t; Table 1). Catches of predatory fishes accounted for 79% (large cod: 61%; redfish: 14%) of the total fishery during the mid-1980s while catches of forage fishes and invertebrates represented only 21%. During the mid-1990s, catches of predatory fishes accounted only for 8% (large cod: 1%; redfish: 4%) of the total landings while catches of forage fishes and invertebrates represented 92%. From the mid-1990s to the early 2000s, there was a slight increase in total landings (from 0.5 to 0.6 t km^{-2} yr^{-1} or from 48,857 to 65,701 t), reflecting increases in each of these groups (Figure 4). Predatory fishes accounted for 15% (large cod: 11%; redfish: 1%) of the total fishery while catches of forage fishes and invertebrates represented 85%. The mean trophic level of the catches decreased from the mid-1980s (TL_c: 3.8) to the mid-1990s (TL_c: 3.1), followed by an increase in the early 2000s (TL_c: 3.4). This recent trend was mainly related to an increase in catches of large cod (from 0.003 t km^{-2} yr^{-1} in the mid-1990s to 0.070 t km^{-2} yr^{-1} in the early 2000s).

Although it is a smaller ecosystem, landings in the sGSL were equal to those in the nGSL during the mid-1980s (180,002 versus 189,819 t, respectively) and were over double those of the nGSL during the 1990s (127,255 versus 48,857 t, respectively), primarily due to the much higher landings of herring and snow crab in the sGSL. On a unit area basis, there was a decrease in total landings in the sGSL of about 30% (from 2.8 t km^{-2} yr^{-1} for the mid-1980s to 2.0 t km^{-2} yr^{-1}). The fishery in the southern Gulf was more diverse during the mid-1980s, with only 39% of total catches represented by cod while forage fishes and invertebrates represented 52% (Figure 4). Thus, the decline in total landings when cod stocks collapsed was lower in the sGSL than in the nGSL. Similar to the nGSL, the loss of predatory fishes resulted in a change in landings from a state where long-lived and predatory groundfish made up about 50% to one where forage fishes and invertebrates (snow crab) comprised > 90% of the landings. The mean trophic level of the catches decreased slightly from the mid-1980s (TL_c: 3.3) to the mid-1990s (TL_c: 3.2).

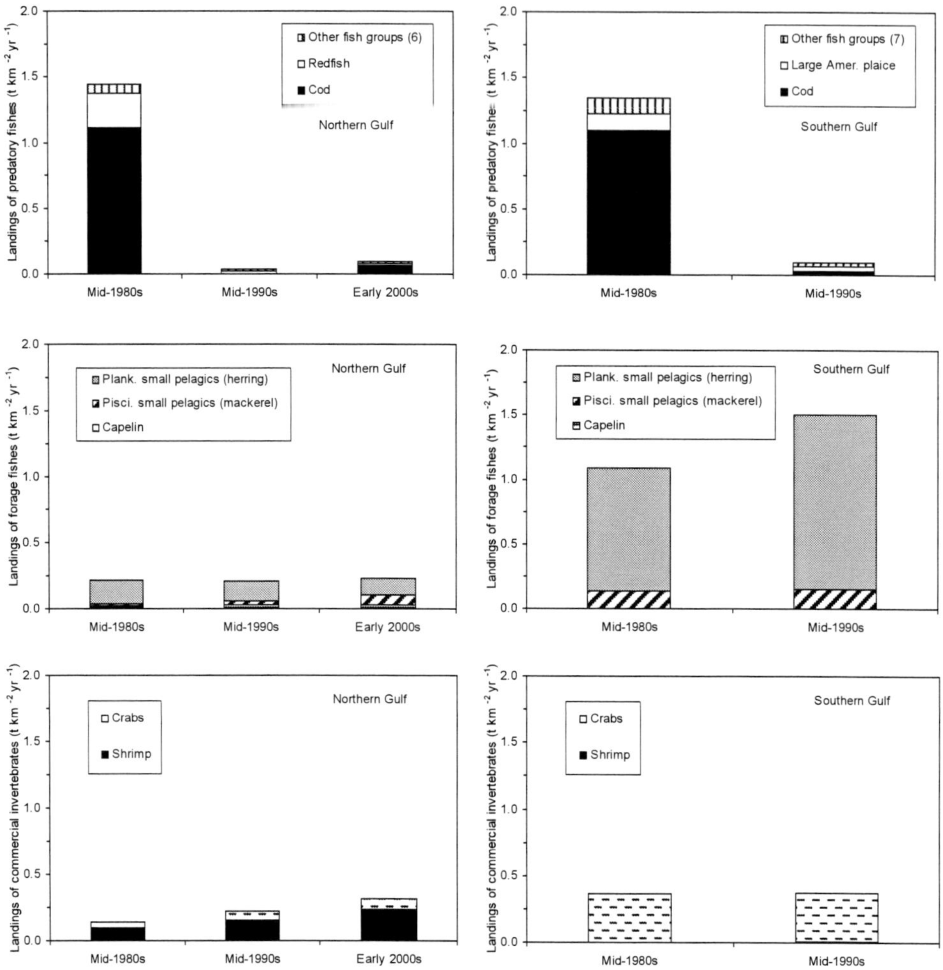

Figure 4. Main fish and commercial invertebrate landings estimated for each time period in the northern (left panels) and southern (right panels) Gulf.

Modelling Results: Comparisons of Ecosystem States between Time Periods

Annual Consumption by Each Species Group

In the nGSL, annual consumption by each species group of marine mammals increased from the mid-1980s to the mid-1990s and again in the early 2000s. With the exception of small Greenland halibut, there was a decrease in food consumption for all fish species between the 1980s and 1990s (Table 3). Except for skates and redfish, annual consumption increased or did not show significant changes between the mid-1990s and the early 2000s. During the mid-1980s, the main predators on fish prey were large cod, followed by redfish and, via cannibalism, capelin. Cetaceans and harp seals were the major predators on fish during the mid-1990s (Figure 5). During the early 2000s, cetaceans and harp seals remained the most important predators on fish, followed by small Greenland halibut (Figure 5).

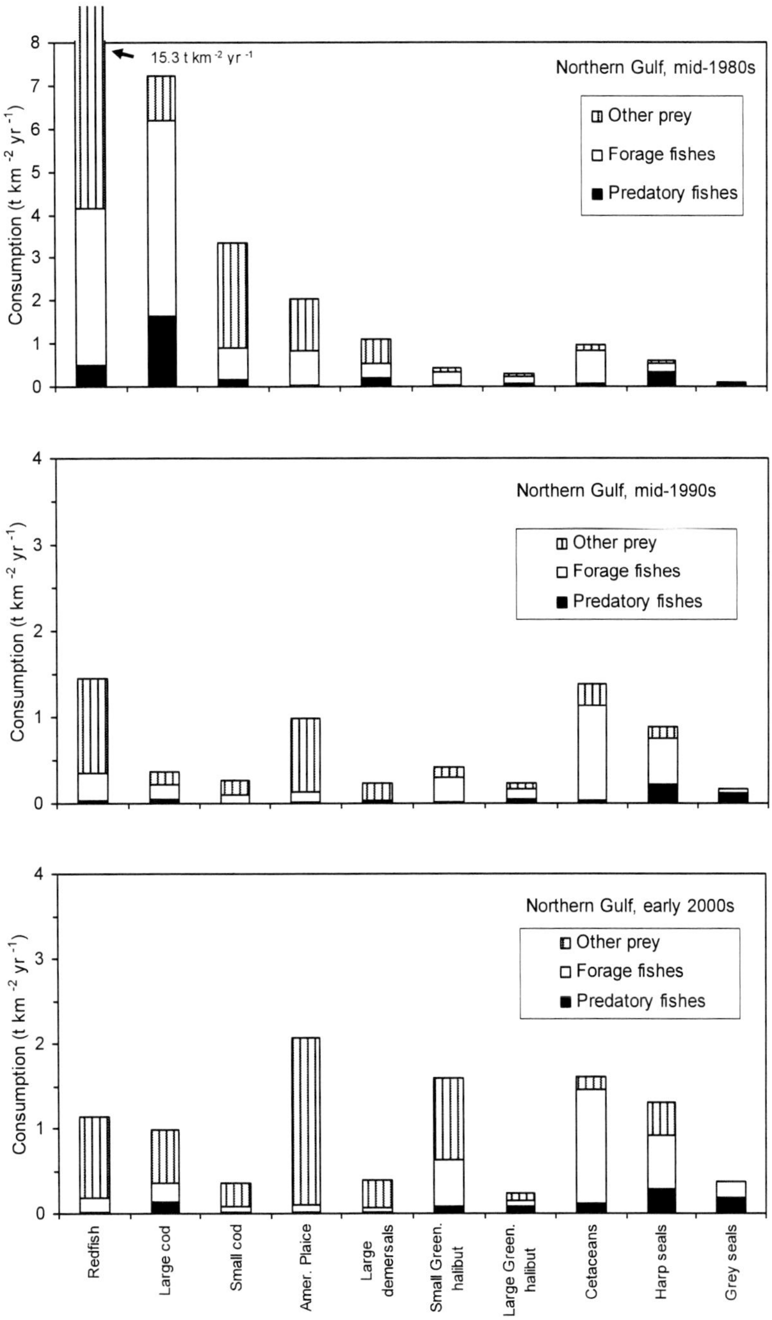

Figure 5. Modelled predation mortality by the main predators for each time period in the northern Gulf. Other prey: pelagic and benthic invertebrate groups and detritus.

Table 3. Comparison of estimated consumption Q (t km^{-2} y^{-1}) for each time period in the northern Gulf. CI: 95% confidence intervals (CI = (SD/√n)*2.042 with n = 31) around the mean. The difference between the two values of a parameter is indicated as significantly higher (+), or lower (−), or else non significant (=) (t-tests).

Group	Mid-1980s		Mid-1990s		Early 2000s		1980s-1990s	1990s-2000s
	Q	±CI95%	Q	±CI95%	Q	±CI95%	change	change
Cetaceans	0.97	0.03	1.39	0.04	1.60	0.14	+	+
Harp seals	0.61	0.06	0.88	0.12	1.30	0.17	+	+
Hooded seals	0.03	0.00	0.05	0.01	0.08	0.01	+	+
Grey and harbour seals	0.12	0.01	0.19	0.01	0.37	0.02	+	+
Seabirds	0.16	0.02	0.28	0.05	0.22	0.05	+	=
Large cod	7.23	1.00	0.37	0.01	0.99	0.07	−	+
Small cod	3.34	0.07	0.26	0.03	0.36	0.05	−	+
Large Green. halibut	0.30	0.02	0.24	0.01	0.24	0.01	−	=
Small Green. halibut	0.42	0.07	0.42	0.07	1.59	0.04	=	+
American plaice	2.05	0.31	0.98	0.11	2.06	0.24	−	+
Flounders	0.74	0.09	0.08	0.02	0.13	0.03	−	=
Skates	0.27	0.03	0.19	0.01	0.08	0.01	−	−
Redfish	15.33	0.97	1.46	0.03	1.13	0.11	−	−
Large demersals	1.10	0.07	0.24	0.03	0.40	0.06	−	+
Small demersals	2.18	0.20	0.34	0.03	1.50	0.13	−	+
Capelin	58.13	3.83	14.33	1.68	16.89	2.11	−	=
Large pelagics	0.09	0.01	0.04	0.01	0.04	0.01	−	=
Pisci. small pelagics	2.71	0.60	0.93	0.25	0.95	0.20	−	=
Plank. small pelagics	16.21	2.08	2.78	0.56	4.69	0.56	−	+

Table 4. Comparison of estimated consumption Q (t km^{-2} y^{-1}) for each time period in the southern Gulf. CI: 95% confidence intervals (CI = (SD/√n)*2.042 with n = 31) around the mean. The difference between the two values of a parameter is indicated as significantly higher (+), or lower (−), or else non significant (=) (t-tests).

Group	Mid-1980s		Mid-1990s		1980s-1990s
	Q	±CI95%	Q	±CI95%	change
Cetaceans	0.21	0.02	0.38	0.03	+
Harp seals	1.80	0.12	1.82	0.13	=
Hooded seals	0.03	0.00	0.04	0.00	+
Grey and harbour seals	0.51	0.03	0.99	0.01	+
Seabirds	0.18	0.03	0.12	0.01	−
Large cod	8.22	0.36	2.99	0.46	−
Small cod	3.57	0.47	1.29	0.22	−
Green. halibut	0.05	0.01	0.03	0.00	−
Large Amer. plaice	1.48	0.19	0.58	0.14	−
Small Amer. plaice	8.93	1.25	9.62	1.97	=
Flounders	2.62	0.34	3.12	0.35	=
Skates	0.13	0.01	0.04	0.00	−
Redfish	0.06	0.01	0.05	0.01	=
Large demersals	0.42	0.02	0.25	0.00	−
Small demersals	1.56	0.09	0.41	0.03	−
Capelin	3.50	0.75	25.22	2.60	+
Large pelagics	0.07	0.02	0.14	0.00	+
Pisci. small pelagics	8.01	1.06	3.36	0.86	−
Plank. small pelagics	19.54	3.34	20.09	3.96	=

Table 5. Estimated diet composition for the main predators and landing proportions (%) in the northern Gulf for each time period. Empty cells indicate that a prey item was never found whereas "0" indicates that it was found in very small amounts.

	Cetaceans			Harp seals			Grey seals			Large cod			Landings		
	1980s	1990s	2000s	1980s	1990s	2000s	1980s	1990s	2000s	1980s	1990s	2000s	1980s	1990s	2000s
Large cod	0	0	1	3	1	2	12	15	10				61	1	11
Small cod	2	0	0	10	1	1	14	10	7	5	4	1	0		
Large Green. halibut				5	2	1	0	0	0				2	2	1
Small Green. halibut				6	5	7	0	0	0	0	0	0			
American plaice	0	0	0	7	6	4	7	11	11	2	3	10	1	0	0
Flounders				5	0	1	11	1	1	1	1	0	0	0	1
Skates				0	0	0	6	10	4	0	0	0	0	0	0
Redfish	0	0	1	8	7	6	0	0	0	14	2	1	14	4	1
Large demersals	3	1	0	10	0	0	14	11	9				1	0	1
Small demersals	2	1	5	2	1	1	7	6	8	0	2	3			
Capelin	50	57	65	18	58	27	9	14	17	59	40	6	1	7	5
Large pelagics	0	0	0	0	0	0	1	1	1				0	0	0
Pisci. small pelagics	19	8	6	4	1	1	3	8	4	2	1	0	1	6	11
Plank. small pelagics	10	14	13	12	1	21	16	15	27	2	7	15	10	31	20
Shrimp	0	0	0	0	1	5	0	0	1	2	9	13	5	33	35
Crabs			0	1	0	2	0	0	0	0	16	5	2	14	13
Echinoderms										3	1	0			
Molluscs									0	1	0	1	2	1	1
Polychaetes										1	0	0			
Other ben. inver.				4	3	5	0	0	0	0	0	0			
Large zooplankton	7	10	8	5	11	18	0	0	0	6	15	44			
Small zooplankton	7	8	0							0	0	0			
Total (%)	100	100	100	100	100	100	100	100	100	100	100	100	100	100	100
Total ($t\ km^{-2}\ yr^{-1}$)	1.0	1.4	1.6	0.6	0.9	1.3	0.1	0.2	0.4	7.2	0.4	1.0	1.8	0.5	0.6

Table 6. Estimated diet composition for the main predators and landing proportions (%) in the southern Gulf for each time period. Empty cells indicate that a prey item was never found whereas "0" indicates that it was found in very small amounts.

	Cetaceans		Harp seals		Grey seals		Large cod		Landings	
	1980s	1990s	1980s	1990s	1980s	1990s	1980s	1990s	1980s	1990s
Large cod	0	0	4	2	24	17			39	2
Small cod	1	1	10	1	21	4	3	2		
Green. halibut			0	0	0	0			0	0
Small Amer. plaice	0	0	7	5	6	6	3	23		
Large Amer. plaice	0	0	4	0	3	5			5	2
Flounders			2	0	14	20	0	1	2	1
Skates					2	1			0	0
Redfish	1	1	0	0					0	0
Large demersals	0	0	1	1	6	5			2	0
Small demersals	0	0	1	1	5	5	2	0		0
Capelin	19	37	13	76	6	18	1	6	0	0
Large pelagics	2	1	0	0	0	1			0	1
Pisci. small pelagics	44	21	6	2	3	4	5	1	5	7
Plank. small pelagics	30	36	22	4	10	13	9	9	34	68
Shrimp	1	1	5	2	0	0	7	9	0	0
Crabs			0	0	0	0	3	1	13	18
Echinoderms							2	2		
Molluscs							13	26	0	1
Polychaetes							13	5		
Other ben. inver.			0	0	0	0	15	4		
Large zooplankton	2	2	25	5	0	0	22	11		
Small zooplankton							0	0		
Total (%)	100	100	100	100	100	100	100	100	100	100
Total ($t\ km^{-2}\ yr^{-1}$)	0.2	0.4	1.8	1.8	0.5	1.0	8.2	3.0	2.8	2.0

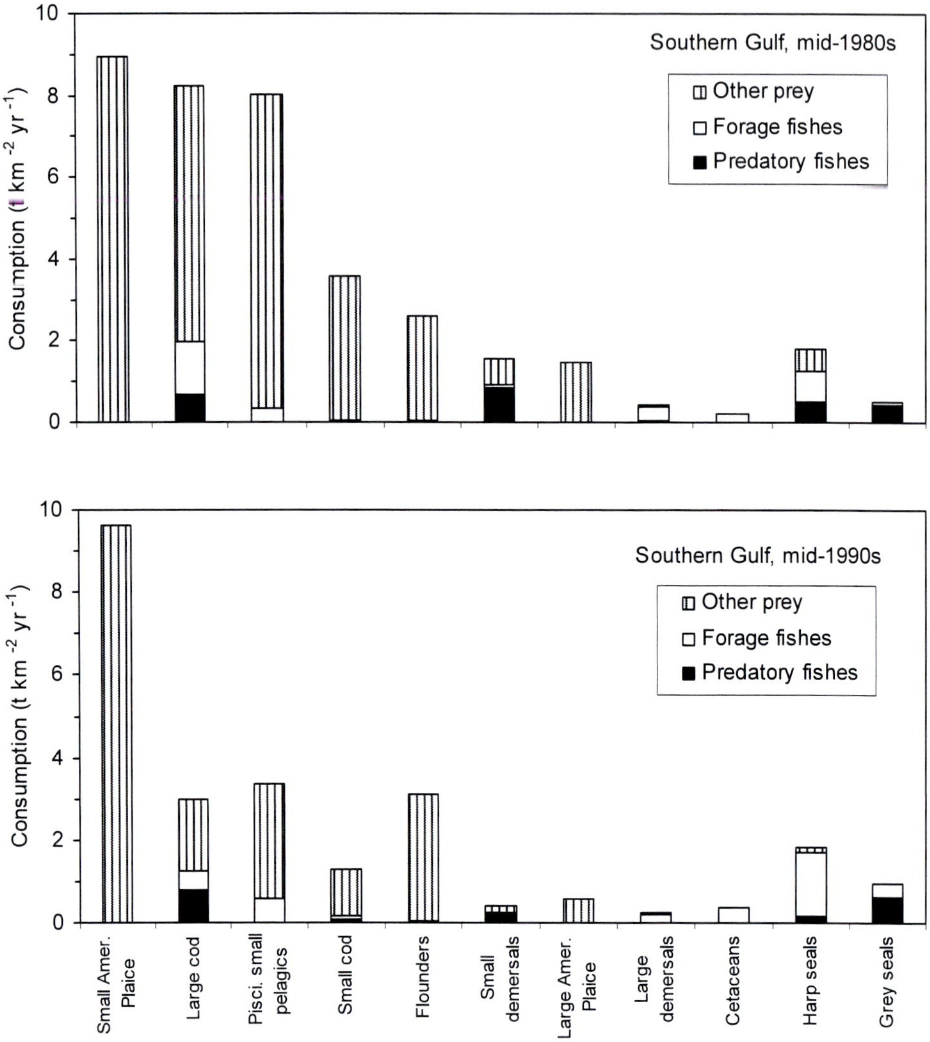

Figure 6. Modelled predation mortality by the main predators for each time period in the southern Gulf. Other prey: pelagic and benthic invertebrate groups and detritus.

In the sGSL, there was also an increase in the total annual consumption by cetaceans and seal species from the mid-1980s to the mid-1990s (Table 4). However, the declines in consumption by fish species were less radical than in the nGSL. There was no change for small American plaice, flounders, redfish, and planktivorous small pelagics while consumption by large pelagic feeders and capelin increased from the mid-1980s to the mid-1990s (Table 4). During the mid-1980s, large cod were the most important predators on fish prey, followed by harp seals and small demersals (Figure 6). During the mid-1990s, harp seals were the most important predators on fish prey, followed by large cod and grey seals (Figure 6).

Main Prey Groups Consumed by Cetaceans

Cetaceans consumed mainly five prey groups (> 5% of the diet) in the nGSL models (Table 5). Fish species, mostly forage fishes, comprised from 81% (1990s) to 91% (2000s) of the diet. Capelin was the most important prey item. Their contribution to modelled cetacean diets increased over time, from 50% in the mid-1980s to 65% in the early 2000s. The other important prey of cetaceans were piscivorous small pelagics (e.g., mackerel, squid, and other cephalopods), planktivorous small pelagics (e.g., herring and sand lance), and large and small zooplankton. Piscivorous small pelagics were the second most important prey item in the mid-1980s (19% of the diet) whereas planktivorous small pelagics were the second most important prey in more recent periods (13-14% of the diet). Large and small zooplankton each contributed 7-10% of the diet in all periods except the early 2000s, when small demersals (5%) replaced small zooplankton as an important prey item. Consumption by cetaceans increased over time, from 0.97 ± 0.07 t km^{-2} yr^{-1} in the mid-1980s to 1.60 ± 0.38 t km^{-2} yr^{-1} in the early 2000s.

Consumption by cetaceans was lower in the sGSL models concurrent with a lower cetacean biomass, increasing from 0.21 ± 0.06 t km^{-2} yr^{-1} during the mid-1980s to 0.38 ± 0.08 t km^{-2} yr^{-1} in the mid-1990s (Table 6). Cetacean diets consisted almost entirely of fish (97-98%), primarily forage fishes (93-95%). As in the nGSL models, the main prey groups were capelin, and piscivorous and planktivorous small pelagics. The most important prey group was piscivorous small pelagics (44%) during the mid-1980s, and capelin (37%) and planktivorous small pelagics (36%) during the mid-1990s. Unlike in the nGSL models, zooplankton were not an important component of cetacean diets in the sGSL models, possibly reflecting the different species composition of the small zooplankton group and an apparent major decrease in availability of euphausiids (large zooplankton group) in the sGSL during the mid-1990s (Hanson and Chouinard 2002).

Main Prey Groups Consumed by Harp Seals

Capelin was the most important prey of harp seals in the nGSL models for all three time periods (Table 5), contributing between 18% (mid-1980s) and 58% (mid-1990s) of their diet. The harp seal diet was most diverse in the mid-1980s model, with nine prey groups contributing 5% or more to the diet (Table 5). After capelin, planktivorous small pelagics (12%), large demersals (10%), and small cod (10%) were the most important prey items of harp seals during the mid-1980s. Redfish, American plaice, small Greenland halibut, large zooplankton, and flounder accounted for between 5% and 8% of the diet. The harp seal diet was least diverse in the 1990s model when a single prey group, capelin, contributed the largest proportion of the diet (58%). Other important prey of harp seals in the 1990s model were large zooplankton (11%), redfish (7%), American plaice (6%) and small Greenland halibut (5%). In the 2000s model, capelin (27%), planktivorous small pelagics (21%) and large zooplankton (18%) were the main prey of harp seals, followed by small Greenland halibut (7%) redfish (6%), and shrimp (5%). The proportion of fish in harp seal diets decreased over time, from 90% in the 1980s and 84% in the 1990s to 70% in the early 2000s. Nonetheless, the amount of fish consumed by harp seals increased over time (Figure 5), reflecting the increasing harp seal biomass and thus consumption. Consumption by harp seals increased from 0.61 ± 0.18 t km^{-2} yr^{-1} in the mid-1980s to 0.88 ± 0.33 t km^{-2} yr^{-1} in the mid-

1990s and 1.30 ± 0.46 t km^{-2} yr^{-1} in the early 2000s. Predatory fishes were more important than forage fishes (55% versus 35%) in the diet of harp seals in the 1980s. The reverse was found during the 1990s (24% versus 60%) and 2000s diets (22% versus 48%).

The main prey of harp seals in the 1980s model of the sGSL were large zooplankton (25%), planktivorous small pelagics (22%), and capelin (13%) (Table 6). Harp seals also heavily consumed small cod (10%), small American plaice (7%), piscivorous small pelagics (6%), and shrimp (5%). Similar to the nGSL models, the diet of harp seals was less diverse in the 1990s model of the sGSL, with capelin the dominant prey (76%), followed by large zooplankton (5%) and small American plaice (5%). Harp seal predation on predatory fishes decreased from the mid-1980s (0.50 ± 0.11 t km^{-2} yr^{-1}, 28% of the diet) to the mid-1990s (0.18 ± 0.05 t km^{-2} yr^{-1}, 10%), while predation on forage fishes increased from 0.74 ± 0.22 t km^{-2} yr^{-1} (41%) to 1.50 ± 0.31 t km^{-2} yr^{-1} (82%) over the same time period. The models showed no change in total consumption by harp seals between the 1980s and 1990s (Table 6) despite their increasing biomass (Figure 3).

Main Prey Groups Consumed by Grey Seals

For the mid-1980s, grey seal consumption was dominated by 10 prey groups in the nGSL (> 5% of the diet; Table 5). The most important prey items were planktivorous small pelagics (16% of the diet), small cod (14%), large demersals (14%), large cod (12%), and flounders (11%). Capelin, American plaice, skates, small demersals, and piscivorous small pelagics accounted for between 6% and 9% of the diet (Table 5). Fish represented 99% of the diet (predatory: 67%; forage: 32%). Grey seal consumption increased from the mid-1980s to the mid-1990s (from 0.12 ± 0.03 to 0.19 ± 0.04 t km^{-2} yr^{-1}) concurrent with the increase in their biomass. For the mid-1990s, grey seals had a large impact on nine prey groups (> 5% of the diet; Table 5). Planktivorous small pelagics and large cod (15% each) were the main prey groups, followed by capelin (14%), large demersals (11%), American plaice (11%), small cod (10%), and skates (10%). The other important prey were piscivorous small pelagics (8%) and small demersals (6%). Grey seal predation on predatory and forage fish groups increased slightly during the mid-1990s (Figure 5). However, the overall proportion of predatory fish species in the diet composition decreased for the mid-1990s (59% vs 40% of the diet for forage fishes). Grey seal consumption increased from the mid-1990s to the early 2000s (from 0.19 ± 0.04 t km^{-2} yr^{-1} to 0.37 ± 0.06 t km^{-2} yr^{-1}). For the early 2000s, seven prey groups were predominant (Table 5). The most important prey items of grey seals were planktivorous small pelagics (27%), capelin (17%), American plaice (11%), and large cod (10%). Large and small demersals as well as small cod accounted for between 7% and 9% of the diet. Grey seal predation on both predatory and forage fish groups still increased (Figure 5) and the proportions of predatory and forage fish groups in the diet composition were similar (nearly 50% each of the diet).

In the sGSL, the most important prey items of grey seals were large cod (24%), small cod (21%), and flounders (14%) in the mid-1980s (Table 6). Grey seals also heavily consumed planktivorous small pelagics (10%), capelin (6%), small American plaice (6%), large demersals (6%), and small demersals (5%). Predatory fishes represented 80% of the diet while forage fishes accounted for 20%. Grey seal consumption increased from the mid-1980s to the mid-1990s (from 0.51 ± 0.08 to 0.99 ± 0.03 t km^{-2} yr^{-1}). Grey seals still had a large impact on eight prey groups (> 5% of the diet) for the mid-1990s (Table 6). Flounders (20%)

were the main prey, followed by capelin (18%), large cod (17%), and planktivorous small pelagics (13%). Small and large American plaice as well as small and large demersals accounted for 5% and 6% of the diet. Grey seal predation on both predatory and forage fish groups increased (Figure 6) and the proportions of predatory and forage fish groups in the diet composition were 64% and 36%, respectively.

Main Prey Groups Consumed by Large (> 35 cm) Cod

The main prey (> 5% of the diet) of large cod included only three to six groups in the nGSL for each time period (Table 5). The most important prey items of large cod in the 1980s model were capelin (59%), redfish (14%), and large zooplankton (6%). Small cod accounted for 5% of the diet. Fish species made up 86% of the large cod diet (predatory: 23%; forage: 63%). Large cod consumption decreased from the mid-1980s to the mid-1990s (from 7.23 ± 2.73 to 0.37 ± 0.03 t km^{-2} yr^{-1}) concurrent with the decrease in large cod biomass. In the 1990s model, capelin (40%) was the main prey, followed by crabs (16%) and large zooplankton (15%) (Table 5). Contributions to the diet of large cod increased for shrimp (from 2% to 9%) and planktivorous small pelagics (from 2% to 7%) and decreased for redfish (from 14% to 2%, concurrent with the decline in the redfish population). The proportion of fish in the diet decreased to 59% (predatory fishes: 12%; forage fishes: 47%). Total prey consumption by large cod increased from the mid-1990s to the early 2000s (from 0.37 ± 0.03 to 0.99 ± 0.19 t km^{-2} yr^{-1}). The most important prey items of large cod in the model for the early 2000s were large zooplankton (44%), planktivorous small pelagics (15%), and shrimp (13%) (Table 5). American plaice, capelin, and small crabs accounted for 10%, 6%, and 5%, respectively, of the diet. The overall proportion of fish in the diet of large cod decreased further to 36% (predatory: 14%; forage: 22%).

In contrast to the model for the nGSL, fish comprised a relatively small proportion (24%) the diet of large cod in the sGSL model for the mid-1980s (Table 6). The main fish prey were planktivorous small pelagics (9%) and piscivorous small pelagics (5%). Unlike in the nGSL, capelin was not an important prey of large cod in the sGSL models, while cannibalism on small cod was small. The main invertebrate prey of large cod in the 1980s model were large zooplankton (22% of the diet, mainly as euphausiids), other benthic invertebrates (15%), molluscs (13%), polychaetes (13%) and shrimp (7%). Consumption by large cod decreased from the mid-1980s to the mid-1990s (from 8.22 ± 0.99 to 2.99 ± 1.26 t km^{-2} yr^{-1}) concurrent with the decrease in their biomass. The proportion of fish in the diet of large cod increased to 42% in the 1990s model, mainly due to a large increase in the consumption of small American plaice (23% of the diet). The other main fish prey of large cod were planktivorous small pelagics (9%) and capelin (6%), with cannibalism contributing only 2% to the diet. The most important invertebrate prey of sGSL large cod in the 1990s model were molluscs (26%), large zooplankton (11%, mainly mysids), shrimp (9%), and polychaetes (5%).

Overlap between Species and Fisheries

In the nGSL models, overlap for food resources between marine mammals and fisheries increased from the mid-1980s to the early 2000s (Table 7). Similar trends in overlap occurred between fisheries and each marine mammal group, except that overlap with cetaceans showed a large increase from the 1980s to 1990s with little subsequent change in overlap between the 1990s and 2000s. During the mid-1980s, fisheries mainly targeted large cod, redfish, planktivorous small pelagics, and shrimp (90% of total landings) that were not heavily eaten by marine mammals (11%, 23%, and 28% of total consumption by cetaceans, harp seals, and grey seals, respectively) during this period. This resulted in relatively weak resource overlap indices between marine mammals and fisheries. Higher overlap occurred between fisheries and all marine mammal groups in the early 2000s (Table 7). Large cod, capelin, small pelagics, shrimp, and crabs accounted for 95% of total landings in the early 2000s. Although crabs were consumed in a lesser amount by marine mammals, the other species accounted for 85%, 57%, and 59% of total consumption by cetaceans, harp seals, and grey seals, respectively. In the nGSL models, resource overlap between large cod and fisheries increased from the mid-1980s to the mid-1990s and then decreased somewhat in the early 2000s (Table 7). The increased overlap with fisheries in the 1990s model mostly reflects the decreased importance of cod and increased importance of small pelagics and shrimp in the landings in the 1990s.

Table 7. Estimated resource overlap index between apex predators and fisheries from the two Gulf ecosystem models.

Ecosystem model and overlap	$\alpha_{j,l}$ mid-1980s	$\alpha_{j,l}$ mid-1990s	$\alpha_{j,l}$ early 2000s
Northern Gulf of St. Lawrence			
Marine mammals & fisheries	0.026	0.030	0.044
Cetaceans & fisheries	0.009	0.045	0.042
Harp seals & fisheries	0.030	0.039	0.096
Grey seals & fisheries	0.018	0.049	0.108
Large cod & fisheries	0.013	0.099	0.082
Southern Gulf of St. Lawrence			
Marine mammals & fisheries	0.128	0.048	
Cetaceans & fisheries	0.025	0.086	
Harp seals & fisheries	0.102	0.015	
Grey seals & fisheries	0.080	0.069	
Large cod & fisheries	0.034	0.046	

Resource overlap between fisheries and marine mammals in the mid-1980s was considerably higher in the sGSL model than in the nGSL model (Table 7). Fisheries in the sGSL mainly targeted large cod, large American plaice, piscivorous and planktivorous small pelagics, and crabs (95% of total landings). Excluding crabs, which were weakly consumed by marine mammals (at most 0.2% of the diet), the other species accounted for 74%, 36%, and 40% of total consumption by cetaceans, harp seals, and grey seals, respectively. Except for cetaceans, resource overlap between fisheries and marine mammals decreased from the mid-1980s to the mid-1990s. This occurred despite an increase in consumption by marine mammals (except for harp seals). In the 1990s model, seals generally consumed food resources that were not the main target of fisheries. Small pelagics and crabs represented 94% of total landings during the mid-1990s, but these species represented only 7% and 17% of the diets of harp and grey seals, respectively. In contrast, small pelagics comprised 57% of the diet of cetaceans, whose resource overlap with fisheries increased in the mid-1990s. Resource overlap of fisheries with large cod was generally lower in the sGSL models than in the nGSL models, especially in the mid-1990s. Fisheries and large cod did not target the same resource in the sGSL (Table 6).

Niche Overlap between Different Species and Cod

In the nGSL models, the niche overlap between both large and small cod and marine mammals increased generally from the mid-1980s to the mid-1990s and then decreased slightly in the early 2000s (Table 8). Overlap was greatest with cetaceans and harp seals and least with grey seals. The greater overlap between large cod and cetaceans and harp seals may reflect the importance of capelin, planktivorous small pelagics, and large zooplankton in the diets of all three groups. Cetaceans and harp seals also competed with small cod for capelin and large zooplankton. Small cod, large cod, and small Greenland halibut share a common niche (i.e., high overlap indices; capelin, shrimp, and large zooplankton) for each time period; the highest values were estimated during the mid-1990s. Cannibalism by large cod was the main source of mortality of small cod (44%) during the mid-1980s and still had a significant effect during the mid-1990s and early 2000s (25% and 13% of total mortality of small cod, respectively), even though small cod constituted a minor component of the diet of large cod (at most 6%). Large and small cod potentially competed even more for the same resources as shown by the increase in the niche overlap indices over time (Table 8).

Table 8. Estimated niche-overlap index between different species from the two Gulf ecosystem models.

Ecosystem model and overlap	O_{jk} mid-1980s	O_{jk} mid-1990s	O_{jk} early 2000s
Northern Gulf of St. Lawrence			
Cetaceans & large cod	0.90	0.86	0.28
Harp seals & large cod	0.57	0.87	0.66
Grey seals & large cod	0.26	0.44	0.33
Small Green. halibut & large cod	0.64	0.85	0.64
Cetaceans & small cod	0.41	0.67	0.31
Harp seals & small cod	0.30	0.69	0.61
Grey seals & small cod	0.09	0.25	0.21
Large cod & small cod	0.41	0.80	0.95
Small Green. halibut & small cod	0.53	0.78	0.71
Southern Gulf of St. Lawrence			
Cetaceans & large cod	0.26	0.24	
Harp seals & large cod	0.64	0.18	
Grey seals & large cod	0.16	0.27	
Small Amer. plaice & large cod	0.77	0.36	
Cetaceans & small cod	0.04	0.11	
Harp seals & small cod	0.54	0.10	
Grey seals & small cod	0.01	0.10	
Large cod & small cod	0.62	0.51	
Small Amer. plaice & small cod	0.90	0.76	

Resource overlap of marine mammals with large cod was generally lower in the sGSL models than in the nGSL models, especially in the mid-1990s (Table 8). This may reflect the greater importance of polychaetes, molluscs and other benthic invertebrates in the diet of cod in the sGSL, species that were not eaten by marine mammals (Table 6). Only the intermediate niche overlap values between harp seals and both large and small cod show partial resource competition during the mid-1980s. For both periods, high niche overlap indices were estimated between large and small cod. Small cod, large cod, and small American plaice potentially competed for the same resource (polychaetes, other benthic invertebrates, and large zooplankton) during the mid-1980s (Table 8). The niche overlap between small cod and small American plaice was still high during the mid-1990s.

Prey Consumption in the Northern Gulf

Table 9. Estimated prey consumption Q (t km^{-2} y^{-1}) and details of consumption (%) for different trophic groups for each time period in the northern Gulf. For comparisons, estimated landings by fisheries are also included. Pred. fish: predatory fish groups; fora. fish: forage fish groups (capelin and small pelagics); SD: standard deviation.

		Mid-1980s			Mid-1990s		
Group	**Prey**	**Q**	**± SD**	**%**	**Q**	**± SD**	**%**
All seals	Pred. fish	0.45	0.09	58%	0.37	0.09	33%
	Fora. fish	0.25	0.14	33%	0.62	0.30	55%
	Other prey	0.06	0.06	8%	0.14	0.11	13%
	Total	0.76	0.18	100%	1.13	0.34	100%
Cetaceans & seabirds	Pred. fish	0.08	0.02	7%	0.04	0.01	2%
	Fora. fish	0.91	0.09	81%	1.33	0.16	80%
	Other prey	0.13	0.04	12%	0.30	0.03	18%
	Total	1.12	0.09	100%	1.67	0.16	100%
Predatory fishes	Pred. fish	2.82	0.51	9%	0.33	0.04	7%
	Fora. fish	10.75	2.75	33%	1.15	0.25	25%
	Other prey	19.46	2.80	59%	3.14	0.31	68%
	Total	33.03	3.96	100%	4.61	0.40	100%
Forage fishes	Pred. fish	0.00	0.00	0%	0.00	0.00	0%
	Fora. fish	2.38	1.04	3%	0.45	0.31	3%
	Other prey	74.67	11.94	97%	17.59	4.86	97%
	Total	77.04	11.98	100%	18.05	4.87	100%
Fisheries	Pred. fish	1.45	0.07	79%	0.04	0.01	8%
	Fora. fish	0.21	0.02	12%	0.20	0.02	44%
	Other prey	0.17	0.01	9%	0.23	0.02	48%
	Total	1.83	0.07	100%	0.47	0.03	100%

Table 9. Continued

Group	Prey	Early 2000s Q	± SD	%
All seals	Pred. fish	0.54	0.07	31%
	Fora. fish	0.82	0.43	47%
	Other prey	0.39	0.16	22%
	Total	1.75	0.46	100%
Cetaceans & seabirds	Pred. fish	0.13	0.04	7%
	Fora. fish	1.53	0.39	84%
	Other prey	0.16	0.10	9%
	Total	1.82	0.41	100%
Predatory fishes	Pred. fish	0.42	0.07	5%
	Fora. fish	1.29	0.28	15%
	Other prey	6.81	0.58	80%
	Total	8.52	0.86	100%
Forage fishes	Pred. fish	0.00	0.00	0%
	Fora. fish	0.31	0.22	1%
	Other prey	22.21	5.98	99%
	Total	22.53	5.99	100%
Fisheries	Pred. fish	0.10	0.00	15%
	Fora. fish	0.23	0.02	36%
	Other prey	0.31	0.02	49%
	Total	0.63	0.03	100%

When species-groups were aggregated into the larger trophic groups, total estimated prey consumption for all the seals combined increased from 0.8 t km^{-2} yr^{-1} for the mid-1980s to 1.1 t km^{-2} yr^{-1} for the mid-1990s (Table 9). There was also an increase in total estimated prey consumption for cetaceans and seabirds from 1.1 t km^{-2} yr^{-1} for the mid-1980s to 1.7 t km^{-2} yr^{-1} for the mid-1990s, while consumption by both predatory and forage fishes decreased over the same interval (predatory fishes: from 33.0 to 4.6 t km^{-2} yr^{-1}; forage fishes: from 77.0 to 18.0 t km^{-2} yr^{-1}). Landings by fisheries decreased from the mid-1980s to the mid-1990s. During the mid-1980s, predatory fishes were the main predators on both predatory and forage fish groups (Table 9). They consumed twice as much of predatory fishes as was landed and six times the amount consumed by seals. Forage fishes were mainly consumed by predatory fishes, with lesser amounts consumed by forage fishes, and cetaceans and seabirds. Other prey were however the main food of both predatory and forage fishes. Overall, seals and fisheries targeted mainly predatory fishes while forage fishes were the main prey of cetaceans and seabirds.

During the mid-1990s, predatory fishes were mainly consumed by seals and predatory fishes. Cetaceans and seabirds were the main predators of forage fishes, followed by predatory fishes. Total consumption increased for all seals combined and cetaceans and seabirds from the mid-1980s to the mid-1990s. In fact, predation on forage fish and invertebrate groups increased while it decreased on predatory fishes. With the net decrease in biomass of predatory fishes, seals and fisheries targeted lower trophic levels during the mid-1990s. The predator structure changed: predatory fishes were the main predators on fish, but cetaceans and seals also became significant predators on fish, together consuming about 60% more fish than the predatory fish group.

Total prey consumption and landings increased for every group from the mid-1990s to the early 2000s. Seals and predatory fishes were the main consumers of predatory fishes for the early 2000s. Forage fishes were mainly consumed by cetaceans and seabirds, predatory fishes, and seals.

Prey Consumption in the Southern Gulf

Total estimated prey consumption increased from the mid-1980s to the mid-1990s for cetaceans and seabirds, seals, and forage fishes (Table 10). Estimated prey consumption by

Table 10. Estimated prey consumption Q (t km^{-2} y^{-1}) and details of consumption (%) for different trophic groups for each time period in the southern Gulf. For comparisons, estimated landings by fisheries are also included. Pred. fish: predatory fish groups; fora. fish: forage fish groups (capelin and small pelagics); SD: standard deviation.

		Mid-1980s			Mid-1990s		
Group	Prey	Q	± SD	%	Q	± SD	%
All seals	Pred. fish	0.94	0.13	40%	0.85	0.06	30%
	Fora. fish	0.85	0.22	36%	1.87	0.31	65%
	Other prey	0.55	0.21	23%	0.14	0.19	5%
	Total	2.34	0.33	100%	2.86	0.37	100%
Cetaceans & seabirds	Pred. fish	0.02	0.00	4%	0.02	0.00	4%
	Fora. fish	0.35	0.11	90%	0.47	0.08	94%
	Other prey	0.02	0.01	5%	0.01	0.00	2%
	Total	0.39	0.11	100%	0.50	0.08	100%
Predatory fishes	Pred. fish	1.65	0.18	6%	1.11	0.25	6%
	Fora. fish	1.93	0.17	7%	0.99	0.32	5%
	Other prey	23.54	3.92	87%	16.41	5.62	89%
	Total	27.11	3.93	100%	18.51	5.63	100%
Forage fishes	Pred. fish	0.00	0.00	0%	0.00	0.00	0%
	Fora. fish	0.32	0.30	1%	0.58	0.53	1%
	Other prey	30.73	9.77	99%	48.09	13.12	99%
	Total	31.05	9.77	100%	48.67	13.13	100%
Fisheries	Pred. fish	1.35	0.04	48%	0.10	0.01	5%
	Fora. fish	1.09	0.12	39%	1.50	0.07	76%
	Other prey	0.37	0.06	13%	0.38	0.02	19%
	Total	2.81	0.14	100%	1.99	0.07	100%

predatory fishes and landings by fisheries decreased over the same time periods. However, except for predatory fishes, consumption of forage fishes increased for all the groups from the mid-1980s to the mid-1990s (Table 10). During the mid-1980s, predatory fishes were the most important consumers of the predatory fish compartment (42%), though fisheries and seals together accounted for the largest proportion (58%) of predatory fish consumption. Forage fishes were mainly consumed by predatory fishes (42%), fisheries (24%), and seals (19%). Other prey represented the main food of both forage and predatory fishes. Overall, seals and fisheries targeted both predatory and forage fishes.

During the mid-1990s, predatory fishes were mainly consumed by predatory fishes themselves (53%) and seals (41%). Seals were the most important predator of forage fishes (35%), followed by fisheries (28%), and predatory fishes (18%). Overall, seals, fisheries, and cetaceans and seabirds all targeted forage fishes during the mid-1990s, while predatory and forage fishes targeted lower trophic levels (other prey).

Discussion

Ecosystem Modelling

A comprehensive understanding of the structure, function, and regulation of major ecosystems is necessary to face the world's growing environmental problems. The past two decades have seen an explosive growth in the number and type of multispecies models directed at fisheries questions (reviewed in Walters *et al.* 1997, Hollowed *et al.* 2000, Whipple *et al.* 2000, Fulton *et al.* 2003, Browman and Stergiou 2004 and references therein).

Fisheries science and management are progressively switching their attention from single species to ecosystems or from single-species models to multispecies models (ICES 2000). By comparison with single-species models, the goal of multispecies models is to improve our understanding of the dynamics of fish populations by considering the ecosystem as a whole, and thus by taking into account trophic interactions between all functional groups (including non-target species) and representing ecological linkages among species, both of which could be altered through fishing (Hollowed *et al.* 2000). However, these authors also concluded that long-term predictions of multispecies models should be considered works in progress owing to their sensitivity to parameter estimation and in particular to assumptions about recruitment levels. It is therefore preferable to accept general rather than specific predictions.

Multispecies models vary from simple extensions of Lotka-Volterra equations to complex simulations of age-size structured interactions (Walters *et al.* 1997). Multispecies models provide a comprehensive representation of predation in the ecosystem and can be used to isolate dominant interactions, but require extensive data input (in particular, accurate diet information) and are unable to provide precise estimates. Also, the relative importance of realism in model structure vs. precision in model output remains a vexing question in modelling predation interactions. In their review, Whipple *et al.* (2000) distinguished between static-flow and dynamic process-based multispecies models in aquatic ecosystems. Static-flow models provide a 'snapshot' of the system at one moment in time and use mass-balance principles to estimate flows of organic matter or energy among components of an ecosystem. This family of models known as mass-balance models is particularly useful for identifying deficiencies in knowledge of the system and is normally used to integrate biomass and rate estimates, and to identify major energy pathways and gaps in the knowledge of an ecosystem. Ecopath is one example of the mass balance modelling approach that has been widely used to explore ecosystem structure and function (Christensen and Pauly 1992, Christensen *et al.* 2005, Bundy 2005). Inverse analysis is an alternate method that is increasingly used in ecosystem modelling to compensate for a lack of data on the ecosystem and to objectively reconstruct a large number of unknown flows or interactions from a small number of observations (Vézina and Platt 1988, Leguerrier *et al.* 2003, Savenkoff *et al.* 2004a, Savenkoff *et al.* 2007a and b). However, since these models assume that energy flows are in balance, they cannot be used to draw conclusions about effects of changes in exploitation rates or predation. Two or more static networks may be compared for different scenarios. In contrast, dynamic models attempt to track variations in population size over time, and some attempt to account for variations in both space and time. The limitation of dynamic models is that they require a high degree of expertise, long time series data, and they are difficult to parameterize. However, they include detailed coupling of physical forcing and its effect on biological interactions and they attempt to model bottom-up and/or top-down forcing in a dynamic framework (Hollowed *et al.* 2000).

Because our empirical databases and scientific understanding of ecological processes will always be incomplete, flow network solutions are not unique. Different approaches (comparisons of different flow networks, random perturbations, sensitivity tests) are needed to develop an explicit implementation and assess uncertainty (or the solution's robustness) to variations in the data (Whipple *et al.* 2000, Savenkoff *et al.* 2004a). Where one modelling approach compromises or simplifies portions of the system, another may provide a realistic and precise representation of the same parts of the system. Modelling of large marine ecosystems is still in its infancy. Our current models, even with 30 compartments, still represent simplifications of the trophic interactions in the northern and southern Gulf of St. Lawrence. The validity of any conclusion regarding the ecosystem under study depends on the input data and the confidence that we have in them. This work is the result of a huge effort to assemble data on the biological characteristics of species occurring in the northern Gulf of St. Lawrence at three time periods and in the southern Gulf at two time periods. Even though most of the data are good estimates for the northern and southern ecosystems, some input values are rough estimates only, meaning that these values are assembled from different literature sources and not from independently measured parameters. Some errors in parameter estimates could significantly alter the system's biomass budget, especially for the most important species of the ecosystem, or produce a totally different balanced solution. The uncertainties remaining in the understanding of the ecosystem may come from: (i) incorrect input values because no data exist; (ii) the confidence limits being too large; (iii) an inaccurate aggregation of species within one functional group; (iv) a poor understanding of the functional relationship between predator and prey; and (v) important processes not accounted for.

Here, the models produced values that were logically consistent with our model structure and constraints. These values were meaningful in the sense that they met some clearly imposed constraints that reflect how we think the system is working. However, even though the models were useful in constraining observations into a coherent picture, it remains that the results are sensitive to choices made regarding the modelling structure and that other equivalent solutions are possible. Moreover, the models were able to derive logical solutions that narrowed the range of possibilities. Several configurations were rejected, e.g., model solutions that included possible parameter values incompatible with the model constraints based on the ranges of input data. Such a model provides a useful starting point to frame hypotheses about the ecosystem, to identify data gaps, and to show where the uncertainties in the food web occur, that could be examined in future studies. Progress is still needed to refine our understanding of the structure of ecosystems and to enhance the quality of future modelling efforts in the Gulf of St. Lawrence as well as in other areas of the world.

Roles of Fishing and Predation in the Gulf Ecosystems during the Mid-1980s

Many of the world's marine ecosystems have undergone major shifts in species composition, which have had severe consequences on their productivity and the derived economic yield (Bax 1998). Regime shifts may be provoked by sudden environmental changes or fishery-induced changes in biological communities (e.g., Cury 2004, Sissenwine and Murawski 2004). The mechanisms underlying these shifts are complex because they often have several causes arising from the interaction of species, fisheries, habitat, and environment. Groundfish populations collapsed in most areas of the Northwest Atlantic in the early 1990s largely as a result of overfishing (Myers *et al.* 1996, Sinclair and Murawski 1997). Most of these populations have failed to recover despite fishing moratoria. This failure to recover reflects a disturbing decline in the productive capacity of these populations (Shelton *et al.* 2006). An important component of this decline in productive capacity is an increase in natural mortality (Shelton *et al.* 2006).

During the mid-1980s, predation by predatory fishes (13.6 t km^{-2} yr^{-1}) was the main cause of fish mortality in the northern Gulf while consumption of fish by seals (0.7 t km^{-2} yr^{-1}) and fishing mortality (1.6 t km^{-2} yr^{-1}) were less important. The relative biomass of forage fishes means that even a small rate of predation (their foraging behaviour targets non-fish prey) can result in a substantial fish removal. During the mid-1980s, forage fishes ate more fish than all marine mammals combined in the nGSL (2.4 vs. 1.7 t km^{-2} yr^{-1}). Among predatory fishes, large cod and redfish were the main predators on fish prey in the nGSL. Large cod mainly consumed forage fishes (63% of total consumption) while redfish ate mostly non-fish prey (73%). Large cod and redfish were also the main species landed (61% and 14%, respectively, of total landings) during the mid-1980s. Predatory fishes and fisheries did not compete for the same resource (Figure 7). Moreover, seals and fisheries, which targeted mainly predatory fishes, could not be considered as competitors of cod during the mid-1980s in the northern Gulf as their overlap indices were the lowest during this time period. Cetaceans and large cod potentially competed for the same resource as the overlap index between them is the highest (Table 8).

In the southern Gulf, fish mortality sources by predation or fishing were more diverse during the mid-1980s (Figure 8). Predatory fishes dominated fish mortality (3.6 t km^{-2} yr^{-1}), while fishing mortality on all fish groups was similar to the predation pressure exerted by marine mammals combined (2.4 vs. 2.2 t km^{-2} yr^{-1}). Except for shrimp and large zooplankton, for which there were no fisheries in the southern Gulf, seals and fisheries competed for cod, American plaice, piscivorous small pelagics, and planktivorous small pelagics, as shown by the highest overlap index between seal species and fisheries in the mid-1980s. Grey and harp seals were important predators of cod, and harp seals were also important potential competitors of both large and small cod for the same resources (high overlap indices).

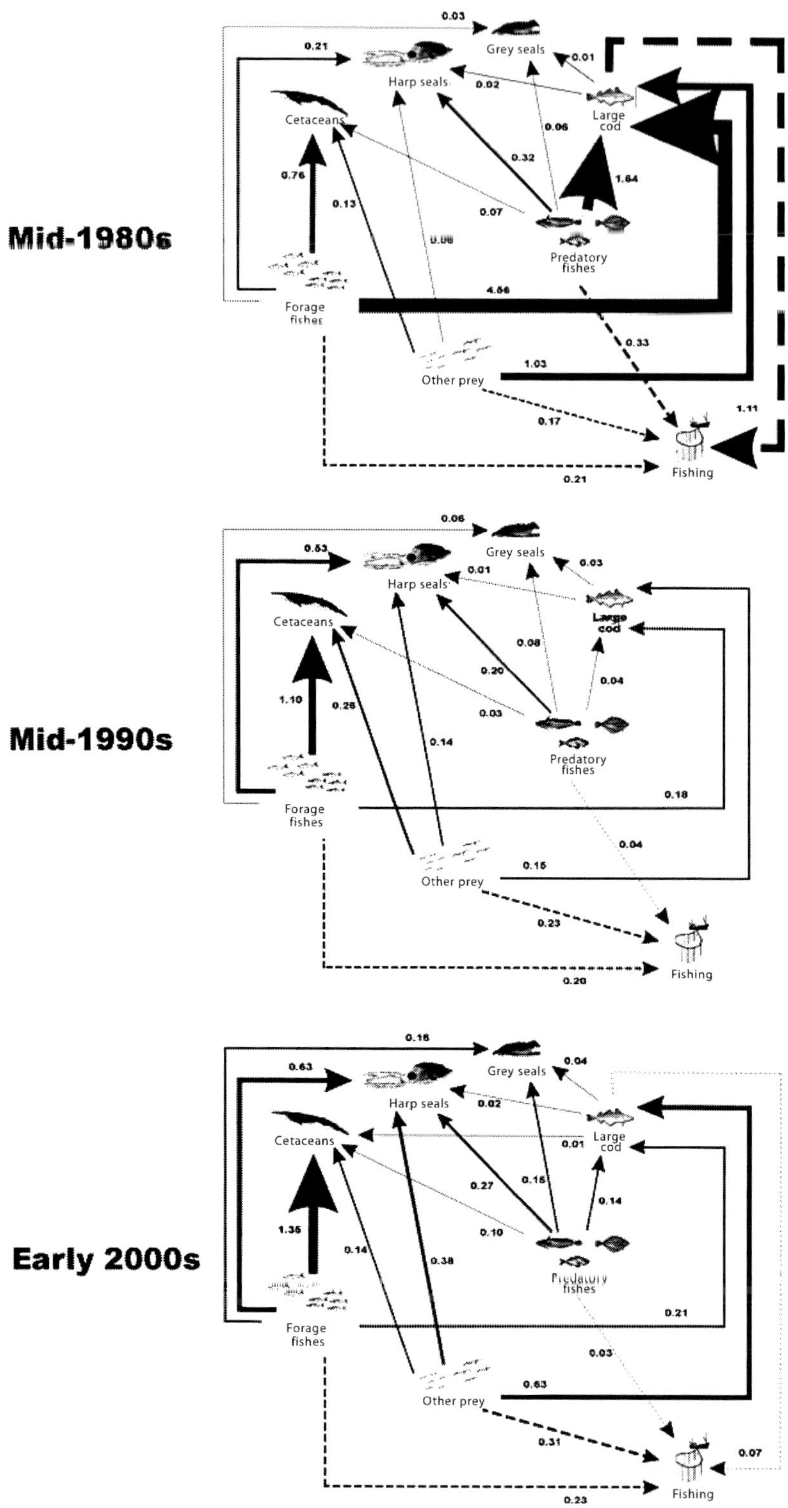

Figure 7. Trophic links related to the main predator species in the northern gulf for the three time periods. Other prey: pelagic and benthic invertebrate groups and detritus.

However, in our model solutions, seal diets contained high proportions of cod (modelling estimates close to the upper limits of the input data ranges). These values (Table 6) were higher than those reported recently from diet studies in the Gulf of St. Lawrence. Cod represented at most 18% of the grey seal diet reported in Hammill and Stenson (2002). Based on quantitative fatty acid signature analysis conducted in other areas, cod represented < 5% to 8.6% of the grey seal diet (Trzcinski *et al.* 2006, Beck *et al.* 2007). Diet estimates derived from the recovery of prey hard parts found in faecal samples or stomach contents often involve substantial biases due to rapid and differential digestion of prey (Trzcinski *et al.* 2006, Beck *et al.* 2007). Thus, in the case of wide-ranging predators that exhibit marked seasonal changes in distribution and foraging effort, the faecal or stomach samples obtained may not be representative of longer-term diet and may overestimate the fraction of prey with large robust otoliths, such as cod, in the diet. This may be the reason that cod represented up to 15% of the grey seal diet when using faecal or stomach samples in Sable Island area (Mohn and Bowen 1996), but only < 5% to 8.6% according to quantitative fatty acid signature analysis (Trzcinski *et al.* 2006). Our input data probably overestimated cod in the diets of harp and grey seals in both ecosystems, and thus our 1980s models represented extreme scenarios of seal predation on cod.

The resource overlap index between marine mammals and fisheries was the highest in the southern Gulf during the mid-1980s. This value is similar to those estimated by Morissette (2007) for the Benguela upwelling system (0.12) and the Gulf of Thailand (0.10), where there are important fisheries for larger fish (Shannon *et al.* 2004). Marine mammals in these ecosystems are also known to be higher trophic-level predators, who mostly feed on these large fish (Dalsgaard *et al.* 1998, Christensen *et al.* 2002, Shannon *et al.* 2004). However, our overlap indices are higher than those found in the North Sea (0.020), or the Strait of Georgia (0.024) (Morissette 2007).

Species interactions are central to ecosystem management considerations. In marine ecosystems, predation can be the major ecological process affecting fish populations and piscivory is often the largest source of fish removals, usually larger than fishing mortality (Bax 1998, Link and Garrison 2002). Also, multispecies predation models indicate that fish can consume more fish than marine mammals do (Overholtz *et al.* 1991, Trites *et al.* 1997). Prior to the collapse of cod and in both Gulf ecosystems, piscivory was the largest source of fish removals and consumption of fish by marine mammals combined was lower than consumption of fish by predatory fishes, but was similar to fish landings.

Roles of Fishing and Predation in the Gulf Ecosystems during Later Time Periods

In both ecosystems, the biomass of demersal and small forage species dominated during the mid-1980s. This structure changed strikingly following the collapse of many groundfish populations, and was dominated by small forage species for the later time periods. The predator structure of both ecosystems has also changed.

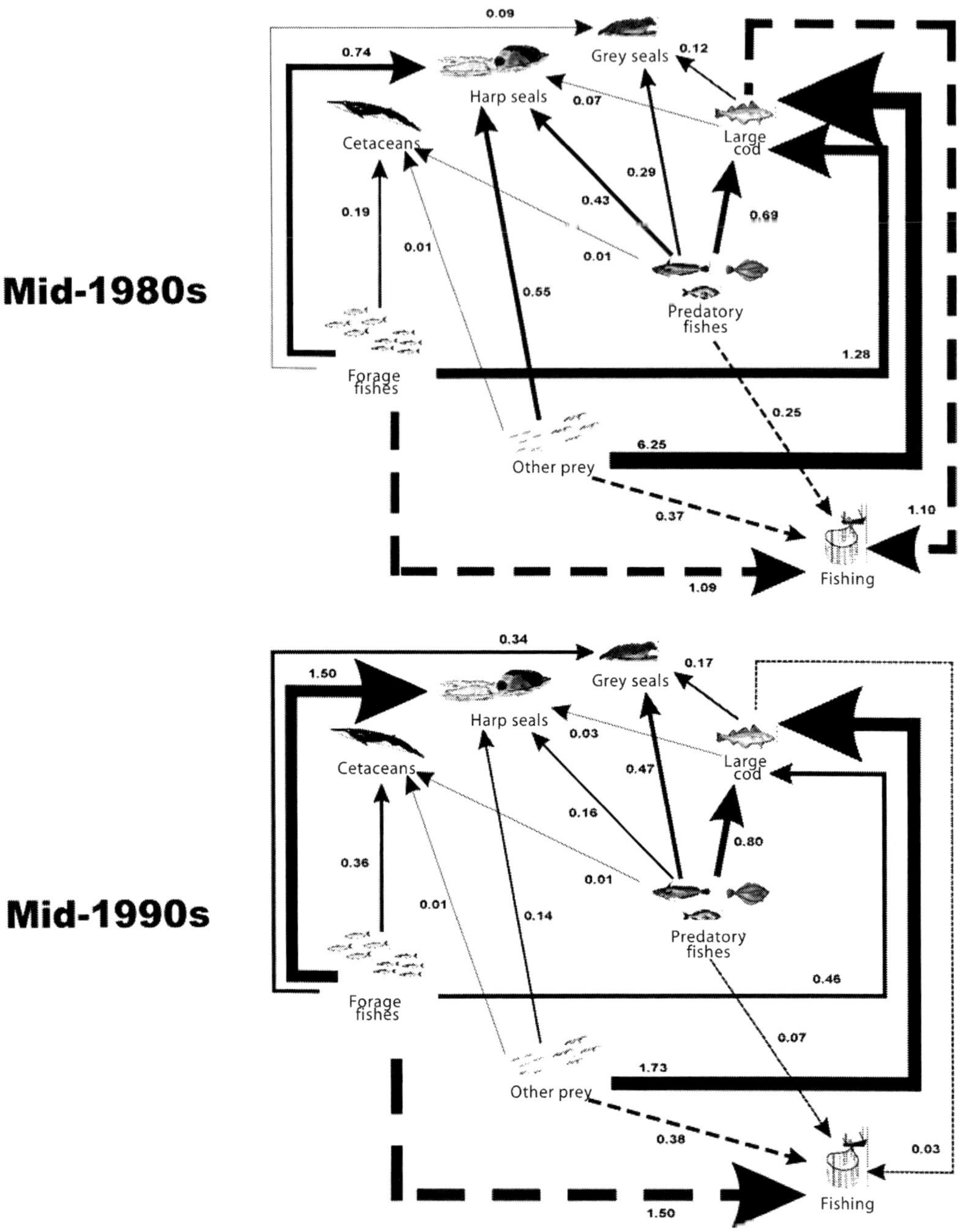

Figure 8. Trophic links related to the main predator species in the southern gulf for the two time periods. Other prey: pelagic and benthic invertebrate groups and detritus.

In both northern and southern Gulf ecosystems, predation mortality exceeded fishing mortality for most groups in recent years because fishing mortality was intentionally reduced by groundfish moratoria in the 1990s (Figures 7 and 8). As expected, fishing mortality decreased significantly from the mid-1980s to the mid-1990s in the northern Gulf, but the decrease in the southern Gulf was weaker (30% vs. 96% in the nGSL, Figures 7 and 8). In both ecosystems, fisheries targeted lower trophic levels during the later time periods. This trend in fishing down the food web has become all too common in the world's fisheries (Pauly *et al.* 1998) and threatens to destabilize ecosystem production and diversity worldwide.

Seals have benefited from the reduction of the seal hunt since the 1970s. The biomass of marine mammals nearly doubled between the 1980s and the 1990s, while forage fish biomass changed little in the two ecosystems. Consumption of fish by marine mammals exceeded both consumption of fish by predatory fishes and fish landings in the two ecosystems in the later periods (Figures 7 and 8). This difference was more pronounced in the northern Gulf for the early 2000s (Figure 8). Frank *et al.* (2005) and Bundy (2005) found a similar pattern in the eastern Scotian Shelf ecosystem off Nova Scotia, Canada, following the decline (and virtual elimination of the ecosystem-structuring role) of the large bodied fish predators. However, Bundy (2005) estimated that the average trophic level of many groups increased, owing to a general increase in piscivory, because an abundant supply of small forage fishes was available. In both Gulf ecosystems, consumption of fish by predatory fishes decreased from the mid-1980s to the mid-1990s, but still accounted for 34% and 36%, respectively, of total consumption of fish during the mid-1990s. Greenland halibut progressively replaced cod as fish predators in the northern Gulf while large cod remained an important predator on fish in the southern Gulf. However, in the northern Gulf, the proportion of fish prey in the diets of large cod, large and small Greenland halibut, and redfish decreased from the mid-1980s to early 2000s, while the proportion of invertebrates increased accordingly. The decline in predatory fishes was less dramatic in the southern Gulf (two-fold vs. six-fold in the northern Gulf). Indeed, in the southern Gulf, piscivory increased from 24% to 42% in the diet of large cod between the 1980s and 1990s, even though large cod consumption decreased between the time periods. Hanson and Chouinard (2002) estimated that fish (mainly herring) had further increased to 60-70% of prey biomass of large cod (>45 cm) diet in 2000. The proportion of fish in cod > 45 cm in the southern Gulf in 2001 was similarly high, 57 to 70%, with herring being the largest component and small mackerel representing 12% to 24% of the diet of large cod (J. M. Hanson, unpublished data).

In the northern Gulf, all resource overlap indices between marine mammal species and fisheries increased from the mid-1980s to the mid-1990s (and to the early 2000s). As shown by the high overlap indices, seals and fisheries competed for the same resources in the early 2000s. Because the biomass of groundfish species declined, our models suggest that consumption by seals shifted towards species at lower trophic level (forage fishes and invertebrates), which were also the main target of fisheries during the later time periods. Also, planktivorous small pelagics and shrimp were sought after both by large cod and fisheries. The highest overlap indices were estimated between large cod and fisheries during the later time periods. Seals were the main predators of large cod and potentially competed with them for the same resource (high overlap indices). Capelin, planktivorous small pelagics, and large zooplankton were the main prey groups targeted by seals and cod.

In the southern Gulf, the overlap index between marine mammals and fisheries decreased from the mid-1980s to the mid-1990s. Fisheries impacted on lower trophic levels (forage fishes and other prey) during the mid-1990s. Fisheries competed mainly with cetaceans and to a lesser extent with grey seals for piscivorous small pelagics and planktivorous small pelagics. In addition to their role as predators, harp and grey seals also potentially competed with large cod for the same resource (capelin and planktivorous small pelagics).

Interactions between Marine Mammals and Fisheries: Implications for Cod Recovery

Marine mammals can be significant consumers for the same resources as humans in marine ecosystems (Bogstad *et al.* 1995, Stenson *et al.* 1997, Trites *et al.* 1997, Hammill and Stenson 2000). In Atlantic Canada, phocid seal populations were estimated to have annually consumed over 3 million tonnes of commercial and non-commercial fish at a time when groundfish population collapses led to closures of commercial fisheries (Hammill and Stenson 2000, Schiermeier 2002). Even though there was an increase in marine mammal biomass and a decrease in the biomass of large predatory fishes in both Gulf ecosystems, the resource overlap indices between marine mammals and fisheries have a reverse pattern. In the nGSL, the low resource overlap between fisheries and marine mammals during the mid-1980s reflects the greater importance of large cod, redfish, planktivorous small pelagics, and shrimp in the landings; species that are not important prey for marine mammals. With the large decrease in cod and redfish biomass, cetaceans, seals, and commercial fisheries all foraged at lower trophic levels (forage fishes and invertebrates) and thus have become important competitors of, and predators on, predatory fishes – possibly slowing their recovery in recent years. In contrast, resource overlap indices between fisheries and marine mammals in the mid-1980s were considerably higher in the sGSL model than in the nGSL model and they decreased from the mid-1980s to the mid-1990s, except for cetaceans. Although marine mammals and fisheries targeted the same species during the mid-1980s (large cod, large American plaice, and piscivorous and planktivorous small pelagics), seals generally consumed food resources that were not the main target of fisheries in the 1990s model (predatory fishes and capelin for seals vs. small pelagics and crabs for fisheries; Tables 6 and 10).

The different seal species preyed on different levels of the food chain. Harp seals preyed on most trophic groups, whereas larger seals, such as grey seals, mainly consumed higher trophic levels. Harp and grey seals consumed cod and also competed with large cod for the same resource (capelin, planktivorous small pelagics, and large zooplankton). Furthermore, other fish predators may have benefited from the decline in cod stocks. The main competitor of small cod and large cod was small Greenland halibut, which was among the main fish predators in the nGSL in the later periods. In the early 2000s, small Greenland halibut were the highest fish biomass (1.3 t km^{-2}), were not heavily consumed (predation mortality: 29% of total mortality), and were not recruited to the fishery in the northern Gulf (44 cm small fish protocol; DFO, 2003). Small Greenland halibut increased in abundance at the same time as cod abundance decreased, resulting in higher potential competition for resources.

Resource overlap of fisheries with large cod and niche overlap between marine mammals and cod were generally lower in the sGSL models than in the nGSL models, especially in the mid-1990s. In the sGSL, cod consumed a large proportion of species largely not targeted by fisheries and marine mammals (polychaetes, molluscs, other benthic invertebrates, and large zooplankton). However, other species (American plaice) potentially competed with cod for this resource. Even though American plaice biomass decreased over time, this value was twice the biomass of both small and large cod during the mid-1990s. In both ecosystems, potential intra-specific and inter-specific competition with cod was high during each time period, particularly with small cod. In our study, the overlap indices are global and do not consider size. In general, different size classes of a species exploit different sizes and species of prey allowing different size classes to use different ecological niches (Polis 1984); there is often greater diet overlap between similar sized individuals of different species than different sized individuals of the same species and they often occupy different habitats. Thus, the overlap index can overestimate the real competition between two predators for the same resources (assuming the resource is in short supply).

Predation on early life history stages (i.e., eggs and larvae) of species is not accounted for in our models. Swain and Sinclair (2000) found a strong negative relationship between recruitment success of southern Gulf cod and the biomass of pelagic fishes, potential predators of cod eggs and larvae. Walters and Kitchell (2001) proposed that large predatory fishes may benefit from "cultivation" effects, with adults cropping down forage species that are potential predators or competitors of early life history stages. Consequently, large predatory fishes may experience reduced recruitment at low abundance if forage fishes increase in abundance when released from their predation.

The predation experienced by a population is influenced by environmental and biological factors that vary spatially, seasonally, annually, and with its own abundance, and with other prey and predators. Predation can directly regulate prey numbers and indirectly regulate their survival via habitat availability, individual growth, and trophic structure (Bax 1998). A species' interaction with other species (and itself) through predation is a potential source of change in the biological environment. Change is direct when a predator kills a prey, and indirect in a multitude of ways, including predation on shared prey resources, effects involving intermediate predators, and predation influencing prey behavior and resource availability (Bax 1998). A model of a Norwegian fjord ecosystem (Bax and Eliassen 1990) demonstrated that although increasing herring abundance directly benefited the cod population at first, the competition for food between adult herring and juvenile cod resulted in higher juvenile cod mortality and a subsequent overall decline in the cod population. It appears that reducing the abundance of a particular predator is unlikely to lead to a proportional reduction in the predation on a targeted prey species; the remaining predators (of the same or different species) will increase their predation as the abundance of the targeted prey species increases. Ontogenetic changes in diet may serve to further compensate (or even reverse) any reduced predation. Given the many potential pathways connecting marine predators and their prey, accurate prediction of the direction and magnitude of predation-driven changes in marine fishery ecosystems are difficult (Bax 1998). Thus, any anthropogenic interference such as reducing the abundance of seals in such already-disturbed and evolving ecosystems could lead to unknown and perhaps undesirable consequences.

Conclusion

The reasons for the lack of recovery of depleted demersal fish stocks are complex and no single factor can explain all observations. Major changes in the structure and functioning of the northern and southern Gulf ecosystems accompanied the collapse of the large-bodied demersal predators in the late 1980s and early 1990s, which was due to overfishing. Overfishing removed a functional group, large predatory fishes, which has not been replaced 15 years after the cessation of heavy fishing (Rice and Rivard 2003). The decline of these predators has left only marine mammals as top predators. Seals were significant predators of juvenile cod during the mid-1990s and early 2000s, and may have played a role in the non-recovery of cod. The small fishery on depleted stocks in the 1990s and 2000s probably also contributed to the lack of recovery (Shelton *et al.* 2006). Other factors may also be involved for the lack of recovery of depleted stocks including declines in stock productivity (Dutil and Brander 2003), increases in starvation linked to poor environmental conditions (Dutil and Lambert 2000), disease, and predation by pelagic fish on different life stages of cod. Reductions of food availability through competition between cod and other species could slow cod recovery, assuming that competitors and prey co-occur, spatially and temporally. The low biomass of cod not only makes them vulnerable to predation, but may also increase competition for prey. Small and large cod compete for their prey with highly abundant competitors, and this may lead to food limitation.

Acknowledgments

This study was carried out as a contribution of CDEENA (Comparative Dynamics of Exploited Ecosystems in the Northwest Atlantic), a program supported by the Canadian Department of Fisheries and Oceans (Science Strategic Fund). We are grateful to Simon-Pierre Despatie, Hugo Bourdages, Red Méthot, Hugues Benoît, and colleagues who made data available for this manuscript. CS particularly thanks Alain Vézina for his inverse method expertise during the past 10 years.

See the CDEENA web site at http://www.osl.gc.ca/cdeena/en/accueil.shtml.

References

Bax, N. J., 1998. The significance and prediction of predation in marine fisheries. *ICES Journal of Marine Science*, **55**, 997-1030.

Bax, N. J., & Eliassen, J. E. (1990). Multispecies analysis in Balsfjord, northern Norway: solution and sensitivity analysis of a simple ecosystem model. *Journal Conseil International pour l'Exploration de la Mer*, **47**, 175-204.

Beck, C. A., Iverson, S. J., Bowen, W. D., & Blanchard, W. (2007). Sex differences in grey seal diet reflect seasonal variation in foraging behaviour and reproductive expenditure: evidence from quantitative fatty acid signature analysis. *Journal of Animal Ecology*, **76**, 490-502.

Bogstad, B., Haug, K. H., & Ulltang, O. (1995). MULTSPEC - a multispecies model for fish and marine mammals in the Barents Sea. *NAFO Scientific Council Research Document*, **95/83**, 47 pp.

Bowen, W. D. (1997). Role of marine mammals in aquatic ecosystems. *Marine Ecology Progress Series*, **158**, 267-274.

Browman, H. I., & Stergiou, K. I. (2004). Perspectives on ecosystem-based approaches to the management of marine resources. *Marine Ecology Progress Series*, **274**, 269-303.

Bundy, A. (2001). Fishing on ecosystems: Interplay of fishing and predation in Newfoundland-Labrador. *Canadian Journal of Fisheries and Aquatic Sciences*, **58**, 1153-1167.

Bundy, A. (2005). Structure and functioning of the eastern Scotian Shelf ecosystem before and after the collapse of groundfish stocks in the early 1990s. *Canadian Journal of Fisheries and Aquatic Sciences*, **62**, 1474-1489.

Bundy, A., & Fanning, L. P. (2005). Can Atlantic cod (*Gadus morhua*) recover? Exploring trophic explanations for the non-recovery of the cod stock on the eastern Scotian Shelf, Canada. *Canadian Journal of Fisheries and Aquatic Sciences*, **62**, 1453-1473.

CAFSAC (1994). Report on the status of groundfish stocks in the Canadian Northwest Atlantic. *DFO Atlantic Fish Stock Status Report*, **94/4**.

Chouinard, G. A., Swain, D. P., Hammill, M. O., & Poirier, G. A. (2005). Covariation between grey seal (*Halichoerus grypus*) abundance and natural mortality of cod (*Gadus morhua*) in the southern Gulf of St. Lawrence. *Canadian Journal of Fisheries and Aquatic Sciences*, **62**, 1991-2000.

Christensen, V., & Pauly, D. (1992). ECOPATH II - A system for balancing steady-state ecosystem models and calculating network characteristics. *Ecological Modelling*, **61**, 169-185.

Christensen, V., Beyer, J. E., Gislason, H., & Vinter, M. (2002). A comparative analysis of the North Sea based on Ecopath with Ecosim and multispecies virtual population analysis. In: V. Christensen, G. Reck, & J. L. Maclean (Eds.). Proceedings of the INCO-DC Conference Placing Fisheries in their Ecosystem Context, Galàpagos Islands, Ecuador, 4-8 December 2000. *ACP-EU Fisheries Research Report*, **39** pp.

Christensen, V., Walters C. J., & Pauly, D. (2005). Ecopath with Ecosim: a user's guide. Fisheries Centre, University of British Columbia, Vancouver. November 2005 edition.

Cury, P. M. (2004). Tuning the ecoscope for the ecosystem approach to fisheries. *Marine Ecology Progress Series*, **274**, 272-275.

Cury, P. M., Shannon, L. J., Roux, J.-P., Daskalov, G. M., Jarre, A., Moloney, C. L., & Pauly, D. (2005). Trophodynamic indicators for an ecosystem approach to fisheries. *ICES Journal of Marine Science*, **62**, 430-442.

Dalsgaard, J., Wallace, S. S., Salas, S., & Preikshot, D. (1998). Mass-balance model reconstructions of the Strait of Georgia: the present, one hundred and five hundred years ago. In: D. Pauly, T. Pitcher, D. Preikshot, & J. Hearne (Eds.). Back to the future: reconstructing the Strait of Georgia ecosystem. *Fisheries Centre Research Reports*, **6(5),** pp. 72-86.

Department of Fisheries and Oceans, DFO (2000). Shrimp of the Estuary and Gulf of St. Lawrence. *DFO – Science, Stock Status Report*, **C4-06**.

Department of Fisheries and Oceans, DFO (2001). Capelin of the Estuary and Gulf of St. Lawrence. *DFO – Science, Stock Status Report*, **B4-03**.

Department of Fisheries and Oceans, DFO (2002). Gulf of St. Lawrence (4RST) Greenland halibut in 2001. *DFO – Science, Stock Status Report*, **A4-03**.

Department of Fisheries and Oceans, DFO (2003). Greenland Halibut in the Gulf of St. Lawrence (4RST) in 2002. *DFO Canadian Science Advisory Secretariat Stock Status Report*, 2003/07.

Department of Fisheries and Oceans, DFO (2004). White hake in he southern Gulf of St. Lawrence (Div. 4T). *DFO Canadian Science Advisory Secretariat Stock Status Report*, 2004/007.

Dutil, J.-D., & Brander, K. (2003). Comparing productivity of North Atlantic cod (Gadus morhua) stocks and limits to growth production. *Fisheries Oceanography*, **12**, 502-512.

Dutil, J.-D., & Lambert, Y. (2000). Natural mortality from poor condition in Atlantic cod (*Gadus morhua*). *Canadian Journal of Fisheries and Aquatic Sciences*, **57**, 826-836.

Frank, K. T., Petrie, B., Choi, J. S., & Leggett, W. C. (2005). Trophic cascades in a formerly cod-dominated ecosystem. *Science*, **308**, 1621-1623.

Fulton, E. A., Smith, A. D. M., & Johnson, C. R. (2003). Effect of complexity on marine ecosystem models. *Marine Ecology Progress Series*, **253**, 1-16.

Gascon, D. (*Editor*) (2003). Redfish multidisciplinary research zonal program (1995-1998): final report. *Canadian Technical Report of Fisheries and Aquatic Sciences*, **2462**, xiii+139 pp.

Gascon, D., Aparicio, M., & Mercille, B. (1990). Estimations du mélange entre les stocks de morue du Nord du Golfe du Saint Laurent (Divisions 3Pn4RS) et les stocks adjacents (2J3KL, 3Ps, et 4TVn [Janvier-Avril]) à partir de résultats de marquage. *CAFSAC Research Document*, **90/61**.

Hammill, M. O., & Stenson, G. B. (2000). Estimated prey consumption by harp seals (*Phoca groenlandica*), hooded seals (*Cystophora cristata*), grey seals (*Halichoerus grypus*) and harbour seals (*Phoca vitulina*) in Atlantic Canada. *Journal of Northwest Atlantic Fishery Science*, **26**, 1-23.

Hammill, M. O., & Stenson, G. B. (2002). Estimated consumption of Atlantic cod (*Gadus morhua*) and some other prey by grey seals (*Halichoerus grypus*) and harp seals (*Phoca groenlandica*) in the southern Gulf of St. Lawrence (NAFO Division 4T). *DFO Canadian Stock Assessment Secretariat Research Document*, 2002/054.

Hammill, M. O., & Stenson, G. B. (2005). Abundance of Northwest Atlantic harp seals (1960 – 2005). *DFO Canadian Science Advisory Secretariat Research Document*, 2005/090.

Hammill, M. O., & Stenson, G. B. (2006). Abundance of Northwest Atlantic hooded seals (1960 – 2005). *DFO Canadian Science Advisory Secretariat Research Document*, 2006/068.

Hanson, J. M., & Chouinard, G. A. (2002). Diet of cod in the southern Gulf of St Lawrence as an index of ecosystem change, 1959-2000. *Journal of Fish Biology*, **60**, 902-922.

Hanson, J. M., & Lanteigne, M. (1999). Results of an experimental shrimp fishery in Chaleur Bay, southern Gulf of St. Lawrence. *North-American Journal of Fisheries Management*, **19**, 713-723.

Hanson, J. M., & Lanteigne, M. (2000). Evaluation of Atlantic cod predation on American lobster in the southern Gulf of St. Lawrence, with comments on other potential fish predators. *Transactions of the American Fisheries Society*, **129**, 13-29.

Hollowed, A. B., Bax, N., Beamish, R., Collie, J., Fogarty, M., Livingston, P., Pope, J., & Rice, J. C. (2000). Are multispecies models an improvement on single-species models for measuring fishing impacts on marine ecosystems? *ICES Journal of Marine Research*, **57**, 707-719.

Hurlbut, T., & Clay, D. (1998). Morphometric and meristic differences between shallow- and deep-water populations of white hake (*Urophycis tenuis*) in the southern Gulf of St. Lawrence. *Canadian Journal of Fisheries and Aquatic Sciences*, **55**, 2274-2282.

ICES (2000). Ecosystem effects of fishing. Proceedings of an ICES/SCOR Symposium held in Montpellier, France, 16-19 March 1999. *ICES Journal of Marine Science*, **57**, 465-792.

Jackson, J. B. C., Kirby, M. X., Berger, W. H., Bjorndal, K. A., Botsford, L. W., Bourque, B. J., Bradbury, R. H., Cooke, R., Erlandson, J., Estes, J. A., Hughes, T. P., Kidwell, S., Lange, C. B., Lenihan, H. S., Pandolfi, J. M., Peterson, C. H., Steneck, R. S., Tegner,

M. J., & Warner, R. R. (2001). Historical overfishing and the recent collapse of coastal ecosystems. *Science*, **293**, 629-638.

Kaschner, K. (2004). *Modelling and mapping resource overlap between marine mammals and fisheries on a global scale*. Ph. D. thesis, University of British Columbia.

Koutitonsky, V. G., & Bugden, G. L. (1991). The physical oceanography of the Gulf of St. Lawrence: a review with emphasis on the synoptic variability of the motion. In: J.-C. Therriault (Ed.). The Gulf of St. Lawrence: small ocean or big estuary? *Canadian Special Publication of Fisheries and Aquatic Sciences*, **113**, pp. 57-90.

Leguerrier, D., Niquil, N., Boileau, N., Rzeznik, J., Sauriau, P.-G., Le Moine, O., & Bacher, C. (2003). Numerical analysis of the food web of an intertidal mudflat ecosystem on the Atlantic coast of France. *Marine Ecology Progress Series*, **246**, 17-37.

Linderman, R. L. (1942). The trophic-dynamic aspect of ecology. *Ecology*, **23**, 399-418.

Link, J. S., & Garrison, L. P. (2002). Changes in piscivory associated with fishing induced changes to the finfish community on Georges Bank. *Fishery Research*, **55**, 71-86.

Mohn, R., & Bowen, W. D. (1996). Grey seal predation on the eastern Scotian Shelf: Modelling the impact of Atlantic cod. *Canadian Journal of Fisheries and Aquatic Sciences*, **53**, 2722-2738.

Molenaar, E. J. (2002). Ecosystem-based fisheries management, commercial fisheries, marine mammals and the 2001 Reykjavik declaration in the context of international law. *International Journal of Marine and Coastal Law*, **17**, 561-595.

Morissette, L. (2007). Complexity, cost and quality of ecosystem models and their impact on resilience: a comparative analysis, with emphasis on marine mammals and the Gulf of St. Lawrence. Ph. D. thesis, University of British Columbia.

Morissette, L., Despatie, S.-P., Savenkoff, C., Hammill, M. O., Bourdages, H., & Chabot, D. (2003). Data gathering and input parameters to construct ecosystem models for the northern Gulf of St. Lawrence (mid-1980s). *Canadian Technical Report of Fisheries and Aquatic Sciences*, **2497**, vi+94 pp.

Morissette, L., Hammill, M. O., & Savenkoff, C. (2006). The trophic role of marine mammals in the northern Gulf of St. Lawrence. *Marine Mammal Science*, **22**, 74-103.

Myers, R. A., Hutchings, J. A., & Barrowman, N. J. (1996). Hypotheses for the decline of cod in the North Atlantic. *Marine Ecology Progress Series*, **138**, 293-308.

Ni, I.-H. (1982). Meristic variation in beaked redfish, *Sebastes mentella* and *S. fasciatus*, in the Northwest Atlantic. *Canadian Journal of Fisheries and Aquatic Sciences*, **39**, 1664-1685.

Odum, W. E., & Heald, E. J. (1975). The detritus-based food web of an estuarine mangrove community. In: M. Wiley (Ed.). Academic Press, New York. *Estuarine research*, Vol. 1, pp. 265-286.

Overholtz, W. J., Murawski, S. A., & Foster, K. L. (1991). Impact of predatory fish, marine mammals, and seabirds on the pelagic fish ecosystem of the northeastern USA. *ICES Marine Science Symposia*, **193**, 198-208.

Pauly, D., & Watson, R. (2005). Background and interpretation of the 'Marine Trophic Index' as a measure of biodiversity. *Philosophical transactions of the royal society: Biological sciences*, **360**, 415-423.

Pauly, D., Christensen, V., Dalsgaard, J., Froese, R., & Torres, F., Jr. (1998). Fishing down marine food webs. *Science*, **279**, 860-863.

Pianka, E .R. (1973). The structure of lizard communities. *Annual Review of Ecology and Systematics*, **4**, 53-74.

Plagányi, E. E., & Butterworth, D. S. (2002). Competition with fisheries. In: W.F. Perrin, B. Würsig, & H. G. M. Thewissen (Eds.). *Encyclopedia of Marine Mammals*. Academic Press, San Diego, pp. 268-273.

Polis, G. A. (1984). Age structure component of niche width and intraspecific resource partitioning: Can age groups function as ecological species? *American Naturalist*, **123**, 541-564.

Rice, J., & Rivard, D. (2003). Proceedings of the zonal assessment meeting – Atlantic cod. *DFO Canadian Science Advisory Secretariat Proceedings Series*, 2003/021, 151 pp.

Roy, S., Silverberg, N., Romero, N., Deibel, D., Klein, B., Savenkoff, C., Vézina, A. F., Tremblay, J.-E., Legendre, L., & Rivkin, R. B. (2000). Importance of mesozooplankton feeding for the downward flux of biogenic carbon in the Gulf of St. Lawrence (Canada). *Deep-Sea Research II*, **47**, 519-544.

Savenkoff, C., Castonguay, M., Vézina, A. F., Despatie, S.-P., Chabot, D., Morissette, L., & Hammill, M. O. (2004a). Inverse modelling of trophic flows through an entire ecosystem: the northern Gulf of St. Lawrence in the mid-1980s. *Canadian Journal of Fisheries and Aquatic Sciences*, **61**, 2194-2214.

Savenkoff, C., Bourdages, H., Castonguay, M., Morissette, L., Chabot, D., & Hammill, M. O. (2004b). Input data and parameter estimates for ecosystem models of the northern Gulf of St. Lawrence (mid-1990s). *Canadian Technical Report of Fisheries and Aquatic Sciences*, **2531**, vi+93 pp.

Savenkoff, C., Bourdages, H., Swain, D. P., Despatie, S.-P., Hanson, J. M., Méthot, R., Morissette, L., & Hammill, M. O. (2004c). Input data and parameter estimates for ecosystem models of the southern Gulf of St. Lawrence (mid-1980s and mid-1990s). *Canadian Technical Report of Fisheries and Aquatic Sciences*, **2529**, vi+105 pp.

Savenkoff, C., Castonguay, M., Méthot, R., Chabot, D., & Hammill, M. O. (2005). Input data and parameter estimates for ecosystem models of the northern Gulf of St. Lawrence (2000–2002). *Canadian Technical Report of Fisheries and Aquatic Sciences*, **2588**, v+91 pp.

Savenkoff, C., Castonguay, M., Chabot, D., Hammill, M.O., Bourdages, H., & Morissette, L. (2007a). Changes in the northern Gulf of St. Lawrence ecosystem estimated by inverse modelling: Evidence of a fishery-induced regime shift? *Estuarine, Coastal, and Shelf Science*, **73**, 711-724.

Savenkoff, C., Swain, D. P., Hanson, J. M., Castonguay, M., Hammill, M. O., Bourdages, H., Morissette, L., & Chabot, D. (2007b). Effects of fishing and predation in a heavily exploited ecosystem: Comparing periods before and after the collapse of groundfish in the southern Gulf of St. Lawrence (Canada). *Ecological modelling*, **204**, 115-128.

Schiermeier, Q. (2002). How many more fish in the sea? *Nature*, **419**, 662-665.

Shannon, L. J., Christensen, V., & Walters, C. J. (2004). Modelling stock dynamics in the southern Benguela ecosystem for the period 1978 – 2002. *African Journal of Marine Science*, **26**, 179-196.

Shelton, P. A., Sinclair, A. F., Chouinard, G. A., Mohn, R., & Duplisea, D. E. (2006). Fishing under low productivity conditions is further delaying recovery of Northwest Atlantic cod (*Gadus morhua*). *Canadian Journal of Fisheries and Aquatic Sciences*, **63**, 235-238.

Sinclair, A. F., & Murawski, S. A. (1997). Why have groundfish stocks declined? In J. Boreman, B. S. Nakashima, J. A. Wilson, & R. L. Kendall (Eds.). *Northwest Atlantic groundfish: perspectives on a fishery collapse*. American Fisheries Society: Bethesda, Md, pp. 71-93.

Sissenwine, M., & Murawski, M. (2004). Moving beyond "intelligent tinkering": advancing an ecosystem approach to fisheries. *Marine Ecology Progress Series*, **274**, 291-295.

Stenson, G. B., Hammill, M. O., & Lawson, J. W. (1997). Predation by harp seals in Atlantic Canada: Preliminary consumption estimates for Arctic cod, capelin and Atlantic cod. *Journal of Northwest Atlantic Fishery Science*, **22**, 137-154.

Swain, D. P., & Castonguay, M. (2000). Final report of the 2000 annual meeting of the Fisheries Oceanography Committee including the report of the workshop on the cod recruitment dilemma. *DFO Canadian Stock Assessment Secretariat Proceedings Series*, 2000/17, 113 pp.

Swain, D. P., & Sinclair, A. F. (2000). Pelagic fishes and the cod recruitment dilemma in the Northwest Atlantic. *Canadian Journal of Fisheries and Aquatic Sciences*, **57**, 1321-1325.

Swain, D. P., Frank, K. T., & Maillet, G. (2001). Delineating stocks of Atlantic cod (*Gadus morhua*) in the Gulf of St. Lawrence and Cabot Strait areas using vertebral number. *ICES Journal of Marine Research*, **58**, 253-269.

Trites, A.W., Christensen, V., & Pauly, D. (1997). Competition between fisheries and marine mammals for prey and primary production in the Pacific Ocean. *Journal of Northwest Atlantic Fishery Science*, **22**, 173-187.

Trzcinski, M. K., Mohn, R., & Bowen, W. D. (2006). Continued decline of an Atlantic cod population: How important is gray seal predation? *Ecological Applications*, **16**, 2276-2292.

Vézina, A. F., & Pahlow, M. (2003). Reconstruction of ecosystem flows using inverse methods: how well do they work? *Journal of Marine Systems*, **40-41**, 55-77.

Vézina, A. F., & Platt, T. (1988). Food web dynamics in the ocean. I. Best-estimates of flow networks using inverse methods. *Marine Ecology Progress Series*, **42**, 269-287.

Walters, C. J., & Kitchell, J. F. (2001). Cultivation/depensation effects on juvenile survival and recruitment: a serious flaw in the theory of fishing? *Canadian Journal of Fisheries and Aquatic Sciences*, **58**, 39-50.

Walters, C., Christensen, V., & Pauly, D. (1997). Structuring dynamic models of exploited ecosystems from trophic mass-balance assessments. *Reviews in Fish Biology and Fisheries*, **7**, 139-172.

Whipple, S. J., Link, J. S., Garrison, L. P., & Fogarty, M. J. (2000). Models of predation and fishing mortality in aquatic ecosystems. *Fish and Fisheries*, **1**, 22-40.

Worm, B., & Myers, R. A. (2003). Meta-analysis of cod-shrimp interactions reveals top-down control in oceanic food webs. *Ecology*, **84**, 162-173.

Yodzis, P. (2001). Must top predators be culled for the sake of fisheries? *Trends in Ecology and Evolution*, **16**, 78-84.

Reviewed by *Dr. Sheila J.J. Heymans*, Scottish Association for Marine Science, Dunstaffnage Marine Laboratory, Dunbeg, PA371QA, Scotland.

In: Ecosystem Ecology Research Trends
Editors: J. Chen and C. Guo, pp. 153-175
ISBN 978-1-60456-183-8

Chapter 4

The Mathematics of Maintenance and Acquisition

Noël Bonneuil[*]
Institut national des études démographiques,
133, bld Davout, 75980, Paris cedex 20, France,
and École des hautes études en sciences sociales

Abstract

Modelling dynamic systems commonly focusses on asymptotic behavior. Is the search for asymptotic patterns appropriate for modelling maintenance and acquisition? In biology, where variability and surprise prevail, moment-by-moment time, not limit sets , matters. Taking the state of a particular system at a particular moment into account requires the substitution and replacement of many of the concepts usual to asymptotic analysis: the replacement of probability by set membership; the replacement of trajectories by attainable sets; the replacement of limit sets by the $\mathcal{C}$-viability kernel or capture domain. Rather than convergence to an attractor and dominance of an optimal or evolutionary stable strategy, the focus here is on identifying the right strategy at the right moment that allows a living system to have at least some possibility of perpetuation or collapse. Basic models of ecosystem and population genetics are revisited from this perspective. Notably, rather than proposing a mechanism and examining how it behaves under various constraints, here the constraints are incorporated in the dynamic system from the start through $\mathcal{C}$-viability multipliers. Case studies reveal that even without using the theoretical tools of $\mathcal{C}$-viability, some experimenters have empirically invoked $\mathcal{C}$-viability multipliers or explored $\mathcal{C}$-viability domains.

Keywords: Dynamical systems — Viability theory — Protected polymorphism – maintenance–acquisition

1. Introduction

Rather than reviewing the principle of maintenance in biological literature, a task done by Chesson (2000), this chapter presents the mathematical handling of maintenance and acquisition, which are two complementary facets of evolution. Chesson distinguishes "stable

[*]E-mail address: bonneuil@ined.fr

coexistence" from "unstable coexistence". In the former, "the densities of the species in the system do not show long-term trends; if densities get low, they tend to recover", while in the latter, "there is no tendency for recovery and species are not maintained in the system on long timescales" (:344). Viewing biological systems through the lens of asymptotic behavior can lead to misleading conclusions and spurious distinctions when the environment so obviously fluctuates in time. Models borrowed from physics promise tremendous precision of measurement and clear articulation of interactions, they often rely on approximations at infinite time which are valid very soon. However, the use of these models, although omnipresent, obscures the complexity of living organisms and the variability in biological systems, where each moment can bring surprise and each state can generate diverse and unpredicted fates. For example, Regan, Colyvan, and Burgman (2002) describe uncertainty as "pervasive in ecology where the difficulties of dealing with sources of uncertainty are exacerbated by variation in the system itself" (:618). Underlying equilibrium models rely on a hope that equilibrium is sufficiently persistent and that disequilibrium is sufficiently transient that robust results can still be obtained. Does this hold true in biology? As pointed out by Law and Morton (1996: 763) "it is not clear that local asymptotic stability is an appropriate condition for coexistence, for species may coexist without tending to an equilibrium point" (Williamson, 1987; Hastings, 1988). Firstly, I argue that more attention should be paid to investigating transient dynamics, the "here and now" in which natural selection operates (Gee, 2002: 611), and present some applications to ecosystems and population genetics of $\mathcal{C}$-viability theory, for which it is specifically designed.

Secondly, in leaving the confines of asymptotic patterns, we leave behind the reduction of behavior of a dynamic system to its limit sets. For example, Law and Morton (1996: 763) admit that the "dynamics close to an interior equilibrium point are not the main issue for coexistence". We also forgo the principle of optimality, which has pervaded the modelling of ecosystems and population genetics. Optimality is a mathematical criterion allowing us to distinguish and identify a particular solution from among many possible solutions. In science, it "has long been an organizing principle" (Buchanan, 2007). In optical geometry, Fermat's least action principle proved dramatically efficient. In biology, where conditions are seldom constant and individuals experience non repeatable events, where "natural selection is directionless with respect to history" (Gee, 2002), we may have to build on a different concept that captures the "temporal structure" of the state space. These concepts are innate to the *$\mathcal{C}$-viability kernel* (the set of states from which there exists at least one trajectory pertaining to a given set of constraints), the *capture domain* (the set of states from which there exists at least one trajectory in the state space attaining a given set), or the *victory domain* of a given player in dynamical games (the set of states from which there exists at least one trajectory leading to the victory of the player over an opponent).

Thirdly, asymptotic arguments have been popular because they exploit the convenient and useful law of large numbers, which allows uncertainty to be described by probability distributions (the stochastic paradigm). One might say that deterministic behavior is a special case of stochasticity, but the converse is also true: probability laws conceal a deterministic way of handling hazard, using the law of large numbers and the central limit theorem to tame the unknown with analytical formulas. In biology and in other situations where time matters, we must question the very existence of an underlying probability distribution. We sidestep probability distributions by building upon the more primal statement that "regulons", for example fitness values or migration or fertility rates, belong to closed sets. This conception of hazard was called "tychastic" (Aubin et al., forthcoming). That regulons belong to closed sets conveys the fact that, to a certain extent, living beings are flexible in an unpredictable way. A consequencc is that concepts of trajectory or confidence interval are replaced by the concept of attainable sets. Does the shape of a trajectory really matter for the survival of an organism, or just the fact that this trajectory allows the survival of this organism (taking indeed a given shape, historically)? Similarly, how important is it that a species be optimal, as long as it perpetuates itself? In a perpetually changing environment, a living fish observed today does not mean that it has ever been optimal, but simply that it has managed to "stay in the game". I argue that in biology the question of the frequencies of the values taken by the regulons may be inadequate because evolution is so full of surprise. What is the probability of the appearance of homo sapiens?

The complementary epistemological strategies of focussing on infinite time behavior, delineating limit sets, and relying on stochastic laws have brought deep insight to the understanding of systems. Revising these strategies in order to address the transient time of biology entails the cost of investing different mathematics. The questions of maintenance and acquisition should be treated at the outset, without the distorting lens of asymptotic modelling, which ignores those states from which a solution exists in finite time. I shall revisit basic models of ecosystems and populations genetics, and argue for the explanatory power of $\mathcal{C}$-viability theory in empirical case studies. I will show that experimenters often appear to think in terms of viability rather than in terms of equilibrium and limit sets.

2. Maintenance

2.1. $\mathcal{C}$-Viability

The analytical or numerical resolution of differential equations has proven effective in uncovering regulatory mechanisms in ecosystems and in population genetics. A dynamical system is expressed as

$$x'(t) = f(x(t)) \tag{1}$$

where $x(t)$ denotes the state in $I\!R^n$ at time t, and f is a functional, traditionally to be guessed or estimated: for example, $x = (x_1, x_2)$ where $x_1(t)$ represents the density of the prey and $x_2(t)$ the density of the predator in Lotka-Volterra models. Parameters and interactions between prey and predator are constant in the simplest model, while they involve endogenous effects, such as frequency-dependence or stochastic variation in more complex ones. However, biologists are aware that interactions within most ecological communities are often poorly known (Mittelbach et al., 1995: 2347). Hassell (2000: 552) showed that "contrary to what is still a popular view", "the density-independent [host-parasitoid interactions] are the more important." Wilbur (1972) pointed out the inadequacies of laboratory ecologists' classical models in omitting trajectories of communities in a changing environment. Even in laboratory settings, arthropod studies of predator-prey interactions describe either sustained oscillations or irregular fluctuations in population size over time (Huffaker, 1958; Pimentel, Nagel and Madden, 1963; Mitchell, Arthur and Farrow, 1992).

Pianka (1978) pointed out that the optimality of a given tactic is restlessly challenged by sources of variation, among which stress or environmental heterogeneity; he also emphasized how the fitness of an organism can "strongly be influenced by the degree of developmental flexibility of a given phenotype trait" (p. 165). For example, Sih et al. (1985: 296) concluded that "environmental stress can either increase or decrease the importance of predation", and that "the outcome depends on the relative effects of stress on predation rates and on prey growth rate". Hickman (1975) showed that plants of annual species *Polygonum cascadense* allocate more resources proportionally to reproduction in harsh open habitats than in more moderate habitats, and that these differences are developmentally plastic, adapting to short-term environmental unpredictability. Giesel (1976: 75), from studies of passerine birds, concluded that variability in fertility is an "important adaptation to account for environmentally induced unpredictability of the major selective forces". Wilbur (1972, 1983, 1987) pointed out that plastic growth and variable size at metamorphosis of amphibians were adaptations to the uncertain environment of temporary ponds, knowing that these morphological variations are induced defenses rather than a genetic polymorphism (McCollum, 1993).

Finally, Brown and Vincent (1985, 1987, 1992) argued that the fitness of an individual in a community is bounded by evolutionary constraints, and that the only thing that can be said with certainty is that interactions vary within a closed set. If $u \in I\!R^p$ is the array of strategies and interactions, the flexibility invoked by Brown and Vincent is expressed as:

$$u(t) \in U(x(t)) \subset I\!R^p \tag{2}$$

where $U(x)$ is a closed set depending on the state x.

Recognizing our ignorance on the variation of strategies and interactions, a simple statement that they can only vary within closed sets alters the dynamic of Eq. (1) to:

$$x'(t) \in \{f(x(t), u(t)), u(t) \in U(x(t))\} := F(x(t)) \tag{3}$$

The direction $x'(t)$ of the system is no longer given by a deterministic or a stochastic function, but by the membership in a set $F(x(t))$ of directions. The set-valued map $x \to F(x)$ is also called a correspondence. We have sidestepped specifying a probability law, of which we generally have no knowledge. As it pertains to the current question of maintenance of a living system, the point is less to predict its fate in terms of probability than it is to depend on at least one chance for success. That chance, if it belongs to the range of possibilities, even if it has never been exploited in the past, is enough for survival. Survival or maintenance are identified by a set of constraints K in which the state is to remain:

$$x(t) \in K \tag{4}$$

In continuous time, remaining in K means that the system whose direction $x'(t)$ belongs to $F(x)$ also enters the set K of constraints. This is possible if there exists a regulon u in the set U of admissible regulons such that the intersection of the set $F(x)$ of directions and the contingent cone at x to K, which extends the concept of tangent space to non smooth sets, is not empty:

Theorem 2.1 (Bebernes and Schuur, 1970) *For $F : Dom(F) \subset X \mapsto X$ an upper semicontinuous correspondence with convex compact values and such that:*

$$\sup_{y \in F(x)} \| y \| \leq c(\| x \| + 1) \tag{5}$$

for some real c, then K is a $\mathcal{C}$-viability domain for F if and only if

$$\forall x \in K,\ F(x) \cap T_K(x) \neq \emptyset \tag{6}$$

where $T_K(x) = \{v \in X$ such that $\liminf_{h \to 0} \frac{1}{h} d_K(x + hv) = 0\}$ *is the contingent cone to K at x.*

If K is not a $\mathcal{C}$-viability domain, is there a closed set $D \subset K$ which is a $\mathcal{C}$-viability domain? Is there also a maximal $\mathcal{C}$-viability domain? Aubin (1991) answered positively:

Theorem 2.2 (Aubin, 1991) *For $F : Dom(F) \subset X \mapsto X$ a upper semi-continuous correspondence with convex compact values and satisfying condition 5, there exists a largest closed $\mathcal{C}$-viability domain for F contained in K. This set, denoted $Viab_F(K)$, is called the* $\mathcal{C}$-Viability Kernel *of K for F. It can be empty.*

Theorem 2.1 is at the heart of the algorithm devised by Saint-Pierre (1994) to compute the $\mathcal{C}$-viability kernel, when F is also Lipschitz. As K is reduced to a discrete grid, this algorithm must be able to update every cell of the grid at each time, which is a daunting task explaining why it is limited in practice to three state dimensions. Bonneuil (2006) addressed the computation of viable states and viability kernel in large state dimension, with possible delay, in minimizing the cost to the set of constraints with stochastic algorithms.

The *regulation correspondence* associates the state $x \in \mathbb{R}^n$ with the set in $U \subset \mathbb{R}^p$ of all $\mathcal{C}$-viable fitness values at x—the values enabling the trajectory $x(.)$ starting from x that remain in K for any given time horizon; that is to say, that maintains the polymorphism above a given threshold ϵ. This regulation map, noted R_K is:

$$R_K(x) := \{u \in U \mid f(x, u) \in T_K(x)\} \tag{7}$$

To be viable in a state x, a viable regulon $u \in R_K(x) \subset U(x)$ must be taken. This procedure endogenizes the regulon, which now depends on the state x through the condition of the existence of at least one trajectory maintaining the system within K. The dependence is, in general, set-valued in the interior of K, sometimes point-to-point on the boundary of the $\mathcal{C}$-viability kernel. Hence, endogeneity results from $\mathcal{C}$-viability, whereas in modelling endogeneity is often postulated. Rather than guessing at a scenario and testing it against constraints, all scenarios are present at the state x through $u \in U(x)$—but only those scenarios for which $u \in R_K(x)$ "answer" the question of maintenance.

2.2. In Ecosystems

For a Lotka-Volterra model opposing prey of density x_1 and predator of density x_2:

$$\begin{cases} x_1'(t) &= x_1(t)(a - u(t)x_1(t) - (v(t) - u(t))x_2(t)) \\ x_2'(t) &= x_2(t)(-b + v(t)x_1(t)) \end{cases} \tag{8}$$

where a and b are parameters, u and v two regulons belonging to closed sets U and V, under constraints

$$x_1 \geq \underline{x}_1, \qquad x_2 \geq \underline{x}_2 \tag{9}$$

where $\underline{x}_1$ and $\underline{x}_2$ are thresholds, Figure 1 shows the $\mathcal{C}$-viability kernel of the prey associated with its set $K_1 = [\underline{x}_1, \infty)$ of constraints, the $\mathcal{C}$-viability kernel of the predator associated with its set $K_2 = [\underline{x}_2, \infty)$ of constraints, and the coexistence kernel associated with the set $K_1 \cap K_2$ of constraints (from each state of the coexistence kernel, there exists at least one trajectory along which both prey and predator have sufficiently high densities). Bonneuil and Müllers (1997) showed that this coexistence kernel is the intersection of the $\mathcal{C}$-viability kernels of prey and predator.

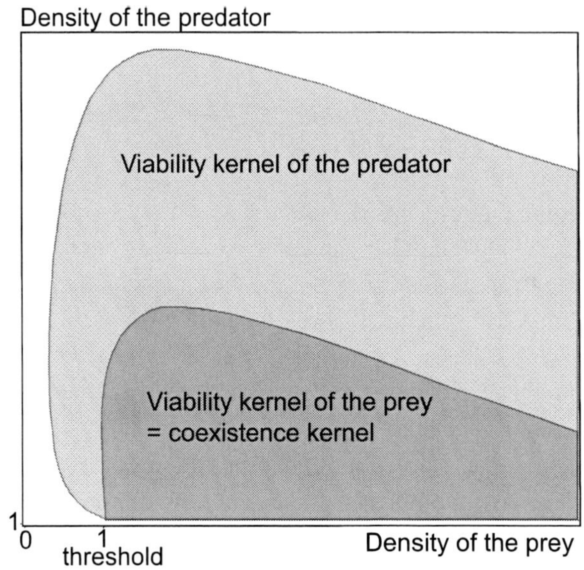

Figure 1. C-Viability kernels and coexistence kernels in prey-predator Lotka-Volterra model. *Source*: Bonneuil and Müllers, 1997.

In the two-species case, the boundaries of the $\mathcal{C}$-viability kernels were obtained as solutions of differential equations, rather than with one of the viability algorithms.

Let me give the context for the concept of $\mathcal{C}$-viability kernel in prey-predator case studies. The $\mathcal{C}$-viability kernel is a set of initial states, from which the question of the existence of a solution perpetuating the system arises. Specifically, Wilbur (1997) designed 144 experimental ponds with varying initial conditions of densities of newts as predators and tadpoles as prey. He followed the course of the system throughout a season. In the absence of predation, or as long as the tadpoles are of sufficient density, they develop a thin tail. Wilbur observed that when the group is on the verge of being totally devored, or when the tadpole density collapses, individuals decrease active foraging or develop morphological adaptations, such as changes in coloration and widening of tails, to accelerate growth by shortening of the larval period. These phenotypic changes enable the survival of the biggest tadpoles, of which there are then few. Tadpoles with wider tails swim faster to escape newt *Anax*, but in the absence of the predator have a higher mortality than tadpoles with narrower tails. Everything happens as stated in the $\mathcal{C}$-viability kernel: far from the boundary, when density is still high enough to satiate the predator, there is no need for anything to change. Near the boundary, changing strategies are necessary to rescue the tadpole population. These strategies (coloration or enlarging of the tail) constitute viable strategies. How likely are these strategies (which increase mortality in the short term!), I don't know, but the important fact is that they do exist and allow the survival of the species. They are the right strategies against predation and they appear at the right time, two characteristics of viable regulons. The rapidity of change displayed by the tadpoles under attack attests to the plasticity of their phenotypes, a property rendered by the regulons as simply measurable controls rather than continuous functions, and the fact that plasticity has certain limits is reflected by the fact that the set within which the regulons vary can be taken as bounded.

The two-dimensional prey-predator case was extended to a three-trophic food chain by Bonneuil and Saint-Pierre (2005). Figure 2 shows the coexistence kernel of prey, predator, and super-predator associated with $K_1 \cap K_2 \cap K_3$ where K_3 is the set of constraints defining the survival of the super-predator.

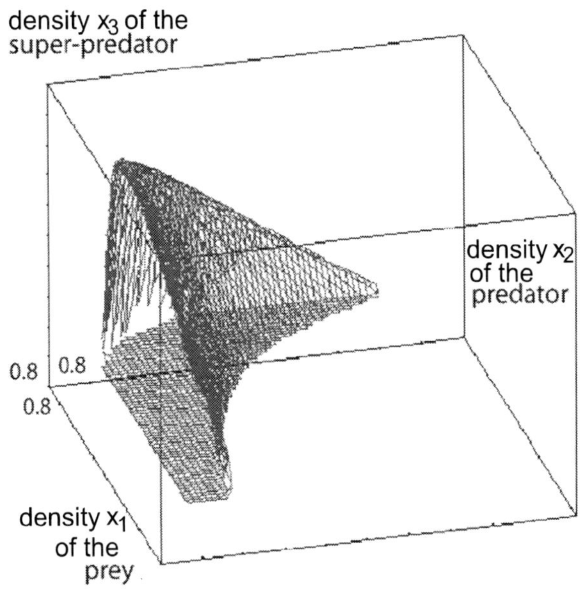

Figure 2. Coexistence kernel in the three-trophic food chain. *Source*: Bonneuil and Saint-Pierre, 2005.

Diverting one's attention from specific trajectories of the system to the benefit of sets of initial states classified with respect to their status of viable (i.e., there exists a trajectory remaining in the set of constraints until the time horizon T) or not viable (i.e., all trajectories leave the set of constraints before T) changes our understanding of eco-systems. The $\mathcal{C}$-viability kernel unifies sparse results from empirical observations. To name a few in the case of the thee-trophic food chain:

- The variation of the volume of the $\mathcal{C}$-viability kernel with respect to the strategies of the three protagonists (Figure 3 of Bonneuil and Saint-Pierre, 2005) emphasizes that coexistence requires that a sufficiently low capture rate is repeatedly taken by the super-predator, and that the capture rate of the predator cannot be too high: predator-prey systems are endangered when predators are too effective, devoring prey until they fall below the capacity to restock. The repetition of sufficiently low values need not be cyclic, and it depends on where the system travels within the C-viability kernel (should it remain viable). The traditional search for cyclic patterns in empirical systems is then finished by the search for repetitions.

- In addition to equilibria within the set of constraints, the C-viability kernel also contains all transient states from which coexistence is possible until T: adding nutrients in sufficient quantity shifts the initial state out of the C viability kernel, where the interactions u are no longer flexible enough to restore the possibility of remaining in the set of constraints. The C-viability kernel then addresses the rarefaction or the extinction of prey or predators resulting from oscillations of high magnitude, as tends to be the case in enriched systems, in the larger case of time varying strategies, and *in transient time* (not only at attractors). Moreover, the C-viability kernel helps identify viable strategies when the system is prone to high magnitude oscillations. If the disturbed system still remains in the C-viability kernel, the species continue to coexist (such as in McCauley and Murdoch, 1990, studying *Daphnia* feeding on algae). A sufficient exogenous addition of predator (or of super-predator) can push the state out of the C-viability kernel, implying the destruction of the system before too long. A case study is provided by predators *Didinium* and *Woodruffia*, which, when added in sufficient quantity and depending on the state of the system at the moment of the addition, can lead either to extinction (Morin and Lawler, 1995), or to coexistence for long periods (Salt, 1967; Luckinbill, 1973, 1974).

- The shape of the C-viability kernel on Figure 2 also implies that adding super-predator does not threaten the conservation of the three species as long as the initial state after the super-predator addition continues to lie within the C-viability kernel. This agrees with Morin and Lawler (1995), who found no consistent effects of omnivores on prey dynamics for two omnivore species feeding on two different prey. When an addition is too high, the initial state is shifted outside of the C-viability kernel and the system is doomed to constraint. Varying the capture rate of the super-predator alters the kernel but does not significantly change the projection onto the prey-predator plane: no nonviable state can be rendered viable by super-predator addition.

2.3. In Population Genetics

In population genetics, the central question of protected polymorphism is the genetic facet of maintenance of a dynamical system within constraints. For example, Feldman (1971) wondered "under what circumstances a stable polymorphism can be maintained" (p. 307). The answer was given in terms of "global asymptotic properties of its equilibrium sets using Lyapunov functions" (Rutschman, 1994, p.167), under five fixed fitness schemes. Equilibrium studies of the two-locus haploid model with recombination are useful indeed to "make statements concerning the time-dependent behavior of the gametic unbalance when the starting conditions are known" (Feldman, 1971, p. 299). Hastings (1981b) demonstrated stable cycling in discrete-time models. Kirzhner, Korol, and Ronin (1995) described "some regularities of the behavior of the sign D under environmental fluctuations in infinite panmictic populations with non-overlapping generations". These fluctuations were generated through a function intended to express specific fitness values and their variations with environment.

Protected polymorphism is a typical problem for which $\mathcal{C}$-viability theory is well-suited. It was discussed by Bonneuil and Saint-Pierre (2000) for unpredictably varying fitness values, as is the case in natural environments. Field experiments have shown that fitness in nature varies significantly, apart from genetic drift (Mueller, Barr and Ayala, 1985; Barker, East, and Weir, 1986; Allard, 1988). A good example is given by García-Martinez et al. (1998) although they work on a single locus. These authors pointed out that varying environment can play a role in protecting polymorphism: they found that two mitochondrial haplotypes, called I and II, of *Drosophila subobscura* are neutral in nature (estimated at x_1 =39.2% and x_2 =56.9% respectively), whereas population cage experiments always led to fixation of haplotype II. They acknowledged that "population cages do not constitute a natural environment" and "that factors promoting positive selection of haplotype II in population cages could be counterbalanced by other factors under natural conditions" (p. 1380) (they first rejected random drift as the force responsible for the observed changes). If the coexistence of haplotypes I and II has been maintained in nature over a certain threshold since y_0 years, then the initial condition (x_1, x_2) is $\mathcal{C}$-viable with time horizon t_0. In population cage experiments, fixing fitness values at constant values drives the distribution of alleles out of the constraint set to an equilibrium located at the edge of the domain of variation. If experimenters wished to maintain the polymorphism of haplotypes I and II over a certain level they would have to alter the environmental conditions prevailing in the cages, and hence the fitnesses, in agreement with the regulation map $u(t) \in R_K(x(t))$ associated with the $\mathcal{C}$-viability kernel.

In this case of two alleles (A and a at one locus, B and b at the other one), time is discrete, denoted by n, the recurrence is given by G, and the $\mathcal{C}$-viability problem is:

$$x^{(n+1)} \in G(x^{(n)}) := \{g(x^{(n)}, u^{(n)}), u^{(n)} \in U\} \quad \text{and} \quad \forall n, x^{(n)} \in K \tag{10}$$

for K a closed set and $G : X \to X$ a locally Lipschitz correspondence with non-empty compact convex values.

Without any assumption about probability distributions or endogenous effects, focussing on change through time rather than dealing with asymptotic properties, $\mathcal{C}$-viability theory provides the conditions for protected polymorphism in an unpredictable environment, or, alternatively, a rule for experimenters to change the environment so as to safeguard a given polymorphism for a given time horizon.

Computations of the $\mathcal{C}$-viability kernel with varying U taken as independent of x emphasizes that in order to remain in K, a trajectory in $x = (x_A, x_a, x_B)$ with $x_b = 1 - x_A - x_a - x_B$ must repeatedly have ab with a fitness close enough to that of AB. "Repeatedly" is not necessarily the same as "regularly"; it depends on where the state travels in the $\mathcal{C}$-viability kernel.

Polymorphism maintained under the joint action of migration and selection has been often studied with respect to equilibria (Bulmer, 1972; Karlin, 1982; Nagylaki, 1992, ch. 6; and references therein for the single diallelic locus; Karlin and Richter-Dyn, 1976; Karlin, 1977; Nagylaki and Lou, 2001, 2007 for multi-allelic patterns). Multiallelic polymorphism maintained by migration and selection involves a large state dimension (equal to the total number of demes times the total number of alleles frequencies minus one) and a large

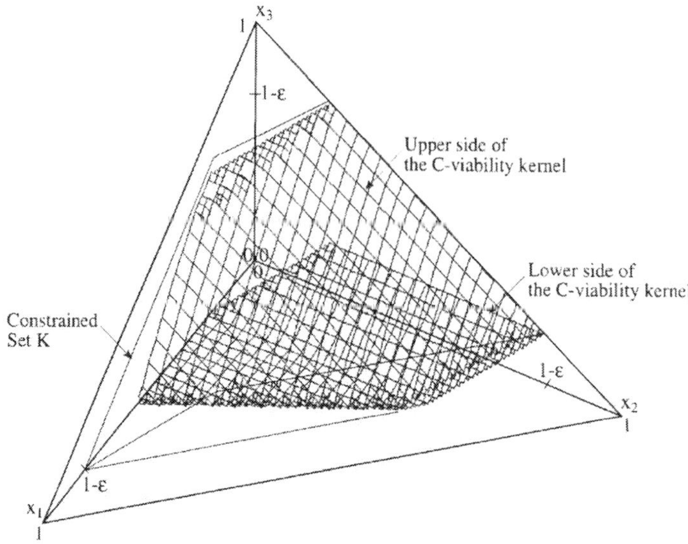

Figure 3. Discrete-time $\mathcal{C}$-viability kernel for fixed recombination rate $r = 0.25$, and all alleles A, B, a, and b over ϵ =20%. *Source*: Bonneuil and Saint-Pierre, 2000.

dimension of regulons (migration and fitness values). Bonneuil (forthcoming) used the $\mathcal{C}$-viability algorithm in large state dimension (Bonneuil, 2006) to delineate $\mathcal{C}$-viability kernels addressing the question of the maintenance of polymorphism in spatial subdivisions with unpredictable fitness values and migration rates.

2.4. In Social Networks

Especially given the importance of the concept of network in eco-systems, much work remains to be done with regard to $\mathcal{C}$-viability in this area of maintenance (in the context of the social sciences, Bonneuil (2000) has characterized viable regulons as connection matrices in a social network case study of 92 nodes). No matter how a network changes, the question that needs to be addressed is this: for $x(t) \in I\!R^n$ the state of the system at time t, what are the measurable connection matrices $W(t) \in \mathcal{L}(I\!R^n, I\!R^n)$ such that there exists at least one solution $x(.)$ to:

$$x'(t) = W(t)g(x(t)) \tag{11}$$

starting from $x(0)$ at $t = 0$ satisfying a collective constraint

$$\forall t \geq 0, h(x(t)) \in M \tag{12}$$

where g and h are functions and M a closed set? This formulation can be generalized to a time-varying set $M(t)$ with no additional difficulty.

Aubin (1997: 159) showed that the regulation map can be written as:

$$R_M(x(t)) = \{W \in \mathcal{L}(I\!R^n, I\!R^n) \mid h'(x(t))^* W g(x(t)) \in T_M(h(x(t)))\} \tag{13}$$

where '*' denotes transposition.

Bonneuil (2000) showed the tight relationship between $\mathcal{C}$-viability of the connection matrix and centrality of the associated graph. Centrality was the property advocated by scholars as the key factor bestowing dominance to one of the nodes, the Medici family. However, what matters for agents taking part in the network is not centrality but dominance itself, a qualitative property which identifies a closed set in which the system must remain. $\mathcal{C}$-viability conditions allow the determination of the whole set of connection matrices that enable the maintenance of the dominance of the Medici family. The historical fact is less the observed network with its particular individual ties than the property that this network was $\mathcal{C}$-viable with regard to the dominance of the specific node. Should a tie to the Medici be broken, the Medici could have established a tie with another node ensuring the same dominance over Florentine society.

2.5. Persistence, Permanence and C-viability

What are the relationships between persistence, permanence, and $\mathcal{C}$-viability? I addressed this question in Bonneuil and Saint-Pierre (2005), which I present here again considering the importance of the concepts of persistence and permanence in current literature. Aubin (1991: 128; 1997: 240) showed that permanence is a specific case involving $\mathcal{C}$-viability: consider a closed set K containing a set Ω in the phase space, such that

$$\bar{\Omega} \subset \mathrm{Int}(K) \quad \text{and} \quad \partial K = \mathrm{Viab}_F(K \setminus \Omega) \tag{14}$$

where $\mathrm{Viab}_F(M)$ denotes the $\mathcal{C}$-viability kernel of the set M under dynamics F, which is the largest set of initial conditions from which there exists at least one solution of $x' \in F(x)$ remaining in M forever. The first condition opens a window of opportunity for the existence of asymptotic states without extinction of any species. The second condition reflects the fact that the system can no longer escape the boundary $x_i = 0,\ i = 1, 2$ or 3 once it is attained: there is no way to recover an extinct species. Permanence occurs in the special case when Ω is *invariant*, that is to say *all* solutions starting from Ω remain in Ω. What is important is that the boundary is not attained from a solution starting from Ω, a property which was simplified, in the statement of permanence, into the existence of a $\delta > 0$ such that $\delta < \liminf_{t\to\infty} x_i(t)$: the system eventually is bounded away from extinction with a "security" ring of width δ. The system can remain under the threshold δ for a long time before the interesting property becomes relevant. In contrast, in $\mathcal{C}$-viability, the question is whether the constraints can be satisfied at each time.

Permanence is a more stringent property than strong persistence, which requires $\delta = 0$, and persistence which requires $0 < \limsup_{t\to\infty} x_i(t)$ (Hofbauer and Sigmund, 1988: 161). As $\liminf \leq \limsup$, permanence implies persistence. However, the concept of $\limsup$ is hard to generalize outside the simplex. Persistence without strong persistence is a key characteristic, meaning that an almost extinct x_i will reappear in the future. This characteristic is innate to the Lotka-Volterra model (Jansen and Sigmund, 1998). Local stability says that there exists a neighborhood V_1, such that for any neighborhood $V_2 \subset V_1$, there exists a neighborhood V_3 such that any solution starting from V_3 returns to V_2. This asymptotic property is also characteristic of the simplex and hard to generalize in general sets.

3. Acquisition

In population genetics, Bonneuil and Saint-Pierre (2002) addressed the concomitant question of the possibility of moving or returning to the set defining polymorphism by the play of natural selection in the one-locus two-allele model: how long will an allele need to recover or, alternatively, when is a deleterious allele doomed to extinction?

For example, the removal of deleterious mutants by pure selection is interesting in the Major Histocompatibility Complex (Meyer and Thomson, 2001): Poland (1998) reviewed several studies showing that human individuals who did not respond to vaccines lacked DRB1*13 and had an excess of DRB1*07 alleles when compared with controls. Rinchik (2000) explored the possibilities of maintaining a mutant stock once one is created. The majority of recessive mutations detrimental to health or to reproductive capacity is hard to maintain. Jones and Probert (1980) advocated habitat selection when contrasting the disappearance in 30 weeks of white-eyed (w) mutants in *Drosophila simulans* in a uniform habitat with the maintenance of w in a heterogenous environment. Barker and East (1980) perturbed the frequencies of alleles at three electrophoretic loci on different chromosomes in cactophilic *Drosophila buzzatii* by releasing adults and injecting larvae into cactus. The release of individuals homozygous for three alleles first results in a large increase of the three alleles, then a rapid decrease of the allelic frequencies to pre-experimental levels.

On the side of theoretical modelling, the one-locus two-allele model was considered by Hadeler and Liberman (1975) in pointing out the existence of polymorphic equilibria, by Clark and Feldman (1986), Christiansen (1989), Hastings (1981a) in studying protected polymorphism, by Doebeli and de Jong (1998) in revealing the existence of stable cycles, or by Gavrilets (1998) in showing "that interactions of maternal and parental selection can result in stable oscillations of genotype frequencies" (:1147). Bonneuil and Saint-Pierre (2002) re-examined this model when fertilities are capable of varying in a bounded but unspecified manner, first from the point of view of protected polymorphism and then the minimal total number of generations spent out of a given set of allele frequencies defining polymorphism.

Fox and Hastings (1992) could "identify the minimum amount of selection necessary to generate any given distribution of allele and genotype frequencies, given any specified initial distribution of these frequencies" (p. 278), without "requir[ing] assumptions that the system is at equilibrium or that selection is constant, or any other knowledge about the form or strength of selection" (p. 278). These authors addressed the specific question of finding the minimal time route leading from a given state $X_0 \in [0,1]^3$ of genotype frequencies to a given state $X_1 \in [0,1]^3$. They studied a problem of "controllability" where they could ignore any additional constraint other than $x(t) \in [0,1]^3$. Bonneuil and Saint-Pierre (2002) expanded the property of remaining in a given closed set K of constraints ($x(t) \in K$) and on the notion of minimal time spent out of these constraints. Crossing the state constraints K defining the desired polymorphism does not necessarily mean that the system collapses, but that polymorphism is relatively "poor": polymorphism can be said to be in "crisis", so the question arises of how long it will need to recover.

The set of constraints combining sufficient density of alleles A and a present in genotypes AA, Aa, or aa is:

$$K_\varepsilon := K_A \cap K_a = \{x \in [0,1]^2 \mid \varepsilon \leq 2x_{AA} + x_{Aa} \leq 2 - \varepsilon\} \tag{15}$$

where x_{ij} is the frequency of genotype ij. The recurrent inclusion according to which the system varies extends Hadeler and Liberman (1975) to the case of set-valued fertilities $u^{(n)} \in U \subset I\!R^4$ of alleles varying in closed sets, in order to reflect unpredictability:

$$x^{(n+1)} \in G(x^{(n)}) \tag{16}$$

where n denotes the number of the generation.

The total number $\tau_{K_\varepsilon,G}$ of generations spent out of K_ε is the key concept as regards the loss of polymorphism: the system may become polymorphic from time to time, yet later one allele can again rarefy. The relevant question is how long will the sojourn out of K_ε last. The $\mathcal{C}$-viability kernel corresponds to those initial states where the time out of K_ε is zero. When no trajectory starting from a state x is able to return to K_ε, the return time is infinite.

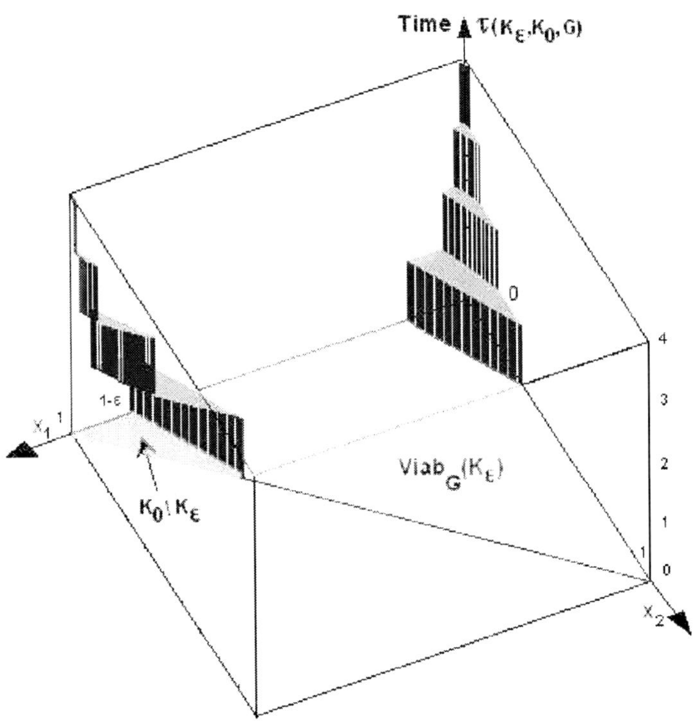

Figure 4. Level surfaces of the minimal total number of generations $\tau_{K_\varepsilon,G}$ to be spent out of constraints K_ε. The level surface for $\tau_{K_\varepsilon,G} = 0$ is also the $\mathcal{C}$-viability kernel of K_ε under G: from each of its states, there exists a trajectory along which the system remains K_ε-polymorphic ($x \in K_\varepsilon$) forever. In this particular case, $\mathrm{Viab}_G(K_\varepsilon) = K_\epsilon$. $\varepsilon = 0.3$, $K_0 = [0,1]^2$. *Source*: Bonneuil and Saint-Pierre, 2002.

Figure 4 presents an example of level surfaces of the minimal total number of generations spent out of K_ε under G. The discrete-time $\mathcal{C}$-viability kernel of K_ε under G is exactly the level surface for $\tau_{K_\varepsilon,G} = 0$. From each state of the level surface n, the minimal total number of generations spent out of K_ε is n. The *regulation map out of* K_ε $\tilde{R}_K$ associates the state x to the set $\tilde{R}_{K_\varepsilon}(x)$ of the fertilities minimizing the total number of generations spent out of polymorphism. On the boundary of K_ε, and if $R_{K_\varepsilon}(x)$ is empty, there are no $\mathcal{C}$-viable fertilities at x and the system leaves the set K_ε of state constraints at the next generation. For $x^{(n)}$ outside of K_ε, if $\tilde{R}_K(x^{(n)})$ is not empty, a $\mathcal{C}$-viable quadruplet $u^{(n)}$ of fertilities, for which $x^{(n+1)}$ has its total number of generations out of K_ε decrease compared to $x^{(n)}$, is necessarily taken from the set $\tilde{R}_{K_\varepsilon}(x^{(n)})$. For $x^{(n)}$ in $K_\varepsilon \backslash \text{Viab}_G(K_\varepsilon)$, if $R_{K_\varepsilon}(x^{(n)})$ is not empty, a $\mathcal{C}$-viable quadruplet $u^{(n)}$ of fertilities has to be picked up from the set $\tilde{R}_{K_\varepsilon}(x^{(n)})$ in order for $x^{(n+1)}$ to have its total number of generations out of K_ε not increase compared to $x^{(n)}$.

One result is that maintaining a chance of regaining polymorphism demands that matings involving heterozygotes repeatedly have sufficiently high fertility, where "repeatedly" depends on the level surfaces and the set K to be attained.

In population genetics, another case study of acquisition is given in Bonneuil (1998). Samples of DNA mitochondrial sequences do not reflect a single unique set of circumstances, due to the variance associated with the mean coalescent time, as Slatkin and Hudson (1991) suspected. However, not all population paths are possible. By recognizing that the mean number of pairwise nucleotide differences is related to the coalescent process as a control problem, Bonneuil (1998) characterized all the possible population paths fulfilling the constraint of a given mean coalescence time. I calculated a 95% confidence interval for the true mean, deduced the capture domain of this confidence interval, and determined the regulation map associating a set of possible population sizes to any state $x(t)$.

Acquiring a property, or attaining a target in finite time T, with controls $u \in U$, where U is a closed set, is expressed as:

$$\left\{ \begin{array}{lcl} x'(t) & \in & F(x(t)) \quad t \in [0,T] \\ x(T) & \in & \Omega \end{array} \right. \tag{17}$$

In the specific case of coalescence, fitting a confidence interval Ω of the coalescence time in a n-gene sample by the coalescent process is well-approximated in continuous time by a two-dimensional problem for $x(t)$ based on $x_1(t)$, the probability that two genes have not a common ancestor t generations ago, and $x_2(t)$, an auxiliary variable:

$$\left\{ \begin{array}{rcl} \lim_{t\to\infty} x_2(t) & \in & \Omega \\ x_1'(t) & = & -u(t)x_1(t) \\ x_2'(t) & = & x_1(t) \\ x_1(t) & \in & [0,1] \\ x_2(t) & \geq & 0 \qquad x_2(0) = 0 \\ u(t) & \in & U := [\frac{1}{N^{\min}}, \frac{1}{N^{\max}}] \end{array} \right. \tag{18}$$

where $N^{\min}$ and $N^{\max}$ are the two extremes of plausible population sizes.

Solving this system yields *the capture domain* represented on Figure 5 of all states visited by all temporal paths $x(.)$ such that the expected coalescence time is in the confidence interval Ω. Pairwise nucleotide differences need not reflect a single unique scenario. Exponential growth scenarios have been proposed to produce the observed data, but they are neither unique nor are then even particularly restrictive.

In anthropology, distributions of human skeletons by age cumulated over a deposition period T have been collected; for example, St. Thomas Anglican Church from 1821 to 1874

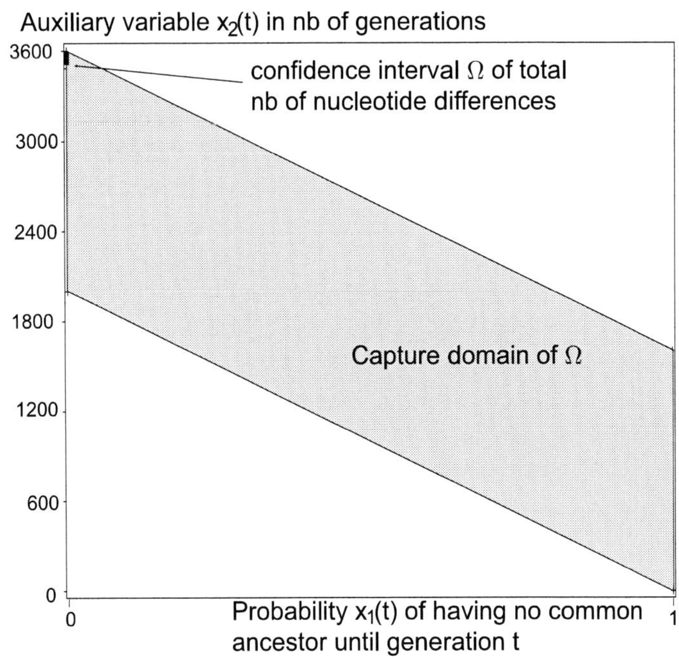

Figure 5. The 'capture domain of the confidence interval containing at least 95% of all the temporal paths leading to the total number of nucleotide differences in the sample of 69 Sardinians mtDNA of DiRienzo and Wilson (1991). *Source*: Bonneuil (1998)

($T = 55$ years) and the Dallas Freedman's Cemeteries from 1869 to 1907 (T =38 years). The complete solution forms the capture basin of a target comprising the distribution of deaths by age cumulated over the deposition time and of the deposition time T itself (since time is a variable of the problem). The dynamic is given by Lotka-McKendrick equations, the regulons are the T life expectancies and the T fertilities; the variables are the age classes of the age pyramids. The solutions that fit the data travel through the capture basin—and they are many. Bonneuil (2005) chose the solution closest to the stable population path, which, in these two case studies, does not belong to the capture basin.

In cancer immunotherapy, Kassara (2005) used viable dosages of effector cells and the therapy zone to expand the tumor-free cell zones to the entire tissue. The dynamic is given by four semi-linear partial differential equations. The target is then to obtain a solution where the entire tissue is free of tumor, and the C-viability problem is to keep the patient healthy. Once again, the problem is to travel within the capture-viability domain, which is the guideline for the protocol. The specific trajectory taken by the system is of secondary importance, because the reaction of the tissue is variable and unpredictable.

4. Maintenance, Acquisition, and Viability Multipliers

Thus far, existing models have been altered to make interactions or strategies set-valued rather than stochastic or deterministic; and in tackling the questions of maintenance or acquisition, rather than interpreting them in terms of limit sets, an approach which ignores transient states and obscures constraints.

However there is no reason why an arbitrary subset under a controlled dynamic should include a nonempty $\mathcal{C}$-viability kernel for given dynamics and constraints. For example, in the absence of constraints, many species would grow exponentially without restraint. Instead of combining a priori physiological growth with counter-effects imposed by interactions with other species and self-limitation ("community ecology has a long tradition of conjectures" (Morin and Lawler, 1995: 511)), the idea is rather to adjust a dynamical system to make it viable, "(i.e. having solutions which do satisfy the constraints), knowing the dynamical behavior of the system without the state constraints" (Křivan, 1991. 379). Because constraints and dynamics are concepts of different status, resource constraints need not be satisfied for any choice of parameters.

If a subset K is not viable under a differential equation $x'(t) \in F(x(t))$, then the additive regulation of the dynamic is obtained by the corrected system:

$$x'(t) \in F(x(t)) - p(t) \tag{19}$$

with

$$p(t) \in F(x(t)) - \overline{co}(T_K(x(t))) \tag{20}$$

where $\overline{co}(T_K(x(t)))$ is the convex hull of the contingent cone $T_K(x(t))$ introduced in Theorem 2.1.

The multiplicative correction is obtained by the corrected dynamic:

$$x'(t) \in W(t)F(x(t)) \tag{21}$$

where W is a $\mathcal{L}(X, X)$ matrix such that:

$$W(t)F(x(t)) \in T_K(x(t)) \,. \tag{22}$$

The $\mathcal{C}$-viability perspective is to identify all viable processes with respect to the known constraints K.

The one-dimensional case of one species feeding on resources is presented in Aubin (1991: 39). In simple terms, Aubin showed that one viable additive correction to the system:

$$\left\{ \begin{array}{lcl} x'(t) & = & x(t) \\ x(t) & \in & K = [x_{\min}, x_{\max}] \end{array} \right. \tag{23}$$

is $p(t) \in N_K(x(t))$ such that the system is corrected to be:

$$\left\{ \begin{array}{lcl} x'(t) & = & x(t) - p(t) \\ x(t) & \in & K = [x_{\min}, x_{\max}] \,. \end{array} \right. \tag{24}$$

The Verhulst correction:

$$p(t) = -\frac{x(t)^2}{x_{\max}} \tag{25}$$

is viable. Elegance guaranteed fame to the Verhulst solution, but it is not the only viable solution of Eq. (24).

In the prey-predator system, the interaction term produces cycling solutions that mimic observations, but it does not stem from the need for the species to survive. On the contrary, *C--viability multipliers* reconcile intrinsic growth and constraints during model-building. In doing so, they extend the method of Lagrange multipliers to non-smooth sets of state constraints. Bonneuil (2003) presented additive and multiplicative corrections of intrinsic growth models. Certain values either of p for the additive correction or of W of the multiplicative correction correspond to Lotka-Volterra or Holling-type (1965) models. However, for certain parameters, Lotka-Volterra or Holling-type models are not viable, and the solutions they generate cross the thresholds of sufficient density. Hence, to sum up, Lotka-Volterra or Holling-type solutions are viable for certain choices of parameters, but not for all; conversely, many other viable processes do not enter into these stylized models.

Even if modelers have not implemented $\mathcal{C}$-viability multipliers, they have long been used by experimenters, as shown in Bonneuil (2003). For example, Gause (1935) confronted prey *Paramecium caudatum* with predator *Didinium nasutum* in a culture tube. The predator soon devored the prey entirely, after which the predator eventually starved. In another experiment, Gause added oat sediment to the culture tube so the paramoecia could find protection, leading the predator to extinction. Gause then experimentally corrected the intrinsic dynamics in weakening the predation rate. In a third experiment, Gause repeatedly introduced new paramoecia into the culture tube containing the paramoecia and the oats sediment whenever the predator *Didinium nasutum* was on the verge of extinction. Both species oscillated. Bonneuil (2003) showed that these immigrations of paramoecia into the milieu correspond exactly to additive viability multipliers. Co-existence is not given by an elegant endogenous mechanism, but results from viable corrections, either inscribed in natural mechanisms, in environmental fluctuations (Schwerdtfeger, 1958), or being brought exogenously.

A second example is due to Luckinbill (1973), who implemented a subtle control in adding methyl cellulose into the culture containing predator *Didinium nasutum* and prey *Paramecium aurelia*, itself feeding on *Aerobacter aerogenes*. Methyl cellulose has the "important effect of reducing the rate of movement of both *P. aurelia* and *D. nasutum* without giving the prey a dispersal advantage" (:1322). The ensuing reduced frequency of encounter between predator and prey enhances the ability of both species to persist, provided that the availability of nutrient is fine-tuned. It is necessary "both that predators be unable to capture all the prey in the system and that some restriction of prey growth exists in addition to predation" (:1326).

A third example was examined by Pimentel and Stone (1968), who introduced parasite *Nasonia vitripennis* to the host *Musca Domestica*. Initial large amplitude fluctuations – attaining up to 400 individuals for each species– eventually collapse to bring both species to the edge of extinction. The survival condition of the host at its threshold condition is not satisfied when the two insects are newly introduced. When the insects have cohabited for two years, the host multiplied to stabilize around 500 individuals, and the parasite oscillated around approximatively 70 individuals. The host had acquired an adaptive resistance to its parasite. Bonneuil (2003) showed that this genetic mutation brings a multiplicative viable correction to the population growth rate of the parasite and subsequently the decreasing fertility of the parasite —from 133 offspring per female among newly associated species versus only 46 among species having cohabited for two years (Pimentel and Stone, 1968).

5. Conclusion

Bare mathematical translation of maintenance or acquisition to apply in transient time requires balancing intrinsic dynamics and constraints, in addition to classical selective arguments. Nagumo (1942) pioneered the use of differential equation models needed to remain within a closed set K. The extension of the concept of tangent space to non smooth sets by Bouligand (1932) allows us to study biological situations often characterized by inequalities. Set-valued analysis was combined with the question of maintenance theoretically by Aubin (1991), and his concept of $\mathcal{C}$-viability kernel sheds light on biological dynamics. I revisited key models of ecosystems and population genetics in advocating the power of these concepts. I showed that field experimenters and observers had come to address the same issues of perpetuation of a living system in transient time, the diversity of initial conditions, the recovery of bio-diversity, and the necessity and potential to alter a system to provide for its continuation. $\mathcal{C}$-Viability theory is a promising candidate to renew the mathematics of maintenance and acquisition.

References

[1] Allard, R. W. (1988). Genetic changes associated with the evolution of adaptedness in cultivated plants and their wild progenitors. *Journal of Heredity,* **79**, 225-238.

[2] Aubin, J.-P. (1991). *Viability Theory.* Boston: Birkhäuser.

[3] Aubin, J.-P. (1997). *Dynamic Economic Theory: A Viability Approach.* Boston, Basel: Springer-Verlag Birkhäuser.

[4] Aubin, J.-P., Bayen, A., Bonneuil, N., & Saint-Pierre, P. (forthcoming). *Viability, Games, Control.* Berlin: Springer Verlag.

[5] Barker, J. S. F., East, P. D. & Weir B. S. (1986) Temporal and microgeographic variation in allozyme frequencies in a natural population of *Drosophila buzzatii. Genetics*, **112**, 577-611.

[6] Barker, J. S. F. & East, P. D. (1980) Evidence for selection following perturbation of allozyme frequencies in a natural population of *Drosophila . Nature*, **284**, 166-68.

[7] Bebernes J. W., & Schuur J. D. (1970). The Wazewski topological method for contingent equations. *Annali di Matematica Pura et Applicata,* **87**, 271-280.

[8] Bonneuil N. (1998). Population paths implied by the mean number of pairwise nucleotide differences among mitochondrial DNA sequences. *Annals of Human Genetics*, **62**, 61-73.

[9] Bonneuil N. (2000). Viability in Dynamic Social Networks. *Journal of Mathematical Sociology,* **24**(3), 175-182.

[10] Bonneuil N. (2003). Making Ecosystem Models Viable. *Bulletin of Mathematical Biology,* **65**, 1081-1094.

[11] Bonneuil N. (2005). Fitting to a distribution of deaths by age with application to paleodemography. *Current Anthropology,* **46**, 29-45.

[12] Bonneuil N. (2006). Computing the Viability Kernel in Large State Dimension. *Journal of Mathematical Analysis and Applications,* **323** (2), 1444-1454.

[13] Bonneuil N. & Saint-Pierre P. (2000). Protected polymorphism in the two-locus haploid model with unpredictable fitnesses. *Journal of Mathematical Biology*, **40**(3), 251-277.

[14] Bonneuil N. & Saint-Pierre P. (2002). Minimal Number of Generations out of Polymorphism in the One-Locus Two-Allele Model with Unpredictable Fertilities. *Journal of Mathematical Biology*, **44-6**, 503-522.

[15] Bonneuil N., & P. Saint-Pierre (2005). Population Viability in Three Trophic-level Food Chains. *Applied Mathematics and Computation*, **169/2**, 1086-1105.

[16] Bonneuil, N., & Müllers, K. (1997). Viable populations in prey-predator systems. *J. Math. Biol.,* **35**, 261-293.

[17] Bouligand G. (1932). Sur la semi-continuité d'inclusions et quelques sujets connexes. *Enseignements mathématiques,* **31**, 14-22.

[18] Brown, J. S., & Vincent, T. L. (1985). The evolutionary game. *Theor. Pop. Biol.*, **31**, 140-166.

[19] Brown, J. S., & Vincent, T. L. (1987). Coevolution as an evolutionary game. *Evolution*, **41-1**, 66-79.

[20] Brown, J. S., & Vincent, T. L. (1992). Organization as an evolutionary game. *Evolution*, **46-5**, 1269-1283.

[21] Buchanan, M. (2007). The best is yet to come. *Nature*, **447**, 39.

[22] Bulmer,M.G. (1972). Multiple niche polymorphism. *Am. Naturalist,* **106**, 254-257.

[23] Chesson, P. (2000). Mechanisms of maintenance of species diversity. *Annu. Rev. Ecol. Syst,* **31**, 343-66.

[24] Christiansen, F. B. (1989). The Multiple-Locus Symmetric Fertility *Model. Theor. Pop. Biol.*, **35-3**, 337-362.

[25] Clark, A. G., & Feldman, M. W. (1986). A numerical simulation of a one-locus, multiple-allele fertility model. *Genetics*, **113**, 161-176.

[26] DiRienzo, A., & Wilson, A.C. (1991). The pattern of mitochondrial DNA variation is consistent with an early expansion of the human population. *Proc. Natl. Acad. Sci. USA*, **88**, 1597-1601.

[27] Doebeli, M., & de Jong, G. (1998). A simple genetic model with non-equilibrium dynamics. *J. Math. Biol.*, **36**, 550-556.

[28] Feldman, M. W. (1971). Equilibrium studies of two-locus haploid populations with recombination. *Theor. Pop. Bio.*, **2**, 299-318.

[29] Fox, G. A., & Hastings, A. (1992). Inferring selective history from multilocus frequency data: Wright meets the Hamiltonian. *Genetics*, **132**, 277-288.

[30] García-Martinez, J., Castro, J. A., Ramón M., Latorre A., & Moya A. (1998). Mitochondrial DNA haplotype frequencies in natural and experimental populations of Drosophila subobscura. *Genetics*, **149**, 1377-1382.

[31] Gause, G. F. (1935). Vérification expérimentale de la théorie mathématique de la lutte pour la vie. *Actual. Sci. Ind.,* **277**. Paris: Hermann.

[32] Gavrilets, S. (1998). One-locus two-allele models with maternal (parental) selection. *Genetics*, **149**, 1147-1152.

[33] Gee, H. (2002). Aspirational thinking. *Nature*, **420**, 611.

[34] Giesel, J. T. (1976). Reproductive strategies as adaptations to life in temporally heterogeneous environments. *Ann. Rev. Ecol. Syst.*, **7**, 57-79.

[35] Hadeler, K. P., & Liberman, U. (1975). Selection models with fertility differences. *J. Math. Biol.,* **2**, 19-32.

[36] Hassell, M. P. (2000). Host-Parasitoid population dynamics. *Journal of Animal Ecology,* **69**, 543-566.

[37] Hastings, A. (1988). Food web theory and stability. *Ecology*, **69**, 1665-1668.

[38] Hastings, A. (1981a). Disequilibrium, selection, and recombination: limits in two-locus, two-allele models. *Genetics*, **98**, 659-668.

[39] Hastings, A. (1981b). Stable cycling in discrete-time genetic models. *Proc. Natl. Acad. Sci.*, **78**, 7224-7225.

[40] Hickman, J. C. (1975). Environmental unpredictability and plastic energy allocation strategies in the annual *Polygonum cascadense* (Polygonaceae). *J. Ecol.,* **63**, 681-701.

[41] Hofbauer, J., & Sigmund, K. (1988). *The theory of evolution and dynamical systems.* Cambridge: Cambridge University Press.

[42] Hofbauer, J., & Sigmund, K. (1998). *Evolutionary Games and Population Dynamics.* Cambridge: Cambridge University Press.

[43] Holling, C. S. (1965). The functional response of predators to prey density and its role in mimicry and population regulation. *Mam. Entomol. Soc. Can.*, **45**, 5-60.

[44] Huffaker, C. B. (1958). Experimental studies on predation, dispersion factors and predator-prey oscillations. *Hilgardia*, **27**, 343-83.

[45] Jansen, V. A. A., & Sigmund, K. (1998). Shaken not stirred, on permanence in ecological communities. *Theor. Popul. Biol.*, **54**, 195-201.

[46] Jones, J. S., & Probert, R. F. (1980). Habitat selection maintains a deleterious allele in a heterogenous environment. *Nature*, **287**, 632-33.

[47] Karlin S. (1982). Classifications of Selection-Migration Structures and Conditions for a Protected Polymorphism, in Max K. Hecht, Bruce Wallace and Ghillean T. Prance (eds), *Evolutionary Biology,* **14** (61-204). Chicago: Plenum Publishing Corporation.

[48] Karlin S. (1977). Gene frequency patterns in the Levene subdivided population model. *Theor. Pop. Biol.,* **11**, 356-385.

[49] Karlin S., & Richter-Dyn, N. (1976). Some theoretical analyses of migration selection interaction in a cline: A generalized two range environment, in S. Karlin and E. Nevo (eds), *Proc. Int. Conf. Pop. Genet. Ecol.* (659-706). New York: Academic Press.

[50] Kassara, K. (2005). Feedback spreading control applied to immunotherapy. *Mathematical Population Studies,* **12-4**, 211-222.

[51] Kirzhner, V. M., Korol, A. B., & Ronin, Y. I. (1995). The dynamics of linkage disequilibrium under temporal environmental fluctuations: two-locus selection. *Theor. Pop. Biol.*, **47**, 257-276.

[52] Křivan, V. (1991). Construction of population growth equations in the presence of viability constraints. *J. Math. Biology,* **29**, 379-387.

[53] Law, R., & Morton, D. (1996). Permanence and the assembly of ecological communities. *Ecology*, **77-3**, 762-775.

[54] Luckinbill, L. S. (1973). Coexistence in laboratory populations of *Paramecium aurelia* and its predator *Didinium nasutum*. *Ecology*, **54**, 1320-27.

[55] Luckinbill, L. S. (1974). The effects of space and enrichment on a predator-prey system. *Ecology*, **60**, 1098-102.

[56] McCauley, E., & Murdoch, W. W. (1990). Predator-prey dynamics in environments rich and poor in nutrients. *Nature*, **343**, 455-57.

[57] McCollum, S. A. (1993). *Ecological consequences of predation-induced polyphenism in larval hylid frogs*. Dissertation. Duke University, Durham, North Carolina, USA.

[58] Meyer D., & Thomson, G. (2001). How selection shapes variation of the human major histocompatibility complex: a review, *Annals of Human Genetics,* **65**, 1-26.

[59] Mitchell, P., Arthur, W., & Farrow, M. (1992). An investigation of population limitation using factorial experiments. *J. of Animal Ecology*, **61**, 591-98.

[60] Mittelbach, G., Turner, A., Hall, D., Rettig, J., & Osenberg, C. (1995). Perturbation and resilience, a long-term, whole-lake study of predator extinction and reintroduction. *Ecology*, **76-8**, 2347-2360.

[61] Morin, P. J., & Lawler, S. P. (1995). Food web architecture and population dynamics, Theory and empirical evidence. *Ann. Rev. Ecol. Syst.*, **26**, 505-529.

[62] Mueller, L. D., Barr, L. G., & Ayala, F. J. (1985). Natural selection vs. random drift: evidence from temporal variation in allele frequencies in nature. *Genetics*, **111**, 517-554.

[63] Nagumo, M. (1942). Über die Lage der integralkurven gewöhnlicher Differentialgleichungen. *Proc. Phys. Math. Soc. Japan,* **24**, 551-559.

[64] Nagylaki T., & Lou Y. (2007). Evolution at a multiallelic locus under migration and uniform selection. *Journal of Mathematical Biology,* **54**, 787-796.

[65] Nagylaki T., & Lou Y. (2001). Patterns of Multiallelic Polymorphism maintained by Migration and Selection. *Theor. Pop. Biol.*, **59**, 297-313.

[66] Nagylaki T. (1992). Introduction to Theoretical Population Genetics. *Biomathematics* **21**. Berlin: Springer-Verlag.

[67] Pianka, E. R. (1978). *Evolutionary Ecology.* New York: Harper and Row Publishers.

[68] Pimentel, D., Nagel, W. P., & Madden, J.L. (1963). Space-time structure of the environment and the survival of parasite-host systems. *Am. Nat.*, **97**, 141-67.

[69] Poland, G. A. (1998). Variability in immune response to pathogens: using measles vaccine to probe immuno-genetic determinants of response. *American Journal of Human Genetics,* **62**, 215-220.

[70] Regan, H. M., Colyvan, M., & Burgman, M. A. (2002). A taxonomy and treatment of uncertainty for ecology and conservation biology. *Ecological Applications*, **12(2)**, 618-628.

[71] Rinchik E. M. (2000). Developing genetic reagents to facilitate recovery, analysis, and maintenance of mouse mutations. *Mammalian Genome,* **11**, 489-499.

[72] Rutschman, D. H. (1994). Dynamics of the two-locus haploid, *Theor. Pop. Biol.,* **45**, 167-176.

[73] Saint-Pierre, P. 1994. Approximation of the viability kernel. *Applied Math. and Optimis.*, **29**, 187-209.

[74] Salt, G. W. (1967). Predation in an experimental protozoan population *Woodruffia-Paramecium. Ecological Monographs,* **37**, 113-44.

[75] Schwerdtfeger, F (1958). Is the density of animal populations regulated by mechanisms or by chance? *Proc. 10th int. Congr. Ent.,* **4**, 115-122.

[76] Sih A., Crowley, P., McPeek, M., Petranka, J., & Strohmeier, K., (1985). Predation, competition, and prey communities, a review of field experiments. *Ann. Rev. Ecol. Syst.*, **16**, 269-311.

[77] M. Slatkin, & R. R. Hudson (1991) Pairwise comparisons of Mitochondrial DNA sequences in Stable and Exponentially Growing Populations. *Genetics*, **129**, 555-562.

[78] Wilbur, H. M. (1997). Experimental ecology of food webs, complex systems in temporary ponds. *Ecology*, **78**(8), 2279-2302.

[79] Wilbur, H. M. (1987). Regulation of structure in complex systems, experimental temporary pond communities. *Ecology*, **68**, 1437-1452.

[80] Wilbur, H. M. (1983). Salamander predation and the structure of experimental communities, anuran responses. *Ecology*, **64**(6), 1423-1429.

[81] Wilbur, H. M. (1972). Competition, predation, and the structure of the *Ambystoma-Rana sylvatica* community. *Ecology*, **53**, 3-21.

[82] Williamson, M. (1987). Are communities ever stable? in A. J. Gray, M. J. Crawley and P. J. Edwards (eds). *Colonization, succession and stability.* Oxford: Blackwell Scientific, 353-371.

Reviewed by - Christian MULLON, Institut pour la recherche et le développement (IRD-GEODES)

In: Ecosystem Ecology Research Trends
Editors: J. Chen and C. Guo, pp. 177-207
ISBN: 978-1-60456-183-8

Chapter 5

ECOLOGICAL SIGNIFICANCE OF COVER CROP AND NO TILLAGE PRACTICES FOR ENSURING SUSTAINABLITY OF AGRICULTURE AND ECO-SYSTEM SERVICE

Masakazu Komatsuzaki*
Collage of Agriculture, Ibaraki University

Abstract

A doubling of the global food demand projected for the next 50 years poses a huge challenge for the sustainability of both food production and global and local environments. Today's agricultural technologies may be increasing productivity to meet world food demand, but they may also be threatening agricultural ecosystems. Cover crops and no-tillage practices may be effective tools for ensuring sustainable agriculture and eco-system service; because these soil management strategies increase soil organic matter and improve soil biological diversities and activities in intensively managed summer field cropping system. This Chapter reviews several aspects of cover crop biology and soil eco-system ecology of no-tillage with cover crop managements in the view point of carbon and nitrogen dynamics, improving soil organic matter and soil biological community dynamics. Recent intensive research revealed that a combined with no-tillage system with cover cropping will be a new soil management strategy because they can scavenge soil residual nitrogen , improve carbon balance and their ecological functions can be utilized to establish an optimal nitrogen cycle due to higher population and diversity of soil fauna organism. Especially, the management technique of subterranean clover reseeding may have the most noticeable economical and ecological benefits, because it can eliminate the cost of cover crop introduction, substitute of fertilizer requirement, and weeds suppression. In order to maximize the potential of no-till and cover crops as a soil management tools, it is essential to obtain and utilize information that expands our ability to manage soil eco-systems and to ensure ecological services.

Key words; cover crop, nitrogen dynamics, carbon dynamics, soil microbe, nematode, soil fauna

* E-mail address: komachan@mx.ibaraki.ac.jp. Tel: +81.298.88.8707; Fax: +81.298.88.8715. Postal address: 3-21-1, Ami, Ibaraki, 300-0393, Japan

Introduction

The exponential growth of the global population has placed increased demands on agriculture to produce more food. The world's population has grown by one billion people in the past 12 years, exceeding 6 billion in 2000, and is projected to reach 9 billion by 2050 (Brown 2004). More than 90% of this growth has taken place in developing countries, in sharp contrast to Western Europe, North America and Japan, where population growth is small or stagnant. Increasing demand for food has resulted in increased soil disturbance, increased fossil fuel consumption to produce agricultural products, and increased biomass burning. The technologies developed since the 1960s to revolutionize agriculture may be increasing productivity to meet world food demand, but they may also be threatening agricultural ecosystems and their services. For instance, chemical-based pesticides and insecticides initially allowed farmers to reduce their losses from harmful insects and disease, although they began to fail as pests developed resistance, and the chemicals left toxic residues in our water, soil, and food (Nierenberg and Halweil 2005).

Japanese farming used to be an environmentally-friendly system because most crop rotations were including both summer cash crops and winter grain crops, such as wheat and barley. However, recent statistical data have revealed that the annual ratio of crop planting area in cultivated fields was 94.4% in 2004, representing a dramatic decrease over the past few decades (Statistics of Agriculture, Forestry and Fisheries 2005). This suggests that Japanese crop rotation systems are being destroyed and intensive farming and single cropping systems are now widely employed. These new farming systems can degrade the quality and health of the soil through the increase of synthetic chemical input and leaching of soil residual nutrients into the groundwater (Kusaba 2001).

Soil health and soil quality are fundamental to the sustained productivity and viability of agricultural systems world-wide, however, soil managers became overdependent on chemical fertilizers to replace or enhance soil nutrients, which also degraded soil and water quality (Stamatiadis *et al.* 1999). As local environmental quality becomes increasingly degraded by agricultural practices, the importance of protecting and restoring soil resources is being recognized by the world community (Lal 1998; Barford *et al.* 2001; Lal 2001). Sustainable management of soil received strong support at the Rio Summit in 1992 as well as in Agenda 21 (UNCED 1992), the UN Framework Convention on Climate Change (UNFCCC 1992), articles 3.3 and 3.4 of the Kyoto Protocol (UNFCCC 1998), and elsewhere. These conventions are indicative of the recognition by the world community of strong linkage between soil degradation and desertification on the one hand and loss of biodiversity, threats to food security, increases in poverty and risks of accelerated greenhouse effects and climate change on the other.

The recent use of policy by the Japanese Government to develop more environmentally-friendly farming practices and the growing awareness of the importance of surplus reduction has led to a widespread interest in organic farming and environmental conservation farming. According to recent statistical data, 75,299 farms in Japan were engaged in environmental conservation farming, accounting for 21.5 % of the total cropping area in the country (Sustainability agriculture office 2001). Under conservation management, traditional agronomic methods are combined with modern farming techniques, and conventional inputs such as synthetic pesticides and fertilizers are excluded or reduced. Instead of synthetic inputs, cover crops, compost, and animal manure are used to build up soil fertility. Cover crops are particularly beneficial ecosystem service in a cropland, as they are instrumental in supplying soil organic matter, adding biological fixed nitrogen, scavenging soil residual nutrients, suppressing weeds, and breaking pest cycles (Peet 1996; Sarrantonio 1998; Magdoff 1998.

No-tillage system is another alternative, that has been increasingly used for crop production in the US, Europe, South America, and Asia during the past decade due to their significant environmental advantage over moldboard plow. For example, no-tillage systems the area under no-tillage has been growing steadily all over the world, the no-tillage technology is now being adopted on more than 95 million ha world wide. Approximately 47% of that area corresponds to South America, 39% to the United States and Canada, 9% to Australia and only 3.9% to the rest of the world (Derpsch 2007). In East Asian countries, including Japan and China, just started to adopt no-tillage farming, however, there are number of research regarding no-tillage practices have great potential to improve soil quality (Robertsons *et al.* 2000; Arrouays *et al.* 2002), and the combination of cover crops and no-tiilage practice increasecd the emvironmetntal benefits such as N leaching reduction, soil organic mater (SOM) increases, soil biological diversity improvements (Komatsuzaki and Ohta 2006).

The soil is a very complex and multi-faceted environment providing the habitat for a diverse array of soil organisms. The activities of this wide range of soil biota contribute to many critical ecosystem services, including: soil formation; organic matter decomposition, and thereby nutrient availability and carbon sequestration; nitrogen fixation and plant nutrient uptake; suppression or induction of plant diseases and pests; and bio-remediation of degraded and contaminated soils.

Healthy soil is defined as a stable system with resilience to stress, high biological diversity, and high levels of internal nutrient cycling (Van Bruggen and Semenov 2000). The major activities of soil microbes include the decomposition of organic matter, mineralization of nutrients, nitrogen fixation, and suppression of crop pests and protection of roots. However, they can also be parasitic and cause injury to plants. Thus, there is a great need to assure that practices introduced in soil management to improve soil quality will also help to maintain the health of the soil. Managing cover crops and no-tillage practice adoption in croplands may be able to contribute to the establishment of a healthy soil, but this depends in large part on understanding their basic biology and eco-system ecology. This Chapter reviews several aspects of cover crop biology and soil eco-system ecology of no-tillage with cover cropping, including carbon and nitrogen dynamics, improving soil organic matter and soil biological community dynamics.

Cover Crop Benefits for Soil Ecosystem

The following are some important ways to evaluate the ecological service aspects: (1) cutting fertilizer input by contributing N to cash crops and by scavenging and mining soil nutrients (Horimoto *et al.* 2002;Komatsuzaki 2002), (2) reducing the need for herbicide because many cover crops effectively suppress weeds as a smother crop that out-competes weeds for water and nutrients, residue or growing leaf canopy that blocks light, alters the frequency of light waves and changes soil surface temperature, and a source of root exudates or compounds that provide natural herbicidal effects (Creamer *et al.* 1996; Fujii *et al.* 2004), (3) improving yields through enhanced soil health resulting from the addition of organic matter that encourages beneficial soil microbial life and enhancing nutrient cycle (Gu *et al.* 2004), (4) preventing soil erosion from wind and rain, because above ground portion of cover crop is quickly spread at the soil surface (Dabney 1998), (5) protecting water quality by slowing erosion and runoff, because cover crops scavenge soil residual nutrients, thereby reducing no-point-source pollution caused by nutrient and agricultural chemicals (Staver and Brinsfield 1998; Delgado 1998), and (6) attracting beneficial insects, serving as a trap for damaging insects, and / or providing food and cover for wildlife (Schutter and Dick 2002). These significant benefits of cover crops vary by location and season (Sarrantonio 1998).

Table 1. Typical winter cover crop dry matter and N accumulation in the several studies.

Cover crop	Dry matter ($Mgha^{-1}$)	N accumulation (kg N ha^{-1})	Sources
Rye (*Secale cereale* L.)	1.5-5.7	17-64	Ranells and Wagger (1996)
	3.4-6.8	44-87	Komatsuzaki *et al.*(2006)
Oat (*Avena sativa* L.)	3.3-4.3	80-82	Dyck and Liebman (1995)
	1.1-1.5	27-45	Komatsuzaki *et al.*(2006)
Wheat (*Triticum aestivum* L.)	4.9-9.8	81-87	Singogo *et al.* (1996)
	1.5-2.0	30-42	Komatsuzaki *et al.* (2006)
Ryegrass (*Lolium multiflorum* Lam.)	0.7-17.5	15-346	Stopes e*t al.* (1996)
Crimson clover (*Trifolium incarnatum L.*)	4.2-5.7	151	Abdul-Baki *et al.* (1996)
Hairy vetch (*Vicia villosa* Roth)	2.9-4.8	125-182	Ranells and Wgger (1996)
	3.0-6.7	104-257	Sainju and Singh (2001)
Subterreanean clover (*Trifolium subterraneum*)	5.1-6.3	51-65	Komatsuzaki (2002)
White clover (*Trifolium repens* L.)	0.6-25.0	17-592	Stopes *et al.* (1996)

Komatsuzaki (2004) summarized the difference crop ecological characters between legumes and cereal cover crops. Table 1 shows typical cover crop species, suggesting that they are involved in a wide range of dry matter accumulation and N accumulation. Winter annual cover crops can be either legumes or cereals. Legumes that are best adapted to the soil and climatic conditions of the Kanto area of Japan, are clovers (*Trifolium*) ,crimson clover (*Trifolium incarnatum* L.), hairy vetch (*Vicia villosa* Roth), and subterranean clover (*Trifolium subterraneum* L.). Cereals or small grains that are best for the Kanto area are rye (*Secale cereale* L.), wheat (*Triticum* spp.), barley (*Hordeum vulgare* L.), triticale (x *Triticosecale*), and oat (*Avena sativa* L). Cereal winter cover crops show significant ability to scavenge soil residual nitrogen, produce a large amount of biomass that can add organic matter to the soil, and have a wide range of planting times in the fall. Leguminous winter cover crops show high nitrogen content and a low C:N ratio, indicating the addition of nitrogen to subsequent crop production; therefore, legume cover crops contribute N to a subsequent crop, relieving the farmer of some of the financial burden of buying fertilizer.

However, most leguminous cover crops have a narrow planting time period, so detailed crop rotation strategies would be required to introduce legume cover crops to a planting rotation.

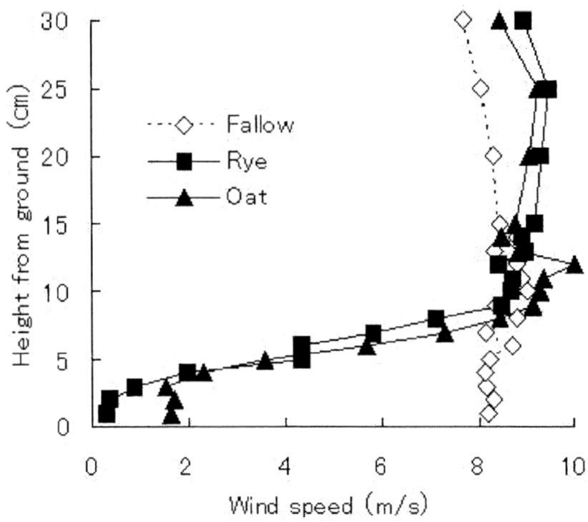

Figure 1. The relationship between cover crop management and wind speed distribution in wind tunnel on January, 2005. Rye and oat cover crop were seeded on 4th November, 2004 in Field research center, Ibaraki University. Data from Suzuki and Komatsuzaki (2006).

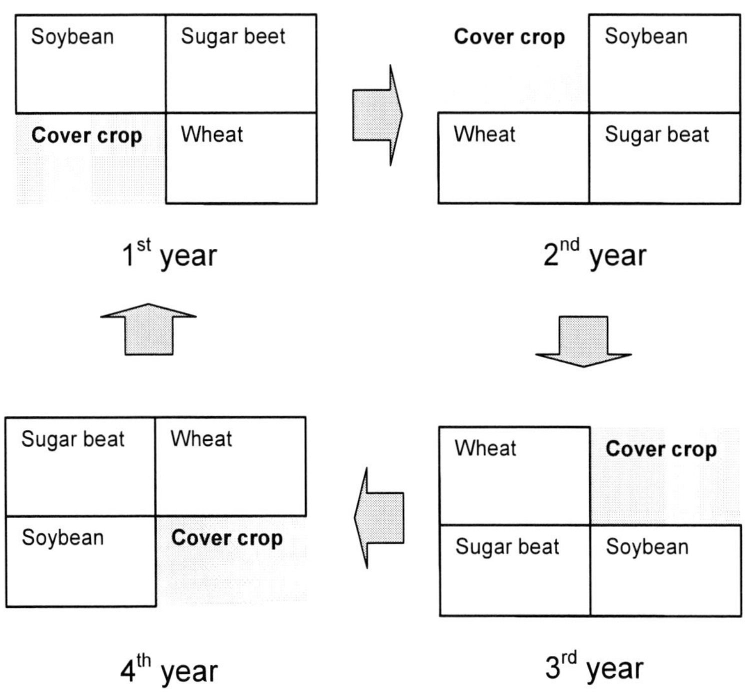

Figure 2. Example of upland field management adopting cover cropping in to their rotation in the Hokkaido region. This schema is proposed in "Policy of Crossing Crop Management" in the Hokkaido area. Modified from Nihon Nougyo Hojin Kyokai, (2005).

Winter tolerance and growth speed strongly affect the abilities to control soil erosion and provide cover crop benefits. The extent of soil erosion control provided by cover crops during the fall, winter, and early spring depends largely on when the crop is established (Figure 1). Timing is particularly important with legumes because late seeding results in small plants with limited root systems. If the legume cover crop is established early enough, adequate growth in the fall can help minimize soil erosion. Because of their rapid growth rates, the non-legumes rye and triticale provide maximum erosion protection during fall and winter. Active cover crop growth in the spring, before the summer crop is planted, offers continued erosion control because mulch is left on the surface.

Cover crop benefits also depend on farming practices, including tillage and cropping systems. Cover crops in a conservation tillage planting system provide erosion control during the winter while the plants are growing and mulch for the soil surface during the summer in the form of crop residue. This surface mulch enhances rainfall infiltration from summer rains, reduces soil water evaporation, and provides weed control by early shading. All these potential benefits are highly dependent on weather and management factors that should be considered when using cover crops. Figure 2 shows an example of a rotation adopted in field management in the Hokkaido region. Such a cropping strategy should provide compatibility between providing economic benefits for farmers and sustaining agro-ecosystems (Nihon Nougyo Houjin Kyokai 2005).

No Tillage Practice with Cover Crop

Conservation tillage systems, including no-till, leave more surface residues because the soil is not turned over. These systems create less potential for soil erosion and therefore conserve SOM. Many studies on ecosystems under long-term management involving conventional tillage and no-tillage practices have demonstrated that tillage causes a substantial decrease of SOM content and mineralization of carbon (Lee *et al.* 1993; Elliott *et al.* 1994; McCarty *et al.* 1995; Six *et al.* 1999). Johnson *et al.* (2005) summarized 44 studies regarding conservation tillage and showed that the rate of SOC storage in NT compared to conventional tillage has been significant, but variable, averaging 0.4 ±0.61 Mg C ha^{-1} $year^{-1}$.

The quality of the no-till system will be defined by the percentage of soil cover with plant residues, by low disturbance seeding, the years of continuous no-till and by the use of crop rotation and cover crops. The use of cover crops with no-till are key factors for the unprecedented growth of no-tillage, because when practiced in monoculture no-tillage is an imperfect and incomplete system, in which diseases, weeds and pests tend to increase and profits tend to decrease (Derpsch 2007). Figure 3 were illustrated an agroecosystem budget model to examine pools and fluxes of C and N in the cover crop fields. Bold box indicate pools where cover crops are predicted to improve. Cover crops have major influence on the soil nutrient cycling process in no-till ecosystems. Cover crops root exudates and litters supply the appropriate food source for soil herbivores , and increasing soil herbivores, also increase predators resulting increase soil biological diversities, in addition, no-till condition provide an appropriate soil condition for their habitat. By ingesting and turning over large amounts of soil and organic matter from cover crop residue and roots, they increase the rate of mineralization on organic matter, converting organic forms of nutrients into inorganic forms that can be taken up more readily by subsequent crops. One of the major disadvantages of no-tillage is increasing soil hardness in plant roots depth layer as long as no-till continued, and increasing soil hardness disturbed plants roots growth and reduced crop yields (Raper *et al.* 2000).Mu *et al.* (2007) reported no-till with cover crop showed significantly soil hardness reduction (Figure 4).

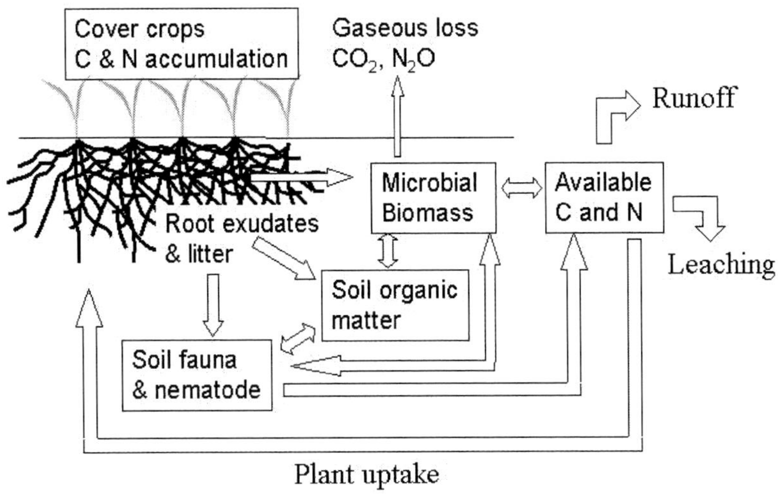

Figure 3. Agroecosystem budget model to examine pools and fluxes of C and N in the cover crop fields. Bold boxes indicate pools where cover crops are predicted to improve.

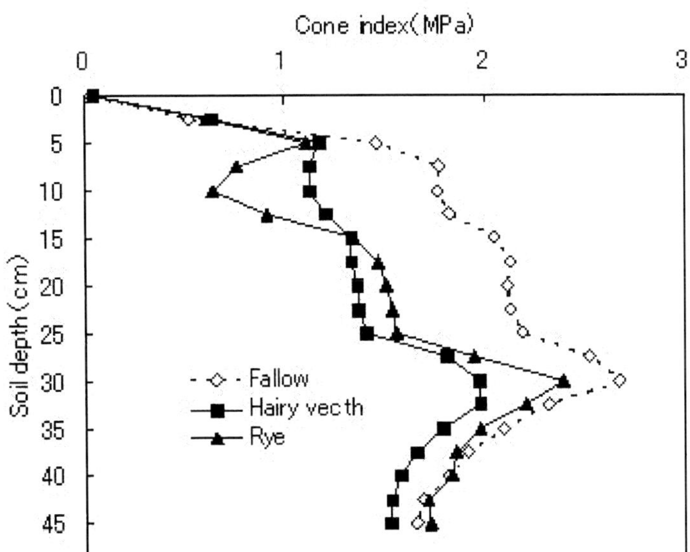

Figure 4. Soil cone index distribution in no-tillage fields in the relation to cover cropping in field rice cultivation were evaluated in field science center, Ibaraki University. No-tillage treatment were continued for 3 years before measuring the soil cone index. Data from Mu *et al.* (2007).

Carbon Dynamics

Soil Carbon Accumulation

The two principal elements regulating carbon dynamics, soil biological activity, and nutrient cycling are carbon (C) and nitrogen (N). In the carbon dynamics of a cropland, crops accumulate carbon resulting in CO_2 fixation by photosynthesis and carbon consumption by respiration. Parts of the net crop carbon accumulation are removed through the harvesting process, while other parts of crop residues, including litter and roots, remain in the soil.

Although the applicability of the options might differ according to soil type and region, one of the recommended management practices for soil C sequestration in cropland was eliminating summer fallow and incorporating legumes and other appropriate cover crops in rotation (Follett *et al.* 2005). Because cover crop decomposition and subsequent nutrient release are mediated by soil microorganisms, it is appropriate to examine C and N inputs to an ecosystem with respect to resource quality. This regulatory effect is manifested primarily in "resource quality", which is a term that refers to the chemical composition and physical characteristics of cover crops.

Cover crop residues replenish organic matter that is naturally decomposed during the year. The more residues can be returned to the soil, the better the effect on soil organic matter. There are many studies regarding the ability of cover crops to increase organic matter in the soil. Magdoff's (1998) investigation of a five-year experiment with clover in California showed that cover crops increased organic matter in the top 5 cm from 1.3 to 2.6 % and in the 5 to 15 cm layer from 1.0 to 1.2 %. In Georgia, Bruce *et al.* (1995) examined the effect of cover crop biomass inputs over a five-year period on SOC and aggregate stability in the top 15 cm, and found that the most dramatic effects occurred when crimson clover was integrated as a cover crop into a no-tillage sorghum production system, as average soil C was more than 100% higher (23.3 vs 10.4 g kg^{-1}) than the other cropping systems and aggregate stability was nearly 50 % higher.

Many scientists have reported that SOC content increased as long as cover cropping was being practiced (Reeves, 1997), but the increase ratio would be very slight in the short term. Gu *et al.* (2004) examined the influence of cover crop species and tillage system on SOC content, noting that cover crops can add organic matter to the soil; however, SOC does not change due to the enhanced soil microbial decomposition that occurs as cover crop residue input increases in short term. On the other hand, many reports have suggested that SOC accumulation depends not only on cover crop residue input but also on the tillage system, including no-tillage treatment. Carbon balance in croplands is used to calculate the difference between the amount of organic matter input and the carbon loss by soil respiration and other losses. In Japan, high humidity and relatively high temperatures in the summer growing season enhance soil respiration which, coupled with SOC breakdown, results in considerable carbon loss from soils.

Ismail *et al.* (1994) evaluated the long-term effect of tillage on continuous corn with a rye cover crop in Kentucky, USA. SOC in the top 30 cm did not change from 1975 to 1980, but substantially increased from 1980 to 1989. In Mu *et al.*'s (2007) evaluation of the effect of tillage on continuous field rice with a rye and a hairy vetch in the Kanto area of Japan, SOC in the top 0 to 2.5 cm increased significantly compared with winter fallow 4 years after adopting cover cropping, but other soil layers did not change (Figure 5). This result suggested that cover crop C accumulation in the soil would increase in short term, if it combined with no-till managements, even in *Andisol* that shows relatively high SOC content initially, and difficult to increase SOC contents.

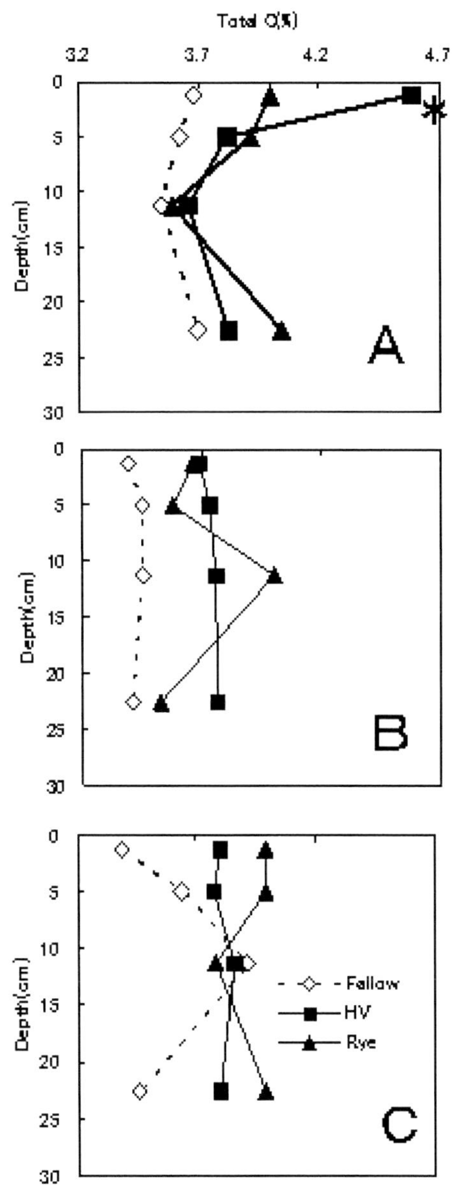

Figure 5. Changes in SOC content in relation to cover cropping and tillage system in continuous field rice cultivation in 2007 at Field research center, Ibaraki University, Japan.The experiment was started in 2002, and each tillage system including No-till (A) plow (B) and rotary (C) was applied in fall and spring in a year. * indicates significant difference (P<0.05) between treatments. Modified from Mu *et al.* (2007).

Cover Crop C Release

Cover crop dry matter (DM) accumulation, which is easily converted to carbon content by multiplying it by 0.4 (Bauer and Reeves 1999), is strongly associated with species and management. For cereal cover crops, the cover crop DM accumulation depends on soil N availability, which itself depends on the amount of residual soil N as well as the mineralization rate, while leguminous cover crops show little difference of DM accumulation regardless of residual soil N availability (Wagger and Mengel 1988 ; Gu *et al.* 2004). Cover crop residues are decomposed into CO_2 (mineralization) or transformed into SOM in the soil by microbial decomposition, with the mineralization strongly dependent on the C:N ratio of cover crop residues.

Cereal cover crop residues show high C:N ratios and yield high amounts of litter, which increase the SOM content in the soil. Leguminous cover crops also produce considerable litter, but their residues have low C:N ratios. Brassica cover crops produce small amounts of litter and the C:N ratio of their residues is low. These low C:N residue-producing crops result in some decrease in SOM content in the soil (Komatsuzaki 1999). Cover crop residue from decomposed crimson clover, hairy vetch, and rye winter cover crops was examined by Wagger (1989), whose results indicated greater C mineralization from residues with narrower C:N ratios and lower concentration of cellulose and lignin.

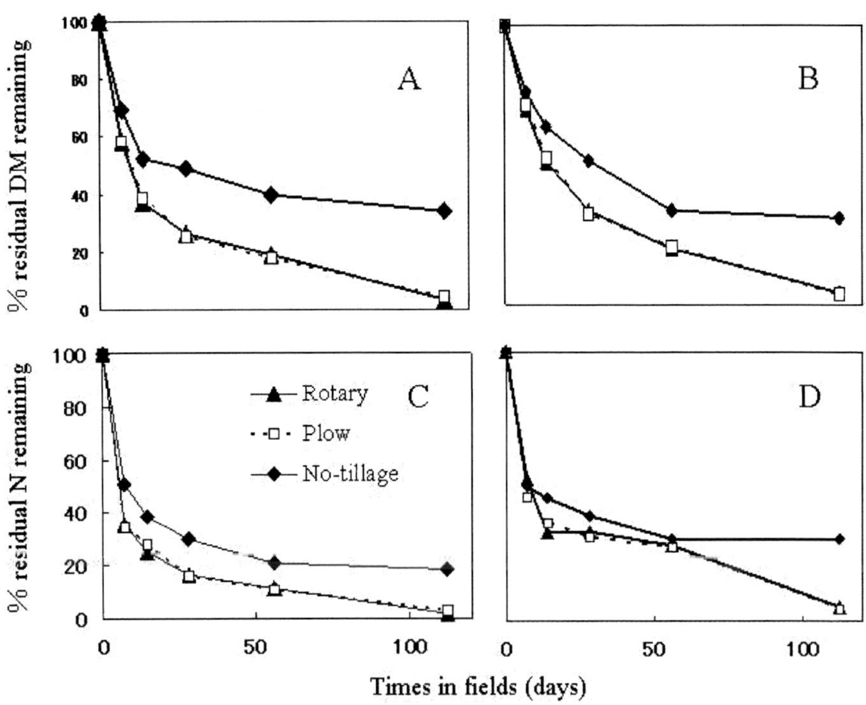

Figure 6. Percentage of initial DM remaining of hairy vetch (A) and rye (B), and N remaining of hairy vetch (C) and rye (D) under field conditions following a early-May desiccation date associated with different tillage systems including rotary, plow and no-tillage in Field Science Center, Ibaraki University, Japan. Data from Komatsuzaki *et al.*(2007).

Concentrations of cellulose and lignin are greater in cereals and more mature plant materials compared with legumes and young plant tissues. The cover crop decomposition process also differs between leaves and shoots. Quemada and Cabrera (1995) reported that in an incubation study, the C/N ratio of crimson clover (leaves and stems) averaged approximately 60 % lower than the mean of the cereals (21 vs. 53). Carbon dioxide emission rates peaked 3 to 4 days after residue application to soil, with the cumulative CO_2 16 days further evolved in leaves than in stems. By the end of the 160-day incubation, differences were less evident among species than plant parts and a mixture of leaves and stems generally resulted in cumulative CO_2 showing values intermediate between those from isolated leaves and stems.

Cover crop decompose ratio strongly associate with tillage system after cover crop termination. Figure 6 shows the differences cove crop DM remaining of hairy vetch and rye cover crop at rotary, plow and no-tillage conditions. Hairy vetch decomposed 62.8% of DM for rotary and 60.9% for plow in first 14 days, although rye decomposed 49.2% for rotary and 47.4 % for plow. On the other hand, cover crop decompose ratio was low in no-tillage condition, decompose ratio of DM in no-till were 47.5% for hairy vetch and 36.3% for rye. at the ends of summer crop cultivation, there were no significant difference of DM remaining between hairy vetch and rye, however, no-tillage showed highest DM remaining compare with rotary and plow. No-tillage provide the soil surface cover during the summer crop growing season, these condition also provide the appropriate condition for soil fauna resulting make a different soil ecosystem between no-till with cover crop and conventional fields (Komatsuzaki *et al.* 2007).

Nitrogen Dynamics

N Contribution and Release

Winter annual cover crops can be an important tool for managing N in various cropping systems. Benefits of cover crops with respect to N dynamics may occur via cover crop recovery of residual soil and/or the cover crop N contribution to a subsequent crop. Soil management influences the amount of nutrients removed from the soil by the cover crops, as well as the availability of nutrients to a following crop as cover crops decompose. Plowing cover crops under early in the spring will increase the decomposition rate of cover plant material, but it also will reduce the amount of nutrients removed from the soil and limit the build-up of soil organic matter from the residue. Allowing the cover crop to mature will improve nutrient accumulation in the cover crop biomass, but it may reduce the ability of microbes to decompose residues for short-term use. Leaving the cover crop on the surface also reduces its decomposition rate, compared to plowing. This, in turn, can benefit subsequent crops by leaving more residual matter to accumulate on the soil surface.

The question of how much nitrogen legume cover crops can supply to subsequent crops is difficult to answer, especially because nitrogen easily changes form in soil. The farming practice and weather following the killing of those legume cover crops will also greatly affect the amount and timing of N release from the legume to the soil. Other environmental factors contribute significantly to how quickly a cover crop releases its N, and many are out of our direct control in the field (Sarrantonio 1998; Magdoff 1998).

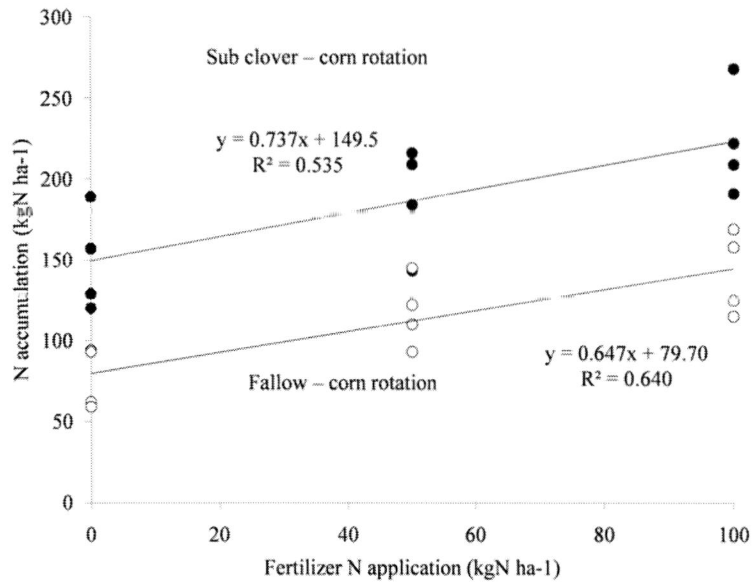

Figure 7. Subterranean clover N contribution to subsequent silage corn compared with fallow treatment. Data from Komatsuzaki (2002).

Regarding the estimated values of the equivalent amount, some scientists have reported that the amount of nitrogen supplied by cover crops to subsequent crops depends on the tillage system (Sarrantonio 1998). A well-established legume cover potentially can supply 85 to 142 kg of N per ha or approximately two thirds of the N required by a corn crop and most of what is needed for grain sorghum and some vegetable crops (Table 2). Komatuszaki (2004) reported that nitrogen released from the residue of cover crops and taken up by corn was estimated by taking the difference in N uptake at corn harvest between a sub clover reseeding system and a fallow system. Based on the results, the amount of nitrogen supplied from subterranean clover to subsequent silage corn as fertilizer was estimated to be 48 - 75 kg N ha^{-1} (Figure 7). This value was somewhat higher than sub clover N accumulation in spring, because subterranean clover N accumulation ranged from 51.1 to 65.1 kg N ha^{-1}. However, these values reflect only the aboveground portion and do not include N in roots.

Table 2. Studies of N supply by cover crop for cash crops (kg N ha^{-1})

Study	Cover crop	Cash crop	Cover crop N (kg N ha^{-1})	N contribution* (kg N ha^{-1})
Wagger, 1989	Hairy vetch	Corn	NA	3.2
Ranells and Wagger, 1997	Hairy vetch	Corn	92	16.5
Ranells and Wagger, 1997	Crimson clover	Corn	30-49	6.3-3.0
Ranells and Wagger, 1997	Rye	Corn	40-111	1.6-5.6
Horimoto *et al*, 2002	Hairy vetch	Tomato	NA	20-60
Komatsuzaki 2002	Subterranean clover	Silage corn	51-65	48-76

*:Contribution of cover crop N to cash crop was calculated by fertilizer input equivalent compared with fallow treatment.

Edelhar *et al.* (1984) estimated that the average amount of fertilizer N replaced by legume cover crops ranged from 90 to 100 kg N ha^{-1} for corn. Toucheton *et al.* (1982) reported that the N provided by crimson clover allowed to reseed was sufficient for maximum grain sorghum yield without the need for supplemental N fertilizer. The estimated values of N contribution to subsequent crops shown in Table 2 suggest a wide range of N contribution associated with cover crop N and subsequent crop types.

The primary chemical factor determining the kinetics of residue decomposition and nutrient release are the C:N ratio and concentrations of lignin and other structural carbohydrates in plant tissues. Nitrogen releases from crimson clover, hairy vetch and rye were examined by Wagger (1989), whose results indicated greater N release from residues with narrower C:N ratios and lower concentrations of cellulose and hemicellulose. For example, 8 weeks after desiccation, 75 to 80 % (81 to 96 kg N ha^{-1}) of hairy vetch and crimson clover residue N was released, compare with 50 % (24 kg N ha^{-1}) from rye residue.

Sainju and Singh (2001) examined the influence of three tillage practices including no till, chisel plowing, and moldboard plowing, with hairy vetch cover crop on soil inorganic N, and silage corn N uptake. After the cover crop was killed and the field was planted in corn, Corn N uptake was higher in no-till than in chisel plowing and moldboard plowing, and soil inorganic N (0- to 30-cm depth) at corn planting was higher in chisel plowing than in no-till (18.4–30.2 vs. 9.9–20.5 mg kg^{-1}) suggesting that plowing tillage practice enhance quickly decompose of cover crop residue by incorporate it compare for no-till.

Varco et al. (1989) reported that only 20 % of the total N in corn at harvest came from hairy vetch, with an additional 7 % recovered the following corn season. In a Wisconsin experiment, 19 % was recovered in microbial biomass, and 38 % in the soil organic N fraction.

All of these examples indicate a relatively low efficiency of N utilization from cover crop residues by the subsequent principal crop. Rye cover crops cause a period of nitrate immobilization in which microbes decomposing the rye outcompete crop plants for soil N, causing N deficiency. The low corn N recovery from the rye monoculture was attributable to the relatively high C:N value of this residue.

With legume cover crops, the potential for significant N contributions may not be realized unless the legume N becomes available during the periods of high N demand by the subsequent crop. Wagger (1989) estimated that corn recovery of legume N was 40 to 45 kg N ha^{-1} (2-yr avg.), representing approximately 36% and 30 % of the total N content of crimson clover and hairy vetch, respectively.

Cover crop N contributions to subsequent crops were significantly associated with their soil ecosystems. Komatsuzaki et al. (2007) examined the cover crop ^{15}N distributions in subsequent crops and soil fauna in between conventional tillage soil ecosystem and no-tiilage soil ecosystem in the field rice cultivation in Japan. There were no significant differences between two soil ecosystems on initial cover crop N release in the fields (Figure 8 A), however, soil fauna in no-tillage accumulate the higher ^{15}N on August compare with conventional tillage system (Figure 8 B), resulting the higher N contribution was observed in no-till soil ecosystem compare with conventional tilled soil ecosystem (Figure 8 C and D).

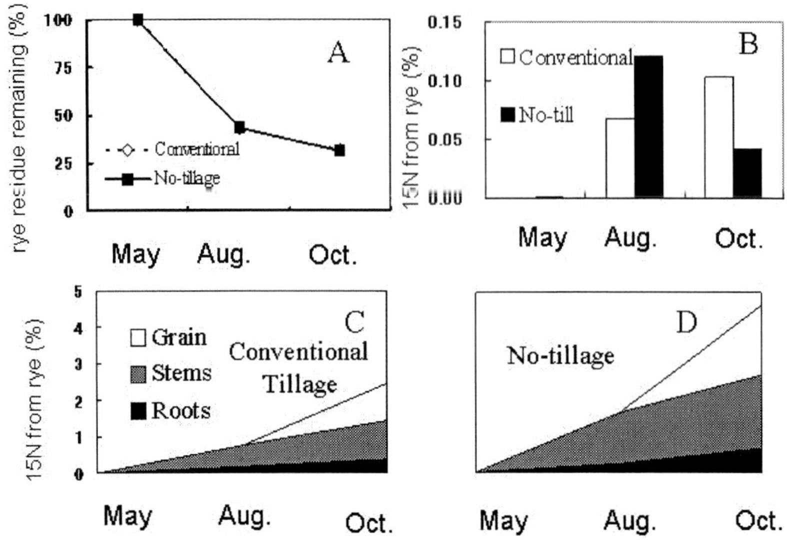

Figure 8. The differences between conventional tillage and no-tillage soil ecosystem on percentages of rye cover crop N release (A), the seasonal changes of rye cover crop N in soil fauna bodies (B), following rice N uptake from rye cover crops between conventional till (C) and no-till (D) were observed in field science center, Ibaraki University, Japan. N contributions were determined by initial ^{15}N contents in rye cover crops. Data from Komatsuzaki *et al.* (2007).

N Recovery

Cover crops can also reduce non-point source pollution caused by nutrients and agricultural chemicals. In particular, cereal cover crops have been shown to prevent N leaching to groundwater by accumulating excess soil N (Wagger and Mengel 1988; Ditsch and Alley 1991; Shipley *et al.* 1992 ; McCrancken *et al.* 1994) . Nitrate losses from cropland represent an economic loss to farmers and a potential environmental contaminant when transported into ground or surface water as NO_3^-, or when denitrified to N_2O. A significant portion of the fall nitrate pool can be lost through leaching and denitrification, especially when summer drought or excess fertilization causes high residual fertilizer nitrogen (Legg and Meisinger 1982).

Several winter cereal species, including rye, wheat and oat, are known to scavenge residual soil N in rotation with summer crops for many years (Wagger and Mengel 1988; Wagger 1996). This system works well following corn, grain sorghum (*S vulgare L.*), tobacco (*Nicotiana tabacum L.*), cotton and soybean, because the range of planting dates for cereal cover crops is wider than for legume cover crops (Wagger 1989).

However, cereal cover crop N recovery also depends on the species. In North Carolina, United States, the order of cover crop ranking for effective N recovery has been rye>oat > wheat, which determined by applying ^{15}N depleted fertilizer. Rye was particularly effective, recovering 35% of the soil residual N fertilizer following corn (Wagger 1996). The potential for N recovery, however, may be strongly influenced by factors related to cover crop species and their management (Ditsch and Alley 1991). The most critical period for recovery of residual N following a summer crop is from immediately after harvest until the cover crop reaches winter growth dormancy. Therefore, early cover crop sowing is essential for maximizing N recovery.

A variety of winter annual cereal and legume cover crops have been evaluated for their ability to assimilate soil inorganic N and thereby reduce potential NO_3 leaching during the winter months. Following the fertilization of a corn crop with ^{15}N-depleted fertilizer applied at a rate of 336 kg N ha^{-1}, Shipley *et al.* (1992) reported a residual level of 94 kg N ha^{-1} in the soil profile (0 to 80 cm) in September following corn harvest. By April of the following spring, rye had recovered 45% of the residual fertilizer N. Greater residual fertilizer N was measured in the top 40cm (90%) in September where more rapid rye root growth occurred compared to crimson clover.

Ditsch and Alley (1991) suggested that the amount of N recovered by the cover crop is largely a function of cover crop plating date, cover crop seeding rate and subsequent plant population, amount of residual mineral N remaining after the harvest of the previous crop, and cover crop growth stage at the time of termination. Compared with the other cover crops, rye had significantly greater total root biomass and the subsequent NO_3^- or inorganic N concentration was lower in the soil compared with legume cover crops, suggesting that root density and above-ground biomass affect the ability to scavenge soil residual N (Sainju *et al.* 1998).

Some information exists on the ability of several cereal cover crops to scavenge residual soil N with respect to their planting and growth termination dates. In Massachusetts, cover crop DM accumulation of September and October planted rye, oat, and wheat resulted in 45 to 62% and 10 to 18%, respectively, of the August planting date when measured the following spring (Herbert *et al* 1986). Wagger (1989) reported that a two-week delay in the desiccation of a rye cover crop resulted in a 45% increase in DM and a corresponding 13% increase in N accumulation.

In addition, new cereal species that may also have potential for N recovery are black oat (*Avena strigosa* Schreb.) and triticale (*Triticosecale*). Black oat is the most widely grown cover crop species on millions of hectares of conservation-tillage soybean fields in southern Brazil, producing above ground biomass, similar to rye. Bauer and Reeves (1999) compared wheat, rye, oat, and black oat as cover crops for cotton in South Carolina, with black oat having greater N concentration than the other three cover crops. Triticale, which has the winter hardiness characteristics of rye and wheat (Martin *et al.* 1976), has been used as a cover crop in West Virginia. Monks *et al.* (1977) and Sheng *et al.* (1991) reported that triticale demonstrated moderate (between rye and wheat) ability to accumulate aboveground biomass.

Soil Ecological Significance

Microbial Community and Structure

Cover crops can enhance SOM associated with soil biological diversity and their activity due to additional cover crop residue in the soil (Magdoff 1998). Quantitative and qualitative changes in the populations of soil microorganisms may reflect changes in soil quality resulting from the use of cover crops. These changes are potentially useful as responsive indicators of the effects of crop and soil management. Zhaorigetu *et al.* (2007) reported that the different soil microbial reactions adopting cover crop residue at different tillage system, leguminous cover crops increased soil microbial populations in no-tillage, after cover crop residues input, however, these effects were moderate in rye cover crop and in all of rotary and plow tillage systems. On the other hands, fungus biomass showed significantly higher in no-till with cover crop on May and October suggesting cover crops with no-tillage practices make a different soil microbial community and structure compare with conventional tilled soil ecosystems (Figure 9).

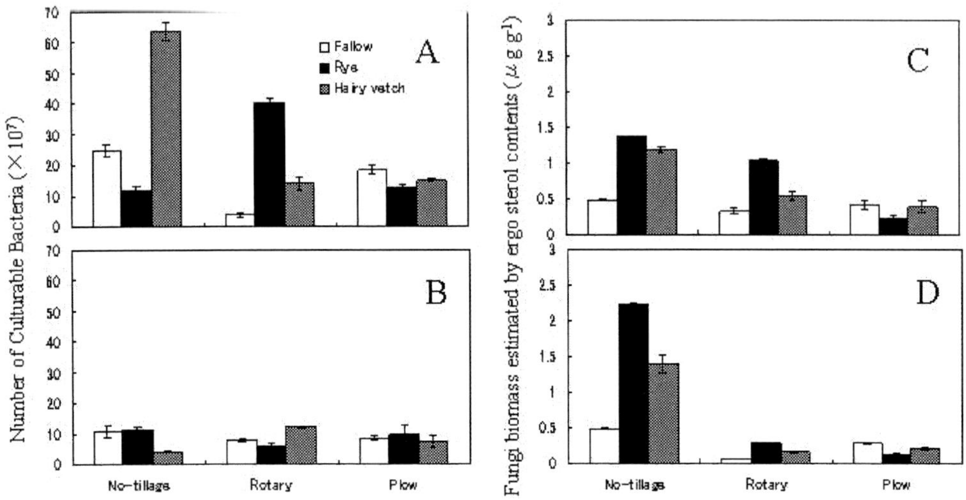

Figure 9. Effects of cover crop and tillage system on soil bacteria population on May (A) and on October (B) and soil fungus biomass on May (C) and on October (D) in field science center, Ibaraki University, Japan. Modified from Zhaorigetu *et al.* (2005).

In Mullen *et al*'s (1998) examination of biological properties in no-till corn plots receiving either 0 or 165 kg N ha^{-1}, with or without cover crop (hairy vetch, wheat, and rye), hairy vetch showed the highest organic C accumulation and the addition of N to the wheat or fallow plots increased C relative to the no-N treatment. In microbial terms, the greatest bacterial population, the highest ß-glucosidase and L-asparaginase activities were observed with hairy vetch with no-N treatment, suggesting cover cropping enhanced the activity of soil microorganisms and provided a better environment for the stabilization of microbally-produced soil enzymes.

Bacterial response (population dynamics) to the input of cover crop residue differed between oligotrophic and copiotrophic bacteria. Hu *et al.* (1999) reported that both copiotrophs and oligotrophs were significantly stimulated by newly added C in the from of cover crop debris, but copiotrophs rapidly peaked at the very early stage of cover crop decomposition while peak populations of oligotropfs occurred at a later stage when available C decreased.

Cover cropping and tillage system strongly affect soil aggregate stabilities and soil microbial and fauna communities, although cover cropping has a lesser effect than tillage system on soil microbial communities. Peixoto *et al.* (2006) examined the effects of conventional tilling and no-tillage in cropping systems with and without cover crops on bacterial community structure, SOC and soil aggregation. SOC was greater in the largest aggregate size class (7.98-19mm), and had greater meanweight diameter under no-tillage than under conventional tillage in the 0-5 cm soil layer; moreover, soil bacterial populations were affected by cultivation, tillage and depth, but not by cover cropping (Peixoto *et al.* 2005). Schutter *et al* (2002) reported, however, that cover crop residues enhance microbial biomass and N mineralization in whole soils and aggregates, and shifted microbial community that was extractable fatty acid methylester profiles although these changes were strongly related to soil aggregate size fractions. Because of the portions of fungi and bacteria differ among aggregate sizes, with lower bacterial:fungal ratios associated more with macroaggregates than microaggregates (Gupta and Germide 1988), microbial community differentiation might be due to the enhancement of soil aggregates through the incorporation of cover crop residues (Schutter *et al.* 2002).

Cover crop N release through microbial decomposition is an important parameter for evaluating the ecological significance of cover crop use. The ability of microbes and plants to assimilate and retain N from cover crop residue was studied by Jackson (2000) in a sandy soil in a field under intensive vegetable production in California. After fresh residue of ^{15}N-labeled phacelia (*Phacelia tanacetifolia* Benth.) (C:N of 19) was added, microbial biomass and NO^-_3–N increased rapidly, then began to decline 2 weeks later. Microbial biomass C declined faster than microbial biomass N. Only a small amount of ^{15}N was ever found in the microbial biomass, but NO_3^-–N was enriched with substantial ^{15}N. Percentage recovery of the added ^{15}N after 4 months was 60.7% as soil organic N; 20.7% in plants; 1.4% as inorganic N; 1.4% in microbial biomass; 4.7% in ion exchange resin bags, leached below a depth of 60 cm; and 11.1% as unexplained loss. Losses of ^{15}N during the first lettuce (*Lactuca sativa* L.) crop after the cover crop were relatively low, most likely due to low rainfall and appropriate scheduling of fertilization and irrigation. Total soil ^{15}N at 0- to 30-cm depth declined for the first 7 months, and thereafter cover crop N was apparently no longer readily mineralizable. Microbes may have assimilated C from the plant residue, met their N demand mainly with soil-derived N, and released cover crop-derived N that was rapidly mineralized and nitrified. The resulting NO_3^-–N was either taken up by plants, leached, or denitrified.

Input of organic matter into the soil has also been emphasized for controlling soil-borne diseases as an agronomic alternative technique; however, fresh organic residues from, for example, a cereal cover crop would enhance the activities of *Pythium*, *Rhizoctonia,* and *Fusarium* in the soil (Ogura 1966). Incorporating fresh clover and rye residues in the soil would also enhance the population of *Pythium* resulting in a low emergence ratio if the following crop were to be sown immediately after the incorporation of these residues (Sawada 1973; Reeleder *et al.* 2006). Dabney *et al.* (1996) reported that legume cover crops increased damage to following sorghum emergence due to *Rhizoctonia solani* kühn; however, residue removal combined with selected in-furrow treatment would allow for successful emergence of seedlings.

Nematode and Protozoa Communities

Nematodes are important participants in the underground energy-transfer system. They consume living plant material, fungi, bacteria, mites, insects, and each other, and they themselves are consumed in turn. Some fungi capture nematodes with traps, sticky knobs, and other specialized strategies (Dropkin 1980). Nematodes and protozoa regulate mineralization processes, for example, Guerena (2006) suggests that between 30 to 50 percent of nitrogen present in crop plants is made available by the activity of bacteria-consuming nematodes.

Many studies have revealed the relationship between cover cropping and nematode ecology. Oat, marigold (*Tagetes* sp.), guinea-grass (*Panucum maximum* Jacq.), and sum-hemp (*Crotalaria juncea* L.) exert a particularly noticeable nematicidal effect, and, when incorporated into the soil, they have been shown to suppress rootknot nematodes (*Meloidogyne arenaria* and *M.chitwoodi*) (Guertal *et al.* 1998). Typically, crops high in nitrogen (N) with a low C:N ratio (Rodríguez-Kábana *et al.* 1987), or those containing tannins, phenols (Mian and Rodríguez-Kábana 1982) or glucosinolates (Gustne and Jung, 1985) have nematicidal activity. To suppress root-lesion nematodes (*Pratylenchus penetrans* Cobb), marigold is grown as a rotation crop, and it also helps to add N to subsequent crops as it decomposes (Ball-Coelho *et al.* 2001).

Hairy vetch grown as a cover crop has been shown to support larger populations of *M. arenaria* and *M. javanica* than fallow plots or those containing rapeseed cover crops (Johnson *et al.* 1992). Rye, wheat and annual ryegrass were evaluated as winter cover crops to suppress nematode populations and increase yield of soybean (Pederson and Rodríguez-Kábana, 1991). These cereal species significantly lowered populations of lesion (*Pratylnchus* spp.) and stunt (*Tylenchorhynchus* spp.) nematodes in both years of the study, but soybean yields were higher in fallow (no cover crop) plots (Pederson and Rodríguez-Kábana 1991).

Van Diepeningen *et al.* (2006) reported that the bactervorous nematode determined to a large extent the differences between the cover crop classes: bacterivorous *Plectus rhizophilus* occurred only in association with a leguminous cover crop, bacterivorous *Acrobeloides* occurred most abundantly in association with graminae cover crops, and, when a cruciferous cover crop was used, the number of carnivorous *Mylonchulus* sigmaturus was highest.

The changes in soil nematode structure are associated with cover crop residue decomposition. However, alleochemicals, which are plant-produced compounds (other than food compounds), also affect the behavior of other organisms in the plant environment. For example, Sudan grass contains a chemical, dhurrin, that degrades into hydrogen cyanide, which is a powerful nematicide (Guerena 2006). Some cover crops have exhibited nematode-suppressive characteristics equivalent to aldicarb, a synthethic chemical pesticide (Diepeningen *et al.* 2006). Bialy *et al.* (1990) found that black mustard (*Brassica nigra*) and brown mustard (*Brassica juncea*) show allelopathic inhibition of other plants. Compounds involved probably include various isothiocyanates, which suppress wheat germination and growth.

Cereal rye produces several compounds that inhibit crops and weeds. The most active compounds are two hydroxamic acids and their breakdown products (Chase *et al.* 1991). Wocjcik-Wojtkowiak *et al.* (1990) reported that residues of tillering plants and rye crop residues contain much lower amounts of allelopathic compounds (various phenolic acids) than do seedlings. Various legumes in the tribe Vicieae (peas, lentils, and vetches) contain Beta-(3-isoxazolinonyl) alanine, which is released into soil as a root exudate, and apparently is an allelopathic compound (Schenk and Werner 1991).

Chen (2007) examined the long-term effects of tillage practices on the soybean cyst nematode, Heterodera glycines, and soybean yields in corn–soybean rotation suggested that tillage is not an option solely for managing SCN population in the northern climate and soil conditions. On the other hand, cover crops and tillage systems showed significant interaction for nematode population and pathogen nematode. Kusumoto *et al.* (2006) reported total nematode population was the highest no tillage with hairy vetch cover crops, however, populations of Heterodera spp. that was one of the pathogen nematode for field rice .were significantly lower in no tillage compares with plow and rotary tillage (Figure 10).

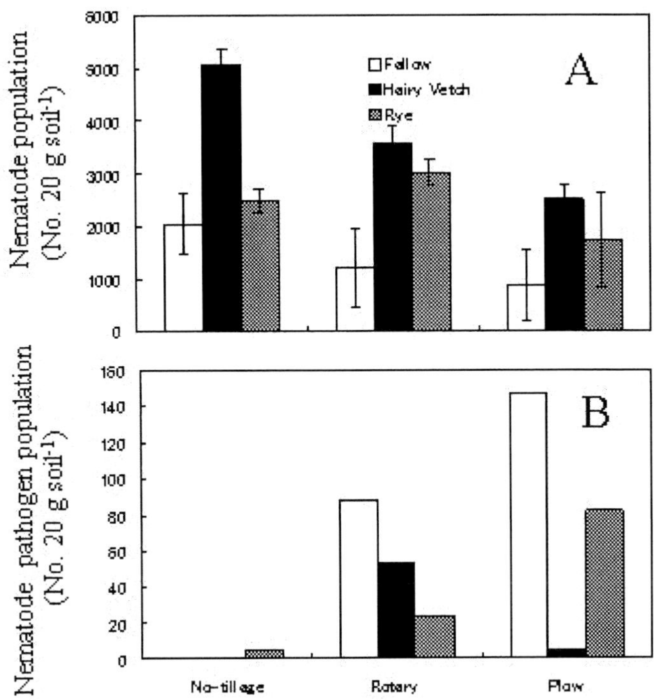

Figure 10. The populations of total nematode (A) and pathgeon nematode (*Heterodera spp.)* (B) in the relation to cover cropping and tillage system were observed in field science center, Ibaraki University, Japan. Soil nematode was extracted by Turuguren methods in 20 g soil, and nematode genera was determined by a microscope. Data from Kusumoto *et al.* (2006).

Protozoa are single-celled animals that feed primarily on bacteria, but also eat other protozoa, soluble organic matter, and sometimes fungi. When they eat bacteria, protozoa release excess nitrogen that can then be used by plants and other members of the food web (Magdoff 1998). While cover crop residue may increase protozoa, that reduces bacterial biomass and increases available N in soil (Clarholm 1985; Kuikman *et al.* 1991). Nematodes and protozoa are by far the most numerous predators of bacteria in soil, but other organisms may also play a role in the N cycling process. Microarthropods and enchytraeids, for example, also regulate bacterial prey populations and release the N sequestered in these organisms (Hansson *et al.* 1990).

Soil Fauna Diversity

Cover cropping and no-tillage practices show a strong association with soil fauna and microbial diversity. No-till with cover crops particularly enhances the population of arthropods (House 1989), and cover crop residue cover on the soil surface also enhances soil micro and macro fauna populations (Wilson-Rummenie *et al.* 1999). On the other hand, Gu *et al.* (2004) reported that soil nematode populations rapidly increased with cover cropping, while soil meso-fauna populations did not show any significant change between cover crop and fallow treatment in several tillage systems.

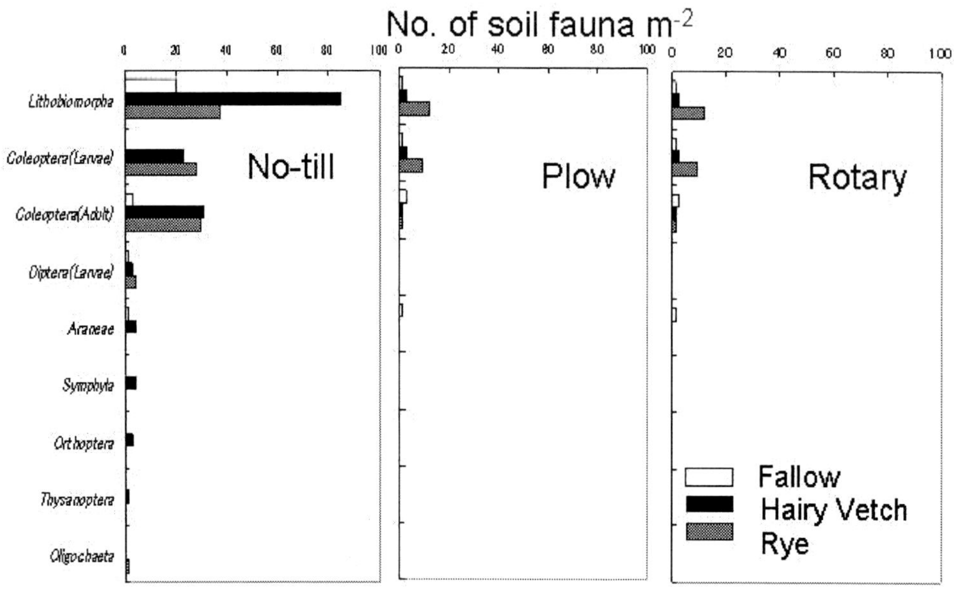

Figure 11. Soil fauna population in the relation to cover cropping and tillage system were obtained in field science center, Ibaraki University, Japan. Soil fauna was extracted by hand sorting methods and genera and families were determined by a microscope. Data from Kusumoto *et al.* (2006).

Cover crops and tillage systems showed significant interaction for soil fauna population and species. Kusumoto *et al.* (2006) examined soil fauna population in the relation to cover cropping and tillage system were obtained in Ibaraki, Japan. Soil macro-fauna populations and species were significantly bigger in no-till with cover crops treatments, however, plow and rotary tillage systems with cover crop did not increased in soil fauna populations (figure 11). Reeledera *et al.* (2006) also examined the impact of tillage regime, cover crop, and nitrogen on various soil organisms in Canada, and reported spring-sampled worm populations were significantly higher in no-till plots than in conventionally tilled plots. Fall-sampled populations were not affected as greatly by tillage, but were generally higher in no-till plots not receiving additional nitrogen or in plots overseeded with a rye cover crop.

Soil macro fauna population is associated with soil meso fauna population. Kusumoto *et al.* (2006) also reported that the soil ecosystem for no-till with rye cover crop showed significant higher co-relationship between the population of predator and soil meso fauna, although these co-relationship was lower in the soil ecosystem for conventional tillage systems. These results suggested that the soil ecosystem in no-tillage with cover crops improve the soil food web structure (Figure 12).

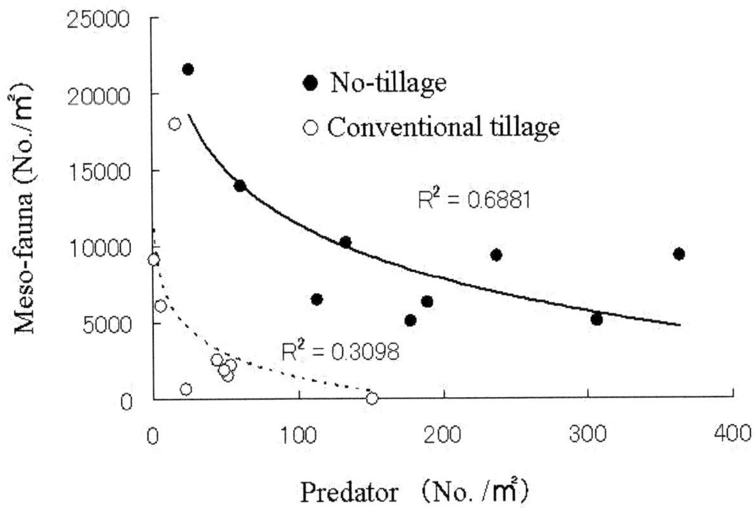

Figure 12. The relationship between the predator population and meso fauna population in no-tillage and conventional tillage soil ecosystem. Data from Kusumoto *et al.* (2006).

Cover crop pollen and nectar can be important food sources for predatory mites and parasitic wasps, which are both important biological controllers of insect feeders that help decrease populations of pests. Reeleder *et al.* (2006) reported that prostigmatid mite populations were generally higher in no-till plots, but spring populations of mesostigmatid mites were higher in plots with a rye cover crop than in plots without an overwintering plant cover. Collembola populations were dominated by the families *Onychiuridae* and *Isotomidae*, but neither was greatly affected by any tillage treatment.

Soil fauna such as earthworms mold the soil, thereby enhancing the accessibility of organic matter to microorganisms. Soil fauna also effects the stabilization of soil organic matter; for example, the earthworm caste has shown long-term stabilization of carbon in agricultural soil (Wolters *et al.* 2000). Using cover crops helps to increase earthworm populations by increasing their food supply and by giving them a longer season to eat and reproduce. Cover crops insulate worms from cold weather in the fall and from warm weather in southern climates. The extra food and ground cover provided by cover crops are especially important where earthworms are removing a high percentage of crop residues (Soil Quality Institute 2001).

Less intensive tillage systems that leave residues on the surface throughout the year improve the environment for earthworms. The residues provide food, insulate earthworms from weather conditions, provide cover to protect them from birds and other surface predators, and protect their burrows. Although a single tillage event will not drastically reduce earthworm populations, repeated tillage over time will cause a decline in earthworm populations. For example, Earthworms were reduced by 70% compared to previously undisturbed sod after five years of plowing (Edwards and Bohlen 1996). In a Georgia experiment, no-till fields had an average of 967 earthworms/m^2 compared to 140 /m^2 in conventionally tilled fields (Coleman and Crossley 1996). There is a strong correlation between earthworm numbers and the amount and quality of residue returned to the soil. Generally, cereal cover crops such as wheat (especially if straw is returned to the soil) encourage earthworms more than crops which leave less residue such as soybean (Kladivko 1993) Therefore, no-tillage and cover crops may enhance the activities of soil ecosystem engineers which in turn may help to establish sustainable crop production.

New Soil Management Strategy Using Cover Crop Reseedings

Cover crops have huge advantages to soil ecosystem, however, there are a number of potential disadvantages to using cover crops. The primary disadvantage in Japan is the cost, particularly for legume seeds, but the expense of legume cover crops might be offset by the value of the legumes' nitrogen fertilizer equivalence (Reeves 1993).

Subterranean clover (*Trifolium subterraneum*) appears to be a suitable winter cover crop for silage corn production based on its relatively early date of full benefit, high production of dry matter, high nitrogen concentration, and reseeding capability. The advantages of subterranean clover include its decumbency and mats of creeping stems that suppress weeds, high nitrogen content and estimated nitrogen fertilizer equivalent of 12–103 kg N/ha (Duke, 1981). Subterranean clover is unique in that its ripening heads turn downward and attempt to bury themselves in the 1 cm depth soil layer where the seeds mature (Enack and Ilnicki 1990). Wheeler and Hill (1957) reported on a pasture where a subterranean clover stand was maintained for 25 years by reseeding; therefore, continuous subterranean clover reseeding could be possible in upland dry fields through several years of proper maintenance.

Komatsuzaki (2006) demonstrated the numerical model regarding the relationship between subterranean clover reseeding and tillage system in terms of weed management. Figure 13 was illustrated the model for predicting reseeding cover crop seedling population included the following sub models: (1) subterranean clover seed production and seed survival; (2) soil and weed management effects; (3) subterranean clover seed bank dynamics; and (4) subterranean clover growth and weed competition.

The appearance of autumn subterranean clover persistency occurred for the following three reasons: many seeds were produced during the period before summer crop planting time; there was some seed persistency under different tillage treatment; and subterranean clover growth competed with weed growth. It is necessary to count the effective number of seed sets and the density of the deposited surviving seed portions under specific conditions of subterranean clover management. Komatsuzaki (2006) summarized that reseeded seedlings were numerous in the case of no-tillage treatment in summer. Thus, if no-till fields are cleared of perennial weeds by contact herbicide spraying, the no-tillage method after subterranean clover seed maturation may have a good effect on the reseeding of subterranean clover. As most silage corn cultures require about 100 days for growth, silage corn should be sown immediately after subterranean clover seed maturation.

Weed management such as contact herbicide spraying and inter-tillage effectively controlled weed growth, and the treatment did not affect the reseeding of subterranean clover. The weed management strategy was also effective in preventing perennial weed invasion in fields, and continuous subterranean clover reseeding may be possible in an upland field using proper cultivation techniques for several years. Therefore, subterranean clover as a winter cover crop appears to be very useful for sustaining crop production in Japan, and the present model will be used widely to realize a sustainable agricultural system utilizing subterranean clover reseeding. Most field crops are planted in April or May, so a field management system for subterranean clover reseeding that takes this into account should be provided to farmers to ensure higher yields and more efficient land use.

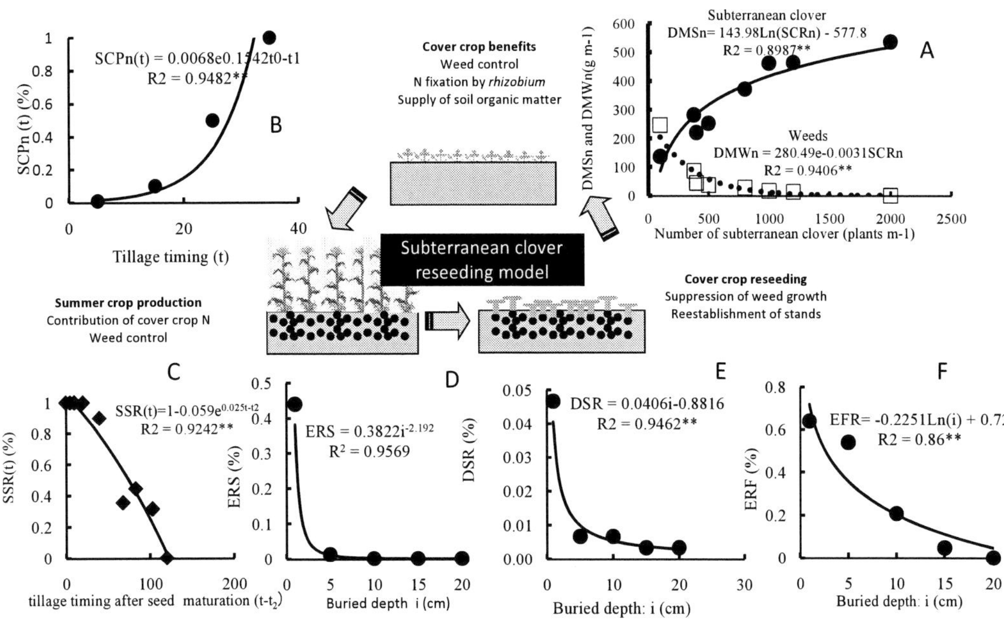

Figure 13. New soil management strategy using subterranean clover reseeding with submodels regarding the relationship between subterranean clover DM and weed DM in the spring (A) , Equations regarding subterranean clover seed production (B) , seed survival ratio (C), and subterranean clover seed bank dynamics including the relationship between buried depth of subterranean seeds and emergence ratio (ERS) in summer (D), the relationship between buried depth of subterranean clover seeds and dormant seed ratio (DSR) (E), and the relationship between buried depth of subterranean clover seeds and emergence ratio in the fall (EFR)(F).The prediction model of subterranean clover reseeding has developed based on these sub models. This management strategy includes several agronomical benefits through agro-ecosystem processes at each stage of land use management. Modified from Komatsuzaki (2007).

These field researches revealed that some economical benefits in introducing the subterranean clover reseeding system to field crop production in Japan, because self-reseeding may eliminate the annual cost of cover crop seeds. On the other hand, there are some countries where chemical fertilizer is more expensive than labor, so the cost saved by not using fertilizer might pay for the cost of the seeds. Reseeding subterranean clover might have other economic benefits in those countries that introduce this system in field production.

Conclusions

Concern for soil management will become more widespread and noticeable because further increases in agricultural output are essential for promoting equity and maintaining global political and social stability. Soil micro- and macro organisms contribute a wide range of essential services to the sustainable function of all ecosystems by acting as the primary driving agents of nutrient cycling; regulating the dynamics of soil organic matter, soil carbon sequestration and greenhouse gas emission; modifying soil physical structure and water regimes, enhancing the amount and efficiency of nutrient acquisition by the vegetation; and by enhancing plant health. To meet the growing demand for and pressures on land and water resources, it will be necessary to develop and adopt eco-specific, eco-friendly and system-based soil management practices. However, for conventional agricultural practices, the planned biodiversity above-ground is reduced with the intention of increasing the economic efficiency of the system. This impacts the associated biodiversity of the ecosystem – micro-organisms and invertebrate animals both above and below ground lowering the biological capacity of the ecosystem for self-regulation and thence leading to further need for substitution of biological functions with agrochemical and petro-energy inputs.

No-tillage with cover cropping will be a critical tool for improving soil ecosystem service increase SOC contents, maintain agriculture productivity, scavenging soil residual nutrients, especially soil fauna organism in no-till with cover crop showed higher population and diversity compared with conventional tilled soil. No-till with cover crops offer the opportunity to enhance soil quality by their influence on C and N dynamics associated with soil microbial and fauna activities. In order to maximize the potential of no-till and cover crops as a soil management tools, it is essential to obtain and utilize information that expands our ability to manage soil eco-systems.

References

[1] Abdul-Baki, A.A., Teasdale, J.R., Korcak, R., Chitwood, D.J. and Huettel, R.N. 1996. Fresh market tomato production in a low-input alternative system using cover crop mulch. *HortScience* **31**:65-69.

[2] Arrouays, D., Balesdent, J., Germon, J.C., Jayet, P.A., Soussana, J.F., and Stengel, P. 2002. Increasing carbon stocks in French agricultural soils? (Available from Scientific Assessment Unit for Expertise, the French Institute for Agricultural Research, Paris at http://www.inra.fr/actualites/rapport-carbone/synthese-anglais.pdf, accessed 26 June 2006)

[3] Ball-Coelho,B. R., Reynolds, L.B., Back, A. J., and Potter, J. W. 2001. Decomposition and Soil Nitrogen are Affected by Mowing and Fertilization of Marigold. *Agron. J.* **93**:207-215.

[4] Bauer, P.J. and Reeves, D.W. 1999. A comparison of winter cereal species and planting dates as residue cover for cotton growth conservation tillage. *Crop. Sci.* **39**:1824-1830.

[5] Barford, C.C., Wofsy, S.C., Goulden, M.L., Munger, J.W., Pyle, E.H., Urbanski, S.P., Saleska, S.R., Fitzjarrald, D. and Moore, K. 2001. Factors controlling long- and short-term sequestration of atmospheric CO_2 in a mid-latitude forest. *Science* **294**:1688–1691

[6] Bialy, Z., Oleszek, W., Lewis, J., and Fenwick, G.R. 1990. Allelopathic potential of glucosinilates (mustard oil glycosides) and their degradation products against wheat. *Plant and Soil* **129**:277-281.

[7] Brown, L.R. 2004. Outgrowing the earth. Earth Policy Institute, New York.

[8] Bruce, R.R., Langdale, G.W., West, L.T., and Miller, W.P. 1995. Surface soil degradation and soil productivity restoration and maintenance. *Soil Sci. Soc. Am. J.* **59**:654–660.

[9] Chase, W.R., Nair, M.G., and Putnam, A.R.1991. 2,2'-oxo-1,1'-azobenzene: selective toxicity of rye (*Secale cereale* L.) allelochemicals to weed and crop species. *J. Chem. Ecol.* **17**:9-19.

[10] Clarholm, M. 1985. Interactions of bacteria, protozoa and plants leading to mineralization,of soil nitrogen. *Soil Biol. Biochem.* **17**:181-187.

[11] Chen, S. 2007. Tillage and Crop Sequence Effects on *Heterodera glycines* and Soybean Yields. *Agron J* **99**:797-807.

[12] Coleman, D.C., and Crossley, Jr. D.A. 1996. *Fundamentals of soil ecology*. Academic Press, San Diego, CA.

[13] Creamer, N.G, Bennett, M.A., Sitinner, B. R., Cardina J., and Regnir, E.E. 1996. Mechanisms of weed suppression in cover crop based production system. *HortScience* **31**:410-413.

[14] Dabney, S.M. 1998. Cover crop impacts on watershed hydrogy. *Soil and Water Cons.* **53**:207-213.

[15] Dabney, S.M., Schreiber, J. D., Rothrock C.S., and Johnson, J.R. 1996. Cover crop affect sorghum seedling growth. *Agron. J.* **88**:961-970.

[16] Delgado, J.A. 1998. Sequential NLEAP simulations to examine effect of early and late planted winter cover crops on nitrogen dynamics. *Soil and Water Cons.* **53**:241-244.

[17] Derpsch,R. 2007. The No-tillage revolution in south America. *Farm Tech 2007 proceedings.* pp54-69. Edmonton, Alberta, Canada.

[18] Diepeningen, A.D., Vos, O.J., Korthals, G.W., and Bruggen, A.H.C. 2006. Effects of organic versus conventional management on chemical and biological parameters in agriculture soils. *Appl. Soil Ecol.* **31**:120-135.

[19] Distch, D.C. and Alley, M.M. 1991. Non leguminous cover crop management for residual N recovery and subsequent crop yields. *J. Fertilizer Issues.* **8**:6-13.

[20] Dropkin, V. H. 1980. *Introduction to Plant Nematology*, p.38–44, p.242–246, p.256. John Wiley & Sons, New York, NY.

[21] Duke, J.A. 1981.Triforlium subterraneum L. In *Handbook of legumes of world economic importance*. Plenum Press, New York.

[22] Dyck, E. and Liebman, M.1995. Crop-weed interface as influenced by a leguminous or synthetic fertilizer nitrogen source: α. Rotation experiment with crimson clover, field corn, and lambsquarters. *Agric. Ecosyst. Environ.* **56**:109-120.

[23] Edwards, C.A. and Bohlen, P.J. 1996. *Biology and ecology of earthworms.* 3rd ed Chapman& Hall, London, UK.

[24] Ebelhar, S.A., Frye, W.W., and Blevins, R.L. 1984. Nitrogen from legume cover crops for notillage corn. *Agron. J.* **76**:51-55.

[25] Elliott, E.T., Burke, I.C., Monz, C.A., Frey, S.D., Lyon, D.J., Paustian, K., Collins, H.P., Halvorson, A.D., Huggins, D.R., Paul, E.A., Turco, R.F., Cole, C.V., Hickman, M.V., Blevins, R.L., and Frye, W.W. 1994. Terrestrial carbon pools: preliminary data from Corn Belt and Great Plains regions. In *Defining soil quality for a sustainable environment.* Doran JW, Coleman DC, Bezdicek DF, Stewart BA (*eds*) *SSSA Special Publication* No. 35. Soil Sci. Soc. Am., Agron. Soc. Am., Madison WI, pp 179–191.

[26] Enach, A.J. and Ilnicki, R.D. 1990. Weed control by subterranean clover used as a living mulch. *Weed Technology*. **4**, 534-538.

[27] Follett, R., Castellanos, J.Z., and Buenger, E.D. 2005. Carbon dynamics and sequestration in an irrigated Vertisol in central Mexico. *Soil Tillage Res*. **83**:148–158..

[28] Fujii, Y., Shibuya, T., Nakatomi, K., Itami, T., Hiradate S., and Paravez, M.M. 2004. Assessment method for allelopathic effect from leaf litter leachates. *Weed Biol. and Manage*. **4**:19-23.

[29] Gu, S., Komatsuzaki, M., Ohta, H., Motobayashi, T., Moriizumi, S., and Yamaguchi, M. 2004.Covcr cropping and tillage system affect soil carbon dynamics and soil organism in field rice cultivation. *Jpn. J. Farm Work Res*. **39**:83-92. (*in Japanese*)

[30] Gu S., Komatsuzaki, M., Moriizumi, S., and Mu, Y. 2004. Soil nitrogen dynamics in relation to cover cropping. *Jpn. J. Farm Work Res*. **39**:9-16. (*in Japanese*)

[31] Guerena, M. 2006. Nematode: Alternative controls. (Available at National sustainable agriculture information service, http://www.attra.org/attra-pub/nematode.html#ingh96. accessed 8 Dec. 2006.)

[32] Guertal, E.A., Sikora, E.J., Hagan, A.K., and Rodrìguez-Kábana, R. 1998. Effects of winter cover crops on populations of southern root-knot and reniform nematodes. *Agric Ecosyst. Environ*. **70**:1-6.

[33] Gupta, V.V.S.R., and Germide, J.J. 1988. Distribution of microbial biomass and its activity in different soil aggregate size classes as affected by cultivation. *Soil Biol. Biochem*. **20**:777-786.

[34] Hansson A. C., Andren, O., Bostrom, S., Bostrom, U., Clarholm, M., Lagerlof, J., Lindberg, T., Paustian, K., Petterson, R., and Sohlenius, B. 1990. Structure of the agroecosystem, p.41-83. In Ecology of Arable Land-Oranisms,Carbon and Nitrogen Cycling. Andern, O. (Ed.). Ecological Bulletin,Copenhagen.

[35] Herbert, S.J., Lithfield, G.V., Liu Zhi-yi.1986. An examination of cover crop seeding dates. *Applied Agricultural Research* vol.1 No.2, 91-95.

[36] Horimoto, S., Araki, H., Ishimoto, M., Ito, M., and Fujii, Y. 2002. Growth and yield of tomatoes grown in hairy vetch incorporated and mulched field. *Jpn. J. Farm Work Res*. **37**:231-240.

[37] Hu, S.J., van Bruggen, A.H.C., and Grunwald, N.J. 1999. Dynamics of bacterial populations in relation to carbon availability in a residue-amended soil. *Appl. Soil Ecol.* **13**:21-30.

[38] Ismail, I., Blevins, R.L., and Frye, W.W. 1994. Long term no-tillage effects on soil properties and continuous corn. *Soil Sci. Soc. Am. J.* **58**:193–198.

[39] Jackson, L.E. 2000. Fates and Losses of Nitrogen from a Nitrogen-15-Labeled Cover Crop in an Intensively Managed Vegetable System. *Soil Sci. Soc. Am. J.* **64**:1404-1412.

[40] Johnson,A.W., Golden, A.M., Auld, D.L., and Sumner, D.R. 1992. Effects of rapeseed and vetch as green manure crops and fallow on nematodes and soil-borne pathogens. *J.Nematol*. **24**:117-126.

[41] Johnson, J.M.F., Reicosky, D.C., Allmaras, R.R., Sauer, T.J., Venterea, R.T., and Dell, C.J. 2005. Greenhouse gas contributions and mitigation potential of agriculture in the central USA. *Soil Till. Res*. **83**:73–94.

[42] Kladivko, E.J. 1993. Earthworms and crop management. Agronomy 279. Purdue Univ. Extension Service. http://www.agcom.purdue.edu/AgCom/Pubs/AY/AY-279.html. (Acsessed on 24 Sep, 2007)

[43] Komatsuzaki, M.1999. Cropping system and rotation. In *Sustainable agriculture system management*. Endo, O. (Ed.) Nourin Toukei Kyoukai, Tokyo (*in Japanese*)

[44] Komatsuzaki, M. 2002. New Cropping Strategy to Reduce Chemical Fertilizer Application to Silage Corn Production Using Subterranean Clover Reseeding. *Jpn. J. Farm Work Res*. **37**: 1-11

[45] Komatsuzaki, M. 2004. Use of cover crops in upland fields. *Jpn. J. Farm Work Res.* 39(3):157–163 (*in Japanese*)

[46] Komatsuzaki, M. 2007. New weed management strategy using subterranean clover reseeding under different tillage systems:Numerical experiments with the subterranean clover-tillage dynamics model. *Weed Biology and Management* **7**, 3–13

[47] Komatsuzaki, M., and Ohta, H. 2006. Soil management practices for sustainable agro-ecosystems. *Sustain Sci*. **2**,103-120.

[48] Komatsuzaki, M., Kusumoto, R., Okiba,C., Asagi, N., Ueno, H., Araki, M., Araki, H., and Matsumura, S. 2007. N Release from a Nitrogen 15 Labeled Cover Crop in Relation to Soil Fauna Activities between No-till and Conventional Tilled Upland Rice Production. *The ASA-CSSA-SSSA International Annual meetings*, **147-7**. New Orleans, USA.

[49] Komatsuzaki, M., Kai, Y., and Nakamura, Y. 2007. Nutrient evaluation of winter cover crop subsequently produced after rice cultivation. *Jpn. J. Farm Work Res*. **42**(2): 74-83. (*in Japanese*)

[50] Kuikman P.J., Jansen, A. G., and Van Veen, J.A. 1991. ^{15}N-Nitorogen mineralization from bacteria by protozoan grazing at different soil moisture regimes. *Soil Biol Biochem*. **23**:193-200.

[51] Kusaba, T. 2001. Development of soil diagnosis research, *J. Agr. Sci.* **56**:7-12. (*in Japanese*)

[52] Kusumoto, R., Komatsuzaki, M., Ymaguchi, S., Araki, M. 2006. Effects of tillage system and cover cropping on soil fauna dynamics. *Jpn. J. Farm Work Res.* **41**(extra issue 1):147-148. (*in Japanese*)

[53] Lal, R. 1998. Soil erosion impact on agronomic productivity and environment quality. *Crit. Rev. Plant Sci.* **17**:319–464

[54] Lal, R. 2001. World cropland soils as source or sink for atmospheric carbon. *Adv. Agron* **71**:145–91

[55] Lee, J.J., Phillips, D.L. and Lui, R. 1993. The effects of trends in tillage practices on erosion and carbon content of soils in the US Cornbelt. *Water Air Soil Pollut.* **70**: 389–401.

[56] Legg, J.O. and Meisinger, J.J. 1982. Soil nitrogen budgets, p. 503-566. In *Nitrogen in agricultural soils.* Stevenson, F.J. (Ed.) American Society of Agronomy, Madison, Wisconsin.

[57] Magdoff, F., 1998. Cover crops, p.73-89. In *Building Soils for better crops*. Magdoff, F. (Ed.) University of Nebraska Press, Lincoln and London.

[58] Martin, J.H., Leonard, W.H., and Stamp, D.L. 1976. *Principles of field crop production, 3rd ed.* Macmillan Publishing Co., Inc., New York.

[59] McCrancken, D.V., Smith, M.S., Grove, J.H., MacKown, C.T., and Blevines, R.L.1994. Nitrate leaching as influenced by cover cropping and nitrogen source. *Soil Sci. Soc. Am. J.* **58**:1467-1483.

[60] McCarty, G.W., Meisinger, J.J. and Jenniskens, F.M.M. 1995. Relationships between total-N, biomass-N and active-N in soil under different tillage and N fertilizer treatments. *Soil Biol. Biochem*. **27**:1245–1250.

[61] Mian, I.H. and Rodríguez-Kábana, R. 1982. Survey of the nematicidal properties to soil organic materials available in Alabama as amendments to soil for control of Meloidogyne arenaria. *Nematropica* **12**:235-246.

[62] Monks, C.D., Basden, T., McFarland, J.L., Poland, S.M., and Rayburn, E.1977. Cover crop response to late season planting and nitrogen application. *J.Prod.Agric*. **10**: 289-293.

[63] Mu, Y., Komatsuzaki, M., Moriizumi, S., Gu, S., Araki, M., Araki, H., and Hirata, S. 2007. Effects of cover cropping and tillage system on soil physical character in upland rice cultivation. *Jpn. J. Farm Work Res*. **42**(1):11-20. (*in Japanese*)

[64] Mu, Y., Komatsuzaki, M., Moriizumi, S., Ryu, E., Araki, H., and Hirata, S. 2007. Effects of tillage system and cover cropping on CO_2 flux and field rice yields. *Jpn. J. Farm Work Res*. **42**(Extra Issue 1):45-46. (*in Japanese*)

[65] Mullen, M.D., Melhorn, C.G., Tyler D.D., and Duck, B.N. 1998. Biological and biochemical soil properties in no-till corn with different cover crops. *Soil and Water Cons*. **53**:219-224.

[66] Nierenberg, D., and Halweil, B. 2005. Cultivating food security. In *State of the world* 2005, Renner, M., French, H. and Assadourian, E. (Eds.) W.W. Norton, New York, pp 62–79.

[67] Nihon Nougyo Houjin kyokai 2005. The proposal of policy that across agricultural products. (Available at Nihon Nougyou Kyokai http: http://www.hojin.or.jp/051012.pdf accessed 8 Dec. 2006.) (*In Japanese*)

[68] Ogura, H. 1966. Studies on the saprophytic behaviour of soil borne pathogenic fungi. Ī. On the existence of *Rhizoctonia solani, Fusarium oxysporum* and *Pyth-ium aphanidermatum* utilizing plant debris in soil. *Ann. Phytopath. Soc. Japan* **32**:236-243.

[69] Peet, M. 1996. Chapter 1: soil management, p. 1–28. In Sustainable practices for vegetable production in the south. Peet, M. (Ed.) R. Pullins Company, MA, USA.

[70] Peixoto, R.S., Coutinho, H.L.C., Madari, B., Machado, P.L.O.A., Rumjanek, N.G., Van Elsas, J.D., Seldin, L., and Rosado, A.S. 2006. Soil aggregation and bacterial community structure as affected by tillage and cover cropping in the Brazilian Cerrados. *Soil and Tillage Res*. **90**: 16-28.

[71] Pederson, J.F., and Rodríguez-Kábana, R. 1991. Winter grass cover crop effects on nematodes and yields of double cropped soybean. *Plant and Soil* **131**:287-291.

[72] Quemada, M., and Cabrera, M.L. 1995. Carbon and nitrogen mineralized from leaves and stems of four cover crops. *Soil Sci. Soc. Am. J.* **59**:71-477.

[73] Ranells, N.N., and Wagger, M.G. 1996. Nitrogen release from grass and legume cover crop monoculture and biculture. *Agron.J.* **88**:777-782.

[74] Ranells, N.N. and Wagger, M.G. 1997. Nitrogen-15 recovery and release by rye and crimson clover cover crops. *Soil Sci.Soc. Am.J.* **61**:943-948.

[75] Raper, R.L., Reeves, D.W., Burmester, C,H,, and Schwab, E,B. 2000. Tillage depth, tillage timing, and cover crop effects on cotton yield, soil strength, and tillage energy requirements. *Applyied Engineering in Agriculture.* **16**(4):379-385.

[76] Reeleder,R.D., Miller, J.J., Ball Coelho, B.R., and Roy, R.C. 2006. Impacts of tillage, cover crop and nitrogen on populations of earthworms, microarthropods, and soil fungi in a cultivated fragile soil. *Applied soil ecology*. **33**:243-257.

[77] Reeves, D.W. 1997. The role of soil organic matter in maintaining soil quality in continuous cropping systems. *Soil Tillage Res.* **43**:131–167.

[78] Reeves, D.W. 1993.Cover crop and rotation, In *Crops Residue Management*. Hatfield, J.L. (Ed.), CRC Press, Boca Raton, FL. 125-172.

[79] Rodríguez-Kábana,R., Morgan-Jones, G., Chet, I. 1987. Biological control of nematodes; Soil amendments and microbial antagonists. *Plant and Soil* **100**:237-247.

[80] Sainju, U. M. and B. P. Singh 2001. Tillage, Cover Crop, and Kill-Planting Date Effects on Corn Yield and Soil Nitrogen. *Agron. J.* **93**:878-886.

[81] Sainju, U. M., Singh, B. P., and Whitehead, W. F. 1998.Cover crop root distribution and its effects on soil nitrogen cycling. *Agron. J.* **90**:511–518.

[82] Sarrantonio, M. 1998. Building fertility and tilth with cover crops, p.16–24. In *Managing cover crops profitably*. Clark A (Ed.) Sustainable agriculture network, Maryland, USA.

[83] Sawada, Y. 1973. The toxic of plant residue and their effects on following crop growth. *Agriculture and Horticulture* **48**:528-532.

[84] Schenk, S.U., and Werner, D. 1991. Beta-(3-isoxazolin-5-on-2yl)-alanine from Pisum: allelopathic properties and antimycotic bioassay. *Phytochemistry* **30**:467-470.

[85] Schutter, M.E., and Dick, R.P. 2002. Microbial community profile and activities among aggregate of winter fallow and cover cropped soil. *Soil Sci.Soc.Am.J.* **66**:142-153.

[86] Sheng, Q., Hunt, L.A. 1991. Shoot and root dry weight and soil water in wheat, triticale and rye. *Can.J.Plant Sci.* **71**:41-49.

[87] Shipley, P.R., Meisingerm J.J., and Decker, A.M. 1992. Conserving residual corn fertilizer nitrogen with winter cover crops. *Agron. J.* **84**:869-876.

[88] Singogo, W., Lamont Jr., W.J., and Marr, C.W. 1996. Fall-planted cover crops support good yields of muskmelons. *HortScience*. **31**:62-64.

[89] Six, J., Elliot, E.T., and Paustian, K. 1999. Aggregate and soil organic matter dynamics under conventional and no-tillage systems. *Soil Sci. Soc. Am. J.* **63**:1350–1358.

[90] Soil Quality Institute 2001. Agricultural management effects on earthworm populations. Agronomy technical note. No. 11 (Available at Soil quality institute, http: http://www.agry.purdue.edu/cti/pdfs/earthworms.pdf accessed 8 Dec, 2006.

[91] Stamatiadis, S., Werner, M. and Buchanan, M. 1999. Field assessment of soil quality as affected by compost and fertilizer application in a broccoli field (San Benito County, California). *Appl. Soil Ecol.* **12**:217–225

[92] Statistics of Agriculture, Forestry and Fishers 2005. Crop land area and field use ratio in 2004.(available at Minster of Agriculture, Forestry and Fishers, http: http://www.maff.go.jp/toukei/sokuhou/data/nobemenseki2004/nobemenseki2004.pdf accessed 5 Nov. 2006) (*in Japanese*)

[93] Staver, K.W., and Brinsfield, R. B.1998. Using cereal grain winter cover crops to reduce groundwater nitrate contamination in the mid-Atlantic coastal plain. *Soil and Water Cons*. **53**:230-240.

[94] Stopes, C., Millington, S., and Woodward, L. 1996. Dry matter and nitrogen accumulation by three leguminous green manure species and the yield of a following wheat crop in an organic production system. *Agric.Ecosyst.Environ.* **57**:189-196.

[95] Sustainable agriculture Office. 2006. Results of certification of ecological friendly farmer, (available at Minster of Agriculture, Forestry and Fishers, http: http://www.maff.go.jp/soshiki/nousan/nousan/kanpo/ecofarmer.pdf (accessed 6 Dec. 2006)

[96] Suzuki, K., and Komatsuzaki, M. 2006. Cover crop affects wind erosion. *Jpn. J. Farm Work Res*. **41**(extra issue 1):151–152 (*in Japanese*)

[97] Toucheton, J.T., Gardner, W.A., Hargrove, W.L., and Duncan, R.R. 1982. Reseeding crimson clover as a N source for no-tillage grain sorghum production. *Agron. J.* **74**: 283-287.

[98] UNCED 1992. Agenda 21: program of action for sustainable development, Rio declaration on environment and development, statement of principles. Final text of agreement negotiated by governments at the United Nations Conference on Environment and Development (UNCED), Rio de Janeiro, Brazil, UNDP, New York, pp 3–14.

[99] UNFCCC 1992. United Nations framework convention on climate change. UNFCCC, Bonn, Germany.

[100] UNFCCC 1998. Kyoto Protocol to the United Nations Framework Convention on Climate Change. (Available from UNFCCC at www.unfccc.int/resource/docs/convkp, accessed 13 September 2006)

[101] Van Bruggen, A.H.C., and Semenov, A.M. 2000. In search of biological indicators for soil health and disease suppression. *Appl. Soil Ecol.* **15**:13-24.

[102] Van Diepeningen, A. D., de Vos, O. J., Korthals, G. W., and van Bruggen, A.H.C. 2006. Effects of organic versus conventional management on chemical and biological parameters in agricultural soils. *Appl.Soil Ecol.* **31**:120-135.

[103] Varco, J.J., Frye, W.W., Smith, M.S., and MacKown, C.T. 1989. Tillage effects on nitrogen recovery by corn from a nitrogen-15 labeled legume cover crop. *Soil Sci. Soc. Am. J.* **53**:822–827.

[104] Wagger, M.G. 1989. Cover crop management and nitrogen rate in relation to growth and yield of no-till corn. *Agron, J.* **81**:533-538.

[105] Wagger, M. G. 1996. Reduction of nitrate leaching in agricultural soils via cover crops, (Report no. 303). Water Resources Research Institute of the University of North Carolina, Raleigh, NC.

[106] Wagger, M.G., and Mengel, D.B. 1988. The role of nonleguminous cover crops in the efficient use of water and nitrogen. p. 115–127. In *Cropping strategies for efficient use of water and nitrogen,* Hargrove WL (Ed.) Special Publication No. 51, American Society of Agronomy. Madison WI.

[107] Wheeler W. A., and Hill, D. D. 1957. Sub clover. In *Grassland seeds*. D. Van Nostland Company, Inc , Princeton, NJ, 388-390.

[108] Wilson-Rummenie, A. C., Radford, B. J., Robertson, L. N., Simpson, G. B., and Bell, K. L. 1999. Reduced Tillage Increases Population Density of Soil Macrofauna in a Semiarid Environment in Central Queensland. *Environ Entomol* **28**:163 – 172.

[109] Wocjcik-Wojtkowiak, D., Politycka, B., Schneider, M., and Perkowski, J. 1990. Phenolic substances as allelopathic agents arising during the degradation of rye (*Secale cereale* L.) tissues. *Plant and Soil* **124**:143-147.

[110] Wolters, V., Silver, W.L., Bignell, D.E., Coleman, D.C., Lavelle, P., van der Putten, W.H., de Ruiter, P., Rusek, J.,. Wall, D.H., Wardle, D.A., Brussaard, L., Dangerfield, J.M., Brown, V.K., Giller, K., Hooper, D.U., Sala, O., Tiedje, J., and van Veen, J.A. 2000. Effects of global changes on above- and belowground biodiversity in terrestrial ecosystems: implications for ecosystem functioning. *BioScience* **50**:1089-1098.

[111] Zhaorigetu, Komatsuzaki, M., Sato, Y., Ohta, H., and Kasahara, Y. 2005. Influence of cover crop on soil microbial community, pp22. In *Proceedings of Japanese society of soil microbiology in 2005*, Japanese society of soil microbiology, Nagoya.

In: Ecosystem Ecology Research Trends
Editors: J. Chen and C. Guo, pp. 209-239
ISBN: 978-1-60456-183-8

Chapter 6

INTEGRATED APPROACH FOR THE ASSESSMENT OF BIOTECH SOYBEAN PESTICIDES IMPACT ON LOW ORDER STREAM ECOSYSTEMS OF THE PAMPASIC REGION

***A.E. Ronco*[1,a], *P. Carriquiriborde*[1], *G.S. Natale*[1], *M.L. Martin*[1], *H. Mugni*[2] *and C. Bonetto*[2,b]**

[1]Centro de Investigaciones del Medio Ambiente, Facultad de Ciencias Exactas, Universidad Nacional de La Plata. Calle 47 y 115, La Plata (1900), Buenos Aires, Argentina. CONICET.

[2]Instituto de Limnología Dr. Raúl A Ringuelet, Universidad Nacional de La Plata-CONICET, Av. Calchaquí km 23.5, Florencio Varela (1888), Buenos Aires, Argentina.

Abstract

In the last decades Argentine farmers have massively adopted the genetically modified glyphosate resistant soybean together with no tillage managerial practices. The area cultured with soybeans increased to 15 million ha with consequent higher pesticide consumption. Although recent publications deal with some detailed aspects of pesticide effects on organisms, the overall impact of such vast agricultural intensification remains unreported. The present contribution review and integrate results from laboratory and field studies within the frame of a major project oriented towards an integrated assessment of soybean agrochemicals impacts on aquatic ecosystems from streams of the region. The main conclusions are: i) chemical analysis of water, sediments and soils indicate low impacts on nutrient levels; ii) all the biota components of the studied ecosystems (flora, insects, fish and amphibians) were affected in different extent after pesticide applications, representing an appreciable stress over the systems. Effects were sometimes low, without noticeable impact on resident populations, as observed in fishes, or being more drastic, populations tend to recover after the episodic toxicity pulses, as observed in the resident

[a] E-mail address: cima@quimica.unlp.edu.ar (Corresponding author)
[b] E-mail address: bonetto@ilpla.edu.ar (Corresponding author)

invertebrate fauna; iii) of the three most commonly used insecticides, endosulphan fumigation caused the most severe effects, followed by the terrestrial chlorpyriphos and cypermethrin applications; iv) regarding spatial extent of impacts, the herbicide glyphosate drift and runoff caused chlorophyll content reduction on adjacent riparian and, at least 250 meters downstream application, on aquatic macrophyte vegetation. Although effects were followed by later recovery, studied sites showed a rather poor ruderal composition, suggesting that reiterated applications, together with the history of land use eventually impoverished the regional community originally described for the area. Populations of non target invertebrates were the most sensitive ecosystem components, showing strongest effects, with total disappearance of ephemeroptera, odonata, amphipoda as a response to pesticide runoff, eventually at long distance from spray areas, while fish and amphibian impacts were generally circumscribed to adjacent sectors; v) *in situ* exposed experimental populations of insects, fish and amphibians showed good correspondence with observed responses of resident populations within the same watercourses; vi) finally, complementation between laboratory and field studies resulted of great significance for detecting erroneous predictions from the former and lighten observations from the latter. An integrated approach analysis seems to be a promising option towards the understanding of the overall impact of intense soybean farming on aquatic related environments. Improving managerial practices and riparian habitat restorations seems feasible goals for long term sustainable production.

Introduction

The Argentine Pampa is an area of mild climate and fertile soils developed over deep loessic sediments. It was originally covered by grasslands and for a long time the farmers employed a mixed system of livestock and crops, mainly wheat and corn. The gentle slopes and the occurrence of heavy showers in coincidence with the seeding period made erosion a matter of concern. Although no tillage managerial practices have long been advised it was not until the soybean resistant to glyphosate became available that the direct seeding technique was adopted by the farmers. Several applications of the economical and efficient glyphosate herbicide are needed in order to implement the no tillage practices. During the last few decades, the traditional production system has been replaced by intensive soybean production. The areas cultured with soybeans increased from less than 40,000 ha in the 70s to 8,300,000 ha at the end of the century (Pengue, 2000), reaching at present 15 million ha. Wheat varieties of short growing period allow two harvests per year, wheat followed by soybean. A three year rotation is now a common practice, soybean is cultured the first year, wheat followed by soybean the second, and corn in the third. Livestock was moved to marginal areas or concentrated in feedlots. The fast implementation of new technological developments transformed soybean related products (oil, flour, pellets) into the main Argentine production. Pesticide consumption increased from 6 to 18 million kg only in the period from 1992 to 1997 (Pengue, 2000), and has continued to increase at slower rates since then, being formulations of the herbicide glyphosate and the insecticide cypermethrin the most commonly used. The pyrethroid cypermethrin represents more than half of the total insecticide consumption, followed by the organophosphate chlorpyrifos and the organochlorinated endosulphan. Cypermethrin is sprayed usually twice during the soybean growing period, at a rate of 25 g/ha. Thus, approximately 500 tons of pesticide is introduced annually to the environment. Pesticides are sold freely to farmers in rural areas and applied without supervision or control. Although some recent publications deal with detailed aspects of pesticide effect on the biota, the overall impact of such vast agricultural intensification remains unreported. The present contribution review and integrates the studies performed on the effect of the typical agricultural management in commercial farms on the biota of low order streams draining intensively cultivated rural basins.

Study Sites

Four stream tributaries from the Arrecifes River basin were studied in the main soybean area of the Pampa, 150 to 190 km northwest from Buenos Aires (34°36'S, 58°30' W) throughout the 2000-2001 soybean growing period (Mugni *et al.*, 2006). They are small second- or third order streams in the ranges 2-6 m width, 0.2-0.5 m depth, and 0.1-0.5 m^3/s discharge, draining basins of 380 -500 ha (Figure 1). Two streams, Helves and Luna, have steep slopes and lack floodplains. The other two, Horqueta and Maguire streams developed a 20 to 30 m wide floodplain containing patches of grasses mixed with macrophytes. Maguire developed a natural small wetland of about 100 m long, 20 m wide and 0.2 m deep, colonized by dense stands macrophytes, mainly bulrush (*Schoenoplectus californicus*), and showing luxuriant periphyton growth. Horqueta has a dam for recreational purposes upstream of the sampling site.

Later on, a small first order stream born within a farm close to Arrecifes, 150 km northwest from Buenos Aires, was sampled at two sites, during two successive growing periods from 2002 to 2004 (see detailed insert of Figure 1). The first site was located within the plot in which the stream is born, the second about 250 m downstream, in another plot. The first studied year, the upstream plot was cultivated with soybean, and the downstream plot was maintained with cattle. The second year the upstream plot was cultured with wheat followed by soybean and the downstream plot was cultivated with soybean.

Another stream tributary from the south-eastern coastal sector of the Pampa plains was selected for comparative studies due to the recent introduction of the soybean transgenic culture in the area. The Pescado stream (see detailed insert of Figure 1) is fed by rainwater and groundwater. The stream is 2-5 m width, with a basal depth of about 20 cm, reaching about a maximum of two metres water depth in the event of heavy rain storms. The longitudinal gradient slope varies from 0.4 to 0.1%, running well-drained and developed soils in the upper and middle course (Ronco *et al.*, 2001; Camilión *et al.*, 2006). The studied stream sector runs through cultivated plots mainly with soybean, intercalated with corn or wheat, depending on the year cycle.

Pesticide concentration levels associated to this region have been published by Peruzzo et al. (2003), Marino and Ronco (2005), and Berkovic et al. (2006). Detectable levels of cypermethrin and clorpyrifos were found in downstream locations of main streams and rivers flowing into the lower Paraná basin during the growing season. Measured concentrations of both insecticides in water from the Arrecifes tributary were mostly below detection limits, occasionally reaching maximum values close to 5 and 10 µg/l, respectively. For the case of sediments, detectable levels were more frequently found, with concentrations within the range of 2 and 1075 µg/kg of cypermethrin; and between 1 and 15 µg/kg of chlorpyrifos. A different mode of cypermethrin partition was observed in Pescado stream. Insecticide application in summer coincided with a drought, resulting in low water levels in the stream. Small pools, some times lacking connection among them, with water containing high levels of suspended material were evident during the field assessments. Minimum and maximum concentrations of cypermethrin here were within the range of 0.3 and 94 μg/l in the water, and below 1 μg/kg in the sediments. The insecticide remained associated to suspended material after reaching the water body from application drifts.

The herbicide levels found in the water streams of the studied areas were often below 1 mg glyphosate/l, in Arrecifes tributary. Although, concentration ranges between 1.8 and 10.9 mg/l, and 2.8 and 10.5 mg/kg were detected during sprays in water and sediments of Pescado tributary.

The climate of the region is mild and humid, with an annual mean temperature of 16°C (mean and extreme temperatures for January until midsummer 24°C and 43°C, and July until midwinter 9°C and –5°C, respectively, and the mean annual rainfall is 1000 mm, with a fairly uniform distribution.

A detail of the type of assessment studies in relation to pesticide inputs from crop management activities and runoff associated to following rains is given in table 1.

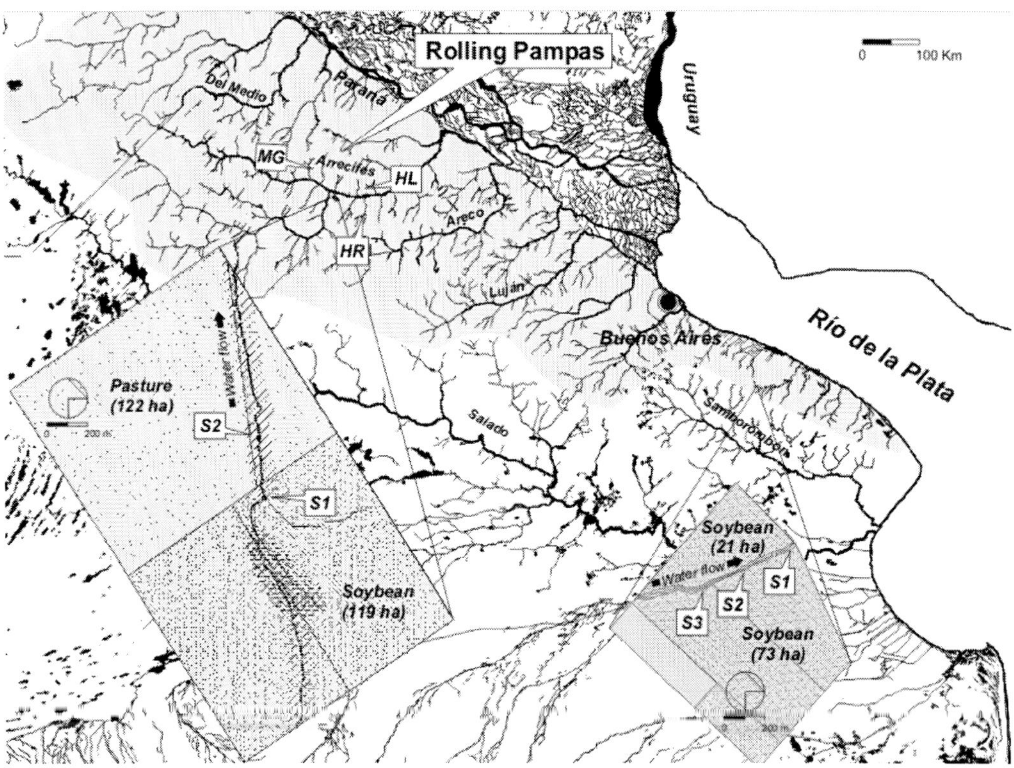

Figure 1. Study areas. MG: Maguire; HL: Helves; HR: Horqueta streams. Insert on the left: Arrecifes stream tributary study area, with sites S1 and S2 locations; insert on the right: Pescado stream tributary study area, with sites S1, S2 and S3 locations.

Table 1. Type of assessment studies performed associated with specific pesticide ground sprays and rain events in stream tributaries. Used pesticide formulations were: Roundup®Max or Roundup®Full for glyphosate; Sherpa for cypermethrin; and Shooter or Lorsban for chlorpyrifos.

Application events in related stream	Pesticide application				Days to the first rain after spray; mm rain	Type of assessment performed
	glyphosate	cypermethrin	chlorpyrifos	endosulphan		
Arrecifes tributary						
First year						
Spring, post-emergence	1500 g/ha	100 ml/ha	-	-	17; 57	Stream and caged fish and amphibian populations. Caged invertebrate.
Summer, pre-fructification	1000 g/ha	-	1000 ml/ha	-	8; 67	Stream and caged flora, fish and amphibians. Caged invertebrate.
Second year						
Summer, post-emergence	1000 g/ha	100 ml/ha	-	-	20; 80	Stream and caged fish and amphibian populations.
Summer, post-fructification	1000 g/ha	100 ml/ha	800 ml/ha	-	9; 34	Stream and caged flora, fish and amphibian populations.
Pescado tributary						
First year						
Spring, pre-emergence	2000 ml/ha				No runoff assessment	Stream and caged fish populations
Summer, post-emergence			800 ml/ha		No runoff assessment	Stream and caged amphibian populations
Summer, post-emergence	1000 g/ha				No runoff assessment	Stream and caged flora.
Summer, post fructification (*)		150 ml/ha		700 ml/ha	No runoff assessment	Stream and caged fish and amphibian populations
Second year						
Summer, post fructification (*)		150 ml/ha		700 ml/ha	No runoff assessment	Stream fish, and stream and caged amphibian populations

(*) corresponds to an aerial application of the pesticide mixture

Nutrients

Conductivity (670-1100 μS/cm), pH (7.2-9.0) and the dominant ions, except nitrates, did not show differences among streams. Conductivity, pH and the dominant ions, including nitrates, decreased in coincidence with rain events and were lower in the runoff than in the streams (Mugni *et al.*, 2005). Soluble reactive phosphorus (SRP) and ammonium showed the opposite trend, being higher in coincidence with the largest discharges after rain events and in the runoff than in the streams. Such higher concentrations were not provided by the rain itself, since nutrient concentrations in rainwater were rather small.

Experiments performed mixing soil samples taken adjacent to each stream and deionized water with the same pH as rain water showed fast pH increase and release of SRP, calcium and bicarbonate. Soil P fractionation showed that the largest soil P pool was organic P, amounting 57-67% of the total P content, followed by iron (19-22%) and calcium bound P (14-21%). The soil which showed the largest pH increase and SRP, calcium and bicarbonate release was the one which showed the largest calcium bound P content. The concentrations of SRP released in the experiments were similar to the concentrations measured in the runoff samples. Since the calcium-bound P fraction is released in response to acidification, it was suggested that the calcium bound P of the surrounding soils contributed most of the observed SRP in the streams after rain events. Fast release of soluble organic P from soil particles seems another P source to the stream water after rain events. Mean SRP showed very little variation among the studied streams attaining an overall mean of 143 μg/l.

Table 2 shows the nutrient concentration in the first order stream at Arrecifes commercial farm during two successive growing periods. Mean SRP concentrations were significantly lower at the plot with soybean than in the downstream plot with cattle on the 2002-2003 growing period. None of them were fertilized. The following year, wheat followed by soybean was cultured at site 1 and soybean was grown at site 2 (Figure 1). Wheat was ploughed in winter and was fertilized with 100 kg/ha of di-ammonium phosphate and 135 kg/ha of urea. SRP concentration at site 1 was significantly higher the second year than the first one. Mean SRP concentration in the Arrecifes first order stream was very similar to mean SRP concentrations previously recorded in the 4 representative streams of the area.

SRP concentrations resulted comparatively high when compared with reported figures from intensively cultivated areas. Schulz & Liess (1999) reported 29 μg/l in Ohebach, (Germany) and Probst (1985) 33 μg /l in Girou (France). Meybeck (1998) compared nutrient concentrations in streams draining agricultural and forested areas in the Seine watershed (France), without observing significant differences between them, and reporting mean SRP concentrations ranging 12-14 μg/l. Similarly, Miller *et al.* (1997) did not observed significant differences between forested and agricultural steams in the Potomac basin (USA), reporting a mean of 20 μg/l. Within the Pampasic streams agricultural loads result comparatively modest, since only a minor proportion of the farmers fertilize the soybean, the culture representing more than half of the cultivated surface, while wheat and corn are fertilized with doses up to 100 kg/ha, but represent a minor proportion of the cultivated land. Breeding cattle represent a still lower proportion of the land surface. However, SRP were significantly higher at the plot with cattle. Although the plot was not fertilized, the roving behaviour of the cattle might have played a role by exploring the grass abundance within the whole plot but concentrating in the stream to drink, and therefore disturbing the riparian strip and increasing the nutrient load.

Serrano *et al.* (1999) showed that leakages from manure increased nutrient loads in a Mediterranean wetland while Nair *et al.* (1995) observed higher labile soil P concentration in grazed pastures than in adjacent forage cultures, suggesting higher leaching to nearby water bodies. Comparatively high SRP concentrations in the Pampasic streams seem more related with the geochemical nature of the rich loessic sediments than with the comparatively modest fertilizer loads.

Table 2. Mean nutrient concentrations in a first order spring along two successive growing periods.

Harvest	N	Site	Culture	Conductivity	PRS	NO_3^-	NH_4^+
				µS/cm	µg/l	µg/l	µg/l
02-03	10	1	Soybean	325-500	144 (75-208)	381 (45-1993)	36 (5-54)
		2	Cattle	430-525	241 (100-322)	1765 (759-3335)	34 (12-61)
03-04	6	1	Wheat/Soybean	190-236	456 (307-775)	142 (31- 410)	69 (25-33)
		2	Soybean	530-600	317 (136-477)	468 (112-998)	60 (33-60)

Nitrate concentration were significantly higher in the two steams lacking floodplain, Helves and Luna, attaining a mean concentration of 3.6 and 3.1 mg N/l, respectively, and lower in the streams provided with a floodplain, Maguire and Horqueta, attaining mean concentrations of 1.4 and 1.1 µg N/l, respectively (Mugni *et al.*, 2005). Nitrate concentrations were higher during dry periods and decreased in coincidence with higher discharges. Nitrate concentrations were significantly lower in runoff than stream samples.

Ammonium stream concentrations were orders of magnitude lower than nitrate concentrations, and were higher in the floodplain streams Maguire and Horqueta attaining means of 30 -31 µg N/l and lower in the Helves and Luna streams lacking floodplain, attaining means of 17 -19 µg N/l. Ammonium concentrations were higher at high water levels and mean concentrations were significantly higher at runoff than stream samples.

Nitrate concentrations at the first order stream at Arrecifes commercial farm were significantly higher at the downstream cattle breeding plot than in the upstream soybean plot (Table 2), during the first studied growing period. None of the plots were fertilized. The following growing period, concentrations were not significantly different at both plots, the upstream one cultivated with wheat followed with soybean and the downstream one cultivated with soybean. As previously stated, wheat was ploughed and fertilized in winter. A lower number of samples concentrated in the soybean growing period were available on the second growing period. Sampling deficiencies might have missed nitrate leaching from the surrounding soils in the first rains after fertilization. Nitrate concentrations at site 2 were significantly lower the second year, when soybean was cultured, than in the first year, when cattle was breaded in the same plot.

Mean nitrate concentrations in the first order stream at the Arrecifes farm was lower than the means of the four previously recorded representative streams of the region, except when cattle was breaded at the downstream plot.

The overall mean nitrate concentrations in the Pampasic streams (2.3 mg N/l) resulted comparatively low when compared with reported figures from intensively cultivated areas in Europe and North America. Mean nitrate concentrations in Ohebach, Germany (Schulz &

Liess, 1999) and Girou, France (Probst, 1985), streams attained 5.9 and 5.7 mg N/l, respectively. Meybeck (1998) compared the nitrate concentrations in forested and agricultural basins of the Seine watershed, the agricultural basins showed an order of magnitude higher nitrate concentration (mean 5 mg N/l) than the forested basins (mean 0.5 mg N/l). Miller *et al.* (1997) studied many streams of the Potomac watershed in Chesapeake (USA) and reported that the highest nitrate concentrations, up to 8 mg N/l, were determined in streams draining basins with intense cropping (mean 3.7 mg N/l), while streams with forested basins showed the lowest (mean 0.4 mg N/l). Boyer & Pasquarell (1995) found that mean nitrate concentrations correlated with the percentage of each basin occupied by agriculture in Appalachian streams (USA). Similarly, Stone *et al.* (1995) and Boyd (1996) found higher nitrates concentrations in streams draining cropland basins in North Carolina (mean 5.6 mg N/l) and Central Nebraska (mean 5 mg N/l), USA, respectively, than in forested (mean 1.1 mg N/l) and rangeland basins. Excess fertilization causes high nitrate concentrations in streams of agricultural basins (Meybec, 1998). It is therefore suggested that low nitrate concentration in Pampasic streams result form the lower fertilizer loads. Soybean is a leguminous symbiotic nitrogen fixer. Since no harvest increase is observed in response to fertilization, it is not fertilized with nitrogen in the Argentine Pampa. However, not all the N present in the biomass pool is attained by fixation, a variable amount is up taken from the soil (Sarandon, 2002). Soybean uptakes available nitrogen from the soil profile resulting in lower nitrate concentrations in the runoff water and adjacent streams, even lower than the concentrations prevailing in cattle breeding plots. Nitrate concentrations were highest at the lowest water levels and became diluted in coincidence with flood events. The opposite trend was observed in Ohebach and Girou streams by Probst (1985) and Schulz and Liess (1999). At high fertilizer loads, nitrate saturates the soil profile and is fast released by rainwater increasing concentrations at flood peaks. In the Pampasic streams the main N contribution comes from the mineralization of the soil organic N pool which is comparatively small and is diluted by the rain water resulting in the lower concentrations observed in the runoff and flood peaks. As previously stated, the roaming behaviour of the cattle likely contributed higher nitrate loads to stream in the corresponding plot. The lower nitrate concentrations in the floodplain streams provided with an upstream marsh (Maguire) and a recreation dam (Horqueta) suggest the influence of these environments on nitrogen transport. Osborne & Kovacic (1993) reviewed the influence that riparian buffer strips might have on the N retention from the adjacent landscape concluding that such environments represent effective N sinks reducing up to 90% of the N leakage from agriculture. N removal by wetlands has repeatedly been reported. Nguyen *et al.* (1999) reported that a riparian wetland removed 54% of the inputs from an adjacent sheep grazed pasture in New Zealand. They also reported higher ammonium downstream leaching consistent with the higher concentrations determined in Maguire and Horqueta streams in the present study. Schulz and Peall (2001) reported that a constructed wetland performed on a small stream of similar characteristics of the Maguire stream reduced the downstream nitrate concentration from a small orchard basin by 70%. El-Habr and Golterman (1990) reported important N losses in marshes and irrigation channels of Camarge (France). Although macrophyte uptake might be important, denitrification seems the main process determining the high observed N sinks in wetlands (Golterman *et al.*, 1988; Minzoni *et al.*, 1988).

Satellite images show that rroughly half of the streams in the area showed stretches provided with floodplains or strips occupied with riparian vegetation of variable width. These stretches represent from 10 to 50% of the total stream length. Differences are often found in neighbor fields along the streams, suggesting that most of them originally counted with riparian vegetation and it was removed in roughly half of them by the agricultural practice. Some farmers leave the natural vegetation along the stream side, while others plough over the small first order stream headwaters in the plots of gentle slopes. Riparian restoration along the streams and maintaining natural vegetation strips along the streams therefore seems a promising and inexpensive option for sustainable long-term improvement of water quality.

Invertebrates

Jergentz *et al.* (2004a) studied the occurrence of toxicity and the effect on the resident invertebrate fauna in the three typical Pampasic streams: Maguire, Horqueta and Helves during the 2000-2001 growing period. Water samples were weekly taken and toxicity assays were performed with *Daphnia magna*. Runoff, flood and suspended matter samples were taken by means of passive devices installed in the surrounding fields (runoff samplers) and within the stream itself (flood and suspended matter samplers). The resident macroinvertebrate fauna associated to the macrophytes was sampled at four occasions from February to April 2001. The macroinvertebrate drift was studied with driftnets installed in the stream bed sampled at five occasions throughout the same period.

No toxicity was found in stream water itself. However, after a rain period of 229 mm occurred from march 1 to march 6 endosulphan was measured in suspended matter from Horqueta and Helves (30 and 40 μg/kg, respectively), Horqueta flood suspended matter (318 μg/kg) and runoff sediments from Helves (10 μg/kg). No pesticide was detected in other samples of Helves and Horqueta streams and no pesticide was ever detected in the Maguire stream at the study site, downstream the marsh. The number of species determined in each stream was reduced from 12 and 9, in Horqueta and Helves respectively before the rain, to 8 and 3 respectively, after the rain, while at Maguire, it remain on 8 throughout the studied period. Ephemeroptera density in Horqueta stream decreased from 350 ind/m^2 before the rain to 50 ind/m^2 after the rain, and from 100 ind/m^2 to none in Helves stream. Odonata density in Horqueta stream decreased from 50 ind/m^2 before the rain to 7 ind/m^2 after the rain, and from 60 ind/m^2 to none in Helves. On the other hand Ephemeroptera density in Maguire did not show significant variations and increased after the rain in Maguire stream. A significantly higher number of individuals were caught in the drift net after the rain at Horqueta, being *Hyallela curvispina* and Ephemeroptera the dominant, while no effect was observed in Maguire. This study showed the occurrence of toxicity pulses in the Pampasic streams associated with intense precipitations and thus, pointing to edge-of -field runoff as the main cause. Although cypermethrin is the most widely used pesticide only endosulphan was detected, plausibly because of its longer environmental persistence (Goebel *et al.*, 1982). Endosulphan was detected in the particulate fraction, pointing that the hydrophobic nature of the pesticide favours its association to soil particles that are later incorporated to the streams by erosion processes. Ensosulphan concentration in runoff suspended matter up to 12 mg/kg has been reported in the Lourens River, South Africa, draining an orchard area, by Schulz (2001). Leonard *et al.* (2000) determined endosulphan concentrations up to 48 μg/kg in sediments from the Namoi River, Australia. No endosulphan was detected in the Maguire stream downstream the small natural riparian wetland. Similarly, Schulz and Peall (2001) showed strong pesticide reduction downstream a small wetland in an affluent of the Lawrence River in South Africa. Pesticide adsorption to particles retained within the marsh seems the cause. Since a high macrophyte biomass is yearly incorporated to the bottom sediments, pesticide association to organic matter likely contributes to the efficient wetland retention.

The observed decrease in density and species number in coincidence with endosulphan contamination points a cause-effect relation. Toxicity associated to the endosulphan particulate fraction was reported by Leonard *et al.* (2001) determining a 10-day LC_{50} concentration of 162 μg/kg for *Jappa cutera* (Ephemeroptera). Furthermore, Leonard *et al.* (2000) inferred that several Ephemeroptera and Trichoptera species were negatively influenced by runoff associated endosulphan exposure in field studies in the Namoi basin. Schulz and Liess (1999) showed increased drift of various invertebrate species at laboratory experiments with fenvalerate inoculated to soil suspensions simulating runoff events.

Horqueta and Maguire streams were also monitored the following growing period, from November 2001 to February 2002, by Jergentz *et al.* (2004b). Maguire was sampled at two sites, up and downstream the wetland. Two widespread and abundant local invertebrate species, *Hyallela cruvispina* (Amphipoda) and *Macrobrachium borelli* (Decapoda) were *in situ* exposed in small cylindrical enclosures with excavated windows covered with plastic nets. After each exposure, lasting 6-8 days, mortality was determined and new organisms were exposed. The local invertebrate population was monthly assessed four times throughout the study period. Stream and runoff suspended matter was retained with passive samplers installed in the stream bottom and in the adjacent field erosion rills respectively. Insecticides were detected in three occasions in runoff suspended matter samples taken after rain events occurred previous to the samples of 8 and 29 January and 7 February, amounting 94, 34 and 85 mm respectively. In the first runoff event 150 μg/kg chlorpyrifos, 46 μg/kg cypermethrin and 8 μg/kg endosulphan were determined in the runoff particulate fraction of Horqueta stream. In the following 29 Jan and 7 Feb runoff particulate samples chlorpyrifos (43 and 15 μg/kg, respectively) and cypermethrin (53 and 13 μg/kg, respectively) were determined. Only chlorpyrifos was determined in runoff water (0.07-0.3 μg/l) and stream suspended matter (64 μg/kg) after a sampling interval without rain lasting from 14 to 23 Jan. Neither endosulphan nor cypermethrin were detected in the stream or runoff water phase or stream suspended matter samples. No pesticide was detected in Maguire stream. *In situ* exposures of *H. curvispina* showed survival rates of 77-85% in both streams throughout the studied period at times when pesticide exposure was not detected. Only one exposure showed severe toxicity; the exposure lasting from 15 to 23 Jan in Horqueta showed 100% mortality of *H. curvispina* and *M. borelli* in coincidence with the detection of chlorpyrifos in stream suspended matter. The resident *H. curvispina* population showed a rather constant density attaining roughly 100 ind/m^2 at the first samplings, and was significantly reduced to zero following the 23 Jan sampling date in coincidence with the 100% mortality in the in situ exposures and chlorpyrifos detection in the stream suspended matter. Simultaneous chlorpyrifos contamination and complete mortality in exposed and resident organisms resulted very consistent. Since there was no rain in the sampling interval it was suggested that pesticide input was related to the drift of air plane fumigations observed in neighbor fields at the sampling area in the previous days. Later *in situ* exposures showed the background survival rates (77-85%) and consistently no pesticide was detected in the stream, pointing that toxicity represented a short pulse-like episode. However, the local invertebrate population did not recover in the last sampling period, until early march. Although the previous and two following runoff episodes contained detectable amounts of pesticides, mortality was not observed in the *in situ* stream exposures but significantly increased drift of small young organisms following the first runoff event of Jan 8. Runoff from a single plot is diluted by the overall flood pulse simultaneously coming from the whole basin. Sublethal exposures might cause increased drift without drastic effects on the resident population. Organism drift has been reported in various studies in relation to transient insecticide pollution (Cuffney *et al.*, 1984; Kreutzweiser and Sibley, 1991). It has been shown that amphipods are present in the organism drift during insecticide exposure, representing an avoidance reaction (Breneman Pontasch, 1994; Schulz and Liess, 1999). It seems likely that the pesticide exposures during the rainfall events caused only a sublethal drift reaction (Taylor *et al.*, 1994), not attaining enough intensity to produce drastic measurable changes in the resident population. Sibley *et al.* (1991) described invertebrate drift caused by permethrin concentrations of 0.35 μg/L, while population effects were evident when concentrations surpassed 1.7 μg/L. Several studies point the sensitivity of amphipods as sentinel organisms for water quality assessment. The amphipod *G. pulex* was the most sensitive organism (Brock *et al.*, 1992a, b) in outdoor experimental ditches contaminated with chlorpyrifos, showing significant population decrease at 0.9 μg/l (Van den Brink *et al.*, 1996). Werner *et al.* (2000) found in stream samples toxicity of chlorpyrifos that caused 100% mortality in 48 h bioassays with *Ceriodaphnia dubia* at a concentration of 0.13 μg/l, from the Sacramento–San Joaquin river delta.

During the following growing period (2002-2003), the Arrecifes tributary was monitored, sampling at two sites, inside the soybean plot (site 1) and at a 250 m downstream plot (site 2) with cattle (Figure 1). On Nov. 27 a mixture of glyphosate and cypermethrin was applied at the beginning of the growing period. Cypermethrin was determined in the bottom sediments attaining 37 and 54 μg/kg at each site, respectively. Eighteen days later (Dec. 15) a 57 mm rain occurred, and soon after (Dec. 17) cypermethrin in the bottom sediments increased to 1,075 and 595 μg/kg at each site, respectively. On Jan. 11 chlorpyrifos was applied to the soybean plot. Although chlorpyrifos was not detected within the stream cypermethrin remained in the bottom sediments on Jan 11, attaining 90 and 120μg/kg at sites 1 and 2, respectively. After the first rain cypermethrin was high and roughly double at the applied plot than the close downstream site. A month later cypermethrin decreased to roughly 10% at the applied plot while concentrations remained in the same order of magnitude at both sites. Present results suggest a comparatively low persistence in the stream environment together with a relatively low downstream transport.

Figure 2 shows the survival rates of *H. curvispina* in the *in situ* exposures at both sites. 100% mortality was observed in coincidence with the exposure from 10 to 17 Dec. when the 57 mm rain caused an important runoff and the maximum cypermethrin concentrations were measured in the bottom sediments. The next exposure, from 17 to 26 Dec. a rainfall event of 80 mm was produced and survival attained 60 and 29% at each site, respectively. The following exposure, from 10 to 12 Jan., included the chlopyrifos application without rain, attaining a survival rate of 45 and 89% respectively. The two following exposures included rains of 17 and 67 mm, with the smaller one, mortality was 100% at site 1 while a 10% survival was attained at site 2. With the later and larger rain, mortality was 100% at both sites. No rains occurred the next two exposures and survival rates increased to 15 and 25% at each site respectively the first one (29 Jan to 5 Feb) and 75-70% the last one. Consistently, almost complete mortality was observed after the first rain following each application while the chlorpyrifos application caused roughly 50% mortality within the applied plot without discernible effect on the downstream plot. Within the managerial practice at the studied farm, edge of the field runoff represents a more important toxicity source than the application drift. Both, cypermethrin and chlorpyrifos are hydrophobic compounds, and therefore are fast retained to the particulate phase. Since the streams produce a fast increase in discharge and water velocity in response to rain events, the presence of pesticides in true solution remains as a sampling challenge. Although brief and sharp ephemeral peaks of dissolved pesticides might eventually occur it would remain extremely difficult to be detected. High concentrations in bottom sediments and the lower effect of drift than runoff inputs on exposed organisms suggest that the applied pesticides are fixed to the soil and introduced to the stream already sorbed to the particulate fraction of the runoff.

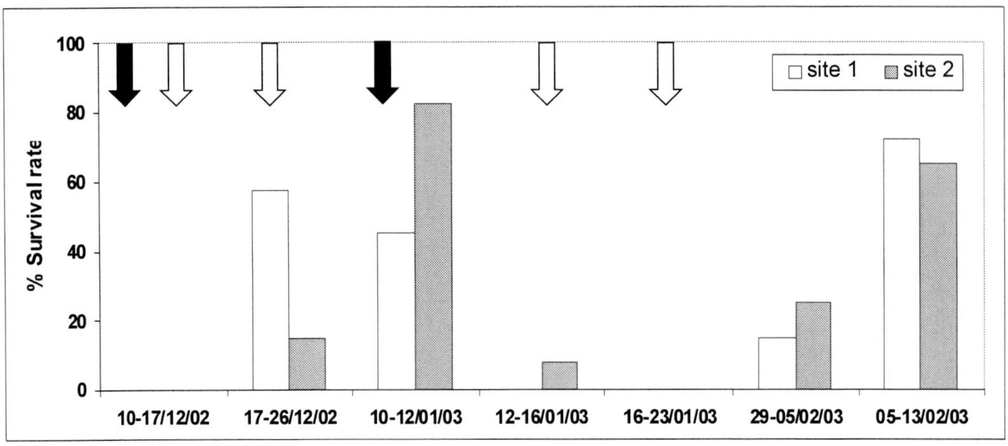

Figure 2. Mortality of *H. curvispina* exposed in field enclosures at two sites of a first order stream born in a soybean plot (site 1) and flowing into a 250 m downstream livestock plot (site 2). Black arrows represent pesticide applications: the left one represent a mixture of cypermethrin and glyphosate applied on 27 Nov. 02, and the right one a mixture of chlorpyrifos and glyphosate applied on 11 Jan. 03. White arrows represent rain events producing runoff.

Vascular Plants

Riparian and aquatic plant communities are essential for maintaining the quality of the aquatic habitat in agro ecosystems. Shoreline vegetation plays critical roles in soil stabilization, and therefore in reduction of erosion (Blackburn and Boutin, 2003). Aquatic macrophytes have shown to be effective in reducing concentrations of pesticides resulting from spray-drift (Dabrowsky *et al*, 2005), regulate water fluxes and provide habitat to aquatic life. Plants of these communities next to agroecosystems are impacted by related activities, particularly the spray of herbicides.

Field assessments from detailed studies of the riparian stream flora carried out monitoring phenological status before and after pesticide (herbicide or herbicide plus insecticides) applications, were performed aiming towards the detection of long term effects on the local flora (Martin *et al.*, 2005). This strategy allowed the assessment of potential changes within reproductive and seasonal cycles, with impacts on the composition of the plant community. Effects assessments on phenological status and richness in Pescado stream from one year crop cycle indicated that the herbicide applications during the soy crop pre-emergence coincided with 12% of the flowering and 10% of the fructification periods of the riparian species. Summer crop development coincided with 50% flowering and 25% fructification. Richness evaluation indicated that of a total of 58 species from the area, 71% corresponded to native species, and 36% of them were weeds. Although the shortness of the studied period, the relevance of potential impacts were clearly associated to the simultaneity of herbicide sprays and seeding time of a high percentage of the species from the area (Martin *et al.*, 2005).

In the Arrecifes tributary no detectable levels of variations in biomass, richness, surface cover and abundance in relation to pesticide applications were seen during a single crop cycle. Three species, *Anthemis cotula* (Compositae), *Trifolium repens* (Leguminosae) and *Echinochloa crusgalli* var. mitis (Gramineae), represented more than 33% of the total plant cover in spring. In summer, *E. crusgalli* replaced *A. cotula* as dominant in the taller stratum, while *T. repens* remained as dominant in the lower stratum. *Cyperus reflexus*, *Bromus unioloides*, *Paspalum dilatatum*, *Setaria geniculata*, *Eleusine tristachya*, *Verbena montevidensis*, *Verbena rigida*, *Verbena gracilescens*, *Melilotus indicus*, *Rumex crispus* and *Polygonum hydropiperoides* represented 10% of the total surface cover. These species coincided with a previous description, performed roughly 10 years ago, and corresponded to the *P. dilatatum* grassland community described by Faggi (1966). However, some characteristic elements of the association were missing (*Stipa hyalina, S. papposa, Bothriochloa laguroides, Bromus catharticus*). A reduction of the species richness was evident, with the presence of only 20% of the 76 species described for the association in the area (Martin *et al.*, 2003). Present evidence suggests an impoverishment in species richness of the assemblage, plausibly related to the agricultural practice. Further work is needed on this subject.

Different type of injuries such as chlorosis, or decreased biomass on riparian and aquatic plants were detected as an immediate response to pesticide spray events on natural and enclosed populations adjacent to the applied culture (Sobrero *et al.*, 2007). Effects on total chlorophyll content (TCC) of a riparian community from the Arrecifes tributary in relation to herbicide sprays indicated significant decrease in TCC for *P. dilatatum* and *R. cripus*. Effects were evident four days after herbicide sprays with 32 and 29% reduction, respectively. Plant recovery was observed 20 and 3 days later, respectively. Measurements of TCC for *T. repens* did not show significant level of variation in relation with spray applications, possibly due to the protection offered by plants from taller strata. For the case of aquatic plants, *Sagittaria montevidensis, Hydrocotyle ranunculoides* and *Pontederia roundifolia* also evidenced TCC reduction at the level of 25, 71 and 12%, respectively. The herbicide reached at least 250 meters downstream application sites, with detectable effects on aquatic macrophyte vegetation. *P. dilatatum, H. ranunculoides, R. crispus* and *S. montevidensis* were sensitive and represented indicators of impacts on riparian and aquatic non-target species in the region, for a rapid assessment of effects at the level of chlorophyll content (Martin *et al.*, 2003). Although no noticeable differences could be observed in biomass in the short term measurement performed, inhibition of TCC gives evidences of potential long term effects on growth and reproduction, depending on the stage of plant development.

Field tests with *L. gibba* or *Spirodella intermedia* within enclosed devices (Sobrero *et al.*, 2007) in stream sectors of the Arrecifes tributary adjacent to soy bean fields, showed inhibition in biomass production (dry weight per frond) and in the TCC of *L. gibba*, with 28 and 75% reduction, respectively, for an application of glyphosate with cypermethrin. Assessments from two later applications also indicated effects on these parameters (35.5 and 75.5% reduction of biomass and TCC, respectively, for an application of glyphosate with chlorpyrifos; and 53.9 and 0% reduction for an application of glyphosate, chlorpyrifos and cypermethrin). Enclosures were protected from the application drift by covering them with transparent nylon. It was shown that the main effect was related to exposure by runoff. Other reports from tests performed in Pescado tributary with *S. intermedia* also indicated significant effects of pesticide runoff on TCC and population growth reduction (multiplication rate and number of frond per colony) between two and seven days after exposure, as observed for *L. gibba* (Martin *et al.*, 2005).

Field glyphosate chemical analyses reported by Peruzzo *et al.* (2003) and Sobrero *et al.* (2007) ranged from 0.5 to 2.1 mg/l, and were consistent with the expected environmental concentrations estimated from recommended doses, following Boutin *et al.* (1995).

Comparisons between the effects actually observed in the field exposures associated to these concentrations with those from laboratory standardized tests (Environment Canada, 1999; Sobrero *et al.*, 2007), indicates that it would be necessary higher concentrations of glyphosate for equivalent effects in laboratory toxicity tests. Taking into account that the studied field sceneries were associated with the application of mixtures of the herbicide with insecticides, a possible explanation for higher effects in the field could be associated to additional or synergistic toxicity in mixtures. Furthermore, limitations in extrapolations between laboratory and field conditions should be also considered.

Martin and Ronco (2006) studied the impact of mixtures of pesticides on vascular plants assessing the effects of single pesticides formulations and their mixtures on the inhibition of germination and seedling root elongation with laboratory bioassays. A comparative assessment of the effects of a glyphosate formulation using a standardized battery of plant seeds showed that sensitivity was largely variable among the assayed families. The IC50 values for root elongation inhibition ranged between 8.89 and 1164.3 mg/L, with a sensitivity rank as follows: *Lactuca sativa*> *Lolium perenne* > *Medicago sativa* > *Allium cepa* > *Brassica napus*. Additionally, a detailed study performed with *L. sativa* showed a slightly higher sensitivity to the Roundup® Max formulation (IC50= 7.3 [6.5-9.4] expressed as mg/L of a.i.), in comparison with glyphosate, the active principle (IC50= 9.8 [8.2-11.8] mg/L). The mixtures of formulated glyphosate, with cypermethrin and chlopyrifos using the most sensitive plant seed, *L. sativa*, indicated higher toxicity of those containing the herbicide with cypermethrin or with both insecticides together when compared to the herbicide alone. Toxicity increased up to seven times when mixed with the insecticide cypermethrin. These results are consistent with the observed higher plant sensitivity in the field than in laboratory exposure.

Results of toxicity tests performed by Martin and Ronco (2006) with a stream sediment elutriate (1:4, sediment: water) sampled after pesticide application (a mixture of the herbicide plus the two insecticides), showed significant effects on root elongation inhibition of *L. sativa* seeds (13 % inhibition at the direct sample elutriate). The detected effects were in agreement with the field study with caged experiments performed simultaneously in the same area (see results previously given in this section), indicating that laboratory tests with environmental samples were capable of detecting toxic effects from exposure to a sediment elutriate sampled at a time and in the place where field tests with aquatic plants were also affected.

We conclude that vascular plants can therefore be affected by pesticide drift or runoff concentrations that are orders of magnitude lower than levels applied in target areas. These results are in accordance with observation done by Ferenec (2001) for low-dose herbicide exposures. Authors also found that adult riparian and aquatic plants could experience effects from a few centimetres distance to as far as 250 m from spray areas. Boutin and Rogers (2003) pointed out that lethal effects on some plants clearly change the composition of and affected community, however, it is unknown to what extent the plant community could be affected when sub-lethal effects are observed (i.e. reduction in the competitiveness or reproductive output of some species). The observed effects on vascular plants from the riparian and aquatic sectors in the Pampasic areas were at the sublethal levels. Long term impacts of sublethal effects on non target plant species growing along margins will strongly depend on their phenological stages (some species may have mature seeds, others may not have reproduced at that moment) during pesticide applications. Since vegetation provide structure to riparian and aquatic communities (including invertebrate, fish and amphibians), changes in plant species may result in an impact on other trophic levels.

Phytoplankton and Periphyton

Pérez *et al.* (2007) studied the effects of Roundup® addition on phytoplankton and periphyton communities in earthen outdoor mescoscosms of 25 m^2 and 1.2 m depth. Effects of added glyphosate at initial concentrations of 5.5 and 12.7 mg/l, by triplicate, were compared with controls. Glyphosate fast disappeared with a mean half life of 6.6 days. No significant differences were observed in nutrient concentration among treatments except for total phosphorus concentrations which increased in the glyphosate treatments. As expected, significant differences in micro and nano phytoplankton densities were observed among treatments. By the end of the experiment, mesocosms treated with both doses of Roundup showed a 2.5-fold decrease in micro- and nanophytoplankton mean abundances. The abundance of the heterotrophic bacteria and picoplanktonic eukaryotes did not exhibit significant differences among treatments. However, picocyanobacteria (Picy) showed the opposite trend. By the end of the experiment, mean Picy abundance increased roughly 40 times in the high-dose treatment as compared to the initial values. Correspondingly, the mesocosms treated with the higher dose experienced a significant two fold increase in phytoplankton primary production.

The periphyton assemblages did not show significant differences in chlorophyll concentration or density abundances among treatments. Similarly as for the phytoplankton assemblage, the periphytic cyanobacteria (particularly *Cyanothece* spp.) appeared to increase after the addition of the herbicide, but observed differences among treatments were not statistically significant.

The most interesting effect observed in this study is the large 40 fold increase in Picy density. Cyanobacteria are known to have a comparatively higher tolerance to glyphosate by either one of two mechanisms: the overproduction of synthases, or the production of glyphosate-tolerant enzymes (Powell *et al.*, 1991, 1992). The total P increase observed in the glyphosate treatments agreed with the 14% contribution of P to the molecular weight of glyphosate. Roundup contains glyphosate, water, and POEA, a complex mixture of fat from cattle or sheep tissues, which is not supposed to contain phosphorus. Since glyphosate is fast metabolized within the water-body its P content enhance the advantage of the comparatively more resistant cyanobacteria species and has important potential implications for water quality, since cyanobacteria represent potential hazard due to bloom forming nuisance and toxic cyanotoxin production and release. It is therefore not only conceivable, but also quite likely that current agricultural practices, which heavily relays upon continual additions of glyphosate at regional scales, may alter the structure and function of many natural aquatic environments leading to enhanced eutrophication levels.

Fish

The fish community of Arrecifes tributary was found to be composed at least by five fish species (*Cheirodon interruptus, Cnesterodon decemmaculatus, Hoplias malabaricus, Rhambdia quelen, Synbranchus marmoratus)* (Carriquiriborde *et al.* (2007). *C. decemmaculatus* represented the most conspicuous and abundant species of the fish community

The effect of cypermethrin-glyphosate mixture (ground applications) on the local resident and caged populations was investigated in two consecutive crop cycles under field-use conditions by Carriquiriborde *et al.* (2007). No fish mortalities or individuals presenting abnormal behaviour were recorded along the stream, and no dead fish were trapped in drift nets after pesticide application or after the first rainfall event during the two consecutive years of study. The total number of fish captured per effort unit, the size -class structures and several other parameters measured (immature: total ratio, male: female ratio and condition index) of *C. decemmaculatus* populations did not show significant differences before and after pesticide application at both sites (Table 3).

A similar trend was observed on population parameters in a later ground application of a mixture containing cypermethrin, chlorpyrifos and glyphosate. However, moderate fish kills in the stream between four and 24 h after spraying were detected in this case. In accordance with these results, observed fish survival within *in situ* caged exposed *C. decemmaculatus* indicated no significant differences in survival rates respect to controls (percent survival: 86.2±4.0) after application of cypermethrin- glyphosate, but a significant reduction (percent survival: 78.8±7.4) after application of the cypermethrin- chlorpyrifos- glyphosate mixture (Carriquiriborde *et al.*, 2005).

Fish community of Pescado stream tributary was composed by the same species as those found in Arrecifes tributary. Additionally, *Corydoras paleatus* was also observed (Carriquiriborde *et al.*, 2005). Studied pesticide spray impacts in this water body included a ground spray of the herbicide and aerial applications of a mixture containing endosulphan and glyphosate (see table 1). The authors found neither fish kills in the stream and no significant mortalities in caged experiments (survival: 88.7±12.5) in relation with the spray of glyphosate. This was also in agreement with the results of laboratory toxicity tests using *C. descemmaculatus* showing that glyphosate was not acutely toxic to the species (Carriquiriborde *et al.*, 2007). The opposite result was observed for the applications of endosulphan in Pescado stream tributary. Field observations and drift nets allowed detecting generalized fish kills 24 and 72 h after fumigation. Only 17.0±5.9 % survival was detected in the exposed *C. descemmaculatus* caged population during the fumigation of the mixture containing endosulphan and glyphosate.

The absence of acute lethal effects observed for cypermethrin under field conditions contrasted with the high toxicity showed by this pesticide under laboratory conditions. The cypermethrin 96 h-LC50 for *C. decemmaculatus* following standardized protocols with dechlorinated laboratory tap water (USEPA, 2002) was 0.46 μg/l (Carriquiriborde *et al.*, 2007), similar to reported figures for other fishes, as *Oncorhynchus mykiss* (0.5 μg/l) and *Scardinius erythropthalmus* (0.4 μg/l), and lower than *Cyprinus carpio* (0.9–1.1 μg/l) and *Tilapia nilotica* (2.2 μg/l) (Stephenson, 1982). The 96 h-LC50 of cypermethrin for C. *decemaculatus* increased to 2 μg/l in filtered stream water and to 5 μg/l when unfiltered stream water was used. A correlation was found between stream organic matter content and cypermethrin LC50.

Although cypermethrin was shown to be very highly toxic to *C. decemmaculatus* at the standardized laboratory tests and that during the second crop cycle the pesticide reached water concentrations of 0.46 and 0.29 μg/l at sites 1 and 2, respectively (Carriquiriborde *et al.*, 2007), higher than the 96 h LC50 determined following standardized protocols, no mortality or behavioral changes were observed in the resident fish population throughout the two successive crop cycles. No differences in the mortality within *in situ* exposed enclosures were observed in both sites comparing the periods previous and following the application.

Reduced toxicity of pyretroid exposure under field conditions has been previously reported (Haya, 1989; Hill, 1989) and attributed to the tendency of pyrethroids to bind to suspended particulate matter, sediments, and aquatic plants. Surface waters in the Pampas are moderately rich in suspended solids with medium to high organic matter contents. Total organic carbon (TOC) values ranged from 2 to 8.2 mg/l in eight typical pampasic streams (Carriquiriborde *et al.*, 2007); showing a mean content close to the mean reported figures for warm temperate-streams (Spitzi and Leenheer, 1991). Total OC concentration in the stream at the Arrecifes farm was the largest of the studied Pampasic streams. Moreover TOC concentrations increased from 4.4 mg C/l at the start of the second crop cycle to 14.6 mg C/l just after the application, and later decreased to 7.0 mg C/l at the end of the sampling period. It seems likely that glyphosate application enhanced organic carbon leakage from the decaying vegetation to the stream, therefore contributing to the observed decrease cypermethrin toxicity for fish.

Risk assessment evaluations relay on toxicity exposure ratios (TER) estimated with standardized procedures for measuring the lethal concentrations. When the procedure was applied to the Arrecifes stream an effect was predicted. If the TER ratios are calculated using the LC50 values actually measured with stream water no effect is expected, in agreement with the lack of effects observed on fish populations in the studied stream. Considering that the main factor explaining the reduction of cypermethrin toxicity in stream water was the TOC, it would be useful to include the TOC for the adjustment of LC50 values, in a similar way as DOC is being used to adjust toxic effects of some metals (Welsh *et al.*, 1996).

Table 3. Effects of soybean pesticides on fish population under current used field conditions in the Pampas region

	Pesticide spraying events			
	glyphosate	**[1]glyphosate+cypermethrin**	**glyphosate+ cypermethrin+ chlorpyriphos**	**glyphosate+ endosulphan**
Field observations and drift nets				
Generalized fish kills in the stream	Not observed	Not observed	Moderate (from 4 to 24 h after spraying)	Severe (from 24 to 72h after spraying)
C. decemmaculatus caged population				
96h-Survival	88.7±12.5 NSD	86.2±4.0 NSD	78.8±7.4 % SD	17.0±5.9 % SD
C. decemmaculatus wild population parameters[†]				
Size class structure	---	NSD	NSD	---
Immature/total ratio	---	NSD	NSD	---
Male/female ratio	---	NSD	NSD	---
Condition index	---	NSD	NSD	---

SD: significantly different respect before spraying; NSD: Not significantly different before and after spraying; ---: not assessed.

[1] Assessed during two consecutive years

Multi Organism Tests

Laboratory studies with mescosm were developed by Mugni *et al.* (2006) in order to detect toxicity thresholds of pesticide exposures at field realistic conditions by simulating runoff events on two representative components of the resident fauna, the amphipod *H. curvispina* and the fish *C. decemaculatus*. Three replicates and controls were used to assess the toxicity of cypermethrin, chlorpyrifos and endosulphan in aquariums containing stream sediment and water. A realistic runoff scenario was simulated by mixing soil and water and adding pesticide to attain a concentration of 10 μg/l in the suspension, giving an initial 1μg/l concentration in the aquariums. Water was taken out of each aquarium at different times and 48 h exposures assays performed.

Chlorpyrifos and cypermethrin were highly toxic for *H. curvispina* attaining 100% mortality at the beginning of the experiment (Figure 3), while endosulphan did not cause any mortality. A significant decrease in water toxicity to *H. curvispina* was observed at the second day for chlorpyrifos and no toxicity was observed from the forth day onward. The treatment exposed to cypermethrin showed decreased mortality from the forth day onward reaching a minimum at the end of the experiment, after nine days incubation. No mortality of fish was observed for any for the three assayed pesticides. A second assay was immediately performed with a higher pesticide concentration of 5 μg/l added in the same way. At such concentration 100% mortality was attained by endosulphan while no mortality was observed by cypermethrin or chlorpyrifos. In spite that 50% lethal concentrations for *C. decemaculatus* following standard laboratory protocols lye in the order of 1 μg/l (Carriquiliborde *et al.*, 2007), our simulated runoff with realistic, field relevant concentrations did not cause any observable effect on the common endemic fish *C. decemaculatus*, in agreement with no observed fish mortality in streams following farm applications. Decreased toxicity due to strong pesticide adsorption to the particulate and dissolved organic mater must be invoked in order to account for such result. *H. curvispina* resulted much more sensitive to cypermethrin and chlorpyrifos exposure than the fish, confirming the suitability of using it as sentinel organism for toxicity assessments.

Toxicity was fast depleted in the local water and sediment environment characterized by alkaline water of comparatively high conductivity, hardness, and organic and suspended matter contents. Toxicity to *H. curvispina* was higher for cypermethrin and chlorpyrifos than endosulphan and decreased faster for chlorpyrifos. *C. decemaulatus* resulted more sensitive to endosulphan than cypermethrin and chlorpyrifos.

Another laboratory experiment was performed in long and narrow aquariums containing sediments and water from the first order stream in the Arrecifes farm, in which a water flux was induced with pumps, and chlorpyrifos was added as a mixture of soil and water, in order to simulate the effect of a runoff event on representative organisms of the local stream fauna (Argemi *et al.*, 2003). Each aquarium contained bottom sediments and water from the stream. A runoff event was simulated by pouring into the aquarium a suspension of soil and water spiked with chlorpyrifos, resulting in a measured initial concentration in the aquariums of 15 and 73 μg/l for the lower and higher dose treatments, assayed by triplicate. Chlorpyrifos disappeared fast from the water in the aquariums attaining a final concentration of 0.6 and 0.4 μg/l after nine days, at the end of the experiment. Bottom sediment concentration at the end of the experiment were conspicuous attaining 83 and 212 μg/kg for each dose respectively, representing 30 and 16% of the total amount of chlorpyrifos initially added. An initial 100% mortality was attained for the amphipod *H. curvispina* and the decapod *Macrobrachium borelli*. However, at the end of the experiment, exposed *H. curvispina* did not show mortality at all. Fishes showed initial symptoms of poisoning such as swallowed gills and spastic swimming movements. Nevertheless, there were no mortality in any replicate and they recover to the end of the experiment.

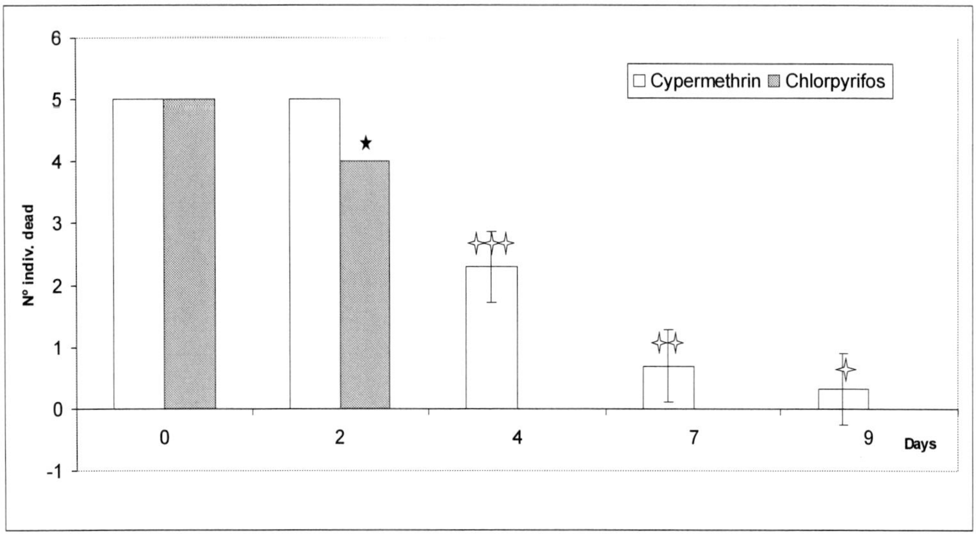

Figure 3. *H. curvispina* mortality in exposure assays following pesticide addition to laboratory aquariums simulating field realistic runoff concentrations. Stars denote significant differences with the previous sampling day, bars represent standard deviation.

Amphibians

Cajade and Natale (2003), reported 15 species of anurans in the study area of Arrecifes (*Chaunus arenarum, Chaunus fernandezae, Hypsiboas pulchellus, Scinax nasicus, Scinax granulatus, Scinax berthae, Leptodactylus latinasus, Leptodactylus gracilis, Leptodactylus ocellatus, Leptodactylus mystacinus, Odontoprhynus americanus, Physalaemus biligonigerus, Physalaemus albonotatus, Pseudopaludicola falcipes, Elachistocleis bicolor*), representing 55.5% of the species present in the Province of Buenos Aires (Gallardo, 1974; Lavilla *et al*., 2000; Lavilla and Cei, 2001). Cajade and Natale (2003) also found higher richness than expected according to vouchers existent in Academic and Scientific herpetological collections for the area. The anuran fauna contained forms belonging to the families Hylidae, Bufonidae, Leptodactylidade and Microhylidae. They have aquatic larvae and different morphological type of adults, with diverse reproductive strategies. Although the anuran community was well represented, only few species reproduced within the study site of the Arrecifes tributary.

Natale and Ronco (2003) and Agostini *et al.* (2005a) reported the impact of several pesticide applications in Arrecifes and Pescado stream tributaries on field sampled anuran larvae. Mortalities within the stream and drift nets located at the sampling sites of the Arrecifes tributary in relation with the applications of a cypemethrin and glyphosate mixture were low (effects were observed only on caged organisms during the first year,

Table 4), pointing to subtle acute effects on tadpoles of this type of ground spray. Sublethal short and long term effects were not assessed in these reports. Associated runoff after the first rain also had an attenuated impact on mortalities (Table 4, Figure 4). Simultaneous field experiments with caged *H. pulchellus* and *C. arenarum* tadpoles detected significant mortalities only in a test group of tadpoles from the former species adjacent to the spray patches in the first year. Mortalities after application and runoff of chlorpyrifos (in its mixture with glyphosate) in the same sites were moderate, with higher mortalities in comparison with the cypermethrin-glyphosate spray (Table 4). Figure 4 shows tadpole mortalities for both types of pesticide spray mixtures before and after application and after the first rain (Natale and Ronco, 2003). Severe effects were observed in the field, assessed by means of observations and drift nets, associated to a ground spray of a mixture containing cypermethrin with chlorpyrifos and glyphosate in the Arrecifes tributary, and also significant mortalities in caged tadpole populations of *L. gracilis* and not significant for *H. pulchellus* (Table 4).

The comparison of the spatial extent of impacts indicate that amphibian and fish impacts were generally circumscribed to adjacent sectors, possibly reaching the water body from pesticide drifts of applications, while the herbicide glyphosate drift and runoff caused chlorophyll content reduction not only on the adjacent riparian vegetation, but at least 250 meters downstream application.

The description of the anuran community from the Pescado stream reported by Agostini *et al.* (2005b) was represented by seven species (*C. arenarum, C. fernandezae, H. pulchellus, L. latinasus, L. gracilis, L. ocellatus, P. falcipes*), constituting 46% of the species present in the region, and only 23% of those reported for the Buenos Aires Province (Gallardo, 1974; Lavilla *et al.*, 2000; Lavilla and Cei, 2001). The field assessments in this study area indicated severe effects of an endosulphan plus glyphosate application, detected immediately after fumigation and throughout the following 24 h. Additionally, high mortality in caged populations of *H. pulchellus* 24 h after fumigation (Agostini *et al.*, 2005a) was detected. The impacts of these aerial applications were also severe on the fish as reported by Carriquiriborde *et al.* (2005). Hence, fumigation of endosulphan determined a high acute impact on these populations, providing additional evidence on the risk associated to the use of this insecticide.

Amphibians occurring in soy bean areas from the province of Entre Ríos and Santa Fe, a few hundred kilometers of distance from the Arrecifes tributary study sector, showed a similar richness as the one found in Arrecifes (Lajmanovich and Peltzer, 2004, Attademo *et al.*, 2005, Peltzer *et al.*, 2006). These authors examined the related landscapes in the studied areas, pointing to the relevance of the presence of a close protected forest, and also at the recent introduction of this type of agricultural activity. The observed higher diversity was compared with another impoverished agricultural area in the province of Córdoba, where agriculture had expanded more extensively. Peltzer *et al.* (2006) proposes that management to prevent recent anuran declines will require defining high-quality habitats for individual species or group of species, followed by efforts to preserve or restore these aquatic habitats. The maintenance of shore vegetation in ponds and hedgerows may increase the number of species and diversity of anurans within agricultural landscapes. Lajmanovich and Peltzer (2004), indicate that these studies will help in our understanding of the status and conservation of amphibians in coastal fluvial areas and will serve to help coordinate research and conservation efforts.

Even though the Arrecifes tributary had a much longer history of soybean cultivation, the presence of several small vegetated patches with tall grasses and typical allochthonous trees could represent eventual refuges for maintaining species richness. Although the community of anurans was well represented, it was only for a few of these species that was possible to detect breeding sites (Cajade and Natale, 2003). It is important to point out, as indicated by Hayes *et al.* (2006) for observations in cornfields in the mid-western of the United States of America, that ponds or streams within agricultural fields could not be predictable year after year, and not even day after day. Results obtained by Piha (2006), pointed on the essentiality of the existence of enough wetlands adjacent to the breeding sites for amphibian survival. The author found that in the studied areas of Sweden increased isolation of breeding ponds from expanding agriculture partly explained the absence of some species.

Although soybean pesticides exhibit low bioaccumulation and short environmental life times, some of them show high toxicity levels, such as the ones referred in this chapter or those reported by other scientists (Smith and Stratton, 1986; Kamrin, 1997). There is an extended amount of reports from the literature on the effects of these insecticides on wildlife, and at present is widely accepted the vulnerability of amphibians to these toxicants. Particularly, sensitivity of early stages of life, lethal and sublethal effects, and different response patterns according to specific life stages have been reported in the last decade (Power *et al.*, 1989; Carey and Bryant, 1995; Hill, 1995; Cowman and Mazanti, 2000; Pauli *et al.*, 2000).

The widespread local agricultural practice makes use of pesticide mixtures in different combinations and proportions, and there is scarce information on their combined effects; additionally, the range of responses observed for anurans from exposure to agricultural mixtures determines a complex approach for undertaking their study. While some extensive reviews about toxicity of pesticides on amphibians are available (Power *et al.*, 1989; Devillers and Exbrayat, 1992), there are there are only few reports of the effects of pesticide mixtures on anurans (Sparling *et al.*, 2000). Since it has been suspected that the effect of chemical pollutants is being underestimated within the multiple stress factors contributing to numerical amphibian decline (Burkhart *et al.*, 2003), there is increasing concern on evidences about the role of organic chemicals involved in retardation of growth and development, time and size to reach metamorphosis, endocrine disruption and related immunosuppressive effects are likely to be highly relevant (Hayes *et al.*, 2006).

The clear impact of some of the pesticides on tadpole mortalities associated to sprays and runoff reported here evidence the toxicity of soybean mixtures of pesticides, associated to different management practices and variable exposure conditions. The observed effects in the field were directly related to pesticide applications reaching the water bodies by drift or runoff. The authors agree that habitat fragmentation, and long term effects of pesticides on populations should be carefully analyzed towards the understanding of the impacts of biotech pesticides on amphibian communities.

Table 4. Effects of soybean pesticides on species of anuran tadpoles under current used field conditions in the Pampas region

	Pesticide spraying events			
	[1]cypermethrin+ glyphosate	[2]chlorpyrifos+ glyphosate	cypermethrin+ chlorpyrifos+ glyphosate	endosulphan+ glyphosate
Field observation and drift nets				
Generalized tadpole kills in the stream	Subtle (through 24h)	Moderate (through 24h)	Severe (through 24h)	Severe (through 24h)
Tadpole caged population				
24h survival after application	*H. pulchellus* 55.0±6.42*/ 100±0 *C. arenarum* 91.7±3.57	*H. pulchellus* 100±0/ 96.5±2.00 *C. arenarum* 90.0±3.87	*H. pulchellus* 83.3±6.80 *L. gracilis* 50.0±8.94*	*H. pulchellus* 27.1±2.80*
24h survival after first rain	*H. pulchellus* 37.0±6.42* / 73.3±8.07 *C. arenarum* 88.3±4.14	*H. pulchellus* 58.3±14.23* *C. arenarum* 73.3±5.71		

Values are given as mean ± standard deviation of percentage of survivals.
*: significantly different ($p<0.05$) before and after spraying.
[1]Assessed during two consecutive years in the same sites; [2]Assessed during two consecutive years in different sites.

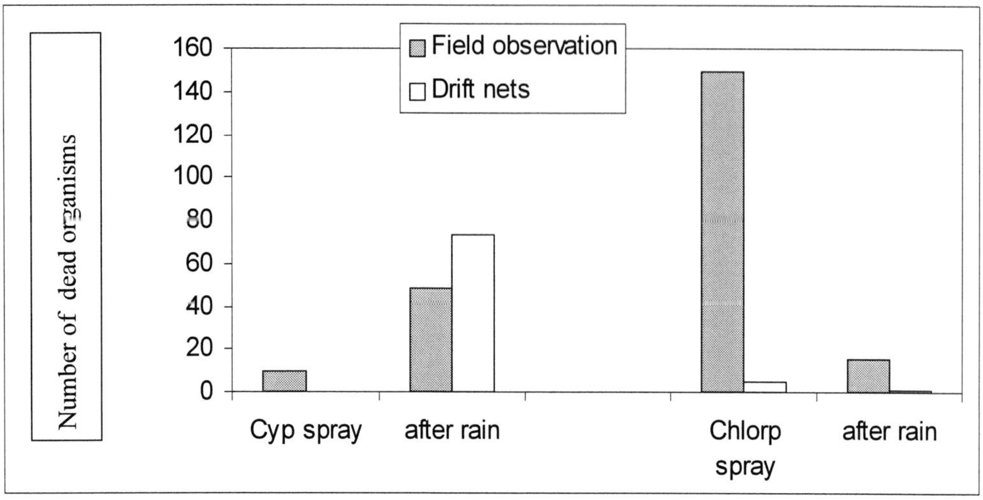

Figure 4. Observed tadpole mortalities in field observations and drift nets in the Arrecifes tributary in relation with cypermethrin (Cyp) and chlorpyrifos (Chlorp) sprays and runoff after the first rain.

Conclusion

Present information, sustained on the studies performed on several typical streams throughout several successive years, suggests that the Pampasic streams contain comparatively low nitrate and comparatively high SRP concentrations. The former results from the relatively low fertilization loads presently employed, since the soybean has no agronomic response to nitrogen fertilization. The latter seems related with the naturally high P content of the deep and rich Argiudol soils developed over the loessic sediments deposited during the Holocene. Available information suggests that agricultural intensification in the Pampas has no evident impact on stream nutrient concentrations at the present fertilizer loads and managerial practices employed. The regional nutrient budget shows that the fertilizer loads do not replace the amount of harvested nutrients. The present agricultural practice is exhausting the soil available pool of nutrients and is therefore not sustainable on the long term. Fertilizer loads, although comparatively low, have shown a significant increase in recent times and it is expectable that such trend will be increased with time leading to enhanced nutrient concentrations in the future.

Results of effects assessments indicate that all the biota components of the studied ecosystems (flora, invertebrates, fish and amphibians) were affected in different extent after pesticide applications, representing an appreciable stress over the systems. Sometimes the effects were comparatively subtle, particularly on the vegetation, and others quite drastic, such as the observed on the invertebrate populations. Evidence from laboratory and field conditions were very consistent, showing high sensitivity of non target invertebrates, resulting in toxicity pulses mainly associated with runoff events that produced mortality in the resident populations with total disappearance of some groups of the resident fauna (ephemeroptera, odonata, amphipoda) in response to toxic pulse events occurring several times per year. Runoff seemed more important than spray drift in the local environment and managerial practice. The toxic pulses were ephemeral as shown by *in situ* exposures, and local populations recover with time. Toxicity persistence in the local environment was quite short. Chlorpyrifos showed faster toxicity dissipation than cypermethrin in laboratory simulated runoff exposures. Chlorpyrifos toxicity disappeared from water-sediment systems within a few days and cypermethrin in about a week. These experiments showed the combined effect of partition between water and sediment and the half life in each phase. Being highly hydrophobic compounds, fast adsorption to the particulate phase and interaction with organic matter seems to contribute the most to the fast observed toxicity disappearance. Soil persistence is longer than that of water and therefore a single field application resulted in several, albeit short, toxic pulses in coincidence with the following rain events.

Simultaneous exposures of the amphipod *H. curvispina* and the fish *C. decemaculatus* showed the larger tolerance of the latter. Of the three most commonly used insecticides, toxic loads associated to endosulphan fumigation caused the most severe effects for amphibians and fishes, followed by the applied mixtures of chlorpyriphos, cypermethrin and glyphoste. Also, severe effects on amphibian mortalities and higher effects on plant chlorophyll were associated to the application of cypermethrin and chlorpyrifos mixture with the herbicide. Information from laboratory toxicity tests with seeds also showed enhanced toxicity of mixtures, in accordance with field observations.

Several environmental aspects influence the impact of each application. Density and height of the riparian and aquatic vegetation, together with the stage of development of the crop, represent protective barriers against pesticide transfer to the water bodies. Larger crop and riparian biomass were coincident with lower effects on the stream biota. Another point to be taken into account in relation to the magnitude of the pesticide impact during application is the water level of the stream. The climatic conditions prevailing at the applications seem also important. The largest effect observed in the studied period was coincident with an endosulphan plane application in the Pescado stream in coincidence with an intense drought. Although endosulphan was more toxic to amphibians and fishes than cypermethrin or chlorpyrifos, drought condition enhanced their effect. The insecticide concentration levels detected in major watercourses of the region, although biological effects were not measured, indicate potential pulsed acute impacts on the local biota.

Complementation between laboratory and field studies resulted of great significance for detecting erroneous predictions from the former and lightens observations from the latter. For example, cypermethrin toxicity in laboratory tests performed with fish in laboratory water point to higher effects than those observed in the field, explained by the protective effect given by the organic matter from stream water. Conversely, lower sublethal effects on vascular plants were detected in the laboratory when compared to those detected in the field for equivalent herbicide concentrations, probably associated to enhanced toxicity of mixtures.

An integrated approach analysis seems to be a promising option towards the understanding of the overall impact of intense soybean farming on associated aquatic ecosystems. Improving managerial practices and riparian habitat restorations seems feasible goals for long term sustainable production.

Streams counting with floodplains and riparian habitats showed less nitrate concentrations and less pesticide contamination, to such an extent that not any single toxic event was observed downstream Maguire wetland. Restoration of riparian habitats and leaving buffer zones between the cultures and the streams would largely improve present hazard condition. Education represents the most important tool for achieving long term environmental improvement.

Acknowledgements

The authors thank the Agencia Nacional de Promoción Científica y Tecnológica (ANPCyT) for financial support.

References

Agostini, M.G., Díaz Seijas, M.E., Natale, G. S.& Ronco, A.E. (2005a). Efectos de la aplicación de pesticidas sobre larvas de anuros utilizando experimentos de campo. *III Congreso Argentino de Limnología*, Chascomús.

Agostini, M.G., Díaz Seijas, M.E., Natale, G. S.& Ronco, A.E. (2005b). Comparación de la diversidad de anuros de dos agroecosistemas del valle de inundación del arroyo El Pescado, provincia de Buenos Aires. *VI Congreso Argentino de Herpetología*, Paraná.

Argemi, F., Jergentz, S., Schulz, R. & Bonetto, C. (2003). Utilización de microcosmos como modelos del comportamiento de pesticidas en ambientes acuáticos. *6ta Reunión Anual SETAC Latinoamericana*. Buenos Aires.

Attademo, A.M., Peltzer, P.M. & Lajmanovich, R.C. (2005). Amphibians occurring in soybean and implications for biological control in Argentina. Agricultura, *Ecosystems and Environment*, **106**, 389-394.

Berkovic, A., Marino, D., Lespade, P. & Ronco, A. (2006). Análisis de pesticidas asociados al cultivo de soja en aguas y sedimentos de un sector productivo de la pampa ondulada. *Anales XXVI Congreso Argentino de Química*, **8-053**, 4 pp. San Luis

Blackburn, L.& Boutin, C.(2003). Subtle effects of herbicide use in the context of genetically modified crops: A case study with glyphosate (Roundup R). *Ecotoxicology*, **12**, 271-285.

Boutin, C. & Rogers, C. A. (2003). Pattern of sensitivity of plant species to various herbicides- An Analysis with two databases. *Ecotoxicology*, **9**, 255-271.

Boutin, C.; Freemark, K & Keddy, C. (1995). Overview and rationale for developing regulatory guidelines for nontarget plant testing with chemical pesticides. *Enviromental Toxicology and Chemistry*, **14**, 1465-1475.

Boyd, R. (1996). Distribution of NO_3 - and orthophosphate in selected streams in central Nebraska. *Water Resource Bulletin*, **32**, 1247-1257.

Boyer, D. & Pasquarell, G. (1995). Nitrate concentrations in karst springs in an extensively grazed area. *Water Resource Bulletin*, **31**, 729-736.

Breneman, D. H. & Pontasch, K.W. (1994). Stream microcosm toxicity tests: Predicting the effects of fenvalerate on riffle insect communities. *Environmental Toxicology and Chemistry*, **13**, 381-387.

Brock, T. C. M., Crum, S. J. H., Van Wijngaarden, R., Budde, B. J., Tijink, J., Zuppelli, A. & Leeuwangh, P.(1992 a). Fate and effects of the insecticide Dursban 4E in indoor Elodea-dominated and macrophyte-free freshwater model ecosytems: I. Fate and primary effects of the active ingredient chlorpyrifos. *Archives of Environmental Contamination and Toxicology*, **23**, 69-84.

Brock, T.C.M., Van den Bogaert, M., Bos, A.R., Van Breukelen, S.W.F., Reiche, R., Terwoert, J., Suykerbuyk, R.E.M. & Roijackers, R.M.M. (1992b). Fate and effects of the insecticide Dursban 4E in indoor Elodea-dominated and macrophyte-free freshwater model ecosystems: II. Secondary effects on community structure. *Archives of Environmental Contamination and Toxicology*, **23**, 391-409.

Burkhart, J.G., Bidwell, J.R., Fort D.J. & Sheffield S.R. (2003). Chemical Stressors. In: Linder, G., Krest, S.K. & Sparling, D.W. (Eds.) *Amphibian Decline: an Integrated Analysis of Multiple Stressors effects*. Pensacola: Society of Environmental Toxicology and Chemistry. p. 111-128.

Cajade, R. & Natale, G. S. (2003). Anfibiofauna asociada a cultivos de soja en el Partido de Arrecifes, Provincia de Buenos Aires. *XVII Reunión de Comunicaciones Herpetológicas*, Centro Nacional Patagónico. Puerto Madryn.

Camilión, M. C., Sayavedra, I., da Silva; M. & Ronco, A. E. (2006). Erosión hídrica en suelos bajo siembra directa. *Anales del XXVI Congreso Argentino de Química*, **8-009**, 4 pp

Carey, C. & Bryant, C.J. (1995). Possible interrelations among environmental toxicants, amphibian development, and decline of amphibian populations. *Environmental Health Perspectives*, **103**, 13-17.

Carriquiriborde, P., Díaz, J. & Ronco, A. E. (2005). Evaluación del impacto de pesticidas asociados al cultivo de soja RR sobre poblaciones de peces mediante estudios de campo. *3er Congreso Argentino de Limnología, CAL III.* Chascomús, Sociedad Argentina de Limnología.

Carriquiriborde, P., Díaz, J., Mugni, H., Bonetto, C.& Ronco, A. E.(2007). Impact of Cypermethrin on Stream Fish Populations under Field Use in Biotech-Soybean Production. *Chemosphere*, **68**, 613-621.

Cowman, D.F. & Mazanti, L.E. (2000). Ecotoxicology of "New Generation" Pesticides to Amphibians. In: Sparling DW, Linder G, Bishop CA, editors. *Ecotoxicology of amphibians and reptiles*. (pp. 233-268) Pensacola: Society of Environmental Toxicology and Chemistry.

Cuffney, T. F., Wallace, J. B., Webster, J. R. (1984). Pesticide manipulation of a headwater stream: invertebrate responses and their significance for ecosystem processes. *Freshwater Invertebrate Biology*, **3**, 153-171.

Dabrowsky, J. M., Bollen, A.; Bennett, E. R. & Schulz, R. (2005). Pesticide interception by emergent aquatic macrophytes: Potential mitigate spray drift input in agricultural streams. Agriculture, *Ecosystems and Environment*, **111**, 340-348.

Devillers, J. & Exbrayat, J.M. (1992). Ecotoxicity of chemicals to amphibians. Volume 1: Effect of chemicals in amphibians. *Hanbooks of ecotoxicological data*. Philadelphia PA: Gordon and Breach.

El-Habr, H. & Golterman, H. (1990). *In vitro* and *in situ* studies on nitrate disappearance in water –sediment systems of the Camargeu (Southern France). *Hydrobiologia*, **192**, 223-232.

Environment Canada. (1999). *Biological test methods: test for measuring the inhibition of growth using the freshwater macrophyte, Lemna minor*. Report EPS 1/RM/37, Ottawa.

Faggi, A. M. (1996) La vegetación espontánea de un área del norte de la Provincia de Buenos Aires. *Parodiana*, **9**, 125-137.

Ferenec, S. (2001). *Impacts of Low-Dose, High-Potency Herbicides on Nontarget and Unintended Plant species*. Pensacola, F.L: Society of Environmental Toxicology and Chemistry.

Gallardo J.M. (1974). *Anfibios de los alrededores de Buenos Aires*. Buenos Aires: EUDEBA.

Goebel, H., Gorbach, S., Knauf, W., Rimpau, R.H. & Hüttenbach, H. (1982). Properties, effects, residues and analytics of the insecticide endosulfan. *Residue Reviews*, **83**, 1-165.

Golterman, H., Bonetto, C. & Minzoni, F. (1988). The nitrogen cycle in shallow water sediment systems of rice fields: Part III: The influence of N-application on the yield of rice. *Hidrobiologia*, **159**, 211-217.

Haya, K. (1989). Toxicity of pyrethroid insecticides to fish. *Environmental Toxicology and Chemistry*, **8**, 381-391.

Hayes, T.B., Case, P., Chui, S., Chung, D., Haeffele, C., Haston, K., Lee, M., Phoung Mai, V., Marjuoa, Y., Parker, J. & Tsui, M. (2006). Pesticide Mixtures, Endocrine Disruption, and Amphibian Declines: Are we underestimating the Impact? *Environmental Health Perspectives*, **114**, S-1.

Hill, I. R. (1989). Aquatic organisms and pyrethroids. *Pesticide Science*, **27**, 429-465.

Hill, E.F. (1995). Organophosphorus and carbamate pesticides. In: D.J. Hoffman, B.A. Rattner, G.A. Burton Jr. & J. Cairns Jr. (Eds). *Handbook of ecotoxicology*. (pp. 243-274) Ann Arbor MI: Lewis.

Jergentz, S., Mugni, H., Bonetto, C. & Schulz, R. (2004 a). Runoff-related endosulfan contamination and aquatic macroinvertebrate response in rural basins near Buenos Aires, Argentina. *Archives of Environmental Contamination and Toxicology*, **46**, 345-352.

Jergentz, S., Pessacq, P., Mugni, H., Bonetto, C & Schulz, R. (2004 b). Linking in situ bioassays and dynamics of macroinvertebrates to assess agricultural contamination in streams of the Argentine pampa. *Environmental Ecotoxicology and Safety*, **59**, 133-141.

Kamrin, M.A. (1997). *Pesticide profiles: toxicity, environmental impact, and fate*. New York: Lewis.

Kreutzweiser, D.P. & Sibley, P.K. (1991). Invertebrate drift in a headwater stream treated with permethrin. *Archives of Environmental Contamination and Toxicology*, **20**, 330-336.

Lajmanovich, R.C. & Peltzer, P.M. (2004). Aportes al Conocimiento de los Anfibios Anuros con Distribución en las Provincias de Santa Fe y Entre Ríos (Biología, Diversidad, Ecotoxicología y Conservación). In: Temas de la Biodiversidad del Litoral Fluvial Argentino. INSUGEO, *Miscelánea*, **12**: 291-302.

Lavilla E.O., Ponssa M.L., Baldo D., Basso N., Bosso A., Céspedez J., Chebez J.C., Faivovich J., Ferrari L., Lajmanovich R., Langone J.A., Peltzer P., Úbeda C., Vaira M., Vera Candiotti F. (2000). Categorización de los anfibios de Argentina. In: E. O. Lavilla, E. Richard & G.J. Scrocchi (Eds.) *Categorización de los anfibios y reptiles de la República Argentina*. (pp. 11-34). Asociación Herpetológica Argentina.

Lavilla E.O., Cei J.M. (2001). Amphibians of Argentina. A second update, 1987-2000. (pp. 177) Turin: Mus. Reg. Scien. Natur. Turin.

Leonard, A.W., Hyne, R.V., Lim, R.P., Pablo, F., Van den Brink, P.J. (2000). Riverine endosulphan concentrations in the Namoi River, Australia: Link to cotton field runoff and macroinvertebrate population densities. *Environmental Toxicology and Chemistry*, **19**, 1540-1551.

Leonard, A. W., Hyne, R. V., Lim, R. P., Leigh, K. A., Le, J.& Beckett, R. (2001). Fate and toxicity of endosulphan in Namoi River water and bottom sediment. *Journal of Environmental Quality*, **30**, 750-759.

Marino, D. & Ronco, A. (2005). Cypermethrin and chlorpiryfos concentration levels in surface water bodies of the Pampa Ondulada, Argentina. *Bulletin of Environmental Contamination and Toxicology*, **75**, 820-826.

Martin, M.L.; Sobrero; C., Rivas, C, Rimoldi; F & Ronco, A. E. (2003). Impacto del uso de pesticidas asociado a la siembra directa sobre especies no-blanco. Flora riparia y acuática. *Memorias Conferencia Internacional Usos del Agua, Agua 2003* (27-31), Cartagena de Indias.

Martin, M. L, Sobrero, M. C. & Ronco, AE. (2005). Fitotoxicidad de herbicidas sobre especies acuaticas y flora riparia en un agroecosistema. *III Congreso Argentino de Limnología*, Chascomús.

Martin, M.L. & Ronco, A. E. (2006). Effects of mixtures of pesticides used in the direct seeding technique on non-target plants. *Bulletin of Environmental Contamination and Toxicology*, **77**, 228-236.

Meybeck, M. (1998). Man and river interface: multiple impacts on water and particulates chemistry illustrated in the Seine river basin. *Hydrobiologia*, **373/374**, 1-20.

Miller, C., Denis, J., Ator, S & Brakebill, J. (1997). Nutrients in streams during baseflow in selected environmental settings of the Potomac River basin. *Journal American Water Research Association*, **33**, 1155-1171.

Minzoni, F., Bonetto, C. & Golterman, H. (1988). The nitrogen cycle in shallow water sediment systems of rice fields: Part I: The denitrification process. *Hydrobiologia*, **159**, 189-202.

Mugni, H., A. Maine & C. Bonetto. (2005). Phosphate and nitrogen transformations in a large floodplain river. In: L. Serrano, & H. Golterman (Eds.) *Phosphate in Sediments* (pp. 139-149). Leiden: Backhuys Publishers.

Mugni, H., S. Jergentz, R. Schulz, A. Maine & C. Bonetto. 2005. Phosphate and nitrogen compounds in streams of Pampean Plain areas under intensive cultivation (Buenos Aires, Argentine). In: L. Serrano, & H. Golterman (Eds.) *Phosphate in Sediments* (pp: 163-170). Leiden: Backhuys Publishers.

Mugni, H., Bonetto,C., Jergentz, S. & Schulz R. (2006). Simulating realistic Runoff scenarios of current used insecticidas in Argentina – Effects on two local species. *Society of Environmental Toxicology and Chemistry*, Landau.

Nair, V., Graetz, D. & Portier, K. (1995). Forms of phosphorus in soil profiles from dairies of South Florida. *Soil Science Soc. Am. J.*, **59**, 1244-1249.

Natale, G. S. & Ronco A. E. (2003). Impacto del uso de pesticidas asociado a la siembra directa sobre especies no-blanco: Anuros autóctonos. *Memorias del Congreso del Agua* **2003**, (pp. 36-39). Cartagena de Indias.

Nguyen, L., Downes, M., Mehlhorn, J. & Stroud, M. (1999). Riparian wetland processing of nitrogen, phosphorus and suspended sediment inputs from a hill Country sheepgrazed catchment in New Zealand (pp. 481-485). *Proceedings of the Second Australian Stream Management Conference.*

Osborne, L. & Kovacic, D. (1993). Riparian vegetated buffer strips in water-quality restoration and stream management. *Freshwater Biology*, **29**, 243-258.

Pauli, B.D., Perrault, J.A. & Money S.L. (2000). RATL: *A Database of Reptile and Amphibian Toxicology Literature*. Technical Report Series N° 357. Hull: Canadian Wildlife Service, Headquarters.

Peltzer, P.M., Lajmanovich, R.C, Attademo, M.A. & Beltzer, A.H. (2006). Anuran diversity across agricultural pond in Argentina. *Biodiversity and Conservation*, **15**, 3499-3513.

Pengue, W. 2000. *Cultivos transgénicos ¿Hacia donde vamos?* Buenos Aires: Lugar Editorial S.A.

Pérez, G. L., Torremorell, A., Mugni, H., Rodríguez, P., Vera, S., do Nascimento, M., Allende, L., Bustingorry, J., Escaray, R., Ferraro, M., Izaguirre, I., Pizarro, H., Bonetto, C., Morris, D. & Zagarese, H. (2007). Effects of Glyphosate based Herbicide on Primary Producer Communities in Aquatic System: A Mesocosm Study. *Ecological Applications*, In press.

Peruzzo, P., Marino, D., Cremonte, C., da Silva, M., Porta, A. & Ronco, A. (2003). Impacto de pesticidas en aguas superficiales y sedimentos asociado a cultivos por siembra directa, *Memorias Conferencia Internacional Usos del Agua, Agua 2003* (pp. 135-142) Cartagena de Indias.

Piha, H. (2006). *Impacts of agriculture on amphibians at multiple scales*. Ecological Genetics Unit, Department of Biological Sciences, Faculty of Biosciences, University of Helsinki, Helsinki.

Powell, H. A., Kerby, N. W. & Rowell, P. (1991). Natural tolerance of cyanobacteria to the herbicide glyphosate. *New Phytologist*, **119**, 421-426.

Powell H. A., Kerby, N. W., Rowell, P., Mousdale, D. M. & Coggins, J. R. (1992). Purification and properties of a glyphosate-tolerant 5-enolpyruvylshikimate- 3-phosphate synthase from the cyanobacterium *Anabaena variabilis*. *Planta,* **188**, 484-490.

Power, T, Clark, K.L., Harfenist, A. & Peakall D.B. (1989). *A Review and evaluation of the amphibian toxicological literature*. Technical Report 61. Ottawa: Canadian Wildlife Service.

Probst, J. (1985). Nitrogen and phosphorus exportation in the Garonne basin (France). *Journal of Hydrology*, **76**, 281-305.

Ronco, A., Camilión, C. & Manassero, M. (2001). Geochemistry of heavy metals in bottom sediments from streams of the western coast of the Río de la Plata estuary, Argentina. *Environmental Geochemistry and Health*, **23**, 89-103.

Sarandon, S. (2002). La agricultura como actividad transformadora del ambiente. El impacto de la agricultura intensiva de la revolución verde, In: S. Sarandón (Ed.) *Agroecología: El camino hacia una agricultura sustentable.* La Plata: Ediciones Científicas Americanas ECA.

Schulz, R. & Liess, M. (1999). A field study of the effects of agriculturally derived insecticide input on stream macroinvertebate dynamics. *Aquatic Toxicology*, **46**, 155-176.

Schulz, R. & .Peall, S. (2001). Effectiveness of a constructed wetland for retention of nonpoint-source pesticide pollution in the Lourens River catchment, South Africa. *Environmental Science and Technology*, **35**, 422-426.

Schulz, R. (2001). Comparison of spray drift- and runoff-related input of azinphos-methyl and endosulphan from fruit orchards into de Lourens River, South Africa. *Chemosphere*, **45**, 543-551.

Serrano, L., Burgos, M., Díaz-Espejo, A. & Toja, J. (1999). Phosphorus inputs to wetlands following storm events after drought. *Wetlands*, **19**, 318-326.

Sibley, P.K., Kaushik, K.N. & Kreutzweiser, D.P. (1991) Impact of a pulse application of permethrin on the macroinvertebrate community of a headwater stream. *Environmental Pollution*, **70**, 35-55.

Smith, T.M. & Stratton, G.W. (1986). Effects of synthetic pyrethroid insecticides on nontarget organisms. *Residue Reviews*, **97**:93-119.

Sobrero, C., Martín, M. L. & Ronco, A. (2007). Efecto del herbicida Roundup® Max sobre plantas acuáticas no blanco: *Lemna gibba* L. *Hidrobiologica*, **17**:1-10.

Sparling, D.W., Bishop, C.A. & Linder, G. (2000). The current status of amphibian and reptile ecotoxicolocal research. In: D.W. Sparling, G. Linder, & C.A. Bishop (Eds). *Ecotoxicology of amphibians and reptiles* (pp. 1-13). Pensacola: SETAC.

Spitzi, A. & Leenheer, J. (1991). Dissolved Organic Carbon in Rivers. In: E.T. Degens, S. Kempe, & J. Richey (Eds.), *Biogeochemistry of Major World Rivers*, Chapter 9. SCOPE 42 (pp. 55-127). Chichester: John Wiley.

Stephenson, R.R. (1982). Aquatic toxicology of cypermethrin. in acute toxicity to some freshwater fish and invertebrates in laboratory tests. *Aquatic Toxicology*, **2**, 175-185.

Stone, K., Hunt, P., Coffey, S. & Matheny,T. (1995). Water quality status of a USDA water quality demonstration project in the Eastern Coastal Plain. *Soil Water Conservation*, **50**, 567-571.

Taylor, E.J., Rees, E.M. & Pascoe, D. (1994). Mortality and a drift-related response of the freshwater amphipod *Gammarus pulex* (L) exposed to natural sediments, acidification and copper. *Aquatic Toxicology*, **29**, 83-101.

USEPA (2002). *Methods for Measuring the Acute Toxicity of Effluents and Receiving Waters to Freshwater and Marine Organisms*, (fifth edition). EPA-821-R-02-012 U.S. Washington, DC: Environmental Protection Agency, Office of Water.

Van den Brink, P.J.; Van Wijngaarden, R.P.A., Lucassen, W.G.H., Brock, T.C.M. & Leeuwangh, P. (1996). Effects of the insecticide Dursbans 4E active ingredient (Chlorpyrifos) in outdoor experimental ditches: II. Invertebrate community response and recovery. *Environmental Toxicology and Chemistry*, **15**, 1143-1153.

Welsh, P.G., Parrott, J.L., Dixon, D.G., Hodson, P.V., Spry, D.J. & Mierle, G. (1996). Estimating acute copper toxicity to larval fathead minnow (Pimephales promelas) in soft water from measurements of dissolved organic carbon, calcium, and pH. *Canadian Journal Fish Aquatic Science*, **53**, 1263-1271.

Werner, I., Deanovic, L.A., Connor, V., De Vlaming, V., Bailey, H.C. & Hinton, D.E. (2000). Insecticide-caused toxicity to Ceriodaphnia dubia (Cladocera) in the Sacramento-San Joaquin River Delta, California, USA. *Environmental Toxicology and Chemistry*, **19**, 215-227.

In: Ecosystem Ecology Research Trends
Editors: J. Chen and C. Guo, pp. 241-274
ISBN: 978-1-60456-183-8

Chapter 7

ECOLOGY AND PREVALENCE OF ACTYNOMYCES-ANTAGONISTS IN SOILS OF KYRGYZSTAN

T.D. Doolotkeldieva[1*] and N.E. Totubaeva[2]
[1]Kyrgyz-Turkish International University, 56, Prospect Mira, 720044, Bishkek, Kyrgyz Republic
[2]State Forestry Service of Kyrgyz Republic 228, Toktogyla street, 720070 , Bishkek, Kyrgyz Republic.

Abstract

The ecological adaptation and the prevalence of actynomyces of Streptomyces class in soils of Kyrgyzstan have been investigated. Various types of soil were served as natural sources for isolation Streptomyces: mountain chernozems, mountain dark-chestnut, bright - brown, mountain meadow subalpine , mountain meadow alpine , mountain-meadow-steppe - alpine soils, mountain – wood dark-colored, mountain meadow chernozem like ,mountain brown typical, mountain - wood black brown.

The phenotypic (taxonomic) characteristic , microscopy and 16S rRNA gene sequencing methods were used to identify the isolates of Streptomyces . Screening and bioassays methods were used for discovery the microbial secondary metabolites of novel Streptomyces isolates.

The results have shown that Streptomyces spp. were particulary abundant and contained a rich biodiversity in the high mountain- meadow-steppe-alpine soils, high mountain alpine soils, mountain chernozems , where they represented 60% of all soil microorganisms. The climatic conditions, agri-physical regime , chemical and mechanical structure of these soils have a favorable influence on the prosperity of Streptomyces groups. The major factors saving stability and developments of Streptomyces populations in these soils are high percent contents of humus , neutral or alkalescent reaction of soils and a significant accumulation of mineral elements as phosphorus, nitrogen, potassium and calcium. In these conditions, Streptomyces ssp are able to participate in decomposition of high-molecular, inaccessible organic compounds, rather than readily available.

The received data also have shown that the occurrence of antagonistic forms of Streptomyces in soils depends on a degree of cultivation of soils, their agrichemical and genetic properties. The isolates with antagonistic properties were seldom found out in soils subjected to agricultural use. The virgin soils, not used in agriculture were allowed much more

[*] E-mail address: tinatin2252@yahoo.com

cultures with antagonistic properties. The soils from non-polluted landscapes contain the greater number of Streptomyces than soils from the territories polluted with industrial wastes. A total 441 Streptomyces strains were isolated from various soil types of Kyrgyzstan, from them 22 strains were selected for the use in agriculture for biotechnological purposes, 105 strains for pharmacology.

1. Introduction

Microorganisms play an important role in functioning and self-regulation of the ecosystem of the Earth. Moreover, microorganisms give a rich variety of molecular and chemical forms in the nature since a single bacterium has 3-4 thousand of genes and approximately 1200 metabolites.

Prokaryotic microbes are already well known as a source of valuable bioactive compounds - anti-infection agents (antibiotics), enzymes, and vitamins. There are over 17,000 antibiotics and other biologically active microbial metabolites in current databases.

Actynomyces always attract scientists' attention in the world because of their strongly expressed ability to synthesize much broader spectrum of compounds possessing antibiotics properties {46, 24].

By their nature, Actynomyces are bacteria that have a peculiar ultra structural and chemical basis like prokaryotes and realize a more complex mycelia plan of the organization like eukaryotes - fungi [16, 12]. The majority of soil Actynomyces known today belongs to aerobic microorganisms with an oxidizing type of metabolism and developed systems of carrying electrons to oxygen [7, 47].

This group of microorganisms has an extensive area of distribution and builds an important part of soil micro flora. Their high adaptation to various conditions of environment allows them to take part in the transformations of a wide group of substances. Actynomyces are very hardy and can grow at low humidity of soils.

There are studies about a quantitative distribution of Actynomyces in soils of Kyrgyzstan [44, 39, 31, 42, and 35]. However, there are no data on the species biodiversity of Actynomyces in these researches. These studies only show the distinctions of species in color of air mycelia; however, depending on conditions the color of mycelia can vary even on the same strain.

Kyrgyzstan, where 93% of the entire territory is occupied by mountains, represents a great complex of unique ecosystems for microbiological and biotechnological researches. In Kyrgyzstan, actynomyces are still a poorly investigated group among soil microorganisms, both from the point of view of biodiversity and from the point of view of biotechnology.

Unique landscapes, original soil - climatic conditions of Kyrgyzstan contribute to the existence of Actynomyces species, which are still unknown for the science. The study will add not only a genetic fund of microorganisms but actynomyces will also be used in biotechnology as producers of effective biological products.

Various types of soil were served: mountain chernozems, mountain dark-chestnut, bright - brown, mountain – meadow sub alpine, mountain – meadow alpine, mountain - wood dark-colored, mountain meadow – blackish, mountain brown typical, mountain - wood black – brown, light grey soil as natural sources to isolate *Streptomyces* – antagonists.

Soil samples were aseptically sampled from virgin layers in depth of 0-5, 0-10, 0-25 cm two times a year (in summer and autumn).

According to A.Mamytov [36], in Kyrgyzstan due to severe and various environment conditions, there are original and unique soils, most of which have no analogues not only in the plains but also in other mountain areas of the world because the formation of them depends on local conditions.

Therefore, selection of soil samples for the research of new strains of *Streptomyces* was accomplished on soil regions and sub regions according to the classification, systematization and provinciality of Kyrgyzstan's soils. The places of selection of soil samples are specified on the map - circuit of geographical division of soil into districts of Kyrgyzstan (fig. 1).

2. The Prevalence of *Streptomyces* in Issyk-Kul Sub Region Soils

Issyk-Kul region has an extremely complex relief in which three basic complexes of the relief are sharply distinctive: mountains, foothills and plains.

Unique natural ecosystems of Issyk-Kul region attracted our attention to research soil microbiological biodiversity and reveal new active producers of antibiotic compounds.

The feature of such ecosystems, which attract scientists' attention, is vertical zones, various unique types of soils, which are distinguished by genesis of both mechanical and biological structures, covered vegetation, temperate, and frequently not exposed to strong anthropogenous influence.

Below-mentioned types of soils of Issyk-Kul sub region were investigated for isolation of *Streptomyces* strains:

Mountain dark-chestnut soils

They are developing under the steppes vegetation. These soils contain in the top horizon from 4, 5 up to 6, and 5 % of humus, up to 0, 3-0, and 45 % of common nitrogen. The relation of carbon to nitrogen changes within the limits of 5, 6-7, 0. In the humus layer collects a significant amount of mobile potassium (18, 3 mg per 100g soil). The contents of mobile phosphorus achieves 2, 5-5, 2 mg on 100 g soil. Mountain dark-chestnut soils are carbonate from a surface, their pH is equal 7, 6-8, 5.

Fir-tree woods' chernozems

They develop mainly under fir and silver fir woods. These soils are characterized by the following morphological attributes: forest bedding of 5 and more cm thickness consists of leaf debris of needle, half-decomposed wood and grass plant remains. The volume of humus horizons A+B1 is 30-35 cm. The mechanical structure of these soils belongs to middle and heavy loams. The content of humus on the upper horizons fluctuates from 5.75 to 17.13 per cent. A narrow proportion of C characterizes these soils: N and nitrogen accumulation. In most cases they are non-carbonaceous. PH fluctuates from 7.0-7.83.

Mid-mountain steppe: Light gray soil

The mechanical structure varies from light loamy to heavy loamy. These soils are characterized by their little humus content (1.0-2.0 per cent), content of CO_2 carbonates is 5-10 per cent, reaction of soil environment is alkaline (pH=8.2-8.8). The upper horizon of these soils contains 0.10-0.25 percent of nitrogen, 0.20-0.30 percent of phosphorus, and 2.0-2.5 per cent of potassium.

Mountain chernozems

Natural vegetation is cereal and various grass meadow steppe. The mechanical structure is mid-loamy and heavy loamy. Humus content is considerable: from 6, 5 to 15, 0 per cent. Total nitrogen is 0.23-0.45 percent. The proportion of carbon to nitrogen equals to 8 11. Reaction of soil environment is neutral and faintly alkaline (pH=7.0-8.4). No solonetzicity. Regarding the amount of exchangeable potassium these types of soil are the richest among other soils. The amount of exchangeable potassium is 23-92 mg on the upper horizon.

Meadow-alpine soils

The soils are formed under alpine meadow-steppes. They have cloddy-granular structure and dark-gray color of the humus horizon. The total volume of the humus horizons is 50-70 cm. These soils are rich with humus (10-11 per cent), the structure of which is predominated by fulvo acids. Nitrogen content reaches 0.7 per cent. On the surface the soils are carbonaceous; the amount of CO_2 in the humus horizon does not exceed 1.5-3.0 per cent. Reaction of soil solution is neutral or faintly alkaline. The mechanical structure of these soils belongs to middle and heavy clayey soils.

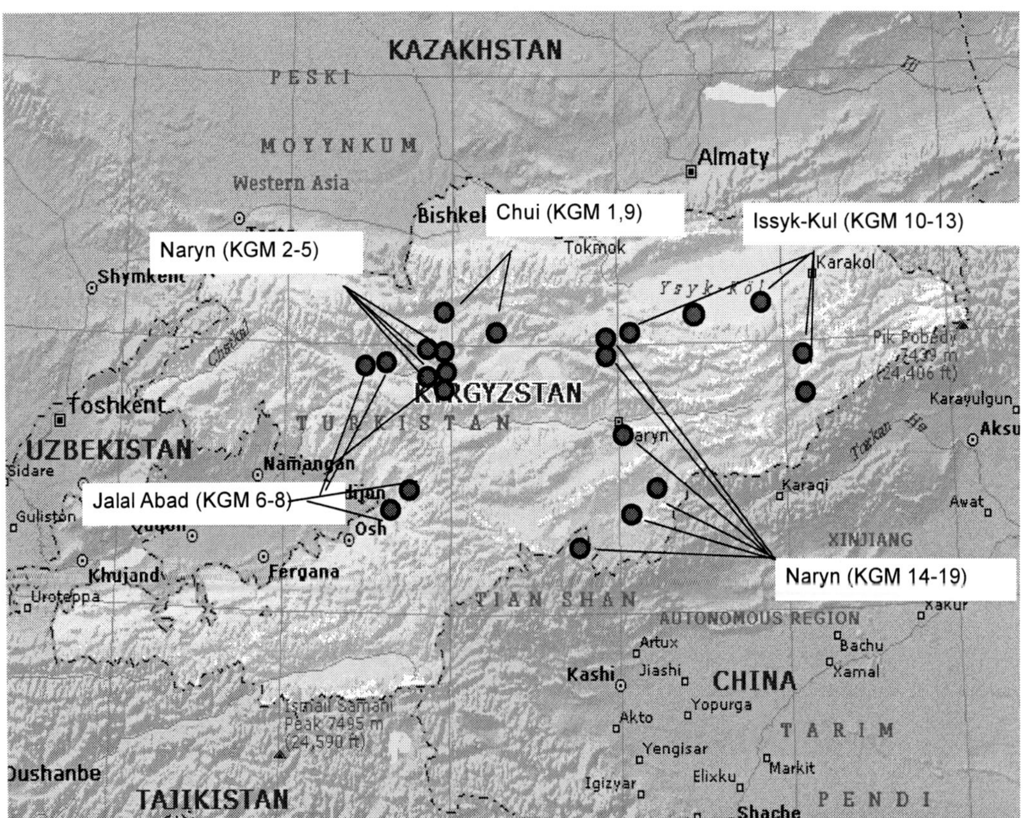

Figure 1. Map of Kyrgyzstan showing the location of the sampling sites (marked with a red sphere).

Mountain - valley northern light, siernozems

Northern fair gray soils are formed under the vegetation of ephemeral desert steppes. Forest-like loams serve as mother rocks. The mechanical structure is dusty-argillaceous. The content of humus is 0.7-1.7 percent, content of common nitrogen is 0.1-0.14 percent. They are carbonaceous on the surface: CO_2 content on the upper horizon is around 2-3 per cent.

Sediment Red

They represent sedimentary rocks of Neogen deposits. Their structure is sandy-clayish.

Sediment White

Neogen- Older Quaternary conglomerates (pebble bed) sandy-clayey deposits of 40-80 and 200-250 m volume.

The data of table 1 show the locality, ecosystems and soil types of Issyk-Kul sub region, where soil samples were selected.

Table 1. Ecosystems and soil types of Issyk-Kul sub region where selected soil samples were taken

Locality	Ecosystem	Soil Type	Elevation	Soil pH
Issyk-Kul/ Aksuu Valley	Spruce forest	Meadow alpine	2202 (m)	6.5
Issyk-Kul/ Aksuu Valley	Spruce forest	Meadow alpine	2206(m)	6.0
Issyk-Kul/ Aksuu Valley	Spruce forest	Fir-tree woods' chernozems	2145 (m)	6.5
Issyk-Kul/ Aksuu Valley	Spruce forest	Fir-tree woods' chernozems	2092 (m	6.5
Issyk-Kul/ Aksuu Valley	Spruce forest	Meadow alpine	2020 (m)	5.5
Issyk-Kul/ North Road (Bishkek – Kara-Kol)	Cultivated land, on vetch field	Light chestnut	1655 (m)	7,0
Issyk-Kul/ North Road (Bishkek – Kara-Kol)	Cultivated land, on Mustard field	Light chestnut	1641 (m)	7.5
Issyk-Kul/ North Road (Bishkek – Kara-Kol)	Cultivated land on Wheat field	Light chestnut	1649 (m)	7.5
Issyk-Kul/ North Road (Bishkek – Kara-Kol)	Cultivated land, on Potato field,	Light chestnut	1747 (m)	6.5
Issyk-Kul\\ Jety-Oguz Valley	Spruce forest	Meadow alpine	2261(m)	6.5
Issyk-Kul\\ Jety-Oguz Valley	Spruce forest	Meadow alpine	2298 (m)	6.5
Issyk-Kul\\ Jety-Oguz Valley	Spruce forest	Meadow alpine	2150 (m)	6.0

Table 1. Continued

Locality	Ecosystem	Soil Type	Elevation	Soil pH
Issyk-Kul\\ Jety-Oguz Valley	Juniper forest/	Dark archa woods' soils	2212 (m)	6.0
Issyk-Kul\\ Jety-Oguz Valley	Juniper forest/	Dark archa woods' soils	2212 (m)	5.5
Issyk-Kul\\ Jety-Oguz Valley	Mountain	Alpine meadow steppe soils;	2062 (m	6.0
Issyk-Kul/ Ton region	Foothill steppe	Meadow-siernozem-saline-dark alkaline in complex with meadow-wetland soils	1629 (m)	7.0
Issyk-Kul/ Ton region	Foothill steppe	Meadow-siernozem-saline-dark alkaline in complex with meadow-	1627 (m)	8.0
Issyk-Kul/ Jety-Oguz region, "Svetlay Polyana" village	Foothill steppe	Mountain - valley northern light,siernozems .	1650 (m)	7,83
Issyk-Kul/ Jety-Oguz region, "Chon-Kizil Suu" village,	Mid-mountain steppe	Red sediment	1800 (m)	7,2
Issyk-Kul/ Jety-Oguz region "Chon-Kizil Suu" village	Mid-mountain steppe	Light siernozem	1800 (m),	8,2
Issyk-Kul/ Ton region "Kok Moynok " village	Foothill desert	Sediment White	1032 (m),	7,9

The obtained data showed that various types of soils of ecosystems of Issyk-Kul region differed in the distribution, specific variety and the ratio of actynomyces species. The data in fig. 2 showed that the soils of mountain ecosystems contain a rich variety of *Streptomyces* - antagonists. So, mountainous meadow alpine soils, mountain chernozems, mountain - forest dark grey soils are rich with *Streptomyces* strains as well as with new antibiotic producing isolates.

Meadow-sierozem-saline-dark alkaline in complex with meadow-wetland soils are rich with species of actynomyces; however, after screening we obtained a small number of new antibiotic containing strains.

Prominent features of these soils, being a suitable ecological niche for actynomyces - they are rich by humus which contents in the top horizon changes from 5 up to 20 % with gradual reduction from top to bottom. They are sated with calcium which makes 80-90 % from the sum of the absorbed bases. The mineralization of fir-trees spreading , which ashes is rich calcium, causes neutral reaction of a soil solution, pH is equal 7,0-7,5.

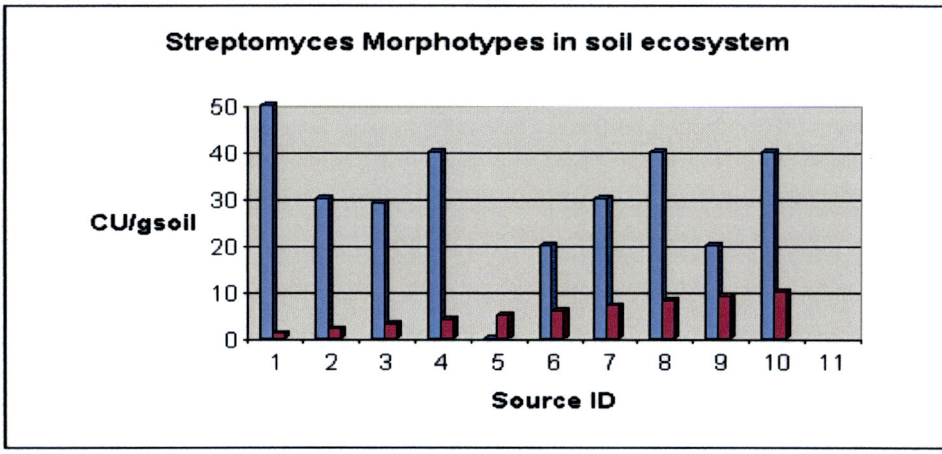

Figure 2. Prevalence *Streptomyces* in soil ecosystems of Issyk-Kul region : 1- mountainous meadow alpine soils, 4- mountain chernozems, 8- mountain -forest dark gray soils ,10- sediment white, 2- light chestnut, 3-light-brown soils,

Table 2. Prevalence of *Streptomyces* antagonistic strains in soils of Issyk-Kul sub region

№	Type of soil	Number of selected strains	Number of strains possessing antagonistic	Occurrence of antagonistic strains, %
1.	Bright - brow	32	12	37,5
2.	Mountain dark-chestnut	56	23	41,1
3.	Mountain chernozems	87	16	18,4
4	Mountainous meadow alpine soils,	104	35	32,4

Eighty-seven strains were revealed in mountain chernozems, 16 cultures of which possessed antagonistic properties, in bright - brown soil -12 cultures, and in mountain dark-chestnut – 23 cultures, in mountainous meadow alpine soils – 35 cultures and among the last the majority showed a wide spectrum of antagonistic actions (tabl.2).

The obtained data also showed that the occurrence of antagonistic forms of actynomyces depends on a degree of cultivation of soils and their agrichemical and genetic properties. Antagonistic forms of Streptomyces were rarely found in soils subjected to agricultural use. So, only 1 culture with antagonistic properties was isolated from chernozems under a fruit orchard whereas no culture was isolated from chernozems under a potato. However, in light-chestnut soils under a mustard plantation it was possible to isolate a big number of Streptomyces with a wide spectrum of antagonistic actions. Virgin soils not used in agriculture allowed far more cultures with antagonistic properties. So, from virgin chernozems soils at the edge of roads and small rivers it was possible to isolate 12 cultures of Streptomyces with strongly expressed antagonistic properties (fig. 3.)

Apparently, physical and chemical regime, mechanical structure, presence of optimum concentration of organic substances favors development of certain groups of *Streptomyces*. Our received data confirm other scientists' opinion. So, according to E.Andreyuk, [3]

mountain soils especially alpine and sub alpine meadow soils are rich with actynomyces whereas according to other researchers, actynomyces predominate in number bacteria in sand-desert soils [19].

The results also determined that *Streptomyces* have a various set of groups and species, a different ratio of their number in different soils. The representatives of *Ginereus* section prevailed in all types of researched soils. The representatives of *Roseus* section were the second in number, and *Albus* and *Azureus* sections were the third in number. The representatives of *Helvolo-Flavus* section were found in insignificant quantity (fig. 4)

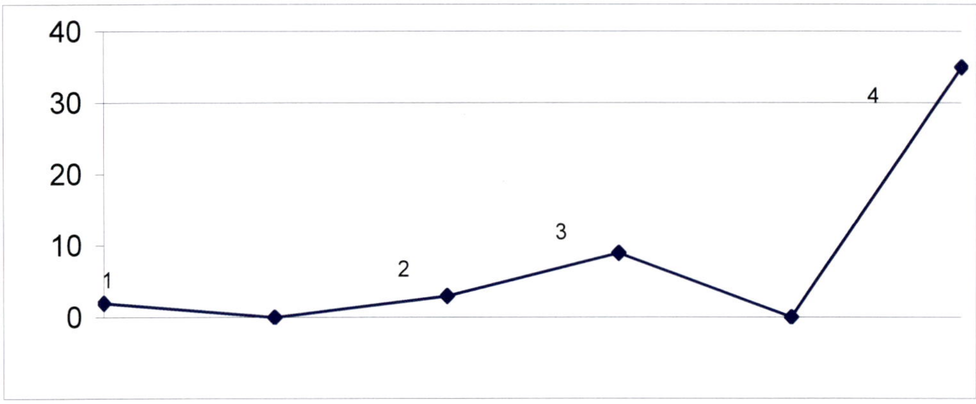

Figure 3. Occurrence of Streptomyces-antagonists in soils of various ecosystems: 1- soils of agri biocenoses, 2 - 3 - virgin soils, 4 – soils of mountain ecosystems.

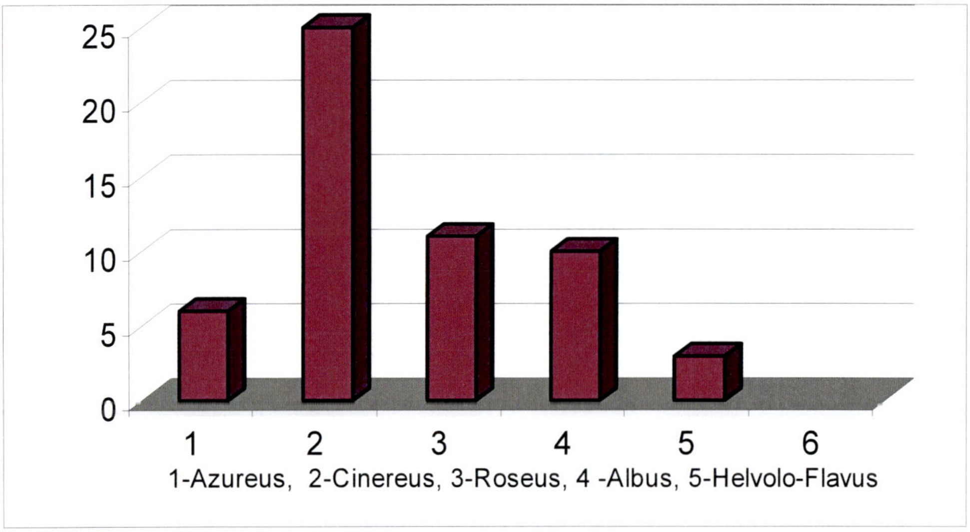

Figure 4. Taxonomic and numerical ratio of Streptomyces section.

Besides, it is determined that climatic conditions, seasonal prevalence, weather conditions and a temperature regime have an essential influence on the prevalence of Streptomyces in soils (fig. 5).

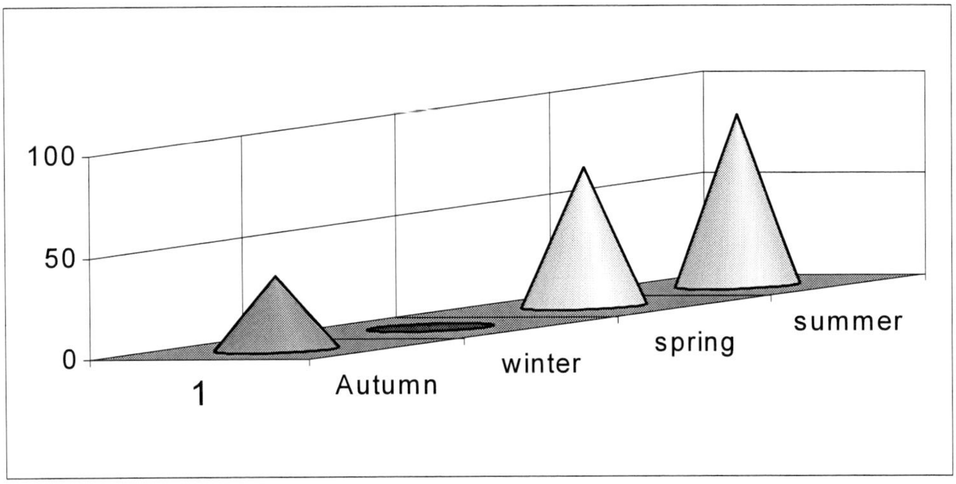

Figure 5. Occurrence of Streptomyces in seasons of the year.

As the data in fig. 5 indicate, in the summer time and when the weather was hot we isolated a larger number of actynomyces than in other seasons of the year. However, the cultures isolated in the spring demonstrated a wide spectrum of antagonistic action. In the winter few cultures of actynomyces was isolated from under the snow cover.

Thus, the soils of Issyk-Kul sub region contained a rich variety of antagonistic forms of Streptomyces, their prevalence depends on the type of soils, its cultivation and a season of the year.

3. Prevalence of Streptomyces in Soils of Northern-Kyrgyz Region (Chuy and Kemin regions)

With a view of studying the degree of prevalence of Streptomyces, we investigated soils of reserved territories of "Ala – Archa" State National Natural Park," Chon –Kemin" State National Natural Park and agricultural landscapes of Chuy region. The following types of soils were subjected to microbiological analyses.

Meadow sub-alpine soils

The climate of this zone is cold, with negative average annual air temperature. Average annual sum of precipitations is 400-500 ml. They are formed under sub alpine meadow vegetation on the altitude of 2500-3000 m. They contain significant amount of humus (8-15 per cent). They are leached out from carbonates and have pH – 6.5-7.0. According to the mechanical structure, these soils are characterized by significant content of sand, gravel, and detritus.

Meadow chernozem like soils

These soils contain up to 15 % of humus. The reaction of these soils changes from neutral up to subacidic. They are leached out from carbonates on the big depth. Their mechanical structure is non-uniform and mid-loamy or heavily loamy

Mountain -forest dark gray soils

They develop mainly under fir and silver fir woods. These soils are characterized by the following morphological attributes: forest bedding of 5 and more cm thickness consists of leaf debris of needle, half-decomposed wood and grass plant remains. The volume of humus horizons A+B1 is 30-35 cm. The mechanical structure of these soils belongs to middle and heavy loams. The content of humus on the upper horizons fluctuates from 5.75 to 17.13 per cent. A narrow proportion of C characterizes these soils: N and nitrogen accumulation. In most cases they are non-carbonaceous. PH fluctuates from 7.0-7.83.

Meadow-siernozem-saline-dark alkaline in complex with meadow-wetland soils

These soils are carbonaceous; according to the type of salt accumulation they are sulphatic and chloride-sulphatic. Content of gypsum is up to 1-2 per cent, pH is 8.8-9.5. Content of humus is 2-4 per cent. Mixed-hydrocarbonaceous and sulphatic subsoil waters lie in the depth of 1-3 meters. Formation of these soils takes place under the ephemeral-cereal and sagebrush vegetation. Mechanical structure varies from skeleton-sandy to grit-like and dust-like loam.

Results of research are shown in table 3.

Table 3. Ecosystems and soil types of North- Kyrgyz region where were selected soil samples

Locality	Ecosystem	Soil Type	Elevation	Soil pH
Chui region, "Sarko" village	Foothill steppe	Meadow siernozems, Meadow-siernozem-saline-dark alkaline in complex with meadow-wetland soils	870 (m)	9,75
Chui, Kara-Balta " gorge	Foothill desert	Sediment Red	1500 (m)	7,84
State National Natural Park "Ala – Archa"	Mountain	Meadow sub-alpine soils	2342 (m)	7,45
State National Natural Park "Chon –Kemin"	Broad-leafed forest	Mountain -forest dark gray soils	1762 (m)	7,17
Ak-Tyuz Ore management	Mountain valley: Urban Industrial waste/mine tailings	Meadow chernozem like soils.	1850 (m)	6,72
Kychy –Kemin region, gorge	Mountain valley: Urban Industrial waste/mine	Dark chestnut	1850 (m)	6,72

Table 4. Prevalence of Streptomyces antagonistic strains in soils of North- Kyrgyz region

Soil Type, Elevation	LOCALITY	Number of selected strains	Number of strains possessing antagonistic activity	Occurrence of antagonistic strains,%
Mountain Meadow sub-alpine soils	State National Natural Park "Ala – Archa"	50	1	2, 0
Mountain -forest dark gray soils	State National Natural Park "Chon –Kemin"	45	6	13,4
Mountain Meadow chernozem like soils.	Ak-Tyuz Ore management	24,2	12	52,2
Meadow-siernozem-saline-dark alkaline in complex	Chui, Jayilskiy region, "Sarko" village	35	10	28,5

The number of Streptomyces in soils of the reserved territories (mountain-meadow sub-alpine soils, mountain-forest dark gray soils) was high in comparison with the soils subjected to human influences (mountain-meadow chernozems), but among them the number of strains with antagonistic properties reached only 7.7% on average. Among the strains subjected to anthropogenesis influence, 52.2 % expressed strongly antagonistic properties, but their total number was minimal and reached 24.2 % from the total number of selected strains (table 4, figure 6).

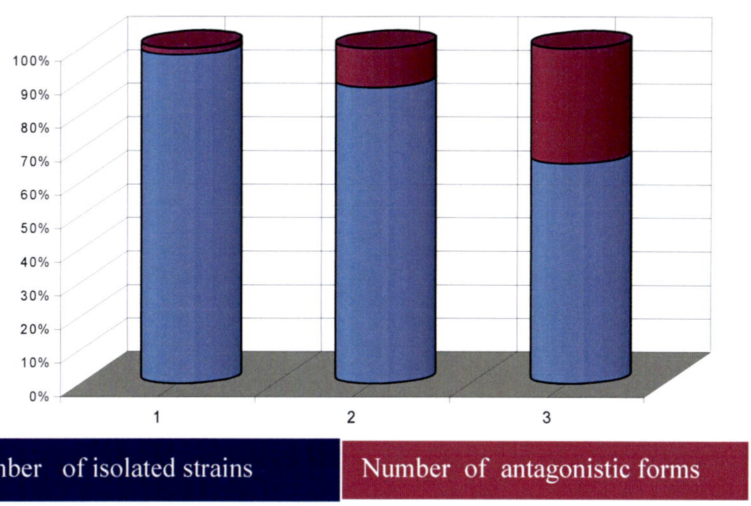

1 – Mountain Meadow sub-alpine soils, 2 – Mountain -forest dark gray soils, 3 – Mountain Meadow chernozems like soils.

Figure 6. Occurrence of Streptomyces strains in *North- Kyrgyz province* soils.

The research ascertained that soils of the reserved territories contained a rich variety of *Streptomyces* cultures. However, there were very few cultures with antagonistic properties among them (fig. 7) whereas the strains isolated from soils of polluted ecosystems (from the zone of Ak-Tyuz ore dressing factory where there has been an emission of heavy metals into soil for several years) possessed strong antagonistic properties. Apparently, the opportunity of prosperity of various microorganisms depends not only on availability of elements necessary for their life but also on the ability of cells to resist to the actions of toxic substances. Researchers mark firmness of Streptomyces in cultures and in soils to the action of toxic agents such as phumigant, insecticides, etc. [4]. Others assume that antibiotics give an advantage to producers in struggle for existence, oppressing the competitors [14]. Demain A.L [9] considers the antibiotics as heterogeneous assemblage of powerful microbe metabolites, the majority of which provide an advantage in struggle for survival.

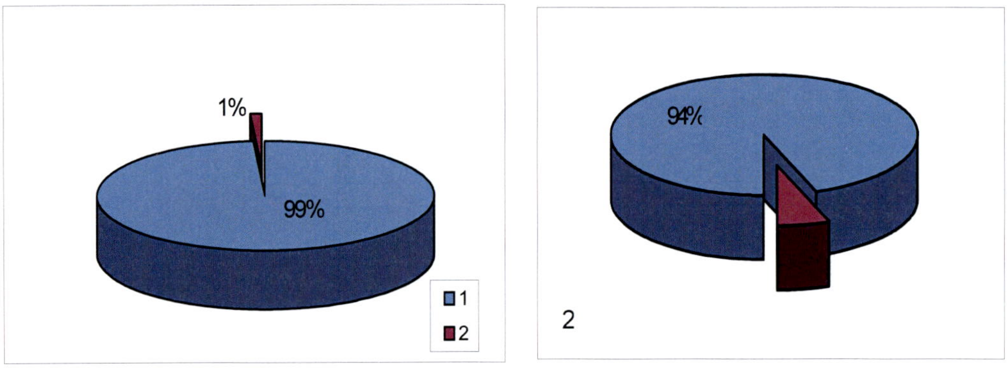

1- Soils of "Ala – Archa" State National Natural Park

2 - Soils of "Chon –Kemin" State National Natural Park

Figure 7. A percentage ratio of antagonists - Streptomyces in soils of reserved territories.

However, further studying of isolated strains showed that the strains isolated from soils subjected to pollution with heavy metals lose their antagonistic properties after the second sowing on agar medium, whereas the strains isolated from soils not subjected to industrial influences do not lose their antagonistic properties for three years (the period of research). Apparently, antagonistic properties of the cultures amplify under the influence of adverse factors, in so-called "stressful" conditions, and then normalize in favorable conditions.

The study of specific variety distribution determined that the representatives of *Roseus*, *Albus*, *Cinereus* sections prevail in soil samples of the reserved territories. The biggest number of antagonists was found among the representatives of *Albus* section. In soils subjected to industrial influences the representatives of *Cinereus* and *Albus* sections were dominating.

Thus, soils from non-polluted landscapes contain a greater number of *Streptomyces* than soils from the territories polluted with industrial wastes. Although the greatest number of strains with antagonistic properties was isolated under "stressful" conditions, they lost these properties in a short time; whereas the strains isolated from soil samples of reserved territories kept their antagonistic properties in laboratory conditions for a long time.

4. Distribution of Streptomyces in Soils of Southern Kyrgyz Region

Southern Kyrgyz region is characterized by its transitive form a temperate to subtropical climate with a sharply expressed contrast of seasonal humidifying, which is of a special interest for a biological research of Streptomyces strains in such conditions.

For this purpose, we researched soils of Ketmen-Tjubo and Fergano-Chatkal districts. Samples were isolated from mountain chernozems, mountain brown typical, mountain - wood black - brown types, brown-gray juniper forest soils, high - mountain meadow-steppe-alpine soils, and skeleton carbonaceous soils (tab. 5).

Mountain - wood black - brown soils of nut -fruit forests

The soil forming breeds of specified soils consist mainly from adjournment with the high contents of carbonic salts (30-35 % CaCO3). Usually they are leached out from carbonates on significant depth. The mechanical structure of these soils is mainly heavy -loamy . The contents of humus in the top horizon reaches 10-15 %, and sometimes 20 %, with depth the amount of it sharply decreases. In connection with an intensive accumulation of fossils the total of nitrogen in the top horizon reaches 0,6-0,85 %, on the depths of 35-40 sm it does not exceed than 0,2-0,4 %. On the structure they are closed to chernozems and characterized with sharp prevalence gumin acids above fulvous acids.

Black – brown soils possess neutral reaction of a soil solution (pH =6,5-7,3). The contents of total phosphorus equally 0,24-0,48 %, potassium - 2,4-3,8 %, mobile P_2O_5-60 mg, and K_2O - 70 mg on 100g soil.

Mountain chernozems

Natural vegetation is cereal and various grass meadow steppe. The mechanical structure is mid-loamy and heavy loamy. Humus content is considerable: from 5.2 to 10.5 per cent. Total nitrogen is 0.23-0.45 percent. The proportion of carbon to nitrogen equals to 8-11. Reaction of soil environment is neutral and faintly alkaline (pH=7.0-8.4). No solonetzicity. Regarding the amount of exchangeable potassium these types of soil are the richest among other soils. The amount of exchangeable potassium is 23-92 mg on the upper horizon. Gumin acids are prevail in the structure of humus . Their ratio to fulvous acids makes 1,8 in the top horizon .

Mountain brown typical soils

These soils belong to dusty- heavy loams on the mechanical structure .

They contain in the top layer from 4,5 up to 6 % of humus , with depth the amount of it decreases. The amount CO2 of carbonates in the top horizon reaches up to 0,5 - 1,5 %, from the top to the bottom it sharply increases and in carbonate- alluvial horizon reaches 10-15 %. Reaction of the soil environment changes within the limits of an alkaline interval (pH = 8,3-8,7). The contents of nitrogen in the humus horizon makes 0,3-0,4 %, the ratio of C:N is rather wide – 7 :10.The absorbing complex of these soils is sated with calcium and magnesium. The structure of humus is fulvous – hum at . Total contents of K2O makes 2,3-2,5 %, P2O5 0,16 0,28 %.

Brown-gray juniper forest soils

They develop under juniper woods. The plant cover is mainly represented by sheep's fescue (Festuca sulcate) with some sub alpine various grasses.

These soils are characterized by the availability of fall of the leaves that transits to brown powder-like mass resembling decayed faintly turf mass. The mechanical structure of these soils is mainly heavily loamy. The content of humus is 12-15 per cent. The volume of absorption on the upper horizon is high and makes up 50 mg-equip per 100 g. The main mass of organic substances is concentrated in the depth of 0-10 and 5-10 cm. These soils differ by their relatively high accumulation of common nitrogen (up to 0.4-1.3 per cent). In most cases they are non-carbonaceous. PH fluctuates from 6.5-7.4.

High-mountain meadow-steppe -alpine soils

The climate of this zone is cold, with negative average annual air temperature, and a short warm period. The soils are formed under alpine meadow-steppes. They have cloddy-granular structure and dark-gray color of the humus horizon. The total volume of the humus horizons is 50-70 cm. These soils are rich with humus (10-11 per cent), the structure of which is predominated by fulvo acids. Nitrogen content reaches 0.7 per cent. On the surface the soils are carbonaceous; the amount of CO_2 in the humus horizon does not exceed 1.5-3.0 per cent. Reaction of soil solution is neutral or faintly alkaline. The mechanical structure of these soils belongs to middle and heavy clayey soils.

High-mountain skeleton carbonaceous soils

The climatic conditions are very severe here: the winter is cold, the summer is cool, strong winds occur all year long. The upper border of their spreading is in the contact with indigenous rocks, talus, and detritus, which go out onto the surface. The humus horizon has a grass covered structure and a gray color with pale-yellow hue. The soil cross-section is shallow and skeletal. The mechanical structure of these soils belongs to light clayey soils. The turf horizon contains from 2.5 to 3.5 per cent of humus with its significant decrease downwards. The amount of bulk nitrogen is 0.2-0.3 percent. The proportion of carbon to nitrogen is from 7 to 11. The soils are alkaline; pH of water extract fluctuates from 8.3 to 8.6. Calcium satiation is significant. Its share is 85-90 per cent. skeleton carbonaceous soils.

The data of table 6 show that the amount of Streptomyces in the investigated samples was not high in comparison with soils samples of Issyk-Kul sub province. Similar results have been received also by the researches of Voronov S.I. [43], he also has noted the low percent of the contents of Streptomyces in soils of Southern Kyrgyzstan .

The greatest amount of Streptomyces has been isolated from mountain chernozem of district Ketmen-Tjubin , high - mountain meadow-steppe -alpine soils, their the least amount contained in black - brown nut woods soils, brown-gray juniper forest soils , in which also we did isolate few strains possessing an antagonistic activity. Opposite the strains isolated from mountain chernozems of Ketmen-Tjubin district possessed strongly expressed antagonistic properties (, fig. 8).

Table 5. Ecosystems and soil types of Southern – Kyrgyz region where selected soil samples were taken

Locality	Ecosystem	Soil Type	Elevation	Soil pH
Komsomol village of. Ketmen-Tjubin district	Mountain valley	Mountain chernozems	2000 (m)	7,45
Masar-Suu village of Ketmen-Tjubin district	Mountain valley	Mountain chernozems	2000 (m)	7,25
Achy village of Ketmen-Tjubin district	Mountain valley	Mountain brown typical soils	1800(m)	8,5.
Village Kochkor-Ata of Jalabad provinse	Nut -fruit forests	Mountain - wood black - brown soils of nut -fruit forests	1700(m)	6.5
"Chychkan" gorge of Ketmen-Tjubin district	Juniper forest	Brown-gray soils of juniper forest	1682 (m),	7,42
"Sharkyratma" place of Ketmen-Tjubin district	High-mountain	Mountain-meadow-steppe - alpine soils.	3015 (m),	7,2
"Ala-Bel" Pass of Ketmen-Tjubin district.	High-mountain	Skeleton carbonaceous soils	3233 (m)	8,4

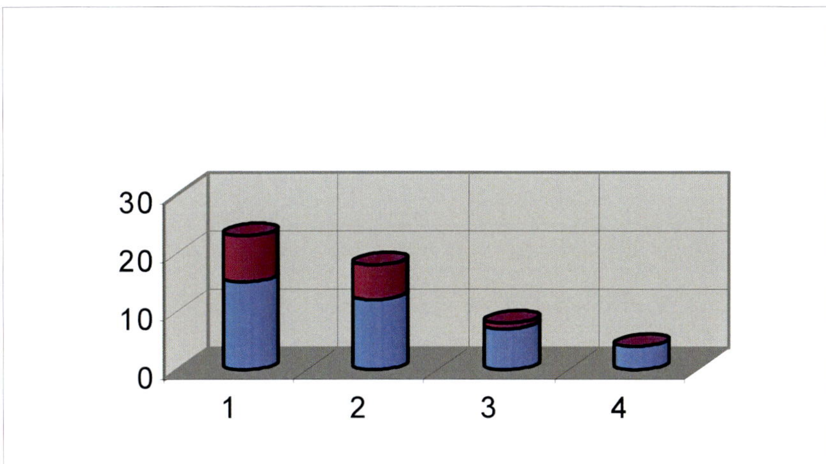

1, 2 Mountain chernozems, 3 – Mountain brown typical soils, 4- Mountain - wood black - brown soils of nut -fruit forests

Figure 8. Prevalence of Streptomyces in soils of the Southern – Kyrgyz province.

Table 6. The prevalence of Streptomyces in soils of the Southern – Kyrgyz province

Soil Type, Elevation	LOCALITY	Number of selected strains	Number of strains possessing an antagonistic activity	Occurrence of antagonistic strains ,%
Mountain chernozems	Village Komsomol of Ketmen-Tjubin district	15	8	53,3
Mountain chernozems	Village Masar-Suu of Ketmen-Tjubin district	12	6	50,0
Mountain brown typical soils	Village Achy of Ketmen-Tjubin district	7	1	14,3
Mountain - wood black - brown soils of nut -fruit forests	Village Kochkor-Ata	4	-	-
Juniper forest: Brown-gray soils of juniper forest.	"Chychkan" gorge of Ketmen-Tjubin district	3	1	33,3
High-mountain: mountain- meadow-steppe -alpine soils.	"Sharkyratma" place.	6	4	66,6
High-mountain: High-mountain Skeleton carbonaceous soils	"Ala-Bel" Pass.	3	-	-

The study of a specific variety of the isolated cultures showed that the strains of Cinereus and Albus sections were mainly isolated from soil samples of mountain chernozem of Ketmen-Tjubo district. Roseus, Helvolo-Flavus sections were mainly isolated from mountain brownish - chestnut soils of Fergano-Chatkal district.

Thus, in soils of the Southern – Kyrgyz region the quantitative distribution of actynomyces was quite low, and among them we revealed an insignificant number of antagonistic forms. Only the soils from Ketmen-Tjubo district contained the greatest number of Streptomyces cultures with sharply expressed antagonistic properties.

5. Distribution of Streptomyces in Soils of Internal Tian-Shan Kyrgyz Region

Climatic conditions are very severe here: the winter is cold and the summer is cool. We researched soils of Narin, Kochkor and At- Bashy districts. Samples were taken from high - mountain meadow-steppe-alpine soils, high-mountain alpine soils, mountain chernozems, mountain brown typical, mountain - wood black - brown types, brown-gray juniper forest soils, skeleton carbonaceous soils, high-mountain sediment, color: yellow, high-mountain light chestnut soils , mid-mountain gray – desert- steppe-stone soils, mid-mountain steppe light –gray - saline- soils , saline sediment.

High-mountain meadow sub-alpine soils

The climate of this zone is cold, with negative average annual air temperature. Average annual sum of precipitations is 400-500 ml. They are formed under sub alpine meadow vegetation on the altitude of 2500-3000 m. They contain significant amount of humus (8-15 per cent). They are leached out from carbonates and have pH – 6.5-7.0. According to the mechanical structure, these soils are characterized by significant content of sand, gravel, and detritus.

High-mountain alpine soils

The climate of this zone is cold, with negative average annual air temperature, and a short warm period, which is broken off by night frost. These soils morphologically are characterized by the availability of turf horizon of dark-gray or brownish and black color, and granular structure. The upper turf horizon contains 10-15 per cent of humus and horizons fertilized with peat contain up to 20 per cent of humus. The upper horizons are distinguished by their structured formation and high water permeability. Fulvo acids prevail in the humus structure. The soils are distinguished by the high content of common nitrogen (0.6-0.8 per cent), general phosphorus (0.25-0.40 per cent), and potassium (2.6-4.0 per cent). These soils belong to lixiviated ones and have sub acidic reaction.

High-mountain: Sediment, Color: Yellow

It represents sedimentary rocks of Neogen deposits. Their structure is sandy-clayish and distinguished by their content of gypsum and salt occurring everywhere especially on the lower part.

High-mountain: Mountain Light Chestnut steppe soils

The climatic conditions of the zone where these soils are spread are characterized by the great dryness of the summer and high continentally. The sum of annual precipitations is approximately 250-350 mm. The mechanical structure of these soils is faintly and mid-loamy. Middle and fine sand as well as large dust-like particles predominate among the fractions.

The upper humus horizon contains from 2.5 to 3.5 per cent of humus, which decreases down the cross-section. The amount of bulk nitrogen is 0.2-0.25 percent. The proportion of carbon to nitrogen equals to 6-9. On the surface the soil is calcareous and contains 1-4 per cent CO_2. The soil has an alkaline reaction of the environment; the pH indicator reaches 8.6-9.0 per cent. The plant cover is mainly represented by Festuca sulcate- wormwood or needle grass-Festuca sulcate-wormwood-various grasses associations.

Mid-mountain desert: Gray – desert- steppe-stone soils

The climate is harshly continental. Strong winds (15-20 meters per second) dominate nearly all year long as a result of which the soils suffer from wind erosion. Average annual air temperature is +5.0-6.6 degrees. Vegetation is very rare. Total plant covering does not exceed 15-20 per cent. The surface of these soils is scattered with cobblestones, large and little pebbles, rock debris with black desert varnish. They are characterized by their faint differentiation of soil cross-section, little humus content (0.5-1.0 per cent) with prevalence of fulvoacids in the humus structure, and low content of main nutrition elements. Reaction of soil environment is alkaline (pH=8.0-8.5).

Mid-mountain steppe: Light gray soils

The climate is harshly continental. Average annual temperature fluctuates from 2.5 to 7 degrees. The mechanical structure varies from light loamy to heavy loamy. They content a little humus (1.0-2.0 per cent). Content of CO_2 carbonates is 5-10 per cent; reaction of soil environment is alkaline (pH=8.2-8.8). The upper horizon of these soils contains 1.10-0.25 per cent of nitrogen, 0.20-0.30 percent of phosphorus, 2.0-2.5 per cent of potassium.

High mountain: - mountain meadow-steppe -alpine soils

They are formed under alpine meadow-steppes. They have dark-gray color of the humus horizon and cloddy-granular structure. Total volume of the humus horizons is 50-70 cm. These soils are rich with humus (10-11 per cent) the structure of which is prevailed by fulvoacids. The content of nitrogen reaches 0.7 per cent. The soils are carbonaceous on the surface. The amount of CO_2 carbonates is 1.5-3.0 per cent. . Reaction of soil solution is neutral and faintly alkaline.

Mid-mountain steppe: Light –gray - saline- soils

They are formed on the tertiary salt-bearing deposits. According to the type of salinization, the soils are chloride-sulphatic or sulphatic-chloride. These types of soil are characterized by the availability of crystal porous layer with a plate-like structure. Little content of humus (1.0-2.0 per cent) with prevalence of fulvoacids in the humus structure, narrow proportion of carbon to nitrogen (4-6). Relatively high amount of common nitrogen, phosphorus and especially potassium and calcareousness. Reaction of soil solution is alkaline (8.2-8.7).

Saline Sediment zonal: Saline Sediment

Red. They represent sedimentary rocks of Neogen deposits. Their structure is sandy-clayish.

Mountain chernozems

Natural vegetation is cereal and various grass meadow steppe. The mechanical structure is mid-loamy and heavy loamy. Humus content is considerable: from 5.2 to 10.5 per cent. Total nitrogen is 0.23-0.45 percent. The proportion of carbon to nitrogen equals to 8-11. Reaction of soil environment is neutral and faintly alkaline (pH=7.0-8.4). No solonetzicity. Regarding the amount of exchangeable potassium these types of soil are the richest among other soils. The amount of exchangeable potassium is 23-92 mg on the upper horizon.

Mountain: Gray-brown desert stony

The surface of these soils is scattered with cobblestones, large and little pebbles, rock debris with black desert varnish. There are few organic substances and nutrition elements in these soils. They contain 0.5-1.5 percent of humus, 0.007-0.008 percent of bulk nitrogen, 0.03-0.04 percent of phosphorus, and 1-3 per cent of potassium. Reaction of soil environment is alkaline (pH=8.0-8.5). In general, these soils are non- salinized and non-alkaline. The date of tab. 7. have shown the sites from them were selected soil samples.

As the tab.7 indicates, 4 classes of different ecosystems and 10 types of soils were analyzed for actynomyces diversity. There is a noticeable presence of desert steppes among ecosystems as well as extremely poor ecosystems – nival and sub-nival. This indicates quite severe conditions of life existence in the significant part of the country.

Thus, by using traditional methods of cultivation of microorganisms it was possible to pick out cultivable forms of prokaryote. The investigated sites differed with yield of actynomyces. From all investigated places such as mountain alpine soils, meadow sub-alpine soils, high-mountain chestnut steppe-brown soils, meadow-sierozem-saline-dark alkaline in complex with meadow-wetland soils, mountain chernozems were rich with Streptomyces, which represents a particular interest as producers of antibiotic substances (fig.9).

Table 7. Ecosystems and soil types of internal Tian- Shan Kyrgyz province where selected soil samples were taken

Locality	Ecosystem	Soil Type	Elevation	Soil pH
Naryn, « Tuy-Ashu » Pass	High-mountain	Meadow sub-alpine soils	2860 (m)	pH- 7,0
Naryn, Descent after of « Tuy Ashu » Pass	High-mountain	Mountain alpine soils	2900 (m)	pH- 7,0
Naryn, Descent after of « Tuy-Ashu » Pass	High-mountain: Intra-zonal	Sediment, Color: Yellow	2642 (m)	pH-7,6
Naryn , "Susamyr" valley	High-mountain	Mountain Light Chestnut steppe soils	2347 (m),	pH-8, 56
Naryn: "Kubakay " passe of Kochkor region	Mid-mountain desert	Gray – desert-steppe-stone soils .	2109 (m),	pH- 8,2
Naryn: "Semizbel " villaje of Kochkor region	Mid-mountain steppe	Light gray soils	2109 (m)	pH- 8,1
Naryn: "Dolon " Pass of Naryn region	High mountain	Mountain meadow-steppe -alpine soils.	3210 (m),	pH-6,9
Naryn: "Kapabuk " place of Naryn region	Mid-mountain steppe	Light –gray - saline- soils	2109 (m),	pH- 8,2
Naryn: "Kysylbel " pass of Atbashy region	Saline Sediment zonal	Saline Sediment Red.	2300 (m)	pH-8,7
Naryn: "Tashrabat place" of Atbashy region	High-mountain steppe	Mountain chernozems	2500 (m)	pH-7,7
Naryn: "Tashrabat place" of Atbashy region	High-mountain steppe	Mountain: Gray-brown desert stony.	2500 (m)	pH-8,6

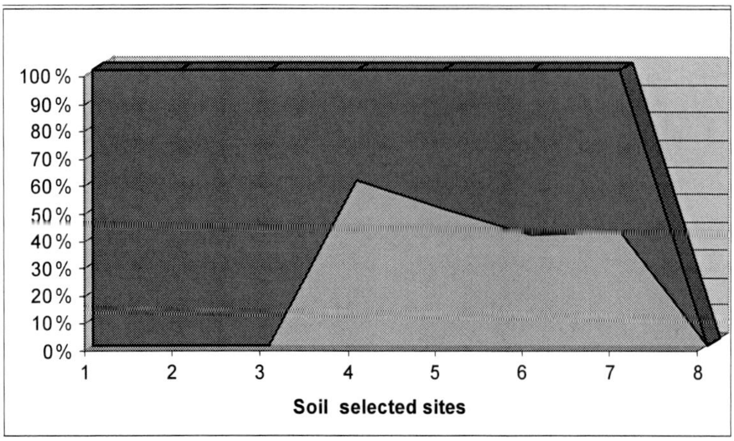

Figure 9. Prevalence of Streptomyces in soils of the Internal Tian- Shan Kyrgyz province: 1- meadow sub-alpine soils ; 2- mountain alpine soils ; 3- high-mountain chestnut steppe- brown soils ; 4- meadow-siernozem-saline-dark alkaline in complex with meadow-wetland soils; 5- mountain chernozems.

Thus , the majority of investigated soils were high mountain-meadow-steppe-alpine soils and high mountain alpine soils, they are formed under alpine meadow vegetation on the altitude of 3000-3800 m. The climate of this zone is cold with negative average annual air temperature and a short warm period, which is disturbed by night frost. These soils are rich with humus (10-20 per cent). The air temperature of these regions in winter decreases till minus 25-30 degrees Celsius, and in summer sometimes goes up to + 25 degrees Celsius. However, the temperature of the soil never reaches 20 degrees Celsius. Such instability of temperature conditions obviously determines the diapason of the growth and development temperatures of microorganisms growing here.

The biodynamic research that we realized showed that mountain soils located at low temperature conditions belong to poor and mid-enriched soils according to their number and biomass. In mountain-meadow alpine soils of a high-mountain zone, such microbiological processes as decomposition of vegetation remain, formation of humus, heterotrophic transformation of substances have a delayed character. We revealed a domination of bacteria adopting organic nitrogen, for instance, species of spore-forming bacteria in these biotopes. It proves the fact that the processes of transformation of organic substances are significantly delayed here.

Actinomyces have a specific place. Their proportion remains high everywhere and is displayed on the profile. Actinomyces usually dominate in soil when there are conditions for using relatively scarcely accessible substrata. They belong to marked aerophilic; therefore, their main mass is concentrated in turf and humus-accumulative subhorizons. Thus, bacterial communities of alpine ecosystems are generally characterized by relatively high abundance of spore-forming bacteria, by bacteria of actinomyces lines, gram-positive coccus, and cellulose-decomposing bacteria as well as mould fungus[17] with low content of nitrogen-fixating bacteria of Azotobacter species, gram-negative bacteria.

Thus, the research give a basis to suppose that soils of Issyk-Kul sub region are characterized by a rich specific variety of Streptomyces and a high content of antagonistic forms in comparison with investigated soil samples from other regions. The soil samples from Northern Kyrgyz region and Internal Tian- Shan Kyrgyz region occupy an intermediate position on the content of Streptomyces antagonistic forms. Southern Kyrgyz region is characterized by a low specific content of Streptomyces -antagonists (fig. 10).

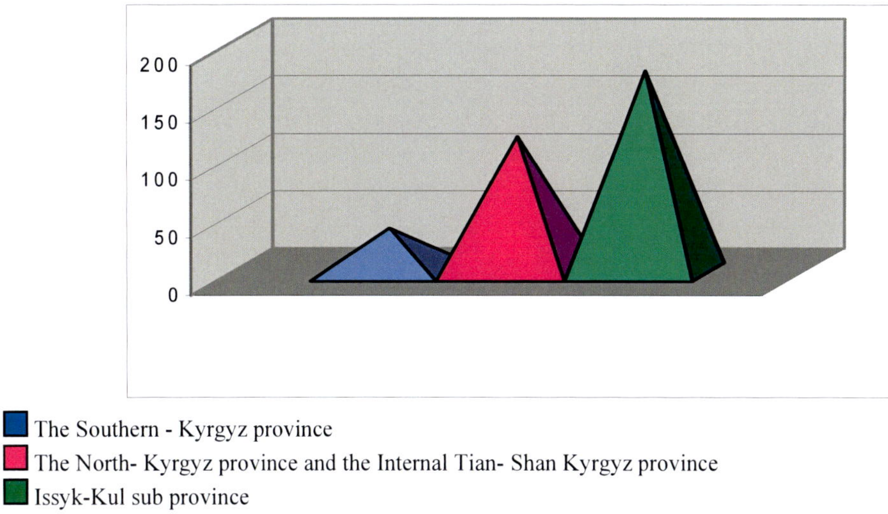

Figure 10. Prevalence of Streptomyces in soils of the Kyrgyz provinces.

The quantitative and qualitative structure (a specific variety and a ratio of species and groups) of actynomyces depends on soil - climatic conditions, elevation of locality , a vegetative cover, cultivation of soils and many other factors.

However, it should be noted that the richness of soils with actynomyces is not always accompanied with abundance of antagonists. There are soils abounding with actynomyces and completely deprived of antagonists, on the contrary, there are soils poor with actynomyces but there are rather rich forms with antibacterial properties.

As the results of our research show, almost in all zones and ecosystems meadow alpine soils, mountain-forest dark gray soils, mountain chernozems are distinguished by high numerical and specific contents of *Streptomyces.*

Characteristic feature of these soils being a suitable ecological niche for actynomyces is that they contain high percent of humus, which on the top horizon changes from 5 up to 20 % with gradual reduction from top to bottom. They are sated with calcium, which makes 80-90 % from the sum of the absorbed bases, and that causes a neutral reaction of soil solution, pH is equal to 7.0-7.5.

According to literary data, a significant number of actynomyces was found in mountain subalpine and alpine meadow soils (about 5 million per 1 g of air – dry soil) and they make 40-50% of all microorganisms growing on starch-ammoniac and starch-casein mediums [50]. Up to 3 million actynomyces per 1 g of soil were revealed in brown mountain soils of Georgia [3]. The data of other authors [38] also testify that entering of organic-mineral and organic fertilizers [10] into soils increase the number of actynomyces. Z.Vaksman's research also [45] showed that the highest percentage of antagonists is found in the composted manure, in the silts and especially in artificial soils enriched with a mixture of different bacteria.

As it is known, in a natural habitat actynomyces complexes participate stage by stage in decomposition of remote organic compounds in soil, taking certain spatial and time position in an ecosystem. Their high adaptation to various conditions of existence allows actynomyces to take part in transformations of a wide group of substances. Actynomyces are very hardy and can develop at low soil humidity. Actynomyces are characterized by a rather small growth rate even in laboratory conditions, in conditions of easily assimilated and well-provided nutrients. Their estimated time of generation in soil can be big [30, 11]. Usually, actynomyces do not appear in the first echelon of "γ--strategists" when readily available for many microorganisms substances are hit upon soils. Local evolution, substantial growth of the number of actynomyces and sometimes domination of their certain representatives is usually observed when there are conditions to use difficult-to-access substrata.

Therefore, in mountain meadow subalpine and alpine meadows soils, mountain-forest dark gray soils, mountain chernozems, where the contents of difficultly decomposed compounds are high, actynomyces are found in a significant number and they make 40-50% of all microorganisms.

Actynomyces population of soils is firmer to drying rather than other bacteria. N.Krasilnikov [22] and R.Tumarkin [40] assume that the optimum humidity for actynomyces is 17-20 % but actynomyces can develop at the humidity of 8-10 %. High humidity (50-60 %) of soils is more favorable for the development of bacteria and the number of actynomyces decreases 2-3 times on all structures [41]. It was noted that during the droughty summer period in soils of Siberia the contents of soils grow 9 times in comparison with damp seasons [18]. Our results also showed that in the summer time and when the weather is hot, the number of isolated actynomyces was higher than in other seasons of the year. However, the cultures isolated in the spring showed a wide spectrum of antagonistic action. In the winter few cultures of actynomyces were isolated from under a snow cover. The similar phenomenon is also marked by other researchers. Therefore, A.Korenjako [20] marks that actynomyces develop plentifully during the summer periods and in the late autumn, i.e. during those moments when the number of population of saprophyte bacteria reduces.

Members of actynomyces groups have a tendency to be prevalent in certain types of soil, so our data showed that predominant members of Streptomyces complex are in Cinereus section of Chromogenus series, which are adapted basically to chestnut and black soil while species of the Roseus section of the Fuscus series are adapted to gray, brown, and chestnut soils. According to G. Zenova [50], actynomyces of three groups - Albus, Griseus and Chromogenes are found in grey wood soils while representatives of three and more color groups are added to them in chestnut and gray soils. It is marked that actynomyces of Chromogenes groups are present almost in all soils [3, 33]. According to L.Sorokina's data [37], a variety of actynomyces in the cultivated soils is more than in wood soils. Representatives of Streptomyces that belong to Cinereus, Chromogenes, Albus, Chrysomallus and Albocoloratus sections dominate in the soils of cultivated lands.

There are literary data that the temperature of soils influences the number of actynomyces [19, 30] but other researchers do not confirm such dependence [49, 34]. They believe that a geographical zone of habitat influences on the optimum growth temperature of actynomyces. So, at 3°C the growth was 48 % of cultures isolated from tundra soils, 28 % - isolated from podsolic soils, 21.4 % - from chernozems. Actynomyces living in the deep layers of soils have a lower temperature optimum [29]. The temperature firmness of Griseus group is higher than groups with other color of mycelium [1]. The temperature optimum for the growth and development of Streptomyces in laboratory conditions is determined in limits of 25-30° C [21, 27].

Under "stressful" conditions it was possible to isolate the biggest number of cultures capable of displaying antagonistic properties, which quickly lose these properties in favorable conditions. The cultures isolated from the reserved territories steadily kept their antagonistic properties for a long term. Apparently, in polluted and extreme conditions, under the mutagen influences of high concentration of heavy metals there is a change of the genom sites that activates special protective mechanisms promoting survival of cells. The literary data also confirm that representatives of actynomyces with certain antibacterial properties are rarely found and only in certain soils [51].

It was determined that growth intensity of Streptomyces colonies on the surface of nutritious agar depends on the sampling conditions and a season of the year. As the data in fig.11 show, from the soil samples taken in May from agribiosenozes of Issyk-Kul area growth of colonies was revealed in 29 days, from soil samples taken in April from the territory of "Ala-Apcha" National Park growth was revealed in 12 days, from soil samples taken in July from "Chon –Kemin" National Park - in 7 days, and from soil samples of Toktogul area taken in November - in 3 days.

Hence, not every type of soils can contain Streptomyces possessing antagonistic properties. Development of antagonistic forms occurs only in certain conditions and types of soils. Apparently, cultivation of soil, meliorative actions, using fertilizers and other agrichemical receptions affect significantly the structure of functioning of microbiozenoses in soil.

Physiological and biochemical abilities of actynomyces are extremely wide and various. Streptomyces are characterized by oligocarbophily that was marked by M. Bejerink [6]. Oligotrophic actynomyces develop well on "hungry agar" being satisfied with insignificant concentration of organic substances on the medium [5].

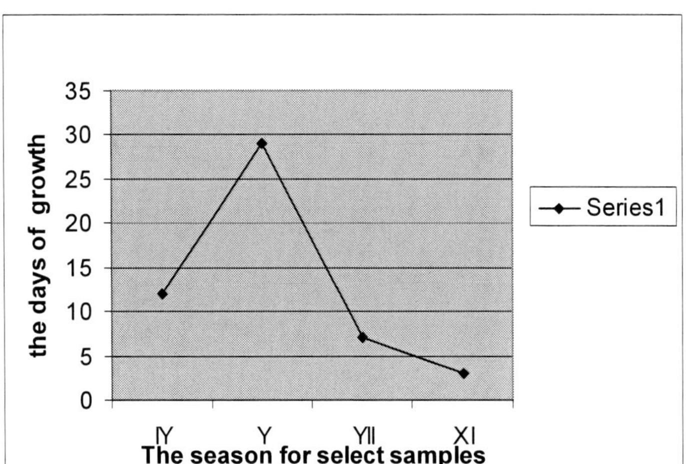

Figure 11. Growth intensity of Streptomyces colonies on the surface of agar medium from terms of season for sampling.

Meanwhile, one of the outstanding features of actynomyces is their big variability. The slightest deviations in the structure of a nutrient medium or in other external conditions cause changes in growth, development or in biochemical activity of actynomyces culture. The variants derived from the same species can be numerous and various in morphological, cultural and physiological properties [25, 21]. It is determined that pH of the environments also influences on the growth of actynomyces. So, for the cultures isolated from tundra and podsolic soils, the optimum pH makes 6.1 while for actynomyces isolated from chernozem – 8.6-9.6 [34].

There are different opinions regarding forms of existence of actynomyces in soils. It is believed that actynomyces in soils along with spores are virtually not distinguished on morphology from bacteria [23]. It is emphasized that Streptomyces exist in soils as a vegetative mycelia but it is defined that the spore is dominate in the population structure of soil actynomyces [26, 15, 28, 30].

Growth of actynomyces in a microzone of non-sterile soils can give up to 80 new spores. Thus, even if 50 % of spores entered into soils in a month destroy, it is enough, because only 1 % of spores developed in soils within a month maintain the population of actynomyces spores at a constant level [13].

Antibiotics are known to be produced by microorganisms in the places of their natural habitat [32]. Antibiotics are adsorbed by soil particles, nevertheless, as A.Zvjagintsev [48] defined, they possess biological activity. Streptomycin is adsorbed on soil particles; penicillin and biomizin oppress the development of both free and adsorbed cells of microorganisms

6. A Brief Review on Biotechnological Potential of *Streptomyces* Strains Obtained from Nature of Kyrgyzstan

To conclude our researches, 331 *Streptomyces* strains from various soil types of Kyrgyzstan were selected. Among them 22 strains possess an antagonistic action concerning with phytopathogenic fungi of agricultural plants and forests, 114 isolates were positive producing potentially novel antibiotics for pharmaceutical industry with tester *E.coli* strains containing resistant plasmids to antibiotics.

We consider it necessary to show some results of the laboratory researches according to biological activity of *Streptomyces* strains.

We used the approach to achieve chemical diversity of soil prokaryotic communities to discover novel antibiotic compounds. This is a traditional novel cultivation method to screen many colony-forming units (CFUs) of actynomyces for activity first, and then isolate only the active ones to exclude familiar antibiotic classes with tester E.coli strains containing resistant plasmids to antibiotics. The results are cited in table 8. In addition, PCR allows us to assay the presence of known and unknown genes and to detect many novel PKS (polyketide syntheses genes). We compared DNA from the environment to the database (GenBan))[2], (fig.12).

Table 8. Results of first screen and 2nd screen of Streptomyces Colonies through *E.coli cultures* containing resistant plasmids to antibiotics

Site	Media type	Colonies	First screen	positives	2nd screen	positives	3rd screen	positives	total positive
KMI _00036	ISP 4	291	E. *coli*	8	*E.coli* pCFF34	6	*e. coli* pMG223	5	5
KMI _00037	ISP 4	233	E. *coli*	8	*E.coli* pCFF34	7	*e. coli* pMG223	6	6
KMI _00038	ISP 4	778	E. *coli*	7	*E.coli* pCFF34	4	*e. coli* pMG223	4	4
KMI _00039	ISP 4	345	E. *coli*	7	*E.coli* pCFF34	5	*e. coli* pMG223	5	5
KMI _00040	ISP 4	187	E. *coli*	7	*E.coli* pCFF34	4	*e. coli* pMG223	4	4
KMI _00041	ISP 4	42	E. *coli*	3	*E.coli* pCFF34	1	*e. coli* pMG223	1	1
KMI _00042	ISP 4	99	E. *coli*	10	*E.coli* pCFF34	7	*e. coli* pMG223	6	6
KMI _00043	ISP 4	H.o.	E. *coli*	-	*E.coli* pCFF34	-	*e. coli* pMG223	-	-
KMI _00044	ISP 4	189	E. *coli*	6	*E.coli* pCFF34	5	*e. coli* pMG223	4	4
KMI _00045	ISP 4	216	E. *coli*	4	*E.coli* pCFF34	3	*e. coli* pMG223	3	3
KMI _00046	ISP 4	324	E. *coli*	7	*E.coli* pCFF34	2	*e. coli* pMG223	2	2
KMI _00047	ISP 4	256	E. *coli*	6	*E.coli* pCFF34	2	*e. coli* pMG223	2	2
KMI _00048	ISP 4	238	E. *coli*	12	*E.coli* pCFF34	6	*e. coli* pMG223	5	5

Table 8. Continued

Site	Media type	Colonies	First screen	positives	2nd screen	positives	3rd screen	positives	total positive
KMI _00049	ISP 4	241	E. *coli*	6	*E.coli* pCFF34	4	*e. coli* pMG223	4	4
KMI _00050	ISP 4	135	E. *coli*	10	*E.coli* pCFF34	5	*e. coli* pMG223	4	4
KMI _00051	ISP 4	281	E. *coli*	4	*E.coli* pCFF34	1	*e. coli* pMG223	1	1
KMI _00052	ISP 4	223	E. *coli*	3	*E.coli* pCFF34	1	*e. coli* pMG223	1	1
Total		4078		108		63		57	57

Table 9. Biological activity of Streptomyces strains against diseases of coniferous species and agricultural plants.

№	Tested Streptomyces strains	Radius of zone lyses of phytopathogenic test-cultures, mm				
		Alternaria	*Fusarium*	*Sclerotinia graminearum. Elen*	*Hypodermella sulsigena .Tub*	*L. pinastri*
1.	*S. bambergiensis К1-3*	20,6	17,8	10,0	10,0	20,0
2.	*S. noursei 2ч-10*	15,0	14,3	8,2	15,2	10,0
3.	*S. griseogromogenes 2ч-8*	15,5	20,0	20,0	20,0	15,0
4.	*S. afghaniensis 1К-6*	0	1,2	0	15,3	0
5.	*S. longispororuber 3К-1*	3,3	12,3	0	1,2	3,0
6.	*S. viridobrunneus 3К-2*	0	15,6	0	18,6	0
7.	*S. rubrogriseus ТК2-5*	18,0	21,0	12,3	21,0	12,5

Table 9. Continued

№	Tested Streptomyces strains	Radius of zone lyses of phytopathogenic test-cultures, mm				
		Alternaria	*Fusarium*	*Sclerotinia graminearum. Elen*	*Hypodermella sulsigena .Tub*	*L. pinastri*
8.	S. fumanus TM2-2	2,5	16,0	1,2	14,8	1,5
9.	*S. aureofuciens 3(2)*	6,4	4,2	1,1	1,6	5,0
10	*S.S.violaceoruber 3(6)*	0	0	0	12,0	0
11	*S.wistariopsis СП3-13*	3,4	16,3	11,2	20,0	4,0
12	*S. rasemochromogenes 6(7)*	1,2	2,3	0,3	8,2	0
13	*S. steffisburgensis 30(a)*	3,2	2,2	0,1	5,1	0
14	*S. fragilis Б1-18*	1,0	2,5	0,5	13,6	0,1
15	*S. longisporus ЧКС3-3*	2,0	2,3	0,3	3,5	0,2
16	*S. albadancus АП3-6*	5,0	15,8	12,5	15,6	0,5
17	*S. heliomycini 2ч-7*	15,0	16,2	14,3	5,4	0
18	*S. olivovariabilis ЧК-33*	1,0	2,1	0,3	0,7	0,1
19	*Control*	growth is abundant	growth is abundant	growth is abundant	growth is abundant	growth is abundant

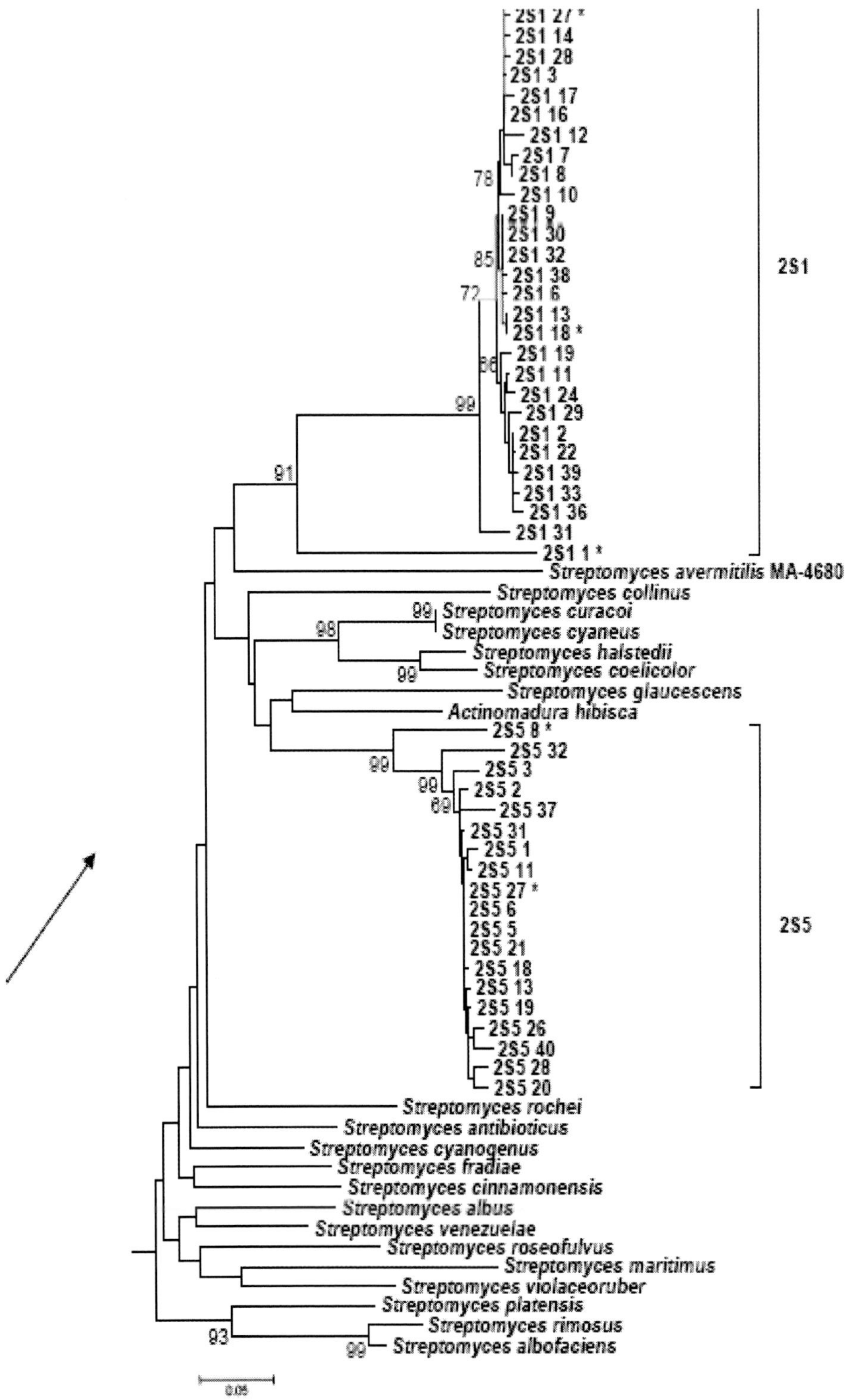

Figure 12. Phylogenetic relationship of the Streptomyses spp. isolates. Accession numbers for the reference strains are shown in parenthesis.

As the data show, 4080 colonies of actynomyces were screened with tester *E.coli* strains containing resistant plasmids to antibiotics, from them 57 isolates were positive producing potentially novel antibiotics(fig.13). These isolates were studied basically for their ability to produce new antibiotic substances with a view of their application in a pharmaceutical industry to develop effective medical products. However, the huge potential incorporated in the genom of these isolates remains not open since one microbe cell can produce a few hundred different biologically active compounds

The biotechnological potential of the obtained strains was discovered during the research of their antibiotic activity to phytopathogenic fungi of agricultural plants and forest. The results are cited in table 9.

Thus, Streptomyces isolated from various soils of ecosystem of Kyrgyzstan have a huge and wide biotechnological potentiality. The results of our research proved the ability of new Streptomyces isolates to produce a complex of biologically active compounds with different useful properties .These isolates have new antibiotic compounds to treat human diseases and an ability to produce antibiotic activity to phytopathogenic fungi of agricultural plants and forest, also growth stimulating substances of agricultural plants.

Figure 13. Continued on next page.

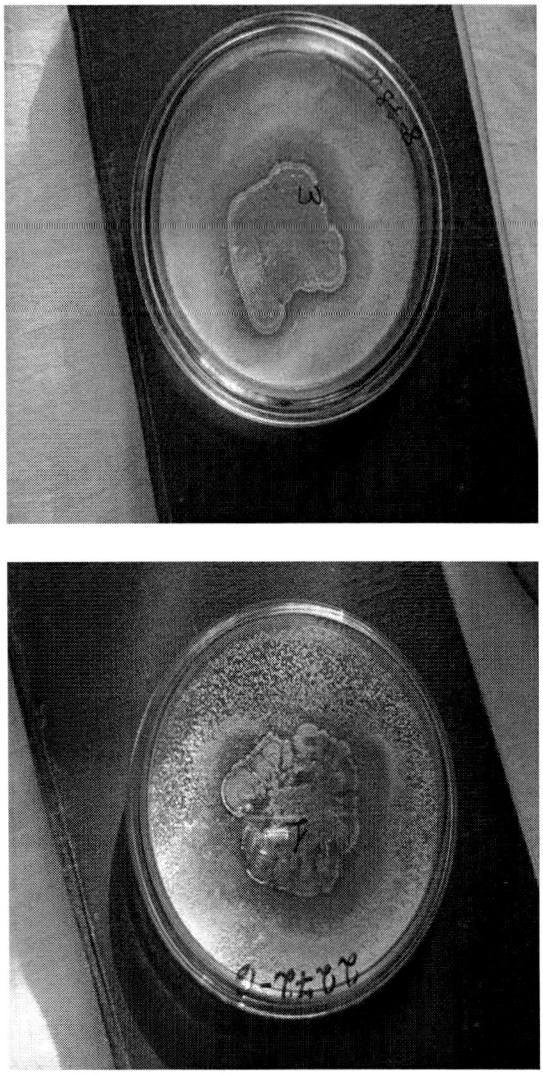

Figure 13. The first, second screening pictures of Streptomyces cultures.

Conclusion

The results have shown that Streptomyces spp. were particularly abundant and contained a rich biodiversity in the high mountain- meadow-steppe-alpine soils, high mountain alpine soils, mountain chernozems , where they represented 60% of all soil microorganisms. The climatic conditions, agri-physical regime , chemical and mechanical structure of these soils have a favorable influence on the prosperity of Streptomyces groups. The major factors saving stability and developments of Streptomyces populations in these soils are high percent contents of humus , neutral or alkalescent reaction of soils and a significant accumulation of mineral elements as phosphorus, nitrogen, potassium and calcium. In these conditions, Streptomyces spp are able to participate in decomposition of high-molecular, inaccessible organic compounds , rather than readily available. The quantitative and qualitative structure (a specific variety and a ratio of species and groups) of actynomyces depends on soil - climatic conditions, elevation of locality, a vegetative cover, cultivation of soils and many other factors.

However, it should be noted that the richness of soils with actynomyces is not always accompanied with abundance of antagonists. There are soils abounding with actynomyces and completely deprived of antagonists, on the contrary, there are soils poor with actynomyces but there are rather rich forms with antibacterial properties.

Members of actynomyces groups have a tendency to be prevalent in certain types of soil, so our data showed that predominant members of Streptomyces complex are in Cınereus section of Chromogenus series, which are adapted basically to chestnut and black soil while species of the Roseus section of the Fuscus series are adapted to gray, brown, and chestnut soils.

Under "stressful" conditions it was possible to isolate the biggest number of cultures capable of displaying antagonistic properties, which quickly lose these properties in favorable conditions. The cultures isolated from the reserved territories steadily kept their antagonistic properties for a long term. Apparently, in polluted and extreme conditions, under the mutagen influences of high concentration of heavy metals there is a change of the genom sites that activates special protective mechanisms promoting survival of cells.

To conclude our researches, 331 *Streptomyces* strains from various soil types of Kyrgyzstan were selected. Among them 22 strains possess an antagonistic action concerning with phytopathogenic fungi of agricultural plants and forests, 114 isolates were positive producing potentially novel antibiotics for pharmaceutical industry with tester *E.coli* strains containing resistant plasmids to antibiotics.

This work was supported by the NIH (USA) – sponsored programmes , ICBG Project.

References

[1] Agra H.C., Umarov J.A. (1973). Temperature borders of growth of cultures of Actinomyces genus. *Scientific reports of the higher school of Biol. Sciences*, #**12**, p. 81-84.

[2] Altschul, S. F., T. L. Madden, A. A. Schaffer, J. Zhang, Z. Zhang, W. Miller, and D. J. Lipman. (1997). Gapped BLAST and PSI-BLAST: a new generation of protein database search programs. *Nucleic Acids Res*. **25**:3389-3402.

[3] Andreyuk E.I., Vladimirova E.V., Kogan S.B. (1974). "Soil's Actinomyces of European Part of the USSR and their Biological Activity"./ Kiyev: "Nauk Dumka", 143 p.

[4] Anderson G.R. (1978). Pesticide effects on non-target soil microorganisms. - Zn: *Pesticide microbiology* \ Ed. G.R. Hill S.G.L. Wright. L.ect: Acad. Press, p.312-533.

[5] Baryshnikova L.M.,Chernova T.P., Agra N. S., (1975). Development and survival of Actinomyces spp. In the starvation conditions. *Biol. Sciences*, #**3**, p. 78-81.

[6] Beijerinck M., (1914). Uber Schroters und Cohns Lakmusmik rokokkus. *Fol. Mikrobiologie*, Delft Bd.2, s.185.

[7] Bergey's Manual of Determinative Bacteriology, (2000). Ninth Edition, John G. Holt et al. Lippincott Williams& Wilkins, p.667-677.

[8] Bondarzov A.S. (1954). The scale of colors - L: *Science* , Leningrad branch.

[9] Demain A.L. (1980). Control of antibiotic biosynthesis. *Microbiol, Rev.*, V.44, p. 230-251.

[10] Flaig W., Kutzner H.(1960). Beitrag zur okologie der Gttung Streptomyces Waksman et Henrici. *Arch. Microbiol.* Bd.35, #3, s.207-228.

[11] Flowers T.H., Williams S.T. (1977). PH influence on the growth and viability of neutrophilic and ascidophilic Streptomycetes. *Microbiol.*, **18.**, p.223-228.

[12] Gauze G.F., Preobrazhenskaya T.P. and others. (1983). "Determinants of Actinomyces class of Streptomyces"/ Moscow: "Nauka"..

[13] Gray R.T.C., Williams S.T., (1971). Microbial productivity in soil. *Microbes and Biological Productivity*. D.E.Hughess, A.H. Rose (Eds), Cambrige Univer. Press., p.225-286.

[14] Gromov B.V., Pavlenko G.V.(1989). "Ecology of Bacteriums"./Leningrad University's publication house.

[15] Jensen H.E. (1943). Observation on the vegetative growth of actinomycetes in soil. *Proc Linn. Soc.N.S.W.*, **68**, p.67-71.

[16] Kalakutskij L.V., Agra N.S. (1977) .The growth of actynomyces M., *Science*, p. 247-256.

[17] Kirilenko T. S. (1978). "Determinants of soil fungus, Kiev.

[18] Klevanskay I.L. (1960). Microflora of chernozems of Siberia - In the book: *Microflora of soils of Northern and Average part of the USSR of M*: *Science*, p. 250-273.

[19] Klevenskay I.L.(1970). Microflora podsolic and turf-podsolic soils. In the book: *Microflora of soils of Western Siberia*, Novosibirsk, *Science*, p. 10-50.

[20] Korenjako A.I., Artamonova O.I., Letunova S.V. (1955). Formation and preservation of antagonistic substances of actinomycetes in soil. // *Microbiology*, V. 24 , #5,

[21] Krasilnikov N.A. (1938). Radiant fungi and related them organisms. M., Publishing house of AS USSR, p.328.

[22] Krasilnikov N.A. (1938). The phenomenon of autolysis of actinomycetes. *Microbiology*, # 7, pp. 708-720, 829-837.

[23] Krasilnikov N.A. (1958). Antagonism of microbes and antibiotic substances, Publishing house « *Soviet science* », Moscow.

[24] Krasilnikov N.A.(1970) .*Radiant fungi* . Publishing house AS of SSSR, M.

[25] Kriss A.E. (1937). *Variability of actinomyces* spp M., Publishing house AS of the USSR

[26] Kubienna W., Renn G.F. (1935) . Micropedological studies of the influence of different organic compounds upon the microflore of the soil. *Zbl. Bakteriol.*, Abt, II,91, p267.

[27] Kuznesov V.D. (1972). Temperature influence of cultivation on the cultural properties of the international species - standards of actynomyces . // *Antibiotics*, V.17, # 3.

[28] Lloyd A.B., J.L.Lockwood. (1966). Lysis of fungal hyphae in soil and its possible relation to autolysis. *Phytopathology*, **56**, p.595-602

[29] Marton M. (1962). Uber die vertikale verbreitung der Strahlenpilz gruppen in einem schwarzerdea-hubichen Ackerbodea. Z.Pfianzenernar, Dung, Bodenk., Bd, 962, H.2, s.105-114

[30] Mayfield C.I., Williams S.T., Ruddick S.I., Hatfield H.L. (1972). Studies on the ecology of actinomycetes in soil. IY Observations on the form and growth of Streptomycetes in soil. *Soil. Biol. Biochem*, **4.** P.79-91.

[31] Mirdzhalilova S.M (1972). Microbiological and biotical characteristic of soils of the Northern Kyrgyzstan //*The works of the Scientific Research Institute of Agriculture*, vol. 5, Frunze.

[32] Mixaylov V.V. (1982). Interaction of actinomyces population in soils - Author's abstract the doctor. dissertation: Moscow, p. 22.

[33] Pavlovicha D.J. (1978). Actinomycetes of Latvia, Riga, Publishing house Zinatne, p. 193.

[34] Rybalkina A.V., Bolshakova V.S.(1952). About physiological features of actinomyces in some types of soils // *Soil science*, # **8**, p. 733-743.

[35] Shamshieva K.T (1969). Enzyme activity of some soils of Kyrgyzstan and their change under the influence of agro technical approaches: Author's abstract of the doctor dissertation, *Frunze*, p. 22.

[36] Soils of Kyrgyz SSR" (1974). Monography under the edition of Kyrgyz Republic's National Academy's academician A.M. Mamytov./ Publishing House "Ilim."

[37] Sorokina L.E. (1990). Structure of complexes soil actynomyces in natural and agri ecosystems: the Author's abstract the doctor. dissertation: 03.00.07 .The Moscow State University name of Lomonosov, M: p. 22.

[38] Steinbrenner K.(1963). Microbiologische veranderungen durch langjahrige Dungungsma Bnahman, *Zbl. Bakteriol*, Abt.2, Bdo 116, #6, s.581-592.

[39] Timofeev V.A. (1960). Microbiological characteristic of some soils of Chuy hollows and northern slopes of the Kirghiz ridge //The works of the Scientific Research Institute of Agriculture, *Frunze*. vol. 3.

[40] Tumarkin R.N. (1961).Specific characteristic and antibiotic properties of actinomycetes isolated from soils of Primorski Krai - *Microbiology*, V.30, issue 1.

[41] Vavulo F.P.(1972). Microflora of the basic soil types of Belarus SSR and their fertility. Minsk, *Crop*, p.232.

[42] Voronov S.I. (1966), Role of microorganisms and physiologically sour fertilizers in mobilization of mineral soil phosphates: the Author's abstract of the doctor dissertation, *Frunze*, p. 22.

[43] Voronov S.I (1961). Microbiological and biotical properties of soils of Kyrgyzstan. - In the monography of Soil of Kyrgyz SSR - *Frunze*, p. 175-217.

[44] Vuxrer G. (1967). Microbiological and biotical activity of soils of Kyrgyz SSR: the Author's abstract the doctor. dissertation, *Vilnius*, p. 22.

[45] Waksman S. (1945). *Microbial antagonism and antibiotic substances*, Agricultural experiment station, 1.

[46] Waksman S., (1959). *The actinomycetes* I Baltimore.

[47] Williams S.T., Bradshaw R.M., Colsterton I.W., Forge A., (1972).Fine structure of the sheath of some Streptomyces species. – *J.Gen. Microbial.*, **72**, p.249-258.

[48] Zvyagintsev D.G. (1964). Influence of the antibiotics adsorbed by soil particles, on the development of free and adsorbed microorganisms. Science reports of high schools of *Biol. Sciences of soils*, #**1**, p. 205-208.

[49] Zenova G.M., Vystina T.P.(1979). Dynamics of actynomyces number in spreading and turf -podsolic soil under a birch and a fir forest - The Bulletin of the Moscow University, A *series of Soils*. # 2, p. 55-58.
[50] Zenova G.M. (1982). *Actinomyces in soils of the USSR.*
[51] Zenova G.M., Zvyagintsev D.G. (1994). Actynomyces in the land ecosystems // *Journal of General biology*. V. 55, # 2.

In: Ecosystem Ecology Research Trends
Editors: J. Chen and C. Guo, pp. 275-297

ISBN: 978-1-60456-183-8

Chapter 8

ASSESSING AND MANAGING WILDFIRE RISK IN THE WILDLAND-URBAN INTERFACE

***Tony Prato*[1], *Bob Keane*[2] *and Dan Fagre*[3]**

[1]Center for Agricultural, Resource and Environmental Systems, University of Missouri-Columbia

[2]Missoula Fire Sciences Lab, USDA Forest Service

[3]USGS Northern Rocky Mountain Science Center

Abstract

Wildfires generate ecosystem benefits and socioecological costs. On the benefit side, wildfire is an essential natural process that stimulates vegetation growth, enhances biodiversity, and sustains ecosystem health. On the cost side, large, intense wildfires often cause human injury/mortality, structural damages, commercial timber losses, natural resource degradation, temporary disruption of ecosystem processes, and high suppression costs. Managing wildfire risk in the wildland-urban interface (WUI) involves balancing the benefits and costs of wildfires. Such balancing is likely to become more challenging due to the convergence of future climate change and future sprawl of residential communities into the WUI. Two conceptual frameworks are proposed for simulating how future climate and land use changes might influence wildfire risk in the WUI, and how stakeholders' can optimally respond to future changes in wildfire risk. Four stakeholders or agents are considered: forest managers and wildfire protection agencies; homeowners; land developers/homebuilders; and community/regional planners. The ex ante framework: (1) specifies alternative futures consisting of combinations of future climate change scenarios and future land use change scenarios; (2) uses biogeophysical process models to simulate how alternative futures influence wildfire risk in the WUI; and (3) employs agent-based modeling and multiple attribute evaluation to determine decision behavior that minimizes wildfire risk to stakeholders under each alternative future. Future climate change scenarios are specified by disaggregating the results of general circulation models. Future land use change scenarios are specified using a land use change model that simulates the expansion of residential communities into the WUI for alternative economic growth rates and land use policies. Potential ecosystem responses to future climate change are evaluated using a GIS-based, biogeophysical model that simulates: (1) carbon, water, and nutrient fluxes at the ecosystem scale; (2) climate change in forest stands; (3) tree growth, vegetation structure, and fuel loads in forest stands; (3) wildfire ignitions; and (4) fire behavior. The ex post framework uses adaptive management, biogeophysical simulation, agent-based modeling, and multiple attribute evaluation to identify ways stakeholders can optimally adjust their decision behavior over time in response to evolving changes in wildfire risk. Both frameworks can be used to evaluate hypotheses, such as: (1) whether convergence of future climate change and economic growth would significantly increase wildfire risk in WUIs; (2) whether limiting future urban sprawl into wildland areas would significantly reduce wildfire risk; and (c) whether lower future economic growth would significantly decrease wildfire risk.

Introduction

US policies for managing wildfires changed substantially through the 20th and early 21st centuries. In 1942, the USDA Forest Service's Cooperative Forest Fire Prevention Program ushered in a new era of forest fire prevention and suppression and introduced the Smokey-the-Bear campaign. For more than 60 years, the program focused on wildfire prevention and suppression despite limited use of prescribed fire and fire for natural resource benefits beginning in the 1970s. A combination of successful fire suppression activities that contributed to high surface and canopy fuel accumulations, sustained drought, trends in long-term climate, increased human development in wildland areas, and greater incidence of insect, disease, and invasive plant infestations contributed to dramatic increases in the number of large wildfires, average area burned, fire suppression costs, and property damages since the 1990s (Covington and Moore 1994; General Accounting Office 1999; Arno and Allison-Bunnel 2002; USDA USDI 2003; Dombeck et al. 2004; Kauffman 2004; Calkin et al. 2005). Property damages were especially acute in the wildland-urban interface (WUI), an area in which homes and other structures are within or adjacent to wildland vegetation that is prone to wildfire (Duryea and Vince 2005; Platt 2006) (Figure 1). The WUI covers 9% of the total land area and contains 39% of all houses in the conterminous US (Radeloff et al. 2005).

Figure 1. Wildfire threatening a residence in the WUI.

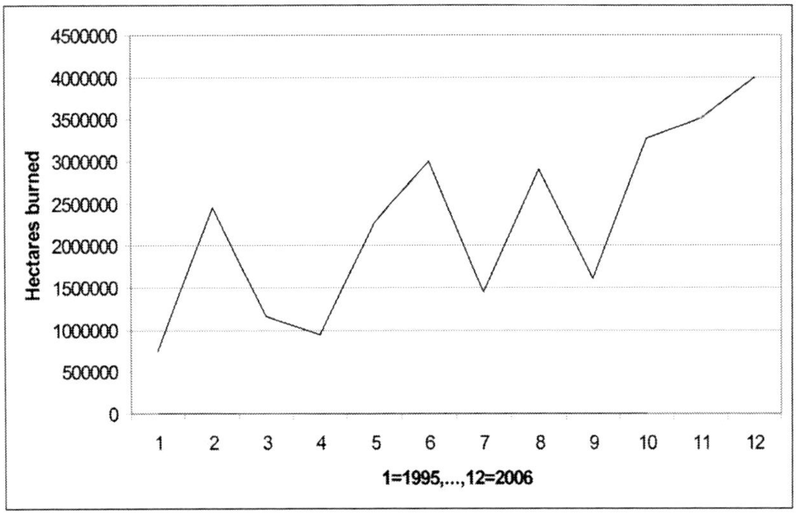

Figure 2. U.S. forest area burned, 1995-2006 Source: National Interagency Fire Center (2007).

The number of major wildfires in the western US has increased fourfold and the area of forest burned has increased sixfold since 1986 (Westerling 2006). During the period 2000-2005, record drought contributed to large, intense wildfires that escalated federal wildfire suppression costs for the period to $7 billion (CIRMOUNT Committee 2006; National Interagency Fire Center 2007). In 2000, 3.4 million ha and in 2002, 2.8 million ha of wildland burned considerably more than the annual average area burned of 1.6 million ha in 1960-1999 (National Interagency Fire Center 2007). There has been an upward trend in forest area burned in the US in the period 1995 to 2006 (Figure 2).

The USDA Forest Service (2007a) notes that "Large wildfires now threaten millions of both public and private acres, particularly where vegetation patterns have been altered by development, land-use practices, and aggressive fire suppression." Dombeck et al. (2004) state that, "New [management] policies are needed that…promote a continuum of treatments with active management and reduction of fuel hazard in wildland-urban interface zones and reintroduction of fire in wildlands." Even if such policies do not significantly alter the frequency and spatial extent of wildfires, they can influence the timing and intensity of wildfires and the adverse impacts of wildfire on natural resources and people (Dellasala et al. 2004).

Wildfires provide ecosystem benefits and impose socioecological costs. On the benefit side, wildfire is an essential natural process that stimulates vegetation growth, enhances biodiversity, and sustains ecosystem health (Miller et al. 2000; US Department of the Agriculture and US Department of the Interior 2003; Arno and Allison-Bunnel 2002; Wuerthner 2006a). In addition, wildfires can reduce fuel accumulations and decrease the occurrence of atypical, intense fires that can damage ecosystems and property (USDA 2001). Wildland fire exclusion in fire-adapted ecosystems can decrease tree growth, homogenize landscapes, and increase the frequency of catastrophic fires that contribute to ecological damages (Keane et al. 2002). Wuerthner (2006b) states that "…if we are to truly preserve wildness as an essential element of the landscape, then we must provide space for wildfire to "live" unimpeded with a minimum interference from humans." On the cost side, large, intense wildfires have caused human injury and mortality, structural damages, commercial timber losses, high suppression costs, and, in some ecosystems, natural resource degradation and temporary disruption of ecosystem processes (Ferry et al. 1995; General Accounting Office 1999; Perkins 2001; Western Governors' Association et al. 2001; Arno and Allison-Bunnel 2002; O'Laughlin 2005).

Managing wildfire risk in the WUI requires balancing the benefits and costs of wildfires, a task that could become more difficult in the years ahead due to the convergence of climate change and expansion of residential development into wildland areas. Balancing the benefits and costs of wildfires will be especially difficult in the western US, which contains approximately 79% of the publicly owned forests in the US (USDA Forest Service 2007a) and is undergoing substantial climate change and substantial residential development in the WUI. On the climate side, the western US is expected to experience more heat, less snowpack, earlier snowmelt and runoff, and more variable weather patterns than in past centuries (Saunders and Maxwell 2005). These climate changes would extend the wildfire season, increase the amplitude and duration of extreme fire weather, and cause some areas of the western US to have more frequent and more severe fires (Price and Rind 1994b; Lenihan et al. 1998; Brown et al. 2004; McKenzie et al. 2004; Littell et al. 2005; Westerling et al. 2006).

On the development side, the population of the Intermountain West doubled and employment tripled between 1969 and 1999 (Regional Economic Information System 2002). In the 1990s, the population of the Rocky Mountain West grew over 25% (Riebsame et al. 1997), with two-thirds of the counties in the region growing at a rate that exceeded the national average (Beyers and Nelson 2000). During this same period, the Rocky Mountain West had five of the 10 fastest growing states and nine of the 15 fastest growing counties in the US (Fagre 2000). Such growth has continued during the first decade of this century and is expected to increase the sprawl of residential communities into wildland areas, which can: (1) reduce fire management options, particularly with respect to prescribed burning (Dellasala et al. 2004); (2) increase fire suppression, thereby decreasing the natural resource benefits of wildfire (Stanionis and Glick 2006); (3) increase the challenges of managing wildfire in and near WUIs (Cortner et al. 1990); (4) raise the probability of human-caused wildfires (Pyne 2001); and (5) possibly increase the risk of wildfire-related losses of human life and property.

Policies for dealing with wildfire have been mixed and controversial due to the complexity of wildfire management in WUIs (Alavalapati 2005). In August 2000, the National Fire Plan was developed "with the intent of actively responding to severe wildland fires and their impacts to communities while ensuring sufficient firefighting capacity for the future" (USDA 2001; USDA USDI 2003). The 2003 Healthy Forests Restoration Act (HFRA) (US House of Representatives 2003) emphasizes controlling wildfire through aggressive logging and hazardous fuels[1] reduction, especially in WUIs. HFRA has been criticized for not accounting for landscape variation in wildfire hazard and not acknowledging the ecological benefits of wildfires (Franklin and Agee 2003; Kauffman 2004). Increased logging, which is one of the strategies in HFRA, has been criticized because it can leave behind combustible slash, which may increase the rate of fire spread in the event of a fire

[1] Hazardous fuels refer to "dry brush and trees that have accumulated and increase the likelihood of unusually large fires" (USDA USDI 2003).

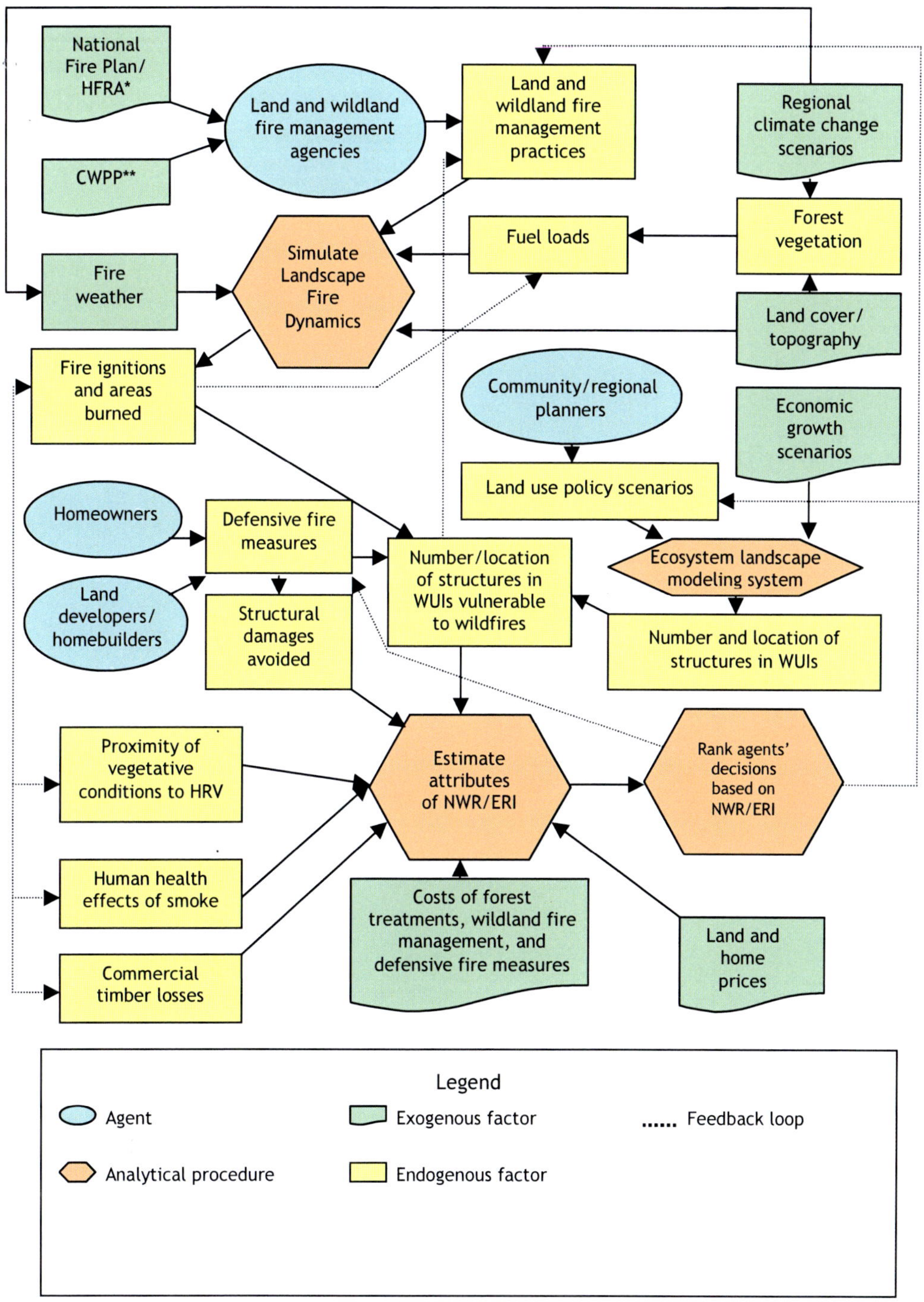

*Healthy Forests Restoration Act
**Community Wildfire Protection Plan

Figure 3. Flow chart for preliminary FIRECLIM model.

(Lindenmayer and Franklin 2002; Dellasala et al. 2004). In addition, there is considerable interest in implementing a new approach to wildfire policy called Appropriate Management Response (AMR). AMR attempts to balance aggressive fire suppression in some areas to minimize potential natural resource damages with non-suppression (a.k.a. wildland fire use) in other areas to achieve natural resource benefits (USDA Office of Inspector General 2006). These changes in wildland fire policy indicate that the US government understands that decades of wildfire suppression have adversely affected forest ecosystems. However, the US government is still reluctant to accept the risk to human life and property of allowing certain wildfires to burn (Arno and Allison-Bunnell 2002; Dombeck et al. 2004; Kauffman 2004).

Several studies have evaluated the causes, consequences, and management of wildfire (e.g., Arno and Allison-Bunnell 2002; Martin et al. 2007), including the effects of climate change on wildfires and wildfire risk (e.g., Cohen 1991, 2000; Miller et al. 2000; Dombeck et al. 2004; Kauffman 2004; McKenzie et al. 2004; Fagre et al. 2005; Platt 2006; Prato and Fagre 2006; Scott 2006; Westerling et al. 2006). Nevertheless, there appears to be a lack of studies that evaluate the potential combined impacts of future climate change and growth in residential development on wildfire risk in the WUI, and that assess how the stakeholders can optimally adapt their decisions to changes in such risk. This is a significant deficiency considering the potential increase in wildfire risk from the convergence of future climate change and the growth of WUIs in the western US.

FIRECLIM Model

The FIRECLIM model incorporates two decision frameworks for assessing and managing future wildfire risk in the WUI (Figure 3). The *ex ante* decision framework is used to determine stakeholders' optimal decisions for managing wildfire risk in WUIs under alternative *cl*imate and *e*conomic *g*rowth (CLEG) futures at the beginning of the evaluation period. The *ex post* decision framework is used to determine how stakeholders' can optimally adapt their wildfire-related decisions to real-time climate change and residential expansion in WUIs. In other words, *ex ante* decisions are made in advance of and *ex post* decisions are made after the resolution of uncertainty about climate and land use changes. Both frameworks have two elements in common. First, they both use an agent-based approach to describe the wildfire-related decisions of four stakeholders or agents (i.e., land and wildland fire management agencies, homeowners, land developers/homebuilders, and community/regional planners). Second, they both simulate how climate change, land use policy, residential development, biophysical conditions, and agents' decisions influence wildfire and wildfire risk. The unique feature of the *ex ante* decision framework is that it determines an agent's optimal *ex ante* decision plan for each CLEG future for the entire evaluation period at the beginning of the evaluation period. The unique feature of the *ex post* decision framework is that it determines an agent's optimal decision plan for each five-year interval in the evaluation period using an adaptive management framework. The remainder of this section describes the components of the FIRECLIM model and the two decision frameworks.

CLEG futures. A CLEG future consists of a particular combination of a future climate change scenario and a future economic growth scenario (this combination is also called an alternative future) for the WUIs of interest. Climate change scenarios for WUIs denote plausible future changes in seasonal and annual patterns of temperature, precipitation, atmospheric CO_2 concentrations, and other climate variables during the evaluation period. They can be developed several ways. For example, the National Oceanic and Atmospheric Administration's (NOAA) Climate Diagnostic Center (Eischeid and Diaz personal comm.) can simulate climate change scenarios for a region by downscaling ensembles of the 23 general circulation models used in the Fourth Assessment Report of the Intergovernmental Panel on Climate Change (Meehl et al. 2007). The resulting regional climate change scenarios can be further downscaled to the 4 km^2 PRISM (*P*arameter-Elevation *R*egressions on *I*ndependent *S*lopes *M*odel) climate mapping system developed by Daly (http://prism.oregonstate.edu/) to provide spatial climate-gridded products that account for the effects of mountain topography on local microclimates in WUIs. Climate change scenarios for WUIs derived in this manner have the dual advantages of being tied to internationally accepted models of climate change and being specific to WUIs in a study area.

Economic growth scenarios specify annual average percentage growth rates during the evaluation period in final outputs of major industries in the economic region that encompasses the WUIs of interest. Growth rates are entered into the IMPLAN model (Minnesota IMPLAN Group, Inc. 2006) to estimate the increase in jobs for each year. The estimated increase in jobs is translated into the estimated increase in the number of new residential housing units required for the additional workers and others (e.g., non-residents). The estimated increase in the number of new housing units required is allocated to developable parcels in the WUIs of interest using a land allocation model, such as the Residential and Commercial, Institutional and Industrial Development model. The latter is a component of the Ecosystem Landscape Modeling System (ELMS) (Prato 2005c; CARES 2007; Prato et al. 2007ab) that allocates the total number of new housing units required for a particular alternative future to developable parcels in an area based on their suitability for six densities of housing and CI&I units subject to the limits imposed on development by three land use policy scenarios (i.e., current, moderately restrictive, and highly restrictive).

Delineation of WUIs. Current WUIs have been delineated for many communities in the US in the process of developing Community Wildfire Protection Plans (CWPPs). For example, the CWPP for Flathead County, Montana defines WUIs as areas in or immediately adjacent to wildland or forest fuels that are located within 2.4 km of residential structures, developments, or private properties suitable for development or residential use (GCS Research 2005). The 2.4 km distance is roughly the maximum distance lofted firebrands can travel from a wildland fire to the roof of a house (SILVIS Lab 2007). Following this definition, future WUIs can be delineated for each CLEG future by first overlaying all grass, shrub land, and forested areas determined from a current land cover map with a parcel boundary map that indicates all developed residential parcels (i.e., those containing residential structures) simulated using ELMS, and second identifying the developed residential parcels that have a majority of their area within 2.4 km of grass, shrub land, and forested areas. The size of WUIs is likely to increase with economic growth and associated residential development.

Simulation of Landscape Fire Dynamics. The effects of climate change and the decisions of land/wildland fire management agencies on fuel loads, area burned, and ecological conditions in WUIs can be simulated at the 30-m pixel resolution using a GIS-based, mechanistic ecological modeling system that integrates FSPro (Fire Spread Probability), Fire-BGCv2 (BGC stands for biogeochemical), FARSITE (Fire Area Simulator), and FlamMap models. In this approach, Fire-BGCv2 (Keane et al. 1996ab; 1997; 1999) is used to simulate tree growth, organic matter decomposition, litterfall, and the effects of fire on ecosystem conditions for several decades. Driving variables for the processes simulated using Fire-BGCv2 are taken from daily weather data extrapolated across the WUI using PRISM or some other method. FireBGCv2 simulated changes in WUI landscape conditions, such as fuel loads, vegetation, and productivity, for each combination of climate change scenario and management practices are then entered into FSPro, FARSITE (USDA Forest Service 2003), and FLAMMAP (Finney 1998, 2006; Farris et al. 1999) to calculate spatially explicit maps of wildfire risk and hazard for WUIs at a 30-m pixel resolution. Growth, spread, and behavior of fire can then be mechanistically computed using the Rothermel (1972) fire spread model as implemented in the FARSITE model. The annual area burned under each CLEG future is estimated by inputting simulated fuel loads and landscape attributes for pixels to the FARSITE and FlamMap models.

One way to assess the effects of climate scenarios and management practices on ecological conditions in WUIs is in terms of deviations from the historical range of variability (HRV) in vegetative conditions (Holsinger et al. 2006). For example, spatial and tabular output from the Fire-BGCv2 model from a 5,000-year simulation parameterized with historical fire regime and weather data can be compared to the simulations of future landscapes under alternative future climate scenarios and management practices to determine how close or departed the future simulated landscapes are from their HRV.

Defining net wildfire risk (NWR). Risk has been defined as: (1) the probability of harm from a hazard (Cohrssen and Covello 1989); (2) the probability of an adverse outcome given a specific set of conditions (PCCRARM 1997); (3) the probability of loss due to wildfire (Rowe 1975; Suter 1993); (4) the expected negative effects of wildfire on human life, property, and air quality (Miller et al. 2000); (5) the undesirable effects resulting from movement away from target conditions (Black and Opperman 2005); and (6) the expected net value change (Finney 2005). The FIRECLIM model assumes that three agents, referred to as the risk-bearing agents (i.e., land and wildland fire management agencies, homeowners, and community/regional planners), minimize a risk-benefit measure called net wildfire risk (NWR), which is unique to each risk-bearing agent. NWR for each risk-bearing agent accounts for the expected losses and expected benefits of wildfires (c.f., Scott 2006). The fourth agent, consisting of land developers/homebuilders, is assumed to maximize expected return on investment (ERI) in defensive fire measures. NWR and ERI are evaluated at three spatial scales depending on the agent: 30-m pixel; parcel; and WUI.

Modeling agent behavior and NWR. FIRECLIM uses an agent-based approach to model the decisions of four agents. Agent-based modeling has been employed in many natural resource management studies (e.g., Berger 2001; Gimblett 2002; Hoffmann 2002; An et al. 2006). Agents typically "have some degree of self-awareness, intelligence, autonomous behaviors, and knowledge of the environment and other agents … and can adjust their own actions in response to environmental changes" (Lim et al. 2002; Parker et al. 2003). The success of the agent-based approach rests on obtaining information from agents about how they make decisions and using that information to model their decisions. This can be accomplished by involving agent panels in the modeling of agent decisions. The impacts of agents' decisions on NWR or ERI and the effects of NWR or ERI on agents' decisions are described below.

Land and wildland fire management agencies select land and wildland fire management practices that minimize annual NWR (i.e., NWR_a) for a particular CLEG future. Examples of management practices include prescribed burning, mechanical thinning, commercial timber harvesting, foaming or wrapping homes in flame resistant material (see Figure 1), herbicide application, livestock grazing, or combinations of these and other practices (USDA USDI 2003; Kline 2004). Annual expected NWR_a is the product of the annual probability that a pixel burns (determined from simulations of wildland fire dynamics) and the conditional annual values of NWR_a. A conditional value is the value of NWR_a given a wildfire occurs in a pixel. NWR_a is a function of four negative attributes of wildfires (i.e., number of structures vulnerable to wildfire, human health effects of smoke, commercial timber losses, and the costs of management practices), and one positive attribute that represents the ecological benefit of allowing wildfires to burn (i.e., the extent to which management practices restore vegetative conditions in a WUI to their HRV). NWR_a decreases (increases) as the negative attributes decrease (increase) and the positive attribute increases (decreases). Conditional values of NWR_a are estimated using the TOPSIS multiple-attribute evaluation procedure described below.

The four attributes of NWR_a are determined as follows. The number of structures in a pixel is determined based on the economic growth scenario being evaluated and the land use policy selected by community/regional planners. Alternative management practices that reduce the probability that a pixel burns, shorten flame length, and/or reduce the delivery of firebrands to ignitable materials on a property will decrease the vulnerability of structures to wildfires. In addition, an increase or decrease in defensive fire measures selected by homeowners and land developers/homebuilders will affect the vulnerability of structures to ignition by wildfires.

Conditional human health effects of smoke are proportional to the spatial extent of a smoke plume, the population of the area covered by a smoke plume, and the possible human health effects of smoke. Epidemiological studies indicate that high levels of $PM_{2.5}$ (i.e., particulate matter that is 2.5 μg or smaller in size) are correlated with substantial adverse health effects, including respiratory problems, pneumonia, chronic obstructive pulmonary disease, heart disease, stroke, and premature mortality (USEPA 2004; Künzli et al. 2006). Human health effects can be estimated based on smoke dispersion, which can be modeled using the Weather Research and Forecasting-Smoke Dispersion System (USDA Forest Service 2007b) and public health standards for particulate matter ($PM_{2.5}$). Death of WUI residents due to wildfires tends to be more common in fast-burning grass and chaparral wildfires. Wildfire-related deaths of firefighters and WUI residents are less frequent in areas dominated by coniferous forests. Moreover, such deaths would be difficult to value in monetary terms. For these reasons, the human deaths from wildfires are not considered in NWR_a.

Conditional commercial timber losses are estimated based on forest type, acreage burned, fire intensity, and market prices for timber. Commercial timber losses should only be considered for pixels in which commercial timber harvesting is permitted and is likely to occur during the evaluation period. Commercial timber losses should be reduced by the expected market value of timber harvested in salvage logging operations, if applicable. The costs of wildland fire management treatments can be estimated using secondary data.

Closeness of vegetative conditions in each pixel in a WUI to their HRV are computed by comparing simulated landscapes to a series of Fire-BGCv2 simulations parameterized to reflect historical climate and fire regimes (Steele et al. 2006). A set of statistics are then used to calculate a similarity index that describes how close future landscapes are to their HRV (Holsinger et al. 2006).

Land developers and homebuilder agents do not directly bear wildfire risk. However, land developers/homebuilders can invest in defensive fire measures that reduce fire spread and hence structural damages should a wildfire occur. Examples of such measures include thinning and removing vegetation to reduce hazardous fuels on a property before it is sold, designing subdivisions to allow adequate road access for firefighting personnel and equipment, and using siding/roofing materials and window types that diminish structure ignitability (e.g., fire resistant roofing, double-pane windows, and deep side yard setbacks, which provide defensive space). Such measures are applied at the time properties are developed and structures are built. Annual ERI in defensive fire measures selected by land developers/homebuilders is the difference between the expected increase in land/home prices from incorporating those measures in a structure class (e.g., high density housing) and the incremental cost of incorporating those measures in a property or structure. The expected increase in land/home prices associated with a defensive fire measure is the expected reduction in structural damages achieved by that measure.

It is impractical to assess NWR for *homeowners* (i.e., NWR_h) for every at-risk structure in a WUI when several WUIs are being evaluated. However, it is feasible to evaluate NWR_h for residential structure classes (e.g., high density, urban, suburban, rural, exurban, and agricultural housing units—the structure classes modeled in ELMS). Structure classes are assumed to have a low, medium, or high combustibility rating in terms of roofing/siding materials and type of windows. Homeowners in WUI communities have been encouraged to take greater responsibility for "creating and maintaining defensible home and landscape conditions" (Long et al. 2005; Power 2006; Firewise Communities 2007). Homeowners can create defensible spaces around their properties by implementing defensive fire measures that decrease structure ignition potential and increase wildfire suppression potential (Dennis 2007). Examples of defensive fire measures for homeowners include clearing vegetation within the home ignition zone, storing wood piles away from the home, improving road access to the home site, and operating a water sprinkler system on flammable roofs and decks during wildfires (Cohen 1991). A lack of fire investigation data for both burned and surviving structures makes it difficult to confirm the efficacy of specific defensive fire measures (Cohen and Saveland 1996). Wildfire suppression potential at home sites is especially difficult to evaluate because it depends in part on whether or not a homeowner is at home when a threatening wildfire occurs or shortly after the fire passes (Deeming et al. 1977; Cohen 1991). The effectiveness of homeowners' defensive fire measures for structure classes is determined using the Structure Ignition Assessment Model (SIAM) (Cohen 1994, 1995)

Expected annual NWR_h is the product of the annual probability that wildfire spreads to a parcel containing a particular structure class and the difference between the amortized average cost of implementing defensive fire measures (net of insurance payments) and structural damages avoided, if any, should a wildfire occur in that parcel (i.e., the latter equals conditional structural damages). Conditional damages for a structure class depend on structure ignition potential, which can be estimated based on: (1) the landscape fire dynamics in a parcel containing that structure class, which depends on the biophysical attributes of the parcel; (2) management practices selected by land and wildland fire management agencies; (3) defensive fire measures selected by homeowners and land developers/homebuilders; (4) the combustibility rating for that structure class; and (5) the land use policy chosen by community/regional planners.

Community/regional planners. NWR for this agent (i.e., NWR_c) can be reduced by implementing specific land use policies, such as zoning ordinances, subdivision regulations, and land use planning measures that decrease structural damages in WUIs in the event of a wildfire. Zoning ordinances and subdivision regulations that are likely to reduce NWR_c include disallowing homes on steep slopes, requiring wider roads in subdivisions to improve access by firefighting equipment and egress of residents in the event of a wildfire, and requiring the establishment and maintenance of firebreaks and other measures that increase the structure survivability (Power 2006). Land use planning measures that can be used to reduce NWR_c include land acquisition, land exchanges, conservation easements, purchase of development rights, transfer of development rights, urban growth boundaries, and others (Alavalapati 2005; Myszewski and Kundell 2005).

For each CLEG future and land use policy, the expected annual NWR_c at the parcel scale is the product of the annual probability that a parcel burns and the difference between the annual policy implementation costs and the annual reduction in conditional structural damages for that parcel (the latter is a benefit). Annual expected NWR_c at the WUI scale are calculated by the weighted average of the annual expected NWR_c for all parcels in that WUI with weights given by the proportion of all structures in the WUI located in that parcel.

Table 1. Attributes of NWR and ERI

Metric	Spatial resolution	Attributes of metric		Method of evaluating metric
		Negative	Positive	
NWR_a[1]	30-m pixel	Number of structures vulnerable to wildfire Adverse health effects of smoke Commercial timber losses Costs of management practices	Index of proximity of vegetative conditions to their HRV	TOPSIS
NWR_h[2]	Parcel/ structure class	Costs of defensive fire measures	Structural damages avoided	Monetary
NWR_c[3]	WUI	Policy implementation costs	Structural damages avoided	Monetary
ERI[4]	Parcel	Costs of implementing defensive fire measures	Expected increases in land and home prices	Monetary

1. Net wildfire risk for land and wildland fire management agencies
2. Net wildfire risk for homeowners
3. Net wildfire risk for community/regional planners
4. Expected return on investment for land developers/homebuilders

Table 1 summarizes the attributes of NWR_a, NWR_h, NWR_c, and ERI. Since the preliminary attributes of NWR_a, NWR_h, and NWR_c are not mutually exclusive, the three values cannot be added together to obtain an overall measure of NWR.

Ex ante Decision Framework

The optimal *ex ante* decisions for each agent under each CLEG future are: (1) the wildland fire management practice that minimizes NWR_a for a pixel; (2) the defensive fire measure for each structure class for a parcel that minimizes NWR_h; (3) the land use policy option that minimizes NWR_c for a WUI; and (4) the defensive fire measure that maximizes ERI for each parcel. Feasible *ex ante* decisions for homeowners, community/regional planners, and land developers/homebuilders can be ranked based on the monetary values of NWR_h, NWR_c, and ERI, respectively (c.f., Scott 2006). The highest ranked *ex ante* decisions for homeowners and community/regional planners have the lowest NWR_h and NWR_c, respectively. The highest ranked *ex ante* decision for land developers/homebuilders is the one with the highest ERI. Since two of the attributes of NWR_a are non-monetary (i.e., number of structures vulnerable to wildfire and landscape proximity to HRV) and it is difficult to monetize these attributes using non-market valuation techniques (Venn and Calkin 2007), decisions for land and wildland fire management agencies are ranked using the TOPSIS multiple attribute evaluation procedure described below.

A fuzzy version of the Technique for Order Preference by Similarity of Ideal Solution—fuzzy TOPSIS for short—can be used to rank alternative decisions for land and wildland fire management agencies (Hwang and Yoon 1981; Chen and Hwang 1992; Chen 2000; Berger 2006). This procedure involves six steps. First, *possible decisions* for land and wildland fire management agencies are determined at the 30-m pixel resolution. Second, *feasible decisions* are determined by eliminating possible decisions that do not satisfy the following chance constraints: p{number of structures vulnerable to wildfire $\geq$ d*} $\leq \beta_d$; p{health effects of smoke $\leq$ m*} $\geq 1 - \beta_m$; p{commercial timber losses $\leq$ t*} $\geq 1 - \beta_t$; p{cost of treatment $\leq$ a*} $\geq 1 - \beta_a$; and p{index of landscape proximity to HRV $\geq$ h*} $\geq 1 - \beta_h$; where d* is the maximum number of structures vulnerable to wildfire on a pixel, m* is the maximum acceptable concentration of particulates in smoke in the affected area, t* is the maximum acceptable level of timber losses, a* is the maximum acceptable level of treatment cost, h* is the minimum acceptable index of landscape proximity to HRV, and $0 \leq \beta_d \leq 1$, $0 \leq \beta_h \leq 1$, $0 \leq \beta_m \leq 1$, $0 \leq \beta_t \leq 1$, and $0 \leq \beta_a \leq 1$ are the reliability levels for the constraints. For example, if t* = \$1 million and $\beta_d = 0.05$, then the probability that commercial timber losses are less than or equal to \$1 million must be greater than or equal to 0.95 for all feasible decisions. The values of the attributes that satisfy the constraints and the corresponding feasible decisions are determined by applying the chance constraints to the probability densities of the expected annual attributes of NWR_a.

Third, the acceptability of the probability densities of expected annual attributes of NWR_a for feasible decisions is determined using the following linguistic variables: very poor; poor; medium poor; fair; medium good; good; and very good. Fourth, the relative importance of the four attributes of NWR_a are rated using the following linguistic variables: very low; low; medium low; medium; medium high; high; and very high. Linguistic variables can be expressed using the positive triangular fuzzy numbers (e.g., Chen 2000). Fifth, the distances of all feasible decisions from the fuzzy positive-ideal solution and the fuzzy negative-ideal solution are determined by applying fuzzy TOPSIS to the overall ratings of feasible decisions and the overall ratings of the relative importance of attributes (Chen 2000). Sixth, the resulting distances are used to calculate annual closeness (to ideal solutions) coefficients for all feasible decisions. The latter are multiplied by the probability that a pixel burns to obtain the expected closeness coefficients for pixels, which are then used to rank the feasible decisions. The optimal feasible decision for land and wildland fire management agencies is the lowest ranked feasible decision for a pixel, or equivalently the feasible decision with the lowest expected annual NWR_a.

Hypothesis Testing

NWRs for agents determined using the *ex ante* decision framework can be used to test various hypothesis about future NWR. Three such hypotheses are described here. The first hypothesis is that the convergence of future climate change and expansion of residential development into wildland areas would significantly increase NWR in WUIs under current land use policy. This hypothesis can be tested using the difference $NWR_{ij} - NWR_{in}$. NWR_{ij} is the annual NWR with the current land use policy for agent i (i = a for land and wildland fire management agencies, i = h for homeowners, and i = c for community/regional planners) and CLEG future j and NWR_{in} is the annual NWR with the current land use policy for agent i with no climate change and economic growth (e.g., climate remains the same as in the 50 years prior to the evaluation period and no economic growth occurs during the evaluation period). The null hypothesis H_o: $NWR_{ij} - NWR_{in} = 0$ is tested against the alternative hypothesis H_a: $NWR_{ij} - NWR_{in} > 0$ using a 'studentized' contrast test (test statistic = student(c), $\alpha = 0.5$) and the expected annual values of $NWR_{ij} - NWR_{in}$ for the evaluation period. If the null hypothesis is rejected, then the first hypothesis is not rejected and the second and third hypotheses are tested.

The second hypothesis is that implementing a more restrictive land use policy would significantly offset some of the increase in future wildfire risk caused by residential expansion into WUIs. This hypothesis can be tested using the difference $NWR_{ijgf} - NWR_{ijgk}$. NWR_{ijgf} is the expected annual NWR for agent i (i = a, h, c) with climate change scenario j, economic growth scenario g, and land use policy f, and NWR_{ijgk} is the expected annual NWR for agent i with climate change scenario j, economic growth scenario g, and land use policy k. Policy f imposes fewer limits on the expansion of residential development into wildland areas than policy k. The null hypothesis H_o: $NWR_{ijgf} - NWR_{ijgk} = 0$ is tested against the alternative hypothesis H_a: $NWR_{ijgf} - NWR_{ijgk} > 0$ for the three economic growth scenarios (i.e., values of g) and three land use polices (i.e., values of f and k) using a 'studentized' contrast test (test statistic = student(c), $\alpha = 0.5$) and the expected annual values of $NWR_{ijgf} - NWR_{ijgk}$. Test results are used to determine the agents and climate change scenarios for which a more restrictive land use policy significantly offsets future increases in NWR from residential expansion into WUIs (i.e., from economic growth).

The third hypothesis is that reducing future residential expansion into WUIs would significantly decrease wildfire risk even under the current land use policy. This hypothesis can be tested using the difference $NWR_{ijgf} - NWR_{ijrf}$. NWR_{ijgf} is the expected annual NWR for agent i (i = a, h, c) with climate change scenario j, economic growth scenario g, and current land use policy f, and NWR_{ijrf} is the expected annual NWR for agent i with climate change scenario j, economic growth scenario r, and current land use policy f. Economic growth scenario g has lower economic growth rates for industries in Flathead County than economic growth scenario r (e.g., scenario g is low growth and scenario r is moderate growth). The null hypothesis H_o: $NWR_{ijgf} - NWR_{ijrf} = 0$ is tested against the alternative hypothesis H_a: $NWR_{ijgf} - NWR_{ijrf} > 0$ for all combinations of growth scenarios g and r for which economic growth scenario g has lower economic growth rates than economic growth scenario r using a 'studentized' contrast test (test statistic = student(c), $\alpha = 0.5$) and the expected annual values of $NWR_{ijgf} - NWR_{ijrf}$. Test results are used to identify the agents and climate change scenarios for which lower future residential expansion into WUIs (i.e., lower future economic growth) significantly reduces future NWR under the current land use policy.

The expected annual values of NWR_a required for the above tests can be approximated by the product of the annual probability that a pixel burns and the annual closeness coefficients for that pixel provided by the TOPSIS procedure. The expected annual values of NWR_h and NWR_c required for the above tests are their corresponding monetary values. For example, if there are 50 years in the evaluation period, then the 'studentized' contrast tests would be based on a sample size of 50.

Ex post Decision Framework

The *ex post* decision framework utilizes the FIRECLIM model to determine how agents can optimally adapt their wildfire-related decisions to future climate change and growth in residential development (i.e., land use change). *Ex post* decisions can be evaluated for successive five-year intervals using an adaptive management approach (Peterman 1977, Holling 1978, Clark et al. 1979, Gunderson et al. 1995; McLain and Lee 1996, Wondolleck and Yaffe 2000; Kiker et al. 2001; Wilhere 2002; Prato 2003, 2005b, 2006; Richter et al. 2003). Helms (1998) states that, "[a]daptive management is a dynamic approach to forest management in which the effects of treatments and decisions are continuously monitored and used, along with research results, to modify management on a continuing basis to ensure that objectives are being met." Dombeck et al. (2004) point out that, "Uncertainty associated with…[planning in fire-adapted ecosystems] can best be addressed through adaptive management."

The *ex post* decision framework incorporates a risk approach and an uncertainty approach to adaptive management. The risk approach is used when scientists can assign subjective prior probabilities to CLEG futures at the beginning of the evaluation period. Posterior probabilities of CLEG futures are estimated at the end of each five-year interval by combining the prior probabilities of CLEG futures at the beginning of that interval and actual changes in climate and land use that occurred during that interval using Bayes theorem (Ellison, 1996; Dorazio and Johnson, 2003; Prato 2000, 2005b). An agent's optimal *ex post* decision at the beginning of a five-year interval is the optimal decision corresponding to the CLEG future having the highest posterior probability at the end of the previous five-year interval. This procedure is repeated for all five-year intervals to obtain each agent's optimal *ex post* decision plan for the entire evaluation period.

Some climate scientists believe it is more rational for experts to provide their best estimates of the subjective probabilities of future climate change events, or equivalently to assign confidence levels to conclusions regarding future climate events and their impacts than to have novice users arbitrarily select the conclusions (Moss and Schneider 2000; Schneider 2001, 2002, 2006). This position maintains that expert knowledge of models and data should be used as a basis for assigning subjective probabilities to future climate events and those probabilities should be revised using Bayesian statistics as more data about the climate system become available. This position is consistent with the risk approach.

The uncertainty approach is used when scientists are unwilling or unable to assign subjective prior probabilities to CLEG futures. An agent's optimal decision for each five-year interval with the uncertainty approach can be determined using the minimax regret criterion (Prato 2005a, 2007). This criterion minimizes the maximum NWR for an agent (or minimizes the maximum loss in ERI) that can occur with feasible decisions. The minimax regret criterion is applied to all five-year intervals to obtain each agent's optimal *ex post* decision plan for the evaluation period.

Stakeholder access to the *ex ante* and *ex post* decision frameworks can be significantly increased by incorporating them in a Web-based, interactive, spatial decision support tool, which is a knowledge-based computer program that integrates data, information, and models for the purpose of helping decision-makers solve complex spatial problems (Dkokic 1993; Loucks 1995). For example, spatial decision support tools have been developed to allow users access to results of a land use change modeling system that simulates locations of future land use changes for alternative futures for Flathead County, Montana (CARES 2007).

Conclusion

This paper describes a FIRECLIM model that can be used to do *ex ante* assessments of the potential impacts of future climate change and economic growth on NWR in the WUI, and implement an *ex post* adaptive management of future wildfire risk in WUIs. The FIRECLIM model integrates: biogeophysical simulation models; a GIS; regional climate change and economic growth scenarios; the IMPLAN economic impact assessment model; the ELMS land use change simulation model; a multiple-attribute, agent-based decision model; Bayesian statistics; decision-making under risk and uncertainty; price and cost data; and information provided by stakeholders. Results from the FIRECLIM model can be used to test hypotheses about the extent to which alternative futures for climate change and economic growth would increase NWR for stakeholders and, if future NWR increases, whether such increases can be reduced by lowering the rate of economic growth and/or limiting the expansion of residential development into WUIs.

Due to its complexity, implementation of the FIRECLIM model should only be undertaken by a multidisciplinary research team with experience in modeling and simulating climate and land use change, and knowledge of forest ecology and management, fire ecology, natural resource management, decision making under risk and uncertainty, statistics, computer programming, interactive GIS, and other areas.

The authors recommend initiating a pilot research project that would apply the FIRECLIM model to one or more WUIs. The project would: (1) assess the availability and adequacy of the geospatial data needed to implement the FIRECLIM model; (2) evaluate the best way to link the sub-models of the FIRECLIM model; (3) assess the estimation and simulation procedures; (4) determine the willingness of stakeholders to participate in agent panels and assess the quality of the information provided by those panels; and (5) identify strengths and weaknesses of the FIRECLIM model. Results of the pilot project could be used to evaluate the feasibility of implementing the FIRECLIM model and ways to improve the components of the model.

References

Alavalapati, J.R.R. 2005. Overview of policies influencing the wildland-urban interface. Pp. 63-76, in S.W. Vince, M.L. Duryea, E.A. Macie, and L.A. Hermansen (eds.), *Forestlands at the Wildland-Urban Interface: Conservation and Management*. CRC Press, Boca Raton, FL.

An, L., M. Linderman, J. Qi, A. Shortridge, and J. Liu. 2006. Exploring complexity in a human–environment system: an agent-based spatial model for multidisciplinary and multiscale integration. *Annals of the Association of American Geographers* **95**:54–79.

Arno, S.F., and S. Allison-Bunnel. 2002. Flames in Our Forest: Disaster or Renewal? Island Press, Washington, DC.

Berger, T. 2001. Agent-based spatial models applied to agriculture: a simulation tool for technology diffusion, resource use changes and policy analysis. *Agricultural Economics* **25**:245–260.

Berger, P.A. 2006. Generating agricultural landscapes for alternative futures analysis: a multiple attribute decision-making model. *Transactions in GIS* **10**:103-120.

Beyers, W.B., and P.B. Nelson. 2000. Contemporary development forces in the nonmetropolitan West: new insights from rapidly growing communities. *Journal of Rural Studies* **16**:459-474.

Black, A.E., and T. Opperman. 2005. Fire effects planning framework: a user's guide. Gen. Tech. Rep. RMRS-GTR-163WWW. USDA Forest Service, Rocky Mountain Research Station, Ft. Collins, CO.

Brown, T.J., B.L. Hall, and A.L. Westerling. 2004. The impact of twenty-first century climate change on wildland fire danger in the western United States: an applications perspective. *Climatic Change* **62**:365-388.

Calkin, D.E., K.M. Gebert, J.G. Jones, and R.P. Neilson. 2005. Forest Service large fire area burned and suppression expenditure trends, 1970-2002. *Journal of Forestry* **103**:179-183.

Center for Agricultural, Resource and Environmental Systems (CARES). 2007. Assessing Ecological Economic Impacts of Landscape Change in Montana's Flathead County. http://cares.missouri.edu/montana/ (accessed March 6, 2007)

Chen, C-T. 2000. Extensions to the TOPSIS for group decision-making under fuzzy environment. *Fuzzy Sets and Systems* **114**:1-9.

Chen, S.J., and C.L. Hwang. 1992. Fuzzy multiple attribute decision making. *Lecture Notes in Economics and Mathematical Systems*, No. **375**, Springer-Verlag, Berlin.

CIRMOUNT Committee. 2006. Mapping New Terrain: Climate Change and America's West. Report of the Consortium for Integrated Climate Research in Western Mountains (CIRMOUNT). Misc. Pub., PSW-MISC-77, Pacific Southwest Research Station, USDA Forest Service, Albany, CA.

Clark, W.C., D.D. Jones, and C.S. Holling. 1979. Lessons for ecological policy design: A case project of ecosystem management. *Ecological Modeling* **7**:1-53.

Cohen, J.D. 1991. A site-specific approach for assessing the fire risk to structures at the wildland/urban interface. *Gen. Tech. Rep. SE-GTR-69*, U.S. Department of Agriculture Forest Service, Southeastern Forest Experiment Station, Knoxville, TN.

______. 1994. Structure ignition assessment model (SIAM). Paper presented at the Biswell Symposium: Fire Issues and Solutions in Urban Interface and Wildland Ecosystems, February 15-17, Walnut Creek, CA.

______. 1995. Structure ignition assessment model (SIAM). Pp. 85-92 in Proceedings of the Biswell symposium: fire issues and solutions in urban interface and wildland ecosystems. GTR-PSW-158. USDA Forest Service, Pacific Southwest Research Station, Albany, CA.

______. 2000. Preventing disaster: home ignitability in the wildland-urban interface. *Journal of Forestry* **98**:15-21.

Cohen, J.D., and J. Saveland. 1996. Structure ignition assessment can help reduce fire damages in the W-UI. Presentation at the Food and Agriculture Organization of the United Nations/Economic Commission for Europe/International Labor Organization Seminar on Forest, Fire, and Global Change, Shushenskoye (Russian Federation).

Cohrssen, J.J., and V.T. Covello. 1989. *Risk Analysis: A Guide to Principles and Methods for Analyzing Health and Environmental Risks*. US Council on Environmental Quality, Washington, DC.

Cortner, H.J., P.D. Gardner, and J.G. Taylor. 1990. Fire hazards at the wildland urban interface: what the public expects. *Environmental Management* **14**:57-62.

Covington, W.W., and M. Moore. 1994. Southwestern ponderosa forest structure. *Journal of Forestry* **92**:39-47.

Deeming, J.E., R.E. Burgan, and J.D. Cohen. 1977. The National Fire Danger Rating System, 1978. Gen. Tech. Rpt. INT-39, USDA Forest Service, Intermountain Forest and Range Experiment Station, Ogden, UT. http://www.wrh.noaa.gov/sew/fire/olm/nfdrs.htm (accessed March 30, 2007)

Dellasala, D.A., J.E. Williams, C.D. Williams, and J.A. Franklin. 2004. Beyond smoke and mirrors: a synthesis of fire policy and science. *Conservation Biology* **18**:976-986.

Dennis, F.C. 2007. Creating Wildfire-Defensible Zones. Colorado State University Extension-Natural Resources. Ft. Collins, CO. http://www.ext.colostate.edu/PUBS/natres/06302.html (accessed September 7, 2007).

Djokic, D. 1993. Towards general purpose spatial decision support systems using existing technologies. *Proceedings of the Second International Conference/Workshop on Integrating Geographic Information Systems and Environmental Modeling. Breckenridge*, CO, September 26-30.

Dombeck, M.P., J.E. Williams, and C.A. Wood. 2004. Wildfire policy and public lands: integrating scientific understanding with social concerns across landscapes. *Conservation Biology* **18**:883-889.

Dorazio R.M., and F.A. Johnson. 2003. Bayesian inference and decision theory-a framework for decision making in natural resource management. *Ecological Applications* **13**: 556-563.

Duryea, M.L., and S.W. Vince. 2005. Introduction: the city is moving to our doorstep. Pp. 3-13 in S.W. Vince, M.L. Duryea, E.A. Macie, and L.A. Hermansen (eds.), *Forestlands at the Wildland-Urban Interface: Conservation and Management*. CRC Press, Boca Raton, FL.

Ellison, M. 1996. An introduction to Bayesian inference for ecological research and environmental decision-making. *Ecological Applications* **6**:1036-1046.

Fagre, D.B. 2000. Changing mountain landscapes in a changing climate: looking into the future. *Changing Landscapes* **2**:31–39.

Fagre, D.B., S.W. Running, R.E. Keane, and D.L. Peterson. 2005. Assessing climate change effects on mountain ecosystems using integrated models: a case project. Pp. 489-500 in U.M. Huber, H.K. Bagman, and M.A. Reasoned (eds.), *Global Change and Mountain Regions: A State of Knowledge Overview*. Springer, Dordrecht, The Netherlands.

Farris, C.A., C. Pezeshki, and L.F. Neuenschwander. 1999. A comparison of fire probability maps derived from GIS modeling and direct simulation techniques. Proceedings from the Joint Fire Science Conference and Workshop *Crossing the Millennium: Integrating Spatial Technologies and Ecological Principles for a New Age in Fire Management* **1**:131-138, June 15-17, Boise, ID.

Ferry, G.W., R.G. Clark, R.E. Montgomery, R.W. Mitch, W.P. Lernout's, and G.T. Zimmerman. 1995. Altered fire regimes within fire-adapted ecosystems. Our living resources: a report to the nation on the distribution, abundance, and health of US plants, animals, and ecosystems. US Department of the Interior, National Biological Survey, Washington, DC.

Finney, M.A. 1998. FARSITE: Fire Area Simulator-model development and evaluation. Res. Pap. RMRS-RP-4, USDA Forest Service, Rocky Mountain Research Station, Ogden, UT. http://www.treesearch.fs.fed.us/pubs/4617 (accessed May 10, 2007).

______. 2005. The challenge of quantitative risk analysis for wildland fire. *Forest Ecology and Management* **211**: 97-108.

______. 2006. An overview of FlamMap fire modeling capabilities. Pp. 213-220, in: P.L. Andrews and B.W. Butler, Fuels Management—How to Measure Success: Conference Proceedings. USDA Forest Service Proceedings RMRS-P-41. March 28-30, Portland, OR.

Firewise Communities. 2007. Resources for the homeowner. http://www.firewise.org/resources/homeowner.htm (accessed April 27, 2007).

Franklin, J.F., and J. Agee. 2003. Scientific issues and national forest policy: forging a science-based national forest fire policy. *Issues in Science and Technology* **20**:59-66.

General Accounting Office. 1999. Western national fires: a cohesive strategy is needed to address catastrophic wildfire threats. US General Accounting Office, House of Representatives, Committee on Resources, Report to the Subcommittee on Forests and Forest Health, RCED-99-65, Washington, DC.

GCS Research. 2005. Flathead County Community Wildfire Fuels Reduction/Mitigation Plan. Missoula, MT.

Gimblett, H.R. (ed.). 2002. *Integrating Geographic Information Systems and Agent-Based Modeling Techniques for Simulating Social and Ecological Processes*. Oxford University Press, New York.

Gunderson, L.H., C.S. Holling, and S.S. Light (eds.). 1995. *Barriers and Bridges to the Renewal of Ecosystems and Institutions*. Columbia University Press, New York.

Helms, J.A. (ed.). 1998. *The Dictionary of Forestry*. Society of American Foresters. Bethesda, MD.

Hoffmann, M.J., H. Kelley, D.C. Parker, and J. Walker. 2002. *Agent-Based Modeling: A Framework for Decision Making*. CIPEC Working Paper CWP-02-01. Bloomington:

Center for the Project of Institutions, Population, and Environmental Change, Indiana University.

Holling, C.S. 1978. *Adaptive Environmental Assessment and Management.* John Wiley & Sons, Chichester, England.

Holsinger, L., R.E. Keane, B. Steele, M. Reeves, and S. Pratt. 2006. Using historical simulations of vegetation to assess departure of current vegetation conditions across large landscapes. Pages 315-366, in M.G. Rollins, and C. Frame (editors), *Assessing Departure of Current Vegetation Conditions from Historical Simulations of Vegetation Across Large Landscapes.* USDA Forest Service Rocky Mountain Research Station General Technical Report RMRS-GTR-175.

Hwang, C.L., and K. Yoon. 1981. Multiple Attribute Decision Making. *Lecture Notes in Economics and Mathematical Systems* No. **186**. Springer-Verlag, Berlin.

Kauffman, J.B. 2004. Death rides the forest: perceptions of fire, land use, and associated ecological restoration of western forests. *Conservation Biology* **18**:878-882.

Keane, R.E., P. Morgan, and S.W. Running. 1996a. A mechanistic ecological process model for simulating fire succession on coniferous forest landscapes. INT Research Paper INT-RP-484. Intermountain Research Station, USDA Forest Service, Ogden, UT.

Keane, R.E., K. Ryan, and S.W. Running. 1996b. Simulating the effect of fires on northern Rocky Mountain landscapes using the ecological process model FIRE-BGCv2. *Tree Physiology* **16**:319-331.

Keane, R.E., C. Hardy, K. Ryan, and M. Finney. 1997. Simulating effects of fire management on gaseous emissions from future landscapes of Glacier National Park, Montana, USA. *World Resource Review* **9:**177-205

Keane, R.E., P. Morgan, and J.D. White. 1999. Temporal pattern of ecosystem processes on simulated landscapes of Glacier National Park, USA. *Landscape Ecology* **14**:311-329.

Keane, R.E., T. Veblen, K.C. Ryan, J. Logan, C. Allen, and B. Hawkes. 2002. The cascading effects of fire exclusion in the Rocky Mountains. Pages 133-153, in J.S. Baron (editor), editor. *Rocky Mountain Futures: An Ecological Perspective.* Island Press, Washington, DC.

Kiker, C.F., J.W. Melon, and A.W. Hodges. 2001. Adaptive learning for science-based policy: the Everglades restoration. *Ecological Economics* **37**:403-416

Kline, J.D. 2004. Issues in evaluating the costs and benefits of fuel treatments to reduce wildfire in the Nation's forests. USDA Forest Service, Research Note PNW-RN-542, Corvallis, OR.

Künzli, N., E. Avon, J. Wu, W.J. Alderman, E. Rapport, J. Millstein, J. Bunion, R. McConnell, F.D. Gilliland, K. Breanne, F. Clurman, A. Winner, and J.M. Peters. 2006. Health effects of the 2003 southern California wildfires on children. *American Journal of Respiratory and Critical Care Medicine* **174**:1221-1228. http://ajrccm.atsjournals.org/cgi/content/abstract/174/11/1221 (accessed March 27, 2007)

Lenihan, J., C. Daly, D. Bachelet, and R.P. Neilson. 1998. Simulating broad-scale fire severity in a dynamic global vegetation model. *Northwest Science* **72**:91-103.

Lim, K., P.J. Deadpan, E. Moran, E. Bromidic, and S. McCracken. 2002. Agent-based simulations of household decision-making and land use change near Altamira, Brazil. Pp. 277–308 in H.R. Gimblett (ed.), *Integrating Geographic Information Systems and Agent-based Techniques for Simulating Social and Ecological Processes.* Oxford University Press, New York.

Lindenmayer, D.B., and J.F. Franklin. 2002. *Conserving Forest Biodiversity: A Comprehensive Multiscale Approach*. Island Press, Washington, DC.

Littell, J.S., D. McKenzie, and D.L. Peterson. 2005. Ecological context of climate impacts on fire: wildfire area burned in the western U.S. 1916-2003. Presentation at MTCLIM Conference 2005, Pray, MT.

Long, A.J., D.D. Wade, and F.C. Beall. 2005. Managing fire risks in the interface: challenges and opportunities. Pp. 201-223 in S.W. Vince, M.L. Duryea, E.A. Macie, and L.A. Hermansen (eds.), *Forestlands at the Wildland-Urban Interface: Conservation and Management*. CRC Press, Boca Raton, FL.

Loucks, D.P. 1995. Developing and implementing decision support systems: a critique and challenge. *Water Resources Bulletin* **31**:571-582.

Martin, W.E., C. Raish, and B. Kent (eds.). 2007. *Wildfire Risk: Human Perceptions and Management Implications*. RFF Press, Washington, DC.

McLain, R.J., and R.G. Lee. 1996. Adaptive management: promises and pitfalls. *Environmental Management* **20**:437-448.

McKenzie, D., Z. Gandalf, D.L. Peterson, and P. Mote. 2004. Climatic change, wildfire, and conservation. *Conservation Biology* **18**:890-902.

Meehl, G.A., T.F. Stocker, W.D. Collins, P. Friedlingstein, A.T. Gaye, J.M. Gregory, A. Kitoh, R. Knutti, J.M. Murphy, A. Noda, S.C.B. Raper, I.G. Watterson, A.J. Weaver and Z.-C. Zhao. 2007. Global climate projections. In S. Solomon, D. Qin, M. Manning, Z. Chen, M. Marquis, K.B. Averyt, M. Tignor and H.L. Miller (eds.), *Climate Change 2007: The Physical Science Basis. Contribution of Working Group I to the Fourth Assessment Report of the Intergovernmental Panel on Climate Change*. Cambridge University Press, Cambridge, United Kingdom and New York, NY.

Miller, C., P.B. Landers, and P.B. Layback. 2000. Evaluating the risks and benefits of wildland fire at landscape scales. Pp 78-87 in L.F. Neuenschwander, and K.C. Ryan (tech. coords.), *Crossing the Millennium: Integrating Spatial Technologies and Ecological Principles for a New Age in Fire Management*, Proceedings of the Joint Fire Science Conference and Workshop, June 15-17, 1999, University of Idaho, Boise, ID.

Minnesota IMPLAN Group, Inc., 2006. IMPLAN System (data and software), Stillwater, MN 55082 www.implan.com (accessed June 21, 2007)

Moss, R.H., and S.H. Schneider. 2000. Uncertainties in the IPCC TAR: recommendations to lead authors for more consistent assessment and reporting. Pp. 33–51 in R. Pachauri, T. Taniguchi, and K. Tanaka (eds.), Guidance papers on the cross cutting issues of the Third Assessment Report of the IPCC, World Meteorological Organization, Geneva, Switzerland.

Myszewski, M., and J.E. Kundell. 2005. Land-use planning and zoning at the wildland-urban interface. Pp. 77-94 in S.W. Vince, M.L. Duryea, E.A. Macie, and L.A. Hermansen (eds.), *Forestlands at the Wildland-Urban Interface: Conservation and Management*. CRC Press, Boca Raton, FL.

National Interagency Fire Center. 2007. Wildland fire statistics. Boise, ID. http://www.nifc.gov/stats/wildlandfirestats.html (accessed March 7, 2007)

O'Laughlin, J. 2005. Relative risk assessments for decision-making related to uncharacteristic wildfire. *Forest Ecology and Management* **211**:3-14.

Parker, D.C., S.M. Manson, M.A. Janssen, M.J. Hoffmann, and P. Deadman. 2003. Multi-agent systems for the simulation of land-use and land-cover change: a review. *Annals of the Association of American Geographers* **93**:314–37.

Perkins, S. 2001. A nation aflame: after last years' conflagration, can we learn to live with wildfires. Science News **159**:120-121.

Peterman, R.M. 1977. Graphical evaluation of environmental management options: Examples from a forest-insect pest system. *Ecological Modelling* **3**:133-148.

Platt, R.V. 2006. A model of exurban land-use change and wildfire mitigation. *Environment and Planning B: Planning and Design* **33**:749-765.

Power, T.M. 2006. Avoiding a new "conspiracy of optimism": the economics of forest fire fuel reduction strategies. Pp. 205-216 in: G. Wuerthner (ed.), *Wildfire: A Century of Failed Forest Policy*. Island Press, Washington, DC.

Presidential/Congressional Commission on Risk Assessment and Risk Management (PCCRARM). 1997. *Framework for environmental health risk management.* Washington, DC.

Prato, T. 2000. Multiple attribute Bayesian analysis of adaptive ecosystem management. *Ecological Modelling* **133**:181-193.

______. 2003. Adaptive management of large rivers with special reference to the Missouri River. *Journal of the American Water Resources Association* **39**:935-946.

_____. 2005a. Accounting for uncertainty in making species protection decisions. *Conservation Biology* **19**:806-814.

______. 2005b. Bayesian adaptive management of ecosystems. *Ecological Modelling* **183**:147-156.

______.2005c. Modeling ecological impacts of landscape change. *Environmental Modelling & Software* **20**:1359-1363.

______. 2006. Adaptive management of national park ecosystems. *George Wright Forum* **23**:72-86.

______. 2007. Evaluating land use policies under uncertainty. *Land Use Policy* **24**:165-174. http://dx.doi.org/10.1016/j.landusepol.2006.02.003 (accessed June 8, 2007)

Prato, T., and D. Fagre. 2006. Coping with climate change. *ActionBioscience* http://www.actionbioscience.org/environment/prato_fagre.html (accessed April 24, 2007).

Prato, T., D. Fagre, and R. Sugumaran. 2007a. Economic growth and landscape change. Pp. 55-66 in T. Prato and D. Fagre (eds.), *Sustaining Rocky Mountain Landscapes: Science, Policy and Management of the Crown of the Continent Ecosystem*, RFF Press, Washington, DC.

Prato, T., A.S. Clark, K. Dolle, and Y. Barnett. 2007b. Evaluating alternative economic growth rates and land use policies for Flathead County, Montana. *Landscape and Urban Planning* **83**:327–339.

Price, C., and D. Rind. 1994. The impact of 2xCO_2 climate on lightning-caused fires. *Journal of Climate* **7**:1484-1494.

Pyne, S.J. 2001. *Year of the Fires: The Story of the Great Fires of 1910*. Viking Press, NY.

Radeloff, V.C., R.B. Hammer, S.I. Stewart, J.S. Fried, S.S. Holcomb, and J.F. McKeefrey. 2005. The wildland-urban interface in the United States. *Ecological Applications* **15**: 799-805.

Regional Economic Information System. Selected years. Geospatial and Statistical Center, University of Virginia. http://fisher.lib.virginia.edu/ (accessed May 22, 2007).

Riebsame, W.E., H. Gosnell, and D.M. Theobald. 1997. Land use and landscape change in the U.S. Rocky Mountains. I. Theory, scale, and pattern. *Mountain Research and Development* **16**:395-405.

Richter, B.D., R. Mathews, D.L. Harrison, and R. Wigington. 2003. Ecologically sustainable water management: managing river flows for ecological integrity. *Ecological Applications* **13**:206-224.

Rothermel, R.C. 1972. A mathematical model for predicting fire spread in wildland fuels. USDA Forest Service, Research Paper INT-115.

Rowe, W.D. 1975. An anatomy of risk. United States Environmental Protection Agency, Washington, DC.

Saunders, S., and M. Maxwell. 2005. Less snow, less water: climate disruption in the West. The Rocky Mountain Climate Organization, Louisville, CO.

Schneider, S.H. 2001. What is 'dangerous' climate change? *Nature* **411**:17-19

______. 2002. Can we estimate the likelihood of climatic changes at 2100? *Climatic Change* **52**:441-451

______. 2006. Global climate change. Newsletter, International Society for Ecological Economics, July.

Scott, J.H. 2006. An analytical framework for quantifying wildland fire risk and fuel treatment benefit. Pp. 169-184 in P.L. Andrews and B.W. Butler (ed.), *Fuels Management-How to Measure Success*, Conference Proceedings, RMRS-P-41, March 28-30, Portland, OR.

SILVIS Lab. 2007. The Wildland-Urban Interface (WUI) defined. Forest Ecology & Management, University of Wisconsin-Madison. http://silvis.forest.wisc.edu/library/ WUIDefinitions2.asp (accessed May 26, 2007)

Stanionis, C., and D. Glick. 2006. Sprawling into disaster: the growing impact of rural residential development on wildland fire management in the Greater Yellowstone area. Pp. 255-259, in G. Wuerthner (ed.), *Wildfire: A Century of Failed Forest Policy.* Island Press, Washington, DC.

Steele, B.M., S.K. Reddy, and R.E. Keane. 2006. A methodology for assessing departure of current plant communities from historical conditions over large landscapes. *Ecological Modelling* **199**:53-63.

Suter, G.W. (ed.). 1993. *Ecological Risk Assessment.* Lewis Publishers, Boca Raton, FL.

US Department of Agriculture (USDA). 2001. The National Fire Plan: preparing for wildland fires and reducing their impacts on people and resources. USDA Forest Service. Washington, DC.

USDA Office of Inspector General. 2006. Audit Report: Forest Service Large Fire Suppression Costs. Washington, DC. http://www.usda.gov/oig/webdocs/08601-44-SF.pdf (accessed July 25, 2007)

US Department of Agriculture and US Department of the Interior (USDA USDI). 2003. National Fire Plan. http://www.fireplan.gov (accessed March 9, 2007)

USDA Forest Service. 2003. FARSITE: Fire Area Simulator. Missoula, MT. (http://www.fire.org/index.php?option=content&task=category&sectionid=2&id=8&Itemid=27 (accessed April 17, 2007)

______. 2007a. Forest Management, Wildland Fire Policy. http://www.fs.fed.us/fire/management/policy.html (accessed February 27, 2007)

US Environmental Protection Agency. 2004. Six common air pollutants: particulate matter. http://www.epa.gov/oar/urbanair/6poll.html (accessed September 14, 2007)

US House of Representatives. 2003. 2003 Healthy Forests Restoration Act H.R. 1904 108th Congress, 1st Session.

Venn, T.J. and Calkin, D.E. 2007. Challenges of socio-economically evaluating wildfire management in and adjacent to non-industrial private forestland in the western United States. Pp. 383-396, in S.R. Harrison, A. Bosch, and J.L. Herbohn (eds.), *Improving the Triple Bottom Line Returns from Small-scale Forestry*, Proceedings from IUFRO Working Groups 3.08 Small-scale Forestry and 6.11.02 Forestry and Rural Development in Industrialized Countries, June 17 to 21, Ormoc City, the Philippines, The University of Queensland, Brisbane.

Westerling, A.L., H.G. Hidalgo, D.R. Cayan, and T.W. Swetnam. 2006. Warming and earlier spring increase western U.S. forest wildfire activity. *Science* **313**:940-943.

Western Governors' Association, the Secretaries of the Departments of Agriculture and the Interior, and their partners. 2001. A Collaborative Approach for Reducing Wildland Fire Risks to Communities and the Environment 10-Year Comprehensive Strategy. www.fireplan.gov/reports/7-19-en.pdf (accessed February 27, 2007)

Wilhere, G.F. 2002. Adaptive management in habitat conservation plans. *Conservation Biology* **16**:20-29.

Wondolleck, J.M., and S.L. Yaffee. 2000. *Making Collaboration Work: Lessons from Innovation in Natural Resource Management*. Island Press, Washington, DC.

Wuerthner, G. 2006a. The Yellowstone fires of 1988: a living wilderness. Pp. 46-61, in G. Wuerthner (ed.), *Wildfire: A Century of Failed Forest Policy*. Island Press, Washington, DC.

Wuerthner, G. (ed.). 2006b. *Wildfire: A Century of Failed Forest Policy*. Island Press, Washington, DC.

In: Ecosystem Ecology Research Trends
Editors: J. Chen and C. Guo, pp. 299-312
ISBN: 978-1-60456-183-8

Chapter 9

SOIL NITROGEN TRANSFORMATION IN DIFFERENT LAND USES IN INDIAN DRY TROPICAL FORESTS

***N. Tripathi*[*] *and R.S. Singh*[**]**
Central Institute of Mining and Fuel Research, Barwa Road, Dhanbad – 826 001
Jharkhand, India

Abstract

The effects of conversion of an Indian tropical forest ecosystem into alternate land uses, such as grassland and cropland on plant available N, N-mineralization, microbial biomass N (MBN) and fine root biomass in soil were studied seasonally and total plant biomass and soil particle size were studied annually for two years. The mean annual mineral N and soil microbial biomass, respectively, in three ecosystems varied significantly with season; the values being highest in summer season and lowest in rainy season, while the net nitrification rate and net N-mineralization rate values were maximum in rainy season and minimum in summer season. The MBN was positively related to root biomass and total plant biomass (i.e., above- and below-ground biomass). Interestingly, MBN and inorganic N are reciprocally and nitrification and N-mineralization are directly related to seasonal soil moisture. The soil microbial biomass, nitrification and N-mineralization are negatively related to clay content. The mean annual net nitrification and N-mineralization rates declined significantly after conversion into grassland and cropland by 43 and 54% and net N mineralization by 33 and 46%, respectively.

Above- and below-ground biomass also have had their impact on MBN, and thereby N-mineralization. Thus, in tropical forests, the conversion into different land-uses affects remarkably the nitrogen transformation process in soil. Therefore, there should be a nitrogen management scheme to prevent the nitrogen losses from the derived ecosystems.

---

[*] E-mail address: nymphaea7@yahoo.com. Telephone No.: 093343-40900, 098351-13131. Fax No. 0091-326-2202429/ 2202774/ 2203156. Address for Correspondence: Central Institute of Mining and Fuel Research, Barwa Road, DHANBAD – 826 001, Jharkhand, INDIA. Address for Reprints: Dr. Nimisha Tripathi, Research Scientist, Central Institute of Mining and Fuel Research, DHANBAD- 826 001. (Jharkhand) INDIA.
[**] E-mail address: rss_60@yahoo.com

Introduction

Land use practices directly affect the distribution and supply of soil nutrients by altering soil properties and by influencing biological transformations in the rooting zone. Conversion of forests to agriculture or other land uses has resulted into the degradation of large areas of soils in tropics (FAO, 2001 and Hurni, 1993).

Land use change is an important human activity and interference in affecting ecosystem processes, particularly the processes of microbial C and N mineralization. Such a change, and particularly conversion of native lands to cultivated fields, strongly modifies land cover, biota, and biogeochemical cycles (Griinzweig et al., 2003).

The impact of conversion of forest to other land uses will depend on how conversion affects N availability and soil water content. Conversion of tropical forest to agriculture or pasture generally leads to decreased rates of N cycling and lower inorganic-N pools (Reiners et al., 1994; Neill et al., 1997; Verchot et al., 1999). Griinzweig et al. (2003) postulated that the conversion of forest into grassland or cropland due to forest harvesting, fertilization, atmospheric depositions and climate change results into significant alterations in soil nitrogen cycling. The total annual deposition of inorganic nitrogen was 26% of the total annual deposition of all ions (Vitousek and Matson, 1988). This conversion leads to several changes in vegetation and soil properties (Sorenson, 1974). Forest clearing and land-use changes have important consequences not only for nutrient availability but also for soil organic matter (SOM) stability Glaser, et al. (2001).

Land use changes, especially cultivation of forest and pasture lands, may instantly lower soil quality. As a result, severe deterioration in soil quality may lead to a lasting degradation of land productivity. Niche also plays an important role in the soil nutrient status (Leuschner et al., 2006).

Nitrogen (N) mineralization is an important component of nitrogen supply for plant growth (Milkha and Rice, 2004) and the nutrient dynamics is controlled by the rate of decomposition of organic matter. N-mineralization is of prime importance in ecosystem productivity and soil nitrogen transformations control N potential availability and losses of nitrogen from the ecosystem (Knoepp and Swank, 2002).

Microbial activity is fundamental in the processes that make energy and nutrients available for recycling in the ecosystem. Microbial biomass can serve as an estimate of microbial N immobilization and the soil microorganisms mediate many of the major processes involved in soil nitrogen cycling. Besides, they are strong competitors with plants for available soil nitrogen. Thus, any change in forest ecosystem, and consequently the microbial N may have significant impacts on nitrogen availability and overall forest nitrogen cycling (Idol et al., 2002). Microbial biomass can serve as an estimate of microbial N immobilization and due to its importance in the functioning of different ecosystems; dynamics of microbial biomass and its role in plant nutrition under different ecosystem conditions is of greater significance (Azam et al., 2003). Microbial biomass acts as a source of plant nutrient in dry tropical forest and savanna ecosystems (Singh and Singh, 2006).

In forest and grassland ecosystems, N-mineralization is affected by the biological factors, such as soil animals, soil microorganisms and plants and the non-biological factors, such as environmental factors and anthropogenic disturbances (Wang et al., 2004, Srivastava and Singh, 1989). Soil macroclimatic conditions vary in different ecosystems, which affect the microbial activity and biomass. Available soil moisture is the key factor in the functioning of the ecosystem in tropical countries (Lodge et al., 1994).

The objectives of this study were to determine the effects of alternative land uses on plant available N, nitrification and N-mineralization rates, and soil microbial biomass N.

Material & Methods

Study Sites

The study sites were located in Similipal Biosphere Tiger Reserve in Mayurbhanj District of Orissa state (India). The Reserve lies between 85°4' to 87°10' E and 20°17' to 22°34' N at an average elevation of 1000 m above mean sea level. Extending over an area of 4374 km^2, Similipal Biosphere Reserve includes the biogeographic zone of Deccan Peninsula, Chhotanagpur and Mahanadian regions. It was constituted by the Government of India as a Biosphere Tiger Reserve for conserving the biodiversity of plants and animals in their natural ecosystem. The forest, grassland and agriculture land cover an area of 845 km^2, 2129 km^2 and 80 km^2, respectively.

Vegetation of Similipal Biosphere Reserve is tropical moist deciduous forest, inhabiting more than one thousand plant species. *Shorea robusta* (Sal) is the dominating species in the studied forest and is considered as one of the best sal forests of India, followed by the other frequently occurring mixed species, such as *Terminalia tomentosa, Anogeissus latifolia, Bauhinia variegata, Sterculia urens, Terminalia arjuna, Murraya koenigii, Schleichera chinensis, Samanea saman, Xylia xylocarpa*, etc. The best elephant species of Central-Indian population are found in Similipal Biosphere Reserve and they constitute the largest elephant population of the country. The area of grasslands comprises of two communities, one dominated by *Imperata cylindrica* and the co-dominant species are *Themada gigantica* and *Saccharum spontaneum*. The other species include *Ergrostis atrovirens, Sporobolos indicus*. The other communities are dominated by *Heteropogon contortus* and the co-dominant species are *Cymbopogon flexosus*, *Bothriochloa bladhii*. The other species include *Bracharia racemosa, Digitaria stricta, Dicanthium caricosum and Panicum notatum*. Tribals are unskilled and constitute the main population of Similipal Biosphere Reserve. Agriculture is mostly rainfed and not well developed. Tribals produce only one paddy crop in a year. So, there is a great pressure on forests for fulfilling the needs of the people, as most of the people derive their income from collection of minor forest products (MFP), sale of firewood and timber. In addition to these, MFP are also collected by them for their own use. 'Akhand Shikar' is another traditional custom which results in large scale killing of wild animals. Both cropfields, which were derived from the sal forest, still have a few standing old trees of *Shorea robusta* (9 ha^{-1}). The cropfields were manured by only 2000 kg ha^{-1} farmyard manure every year in the month of March.

Two forest sites (sal forest and mixed forest), two grassland sites (*Imperata* and *Heteropogon* grasslands) and two cropland sites (10- and 15-yrs. old), all derived from the forest ecosystem, were studied for three seasons for two years in the years May 2003-December 2004.

Climatic Conditions and Soil Characteristics

The climate of Similipal forest reserve is tropical. The year is divided into three seasons; summer (March to June), rainy (July to October) and winter (November to February). Mean maximum temperature varies from 16°C in December to 35°C in May-June and mean minimum from 6°C in January to 22°C in June. The mean annual rainfall varies from 28-396 mm, of which 79% occurs in the rainy season. Relative humidity during morning hour accounts upto 70-100% and 30% during midday (Srivastava and Singh, 1998, Tripathi and Singh, 2007).

Rocks are metamorphic, sedimentary and igneous. The soil is slightly acidic, reddish in color, loam to sandy loam in texture and derived from haematite rocks or weathering of shales. The outcrop of submetamorphic sand stone and quartzite hematite's, on disintegration, produce reddish sandy soil. Soil erosion is common around the valley (Srivastava and Singh, 1998). The organic carbon contents are high in the soil, but the nitrogen and phosphorus contents are very low (Tripathi and Singh, 2007).

Sample Collection

Three uniform plots of one ha each were marked from each site for the sample collection and from each plot twenty soil samples were randomly collected in 5X5X10 cm block from the upper layer and composited to get one sample per plot in the months of May, for summer season, August, for rainy season and December, for winter season in the years 2003 and 2004. Large pieces of plant materials were picked out by hand for removing the large plant materials to check the nutrient immobilization by the microbes (Singh and Singh, 2006, Tripathi and Singh, 2007) and the soils were sieved through 2-mm sieve. Root sampling was confined to 10 cm., as the majority (70%) of roots were located in this layer. Each sample was divided into two halves. One half in the field moist condition was used for determination of mineral N, microbial nutrients, and the other part was air-dried (<0.2% moisture) for all the other analyses.

About 150-200 g. of soil from five pits was incubated in each ecosystem for 30 days in the respective pits to quantify nitrification and N-mineralization. Rest of the samples was brought to the laboratory for the analysis of plant available nutrients.

Soil Analyses

Soil sampling for texture analysis was done once in winter (December) 2004, and physico-chemical characterization, organic C, N, and P, total N and P, mineral N and microbial N were done three times, i.e., seasonally (winter, summer and rainy season). All the data reported in this paper are means for two sites. Estimations of pH, nitrate-N and ammonia-N were done in the field-moist condition of forest, grassland and cropland soils. Water holding capacity, bulk density, particle size distribution, organic carbon, total N and total P were analyzed after air drying the soil.

Soil pH was analyzed by ORION ion analyzer, where soil:water ratio was 1:2. Bulk density and water holding capacity were determined following (Piper, 1994). Organic carbon of soil was determined by Walkley Black's method and total N by modified Kjeldahl method (Idol et al., 2002). For the determination of total phosphorus, the perchloric acid digestion method was followed (Mehta, et al., 1954). Ca^{++} in soil was determined following the standard EDTA method and for estimating K^{+} the ORION ion analyzer was used.

Analyses for Plant Available N (Nitrate- and Ammonium-N), Nitrogen Mineralization and Microbial Biomass N

Nitrogen mineralization in soil was quantified by the *in situ* buried bag technique (Eno, 1960). Freshly collected soil samples (ca. 150 g) enclosed in polyethylene bags were buried at 0-10 cm depth and incubated for 30 days. Nitrate-N and ammonia-N were determined at the day of fresh sample collection and after 30 days of field incubation. Phenol-di-sulfonic acid method (Jackson, 1958) was followed for the estimation of nitrate and phenate method (Wetzel and Likens, 1979) for analyzing ammonium-N. Nitrification was indicated by an increase in nitrate-N during the course of incubation. Net N-mineralization was calculated by taking into account the increase in the concentration of ammonium-N plus nitrate-N in between the period zero day and 30 days of incubation (Singh and Singh, 2006).

Microbial biomass N in all soils was determined by the chloroform fumigation extraction method (Brookes et al., 1985). All the samples were stored in the field-moist condition at room temperature without adjusting the soil moisture, to settle down the respiration (Srivastava and Singh, 1989). The flush of total N was obtained by subtracting the K_2SO_4-extractable N in unfumigated soil from that of the fumigated soil, and is divided by a fraction value (K_N) of 0.54 (Brookes, et al., 1985) of biomass N extracted after chloroform fumigation.

Plant Biomass Analysis

Above ground litter biomass in forests, canopy biomass in grasslands and minespoils and shoot residue in cultivated lands were measured in randomly placed five 1X1 m quadrats. For belowground biomass (live + dead root) (<5 mm diameter) five replicate monoliths of 15X15X10 cm size were sampled. The monoliths were washed with a fine jet of water using a 0.5 mm sieve. Both above- and below-ground samples were oven dried at 80°C till the constant weight.

The above- and below-ground biomass of forest was measured in summer (May 2003 and 2004). The litter layer was greatest on ground at this time. The determination of above-and belowground biomass of cropfield was done twice in the annual cycle, once in May 2003 and 2004 and then in September 2003 and 2004 after the crop harvest. The canopy biomass and root biomass of grassland were measured in September 2003 and 2004, when the canopy biomass was at its maximum.

Statistical Analysis

The statistical analyses (analysis of variance, least significant difference) of the data were analyzed following SPSS (1997).

Results

Physico-Chemical Characterization

Physico-chemical characteristics of soil are given in Table 1. The soils of forest, grassland and cropland were acidic, showing a pH range of 4.68 to 5.88. In the present study, the range of bulk density from forest to cropland was 0.97-1.2 g cm^{-2}, indicating that the conversion of forest into grassland and cropland resulted into an increase in bulk density by 23.7% and 13.4%, respectively. The organic carbon content in soil ranged from 0.82-1.91%, showing a decline of 39% and 57% due to conversion into grassland and cropland, respectively. Similarly, there was a decline of 31% and 48% in total nitrogen content after conversion into grassland and cultivation, respectively, as shown by the range of total N from cropland to forest (i.e., 0.098-1.88%). Total P from cropland to forest ecosystems ranged from 0.03-0.042%, showing a percent decline of 13% and 28.6%, due to conversion of forest into grassland and cropland, respectively.

Table 1. Physical characteristics of soils of different land uses (means ±1 S.E.).

Parameters		Forest	Grassland	Cropland
pH		5.885	4.68	5.82
Temperature°C		21.35	23.5	25.95
Soil Particle Size	>0.2 mm (%)	67±2.2	73.5±2.3	77±3.7
	0.2-0.1mm (%)	19±1.9	16.5±1.6	15.5±1.37
	<0.1 mm (%)	14.5±1.5	9±1.2	7.5±1.0
Bulk Density (g cm^{-2})		0.97±0.26	1.1±0.95	1.2±0.12
Water Holding Capacity (%)		38.96±2.4	34.33±2.0	30.2±2.0
Organic Carbon (%)		1.908±0.2	1.17±0.3	0.82±0.14
Total N (%)		0.188±0.02	0.13±0.01	0.098±0.004
Total P (%)		0.042±0.01	0.0365±0.01	0.03±0.01
Organic C/N Ratio		10.10	9	8.36

Soil Moisture

The mean annual soil moisture content in forest, grassland and cropland varied from 5.43-11.38% (Table 2), showing a decline of 39% and 52% after conversion of forest into grassland and cropland, respectively. The seasonal pattern has shown the highest values of

Table 2. Mean annual variations of various parameters in different land uses.

Ecosystem	Parameters					
	Moisture (%)	Nitrate nitrogen (µg g^{-1})	Ammonia nitrogen (µg g^{-1})	Aboveground biomass (kg ha^{-1})	Belowground biomass (kg ha^{-1})	Total (kg ha^{-1})
Forest	11.38±0.5	9.98±0.33	7.42±0.4	3575±15	6420±18	9995±28
Grassland	6.91±0.36	8.0±0.3	10.16±0.43	3431±11	3725±13	7156±22
Cropland	5.43±0.32	6.69±0.27	11.47±0.48	1130±9	3365±11	4490±16

soil moisture during rainy season and lowest during summer season. Differences due to site and season were significant ($P<0.05$).

Above- and Below-Ground Biomass

The mean annual values of above- and below (up to 10 cm depth) ground biomasses for different ecosystems are given in Table 2. Above (litter) and below ground (live + dead roots) biomass in cropland to forest ecosystems ranged from 1130 to 3575 kg ha^{-1}, respectively and 3365 to 6420 kg ha^{-1}. Thus, the total biomass in cropland to forest ecosystems ranged from 4490-9995 kg ha^{-1}. The conversion of forest into grassland and cropland caused a decline of 4% and 68% in above-ground, 42% and 46% in below-ground and 28.4 and 55% in total biomass, respectively.

Inorganic Nutrient Status

The seasonal data of nitrate-N and ammonia-N show the maximum values during summer season and minimum during the rainy season. The mean annual values of nitrate-N in different land uses ranged between 6.69 to 9.98 μg g^{-1}. μg g^{-1}, respectively (Table 2). Nitrate-N was found to be higher in forest ecosystem followed by grassland and cropland ecosystems. On the other hand, ammonia-N was higher in cropland followed by grassland and forest ecosystems, ranging from forest to cropland from 7.42-11.47 μg g^{-1}, respectively (Table 5). There was a decline of 19.8% and 33% in nitrate-N concentrations after the conversion of forest into grassland and cropland ecosystems, respectively. Ammonia-N was increased after the conversion of forest into grassland and cropland by 36.9% and 54.6%, respectively. Differences due to site and season were significant ($p<0.05$) but not significant due to years.

Net Nitrification and Net N-Mineralization

The mean annual values of net nitrification and N-mineralization rates for cropland to forest ecosystems ranged from 7.93 to 18.77 μg g^{-1}mo^{-1} and 11.74-21.86 μg g^{-1}mo^{-1}, respectively (Table 3), with a higher value in forest ecosystem followed by grassland and cropland ecosystems. On a seasonal basis, net nitrification values were higher in rainy season followed by winter and summer seasons. Differences due to site and season were significant ($P<0.05$).

Table 3. Mean annual variations of nitrification, N-mineralization (μg g^{-1} mo^{-1}) and MBN (μg g^{-1}) in different land uses.

Parameters	Forest	Grassland	Cropland
Net nitrification	18.77±1.6	10.74±1.0	7.93±1.1
Net N-mineralization	21.86±2.0	14.7±1.2	11.74±1.4
MBN	50.05±3.5	35.74±2.2	29.15±2.4

The conversion of forest into grassland and cropland resulted into a decline in mean annual values of net nitrification by 42.8% and 57.75% and net N-mineralization by 32.7%

and 37.45%, respectively. Net nitrification and net N-mineralization values show a significant difference (P<0.05) due to sites as well as the seasons but not significant due to years.

Soil Microbial N

Mean annual values of MBN ranged between 29.15-50.05 $\mu g\ g^{-1}$ from cropland to forest sites; the values being highest in forest ecosystem followed by grassland and cropland ecosystems with a similar seasonal trend as shown by inorganic nutrients. The MBN values were higher during summer followed by winter and rainy seasons (Table 3).

In forest ecosystem, the values are highest followed by grassland and cropland. The conversion of forest into grassland and cropland resulted into a decline in mean annual values of microbial-N by 28.6% and 41.4%, respectively. The microbial-N values show a significant difference (P<0.05) due to sites as well as the seasons but not significant due to years.

Statistical Analysis

Mean annual MBN was positively correlated with inorganic nitrogen concentration ($r^2 = 0.99, P<0.05$), belowground biomass ($r^2 = 0.83, P<0.05$), net nitrification ($r^2 = 0.99, P<0.05$) and net N-mineralization ($r^2 = 0.90, P<0.05$). Microbial biomass is negatively related to soil moisture ($r^2 = 0.84$, P<0.1) while net nitrification ($r^2 = 0.82, P<0.05$) and net N-mineralization rates are positively correlated ($r^2 = 0.86, P<0.01$) to it. A positive relation of total N with microbial N ($r^2 = 0.98, P<0.05$), net nitrification ($r^2 = 0.97, P<0.05$) and net N-mineralization ($r^2 = 0.98, P<0.05$) has been found. In the present study, a strong correlation of organic carbon content with microbial N ($r^2 = 0.88, P<0.05$), net nitrification ($r^2 = 0.94, P<0.05$) and net N-mineralization ($r^2 = 0.88, P<0.05$) has been found.

Discussion

Inorganic Nutrient Status

The seasonal pattern for nitrate-N and ammonium-N was found maximum in summer and minimum in the rainy season in all sites. The decrease in inorganic N during the rainy season is mostly due to the strong demand of these nutrients by the vigorously growing higher plants, during the period (Raghubanshi et al., 1990). However, leaching and denitrification may also be responsible for this decline. A decline of nitrate was also recorded in flooded soil (Alaoui-Sosse et al., 2005). Conversely, an increase in summer season is associated with a decreased demand by the plants. According to Sanchez (1976), in tropical soils the previously present or recently formed nitrate in the sub-soil may also move up and accumulate in the topsoil. Birch (1958) and Singh et al. (1991) opined that greater amount of ammonium N during the dry period may be partially due to the release of free ammonium and amino acids due to soil drying.

In the present study, mean annual nitrate-N decreased by 19.81% and 33%, respectively due to conversion of forest into grassland and cropland. Interestingly, the mean annual ammonium-N was increased by 36.92% and 54.58%, respectively in grassland and cropland ecosystems as compared to that in forest ecosystem (Table 2). The decrease in nitrate-N in different land uses and increases in ammonium N in the grassland and agricultural soils may be due to higher percentage of net ammonification rates in grassland and cropland as compared to forest. This may be one of the conserving mechanisms of nitrogen in dry tropical ecosystem to check easily mobile nitrate nitrogen loss from the open derived ecosystems. Reduced plant cover may also reduce nutrient uptake, thereby causing accumulation of inorganic N in the soil.

Net Nitrification and Net N-Mineralization

The seasonal patterns of net nitrification and net N-mineralization were similar; both being highest in rainy season and lowest in the summer season. In dry tropical forest ecosystems, nitrification and N-mineralization are moisture limited. During dry periods plant uptake of nutrients is greatly reduced and the N-mineralization and nitrification are either immobilized in microbial biomass or accumulate in the soil as inorganic N (Singh and Singh, 2006). It was reported (Sorenson, 1974) that N-mineralization of dry soil increases after rewetting. In arid conditions, the high stock of nutrients is rapidly released when rainfall occurs. Consequently, the net N-mineralization enhanced significantly in all the sites of dry tropical soils during rainy season due to high soil moisture, which is limiting in the tropical deciduous forest. Maly, et al. (2002) also support that under more intense fluctuation in soil moisture content microbial activity stimulated the mineralization of organic N. Hence, a strong seasonal variation of nitrification and net N-mineralization can be seen in the dry tropical deciduous forest of the present study. Mean annual nitrification and N-mineralization rates in forest, grassland and cropland are 700, 470 and 330 kg $ha^{-1}yr^{-1}$, and 816, 644 and 450 kg ha^{-1} yr^{-1}, respectively.

The decline in mean annual net nitrification and net N-mineralization rates in grassland and cropland by 43 and 54%, and 33 and 46%, respectively (Table 3) is equivalent to a loss of inorganic nitrogen by 230, 370 and 517 kg ha^{-1} yr^{-1}, and 172, 366 and 523 kg ha^{-1} yr^{-1}, respectively. The reason for this nitrogen loss may be the lower plant biomass and loss of organic matter content from the soil of these ecosystems. The high N mineralization rates found in dry-forest soils are in agreement with other studies that lowland tropical forests are characterized by high rates of N mineralization (Vitousek and Matson, 1988).

Cultivation of native forest and grassland results in marked changes in the soil organic matter and available nutrients. Whenever the forests are cleared, the thin layer of humus with nutrients readily washes away. Haas et al (1957) compared organic C and total nitrogen concentration in cropland soils with that of virgin grassland in Great Plains in North Dakota and reported 41 and 31% loss in C and N, respectively. Lobe et al (2001) and Mrabet (2002) have also reported reduction in organic matter and total N levels by soil cultivation. Recent studies reveal that the N losses from undisturbed forest are much lower and mainly in the form of dissolved organic compounds (Perakis and Hedin, 2002). Srivastava and Singh (1989) reported 30 kg ha^{-1} yr^{-1} less contribution of microbial N in comparison to forest in annual flux of nitrogen in the cropland of tropical soil. Consequently, the net N mineralization has declined in other land uses.

Soil Microbial N

A marked seasonal variation was observed in soil microbial biomass N similar to those of inorganic N. MBN was highest in summer season and lowest in rainy season. Singh and Singh (2006) observed a reciprocal relationship between the plant growth rate and microbial biomass in Indian tropical ecosystems. Microbial biomass of dry tropical soil is pre-adapted to moisture stress and accumulates intracellular solute under conditions of low water potential (Sparling et al., 1987). Sanchez (1976) proposed that the microbial activity continues during the summer season, because the microbial activity is less sensitive to soil water potential than is water uptake by plants and the increase in soil water potential in dry tropics in the rainy season may induce plasmoptysis. In wet period, the microbivore population is greatly enhanced. Feeding by this microbivore population may cause reduced microbial biomass and increased microbial turnover in the wet period (Singh et al., 1991).

Table 4. Mean annual soil microbial biomass N and flux of N through microbial biomass.

Ecosystem	MBN (μg g^{-1})	Kg ha^{-1}	Annual flux (kg/ha/yr)
Forest	50.05	51.05	40.84
Grassland	35.74	42.9	34.32
Cropland	29.15	30.6	24.48

Assuming a biomass turnover time of 1.25 year, as suggested [12] for warmer countries.

The study reveals that the conversion of forest into grassland and cropland affects MBN by reducing its values by 29 and 41%, respectively (Table 3). Cultivation of forests not only diminishes soil carbon (C) within a few years of initial conversion (Murty et al., 2002), but also substantially lowers mineralizable nitrogen (N) (Raghubanshi et al., 1990). Srivastava and Singh (1989) have also shown in their study a decline in microbial biomass due to cultivation in tropical deciduous forest. They further suggested that in natural ecosystems most of the organic matter produced by vegetation is returned to the soil, but after cultivation much of it is removed for human and animal consumption and relatively very less amount is returned back to the soil. In cropland, the microbial biomass development is limited by C availability. The decline in soil organic matter and microbial nutrients may be due to lower inputs of organic matter into soil. Consequently, the microbial biomass is decreased along organic matter gradient. In the present study, microbial biomass N reflected in total nitrogen ranged from 2.12-4.1%. Different land uses caused increase in biomass N contribution with maximum values shown by cropland (2.9%) followed by grassland (2.75%). The minimum value (2.12%) reflected in biomass N was exhibited by forest ecosystems. By assuming a turnover time of 1.25 years for microbial nutrients in the present study, the flux of N was calculated. The calculated flux of N through the microbial biomass in forest (40.84 kg ha^{-1}) in upper 10 cm soil layer was nearly 1.66 times as large as N flux in the cultivated soil (24.48 kg ha^{-1}) (Table 4). There are other studies, which also indicate a significant amount of nutrient flux through microbial biomass. Shioda (1984) estimated that N released from the decomposed microbial biomass in paddy soils was about 50 kg $ha^{-1}yr^{-1}$. Similarly, Marumoto et al. (1982) showed 40 kg of N ha^{-1} flush mineralization in paddy soils. This indicates that different land uses caused increased immobilization of nitrogen by microbial biomass. Thus, soil microbial activity and levels of soil microbial biomass play an important role in nutrient cycling. Soil microbial biomass is a critical factor in the recovery of drastically disturbed sites as it aids in the re-establishment of nutrient cycling (Schnürer et al., 1985). However, the reduction in total microbial biomass N in grassland and cropland is due to reduced leaf litter inputs, fine root production and plant biomass accumulation. Singh and Singh (2006) postulated that there is a relationship between the soil N and the quantity of N deposited by litter fall indicating that soil is indeed a function of N returned by the vegetation to the soil. Evidently, the species, which allocate a greater proportion of biomass to foliage and produce N-rich litter play a major role in accelerating the fertility of soil.

In this study, annual mean MBN was positively related to the annual mean net N-mineralization ($P<0.05$) across the sites according to the regression equations:

$$\text{Microbial N } (\mu g\ g^{-1}) = 6.5537 + 0.554 \text{ Net N-mineralization } (\mu g\ g^{-1}\ mo^{-1})$$

These regression equations explained about 95% variability in the microbial N due to variability in net N-mineralization. This may be due to higher immobilization of nutrients by the microbial biomass during the dry periods, which contributes higher annual flux of nutrients during the rainy season.

Similarly, annual mean microbial N was positively related to the annual mean belowground biomass and total plant biomass ($P<0.05$) according to the regression equations:

$$\text{Microbial N } (\mu g\ g^{-1}) = 7.741 + 0.006 \text{ belowground biomass } (g\ m^{-2})$$

and

$$\text{Microbial N } (\mu g\ g^{-1}) = 10.855 + 0.003 \text{ total biomass } (g\ m^{-2})$$

These regression equations explained about 84% and 68% variability in the microbial N due to variability in belowground biomass and total biomass, respectively. Root biomass and aboveground biomass are considered to be the main source of soil organic matter and are highly correlated with microbial biomass (Singh and Singh, 2006). Root biomass provides a direct input of organic matter and mineral nutrients through exudation and upon mortality. Srivastava and Singh (1991) also found a positive relation between microbial biomass N and belowground biomass, and microbial biomass N and total plant biomass in Indian dry tropical soils.

Conclusion

There is direct impact of conversion of a natural dry tropical forest ecosystem into different land uses, such as grassland or cropland, as evident by the changes in plant available inorganic nitrogen concentration, nitrification, N-mineralization and MBN. We opine that these parameters are the critical soil functions indicating the influence of conversion of a native dry tropical forest ecosystem into different land uses. These soil functions play a very crucial role in re-establishment of nutrient cycling and also soil redevelopment. Particularly, the loss of nitrogen can be taken as function indices of ecosystem stability.

In dry tropical forests, soil moisture and microbial biomass regulate the nitrogen availability, nitrification and nitrogen mineralization rates. The conversion of forest into different land use ecosystems cause a loss of soil nitrogen and soil microbial biomass and N transformation rates. So, the loss must be checked by adopting nutrient, particularly N, management strategies through seeding of leguminous plant species.

Acknowledgements

Funding support by the Ministry of Environment & Forests, New Delhi (India) and Council of Scientific & Industrial Research (CSIR), New Delhi are acknowledged.

References

Alaoui-Sosse, B., Gerard, B., Binet, P, Toussaint, M. and Badot, P. (2005). Influence of flooding on growth, nitorgen availability in soil and nitrate reduction of young oak seedlings (*Quercus robur*). *Annals of Forest Science* **62**, 593-600.

Anderson, J.P.E. and Domsch, K.H. (1980). Quantities of plant nutrients in the microbial biomass of selected soils. *Soil Science* **130**, 211-216.

Azam, F.A., Farooq, S. and Lodhi, A. (2003). Microbial biomass in agricultural soils-determination, synthesis, dynamics and role in plant nutrition. Pakistan Journal of *Biological Sciences* **6(7),** 629-639.

Birch, H.F. (1958). The effect of soil drying on decomposition and nitrogen availability. *Plant and Soil* **10**, 9-31.

Brookes, P.C., Landman, A., Pruden, G. and Jenkinson, D.S. (1985). Chloroform fumigation and release of soil N: A rapid direct extraction method to measure microbial biomass N in soil. *Soil Biology & Biochemistry* **17**, 837-842.

Eno, C.F. (1960). Nitrate production in the field by incubating the soil in polyethylene bags. *Soil Science Society of America Proceedings* **24**, 277-279.

FAO, Highland Reclamation Study, Ethiopia, FAO, Rome (1986).

Glaser, B., Johannes, L. Führböter, M., Solomon, D. and Zech, W. (2001). Carbon and nitrogen mineralization in cultivated and natural savanna soil of Northern Tanzania. *Biology and Fertility of Soils* **33**, 301-309.

Grünzweig, J.M., Sparrow, S.D. and Chapin, F.S. (2003). Impact of forest conversion to agriculture on carbon and nitrogen mineralization in subarctic Alaska. *Biogeochemistry* **64(2)**, 271-296.

Haas, H.J., Evans, C.E. and Miles, E.F. (1957). Nitrogen and carbon changes in great plain soils as influenced by cropping and soil treatments. *Technical Bulletin* No. **1164**, *USDA*, U.S. Government Printing Office, Washington, D.C.

Hurni, H. (1993), Land degradation, famine and land resource scenarios in Ethiopia. In: D. Pimentel, Editor, *World Soil Erosion and Conservation*, Press Syndicate, University of Cambridge, Cambridge, pp. 27–61.

Idol, Travis W., Pope Phillip, E. and Ponder, Felix, Jr. (2002). Changes in Microbial Nitrogen Across a 100-Year Chronosequence of Upland Hardwood Forests. *Soil Science Society of America Journal* **66**, 1662-1668.

Jackson, M.L. *Soil Chemical Analysis* (1958). Prentice Hall, Englewood Cliffs.

Knoepp, J.D.and Swank, W.T. (2002). Using soil temperature and moisture to predict forest soil nitrogen mineralization. *Biology and Fertility of Soils* **36**, 177-182.

Lobe, I. Amelung, W. Preez C C Du (2001). Losses of carbon and nitrogen with prolonged arable crppping from sandy soils of the South African Highveld. *Eur. J. Soil Science* **52**, 93-101.

Leuschner, C. Meier, I.C. and Hertel D. (2006). On the niche breadth of Fagus sylvatica: soil nutrient status in 50 Central European beech stands brad range of bedrock types. *Annals of Forest Science* **63**, 355-368

Lodge, D.J., McDowell, W.H. and McSwiney, C.P. (1994). The importance of nutrient pulses in tropical forest, *Trends in Ecol. E.* **9**, 384-387.

Maly, S., Tarapatka, B. and Krtkova, M. (2002). Seasonal variability in soil N mineralization and nitrification as influenced by N-fertilization. *Rostlinna Výobra* **9**, 389-396.

Marumoto, T., Anderson, J.P.E. and Domsch, K.H. (1982). Decomposition of ^{14}C and ^{15}N labeled microbial cells in soil. *Soil Biology & Biochemistry* **14**, 461-467.

Mehta, N.C., Leg, J.O., Goring, C.A.I. and Black, C.A. (1954). Determination of Organic Phosphorus in Soil. I. Extraction method. Proceedings of the *Soil Science Society of America* **18**, 443-449.

Milkha, M.M. and Rice, C.W. (2004). Nitrogen mineralization from a soil with differing N supply capacities. Agronomy Abstracts. Presented at the 2004 Asa-Cssa-Sssa *International Annual Meeting,* Oct. 31-Nov. 4 Seattle. Wa.

Mrabet, R. (2002). Stratification of soil aggregation and organic matter under conservation tillage systems in Africa. *Soil Tillage Research* **66**, 119-128.

Murty, D., M. U. F. Kirschbaum, R. E. McMurtrie, and H. McGilvray (2002). Does conversion of forest to agricultural land change soil carbon and nitrogen? A review of the literature. *Global Change Biology* **8**, 105–123.

Neill, C., Piccolo, M.C., Cerri, C.C., Steudler, P.A., Mellilo, J.M., Brito, M. (1997). Net nitrogen mineralization and net nitrification rates in soils following deforestation for pasture across the south-western Brazilian Amazon Basin landscape. *Oecologia* **110**, 243-252.

Perakis, S.S. and Hedin, L.O. (2002). Nitrogen loss from unpolluted South American forest mainly via dissolved organic compounds. *Nature* **415**, 416–419.

Piper, C.S. (1994). *Soil and Plant Analysis*, Inter Science, Adelaide.

Raghubanshi, A.S., Srivastava, S.C., Singh, R.S. and Singh, J.S. (1990). Nutrient release in leaf litter. *Nature* **346,** 227.

Reiners, W.A., Bouwman, A.F., Parsons, W.F.J., Keller, M. (1994). Tropical rain forest conversion to pasture: changes in vegetation and soil properties. *Ecological Applications* **4**, 363-377.

Sanchez, P.A.(1976). *Properties and management of soils in the tropics*, John Wiley, New York.

Schnürer, J. Clarholm, M. and Rosswall, T. (1985). Microbial biomass and activity in agricultural soils with different organic matter contents, *Soil Biology & Biochemistry* **17**, 611-618.

Shioda, Y. (1984). Microbial biomass in manured paddy field. *In Biomass in soil quality and turnover of soil organism*, Hakuyusha, Tokyo. 141-167.

Singh, A.N and Singh, J.S. (2006). Experiments on ecological restoration of coalmine spoil using native trees in a dry tropical environment, India: a synthesis, *New Forests*, 25-39.

Singh, R.S., Raghubanshi, A.S. and Singh, J.S. (1991). Nitrogen mineralization in dry tropical savanna: Effects of burning and grazing. *Soil Biology & Biochemistry* **23**(3), 269-273.

SPSS (1997). *SPSS Base 7.5 Application Guide*. SPSS Inc., Chicago.

Sorenson, L.J. (1974). Rate of decomposition of organic matter in soil as influenced by repeated air drying-rewetting and repeated additions of organic matter. *Soil Biology and Biochemistry* **6**, 287-292.

Sparling, G.P., Milne, J.G. and Vincent, K.W. (1987). Effect of moisture regime on the microbial contribution to Olsen P values. *New Zealand Journal of Agricultural Research* **30**, 79-84.

Srivastava, S.S. and Singh, A.K. (1998). Simlipal Bioshere Reserve. In *Biosphere reserves and management in India.* Eds by R.K Maikhuri, K.S. Rao, and R.K Rai. Himavikas Occassional Publication No. 12, G.B. Pant Institute of Himalayan Environment and Development. India. 65-70.

Srivastava, S.C. and Singh, J.S. (1989). Effect of cultivation on microbial carbon and nitrogen in dry tropical forest soil. *Biology and Fertility of Soils* **8**, 343-348.

Srivastava, S.C. and Singh, J.S. (1991). Microbial C, N and P in dry tropical forest soil: effects of alternate land uses and nutrient flux. *Soil Biology and Biochemistry* **23**(2), 117-124.

Tripathi, N. and Singh, R.S. (2007). Cultivation impacts nitrogen transformation in Indian forest ecosystems. *Nutrient Cycling in Agroecosystems* **77**, 233-243.

Verchot, L.V., Davidson, E.A., Cattânio, J.H., Ackerman, I.L., Erickson, H.E. and Keller, M. (1999). Land-use change and biogeochemical controls of nitrogen oxide emissions from soils in eastern Amazonia. *Global Biogeochemical Cycles* **13**, 31-46.

Vitousek P.K. and Matson P.A. (1988). Nitrogen transformations in a range of tropical forest soils. *Soil Biology & Biochemistry* **20**, 361-367.

Wang, C., Xing, X. and Han, X. (2004). Advances in study of factors affecting soil N mineralization in grassland ecosystems. Ying Yong Sheng Tai Xue Bao. *Soil Biology & Biochemistry* **15** (11). 2184-8.

Wetzel, R.G. and Likens, G.E. (1979). *Limnological Analyses*, W.B. Saunders Co. Philadelphia. 357.

In: Ecosystem Ecology Research Trends
Editors: J. Chen and C. Guo, pp. 313-325
ISBN: 978-1-60456-183-8

Chapter 10

Temperature and Soil Moisture Effects on the Respiration of a Grassland Area

***A.F.G. Jacobs*[1], *B.G. Heusinkveld*[1] *and C.M.J. Jacobs*[2]**

[1] Meteorology and Air Quality Group, Wageningen University, P.O. Box 47, 6700 AA Wageningen, The Netherlands

[2] Alterra, Wageningen University, P.O. Box 47, 6700 AA Wageningen, The Netherlands

Abstract

In an ecosystem the respiration is affected by many biological and physico-chemical factors, of which year round the soil temperature and soil moisture are the most important ones. During six months in 2003, the net ecosystem exchange (*NEE*) of CO_2 was measured with the eddy-covariance technique along with soil temperature and soil moisture profiles. Nighttime surface respiration measurements were analyzed and compared with model calculations using an Arrhenius relation. Not the soil temperature but the surface temperature of the grass cover, fitted best the experimental data. For dry soil wetness conditions but also for very wet soil water conditions a correction function is introduced, which appears to work well. For the soil wetness correction function, the soil water content nearest to the soil atmosphere interface appeared to work best.

Introduction

Carbon dioxide (CO_2) exchange involves a flux from atmosphere to ecosystems, due to photosynthetic production, and an opposite flux, due to respiration. A possible shift in the balance of these two fluxes, caused by global climate change, can be very significant due to the large amount of carbon stored as soil organic matter.

Soils contain more than two-thirds of the total carbon on earth which means that soils form the largest reservoir of carbon in our terrestrial ecosystems. A small change in the soil respiration rate due to changes in physico-chemical or biological parameters in the soil can thus have a major impact on the atmospheric CO_2 budget and on the global ecosystem [Hunt et al., 2002; Reichstein et al., 2005]. Soil temperature and soil moisture are considered to be two main drivers of respiration. So far there is still no consensus of the effect of temperature and soil moisture on respiration.

Atmospheric CO_2 enrichment generally increases photosynthetic rates and biomass, particularly under conditions of optimal soil fertility and soil moisture condition. As the air's CO_2 content continues to rise, therefore, earth's plant species will likely respond by increasing their photosynthetic rates and biomass production, even under conditions of very low soil nitrogen availability. Thus, whether climate changes or not, an increase in CO_2 will directly influence growth of all plants. It will also cause higher decomposition rates: a so-called feed-forward cycle.

This paper focuses on the respiration of a grassland area. The respiration of such areas is the sum of plant respiration and soil respiration. Plant respiration is the sum of growth respiration and maintenance respiration. Growth respiration means the conversion of assimilatory products into structural plant material, such as cellulose, lignin, proteins, organic acids and fats. The CO_2 production rate related to this process is proportional to temperature. Maintenance respiration is needed to support plant functioning [Rabbinge et al., 1995; Campbell and Norman, 1997]. Soil respiration is mainly the result of microbial and root activities in the soil. It includes processes like root respiration, root decomposition, soil organic matter decomposition and litter decomposition. The first three processes occur in the root zone, where soil organic matter accumulates and the latter process at the surface. The rate at which CO_2 is transferred from the soil to the atmosphere is controlled by the rate of CO_2 production in the soil by root and microbial decomposition, the CO_2 concentration gradient between the soil and the atmosphere and the gas diffusivity within and above the soil. These processes are affected by many factors like soil moisture and organic matter contents, litter depth, aeration, pH, temperature, salinity, soil nutrition, soil texture and wind speed [Raich and Schlesinger, 1992]. The microbial activity, and therefore the soil respiration, can be related to soil moisture and organic matter contents, soil texture, aeration, crop type and age, pH, season, time of day, temperature, salinity, fertility level and (de)nitrification.

A lot of factors affect the microbial activity and photosynthetic rate, and both are important in the respiration process. Because in this study respiration at one specific site for duration of six months will be analyzed, some of the factors that affect microbial activity, plant respiration and photosynthetic uptake of CO_2 can be taken constant. Year round the soil organic matter content, soil texture, nitrification and denitrification, crop type, age, salinity, pH and fertility level can be considered constant for the study site. This is not the case for the soil moisture content, aeration status, soil-surface and air temperature, season and time of day. The present research will be focused on these parameters. The aim is to find a clear and relatively simple relationship between respiration and soil temperature, air temperature and soil moisture content that can be used compute annual carbon budgets of grassland areas with reasonable accuracy.

Traditionally, soil CO_2 fluxes are measured with open or closed chambers covering small patches of soil-plant system [De Jong et al., 1978]. In recent years, the eddy-covariance technique has been used to measure CO_2 fluxes above the surface [Xu and Baldocchi, 2004; Suyker et al., 2003]. An advantage of this technique is that it covers not only a restricted patch but that it averages the CO_2 flux over a larger area depending on its footprint. In the present study we used the eddy-covariance technique to estimate the respiration over grassland. The eddy-covariance technique, however, cannot distinguish between the respiration and assimilation fluxes. Since at night no assimilation can occur, we will only use the nighttime observations to derive the respiration.

The majority of previous direct observations of carbon exchange by respiration over grasslands has been executed during short-term campaigns by different techniques. For example, Norman et al. (1992) and Dugas et al. (1997) used the chamber technique over grassland. Dugas et al. (1997) and Ham and Knapp (1998) used the Bowen Ratio Energy Balance (BREB) technique. Kim et al. (1992) applied the eddy-covariance technique over various natural prairie grasslands (tall grass, mixed grass and short grass) and Hunt et al. (2002) used the same technique over a tussock grassland during a summer drought. Only a few long-term studies have been conducted over grasslands [e.g. Suyker et al, 2003; Meyers, 2001]. Although many studies were carried out, there is still no consensus on the relation between respiration and temperature and soil moisture.

About 60% of the agricultural land in the Netherlands consists of grasslands, where rotational grazing is the most common land use. These grasslands are mostly relatively wet and characterized with a water table ranging between 0.2 and 1 m. The objective of the present research is to find a relatively simple expression for the respiration process for a grassland area. To achieve our objective, data of our agro-meteorological observatory have been used in combination with additional observations within the soil and at the soil surface. The observatory is mainly surrounded by other grasslands with rotational grazing.

Experimental Set-up

Year round in 2003 experiments were made at a grassland area of the Wageningen University agro-meteorological observatory, Haarweg Station, in the centre of the Netherlands (lat. 51° 58' N, long. 5° 38' W, altitude +7 msl). The dominating plant species in this perennial grassland are *Lolium perenne* and *Poa trivialis*. The Leaf Area Index, *LAI*, of the grass cover was estimated with a Digital Plant Canopy Imager (Licor, type CI-110). During the experimental period, the grass was weekly mowed, had a height of 10 cm and had a mean *LAI* of 2.9 [Snel, 2004]. The *LAI* was an average of measurements obtained at 10 plots within the observation site. At the measurement site the mowed grass is not removed.

The soil at the site is predominantly heavy basin clay resulting from the back-swamps of the river Rhine. The accumulated soil organic matter, especially in the top soil (first 300 mm), is a storehouse of plant nutrients and is derived mainly by plant residues. It also increases the water holding capacity and the transmission of soil water, largely as a result of its influence on soil aggregation and associated pore space [Hillel, 1998]. During the experimental period the organic matter was estimated by taking 15 soil samples over the area, from 5 different spots and at 3 depths. It appeared that the overall mean organic matter is 0.067 ± 0.007 kg kg^{-1} [Snel, 2004].

A lattice tower was instrumented with an eddy-covariance system installed at a height of 4 m. This system included: a 3-Dimensional sonic anemometer (3-D Solent Res. Gill Instruments Ltd., model A1012R2), a fine wire thermocouple (home made) and an open path infrared CO_2 and H_2O gas analyzer (IRGA) (LI-COR Inc. Lincoln Ne, model LI-7500). The 3-D sonic anemometer and the IRGA were placed together, at 0.05 m apart.

In a thermometer screen an aspirated psychrometer was installed to estimate the air temperature as well as the air humidity. At 10 cm height, the grass temperature, *Ta(10cm)*, was measured with a shielded Pt-100 thermometer (homemade).The incoming short-wave radiation, R_g, was measured with an aspirated pyranometer (Kipp & Zonen, model CM11). The short-wave radiometer was used to distinct between day and night. The outgoing long-wave radiation, R_{Lo}, was measured with a pyrgeometer (Kipp & Zonen, model CG 1). The outgoing long wave radiometer was used to evaluate the surface temperature T_{surf}, ($T_{surf} = \sqrt[4]{\frac{R_{lo}}{\varepsilon\sigma}}$).

Soil temperatures, T_s, were measured by Pt-100 element at depths 0.01, 0.02, 0.05, 0.075 and 0.10 m. These soil temperatures along with the surface temperature, T_{surf}, and grass temperature $T_a(10cm)$, were used to estimate the soil respiration term. At the same soil depths the soil moisture content was measured with a TDR (Time Domain Reflectrometry) system. Details about this technique and analysis can be found elsewhere [De Jeu and Owe, 2004]. Soil moisture measurements were made only during six months in 2003 (April – September).

The slow response meteorological instruments were sampled at 0.25 Hz. At 30 minutes intervals, data were averaged and stored in data loggers for later processing. The fast response sonic anemometer, the IRGA system and the fine wire thermocouple were sampled at 20.8 Hz.

The raw data of the eddy-covariance system were stored on a PC and processed later, using a first order recursive digital filter with a time constant of 200 seconds [McMillen, 1988]. Here a moving average was subtracted from every sample to get the fluctuating value of all the measured components. A software program [Van den Hurk, 1996] performed the necessary corrections, including: co-ordinate rotation [McMillen, 1986], Webb corrections [Webb et al., 1980] and frequency response corrections [Moore, 1986], required for calculation of the half-hour averaged flux densities More details about the experiments and area can be found elsewhere [Jacobs et al., 2003b].

Methodology

Soil respiration contributes for a large part to the total exchange of CO_2 between atmosphere and ecosystems. This part of the total respiration can be divided into litter decomposition, soil organic matter decomposition and root respiration.

Attempts have been made to describe these processes as chemical reactions in biological systems. A well-known approach is to use an Arrhenius equation. Arrhenius (1889) described the temperature dependence of a reaction rate for gas reactions at constant pressure:

$$k = Ae^{\frac{-E_a}{\Re T}} \tag{1}$$

in which k is reaction rate (mol m^{-3} s^{-1}), A is frequency factor (mol m^{-3} s^{-1}), E_a is activation energy (J mol^{-1}), $\Re$ is the universal gas constant (8.314 J mol^{-1}K^{-1}) and T is the absolute temperature (K). Eq. (1) can be rewritten for the reaction rate and relative to the rate at some standard temperature, in this case 283.15 K, yielding [Lloyd and Taylor, 1994]:

$$R = R_{10} e^{\left(\frac{E_a}{283.15\Re}\right)\left(1-\frac{283.15}{T}\right)} \tag{2}$$

where R_{10} is a constant, representing the reaction rate at a reference temperature of 10 °C. This constant is a data-set dependent variable. For our grassland area it appeared that the numerical value for this constant under no water and air limited conditions is R_{10} = 0.270 mg m^{-2} s^{-1} (see later). This numerical value is used in the present study.

Carbon dioxide produced by plant respiration and litter decomposition occurs at the earth's surface. Carbon dioxide, released from these sources will directly be dispersed into the turbulent atmosphere, and consequently can directly be estimated. Carbon dioxide produced by decomposition in the soil and root respiration, does not contribute to the atmospheric flux directly, since the gas needs to diffuse through the soil to the surface. This is a slow process, which can be delayed by small pores and blocking of water. To first order, the production of CO_2 by the litter and plants may be expected to follow the Arrhenius equation. Dispersion of CO_2 in the air by turbulence, generated by wind and buoyancy will certainly not follow an Arrhenius behavior and will cause scatter of the measured CO_2 fluxes relative to the Arrhenius function.

The atmospheric flux resulting from the CO_2 production processes can be measured directly by the eddy-covariance method. An exception occurs under nighttime conditions if the atmosphere is extremely stable. Then atmospheric turbulence is suppressed and respired CO_2 will accumulate below the measurement level of the eddy-covariance system [Baldocchi et al., 1996; Jacobs et al., 2003a]. To avoid very stable conditions, only observation have been analyzed where the friction velocity, u_*, is higher than a certain critical value, u_{c*}. Here we have chosen the critical level u_{c*} = 0.1 m s^{-1} [Hunt et al., 2002; Pattey et al., 2002; Jacobs et al., 2003b].

Results

First a period is analyzed in which the soil moisture content is more or less at field capacity, thus no serious moisture reduction on the respiration can be expected. May 2003, appeared to be the best acceptable period for this analysis. Precipitation, freatic surface and some soil moisture content data for this period are depicted in Figure 1.

To find out what temperature in the Arrhenius relation (Eq. (2)) gives the best model results, the experimental results have been plotted in Figure 2 versus the model results calculated with different temperatures. The data in Figure 2 are mean nighttime values in order to reduce the scatter. Moreover in Figure 2, the unbiased linear regressions and the standard errors of estimate (see) have been plotted.

Figure 2 shows that the model calculations with the surface temperature, T_{surf}, give the best fit and the deeper the soil temperature is taken in the Arrhenius' equation, the larger the deviation from the 1:1 line and the larger the standard error of estimate. It also was found (not presented here) that the 10 cm grass temperature, $T_a(10cm)$, agreed well with the surface temperature, T_{surf}, and consequently did the model calculations with this temperature. The surface temperature, T_{surf}, fits best suggests that in our grassland area the CO_2 efflux from the earth's surface must mainly originate from plant respiration and from litter decomposition. Also it can be shown that the diffusivity of CO_2 in our heavy basin clay is small, which limits the CO_2 efflux from the deeper soil and roots.

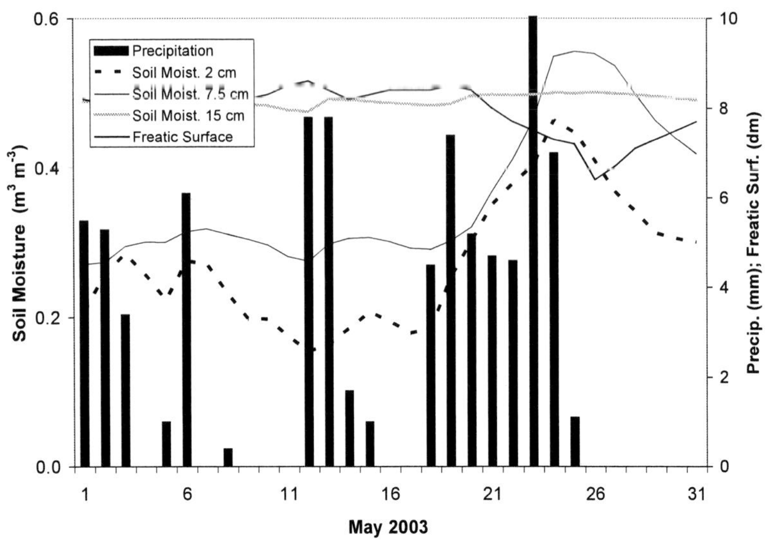

Figure 1. Precipitation and moisture conditions of the upper soil during the reference month of May.

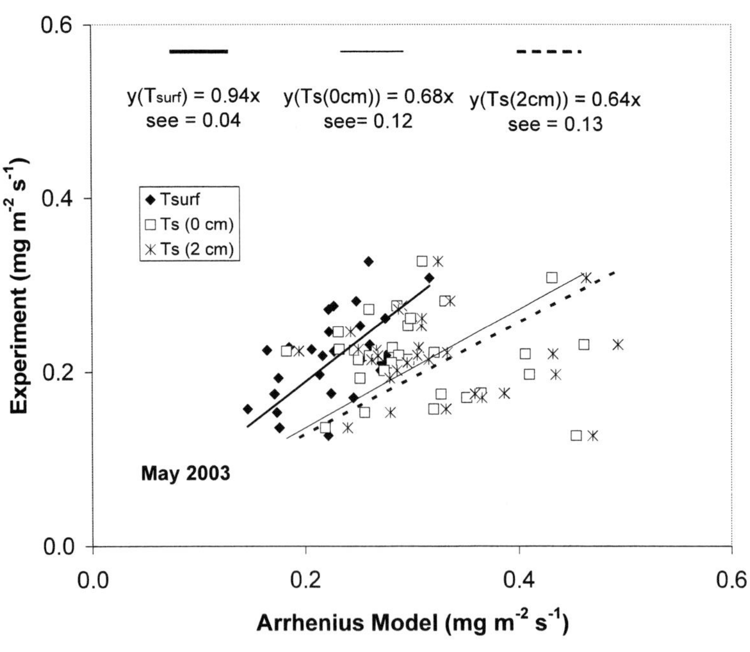

Figure 2. The mean night-time measured respirations versus the mean night-time calculations according to Arrhenius' Eq. (2) for surface temperature T_{surf} and soil temperatures at a depth of 0 and 2 cm.

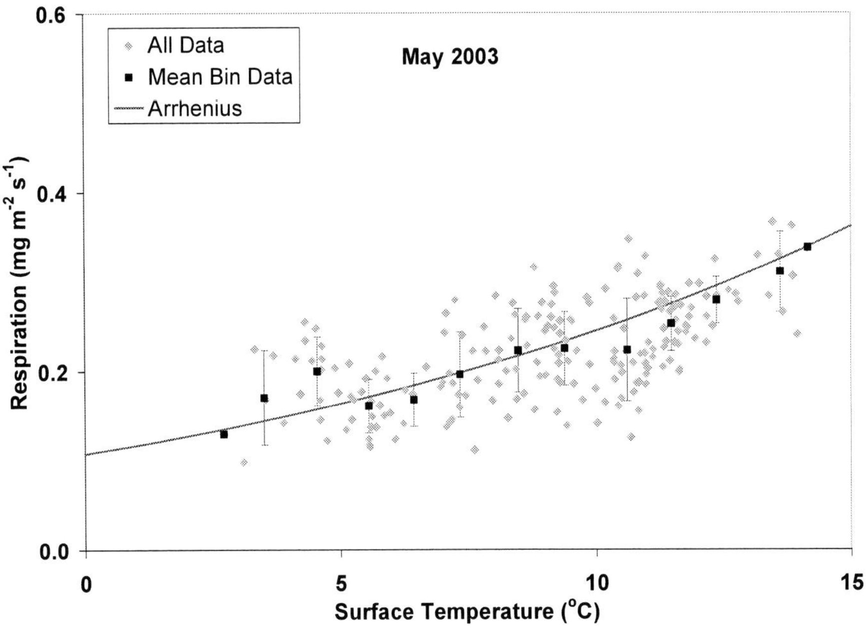

Figure 3. The 30-min averaged surface respirations as a function of the surface temperature, T_{surf}, along with model calculations according to Arrhenius' Eq. (2).

In Figure 3, all nighttime observed respiration data for May 2003, averaged over 30 minutes, have been plotted versus the surface temperature, T_{surf}, along with the Arrhenius model results according Eq. (2). The observed data in Figure 3 are selected under the condition that $u_* > u_{c*} = 0.1$ m s^{-1} since the eddy-covariances become ill-defined if $u_* < u_{c*}$ since then the atmospheric condition is non-stationary and non-homogeneous. Also rain periods are skipped since rain events can seriously affect the reliability of the measurement instruments. Due to the nature of nighttime eddy-covariance measurements and varying contribution of the plant respiration to the total respiration [Nieveen et al., 2004], there is a considerable amount of scatter in the nighttime measurement data. To deal with this scatter, the data were averaged in 1 °C bins and plotted in Figure 3 as well along with its standard deviation. From the latter results the R_{10} was calculated to be 2.70 ± 0.05 mg m^{-2} s^{-1}.

In order to study the soil moisture dependency of the respiration, all relative mean nighttime CO_2 deviations, $\Delta R/R_A$, have been plotted in Figure 4 versus the soil moisture content at 1 cm depth, where R_A is the mean nighttime calculated respiration according to Arrhenius (Eq. 2) and $\Delta R = (R_A - R_O)$ where R_O is the observed mean nighttime respiration. To deal with the scatter in Figure 4, the data were averaged in soil moisture bins with about equal number of data points and plotted in Figure 4 as well along with its standard deviation. A simple rectangular hyperbolic function, $f1(\theta)$,

$$f1(\theta) = \frac{a_1}{\theta - b_1} \quad (3)$$

is fitted through the data points, where $f1(\theta)$ has two data-set dependent constants, a_l (= 0.007) and b_l (= 0.008) and is plotted in the upper frame (A) of Figure 4 as well. The same calculations were executed with deeper soil moisture contents, however, the above choice gave the best fit.

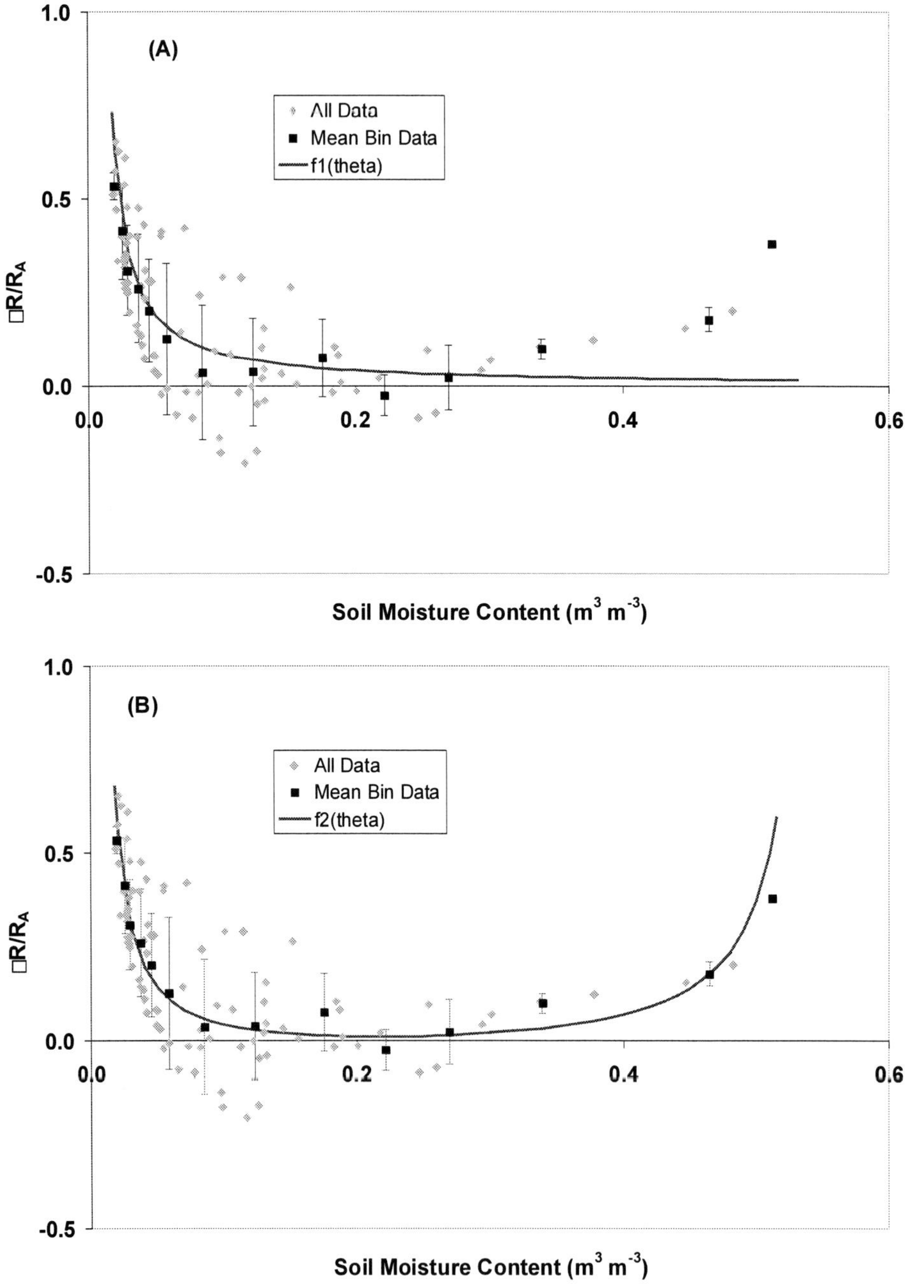

Figure 4. Relative respiration deviation for mean night-time CO_2 fluxes, along with a rectangular hyperbolic regression, versus the soil moisture of the uppermost soil layer.

It is interesting to note that the soil moisture correction function in the Arrhenius relation, *(1 – f1(θ))* agrees exactly with the soil moisture correction function found earlier by Hunt et al. (2002) defined by:

$$\frac{R_{10}}{R_{10\,max}} = \frac{\theta - a}{(\theta - a) - (\theta_{50} - \mathbf{a})} = 1 - \frac{\theta_{50x} - a}{\theta - (2a - \theta_{50})} \tag{4}$$

where $a = (a_1 + b_1)$ is the x-axis intercept of Eq. (4) and θ_{50} $(=2a_1 + b_1)$ is θ when R_{10}/R_{10max} = 0.5. It must be noted that in Hunt et al.'s original paper the bold constant a in Eq. (4) was missing but mathematically must be inserted (personal communication John Hunt).

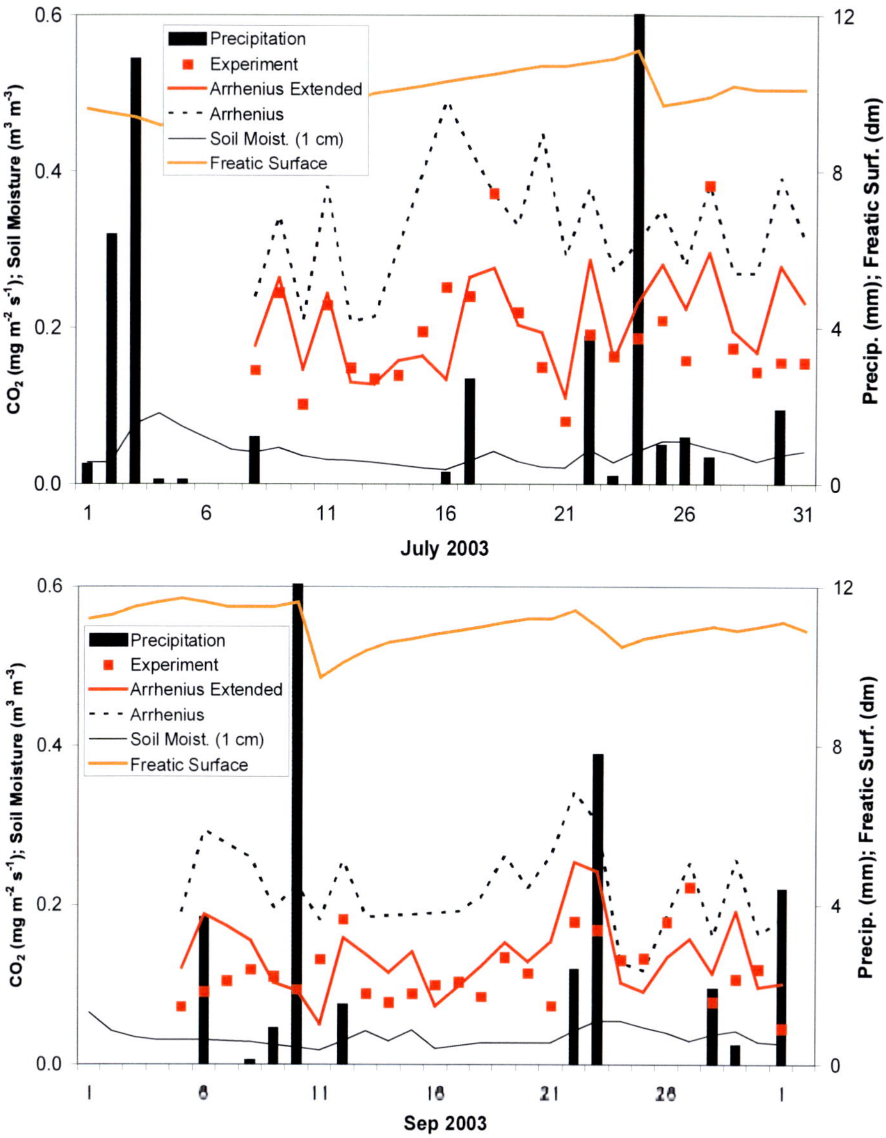

Figure 5. Mean measured night-time respiration fluxes along with model calculations, weather, and soil variables for two months in 2003.

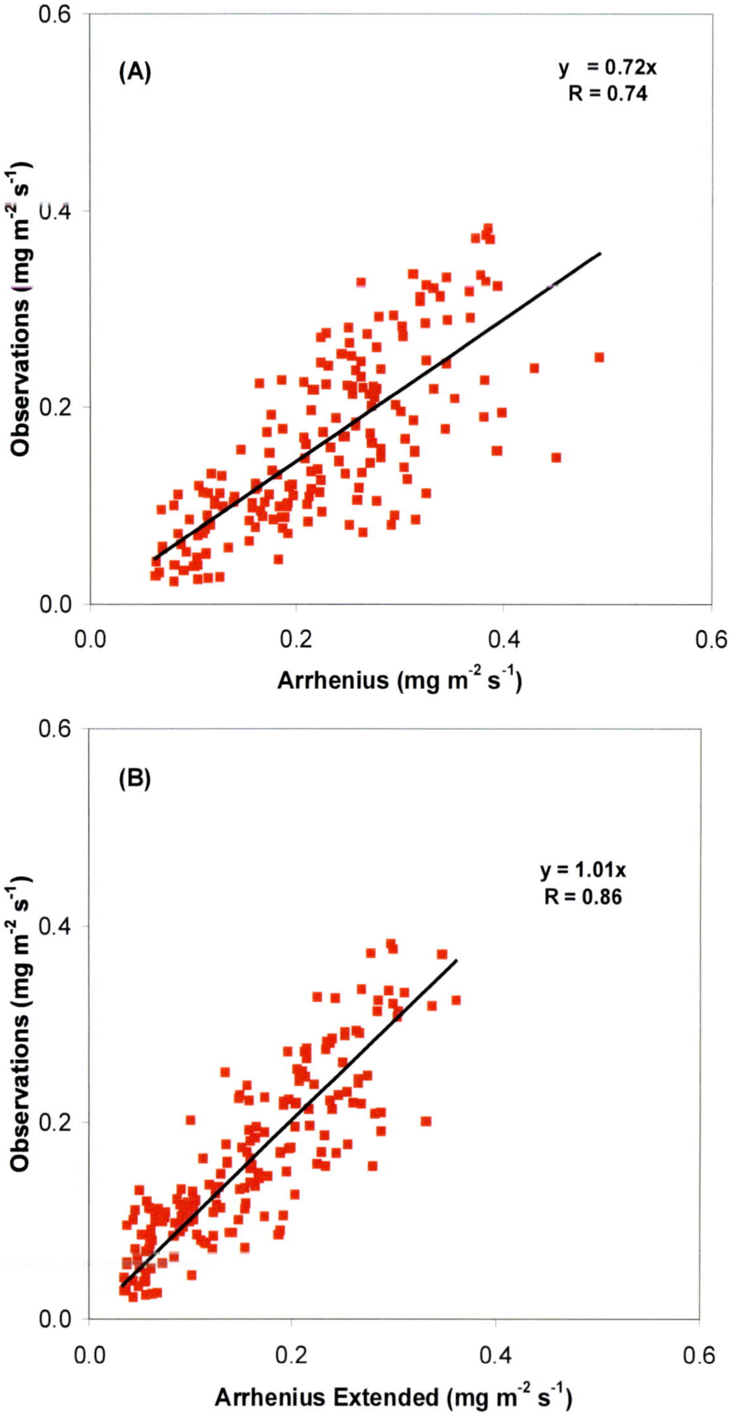

Figure 6. Observed mean night-time respiration fluxes versus the uncorrected Arrhenius equation (panel A), and versus the extended Arrhenius equation (panel B).

As mentioned earlier, the respiration is reduced under dry soil conditions but the respiration is also reduced if hardly any air is available in the soil pores. This characteristic is clearly shown in Figure 4 for soil moisture contents, $\theta > 0.35$. In order to correct the Arrhenius equation also for wet conditions an extended correction function, $f2(\theta)$ is fitted through the data points with:

$$f2(\theta) = \frac{a_1}{\theta - b_1} - \frac{a_2}{\theta - b_2} - c_1 \tag{5}$$

where a_1 and b_1 are the same constants as mentioned before and a_2 (= 0.020), b_2 (=0.545) and c_1 (=0.089). Note, that the constant b_2 agrees with the saturated soil moisture content. The function $f2(\theta)$ is plotted in the bottom frame (B) of Figure 4. Function $f2(\theta)$ is applied in the further analysis with:

$$R_E = (1 - f2(\theta))R_{10} e^{\left(\frac{E_a}{283.15\Re}\right)\left(1 - \frac{283.15}{T}\right)} \tag{6}$$

where R_E is the extended Arrhenius relation and θ is the soil moisture content of the upper soil.

For two months the courses of the daily mean measured nighttime respiration fluxes have been depicted in Figure 5, along with Arrhenius' model calculations and some important weather and soil variables. In Figure 5, the original Arrhenius' equation (Eq. 2) as well as the extended Arrhenius' equation (Eq. 6) have been depicted in order to see the effect of the suggested moisture correction. From Figure 5 we clearly can observe the precipitation effect on the soil moisture content and the consequent effect of this on the surface respiration. Also it can be noticed that for mean nighttime values the soil moisture correction on the extended Arrhenius' equation works reasonably well.

All observed mean nighttime fluxes, from April until October, have been plotted in the scatter diagram of Figure 6 versus the original (upper frame A) as well as the extended Arrhenius' equation (lower frame B), along with the unbiased linear regressions. Figure 6 shows that the original Arrhenius' equation deviates about 30% from the measurements while the extended Arrhenius' relation deviates only 1 % which is very close to the 1:1 line. Moreover, there is a considerable improvement in the correlation coefficient (original: R = 0.74 and the extended: R = 0.86).

Conclusion

The respiration of a grassland area in the centre of the Netherlands was measured by using nighttime observations of CO_2 fluxes with the eddy-covariance technique. An Arrhenius type of relation was used to calculate this CO_2 efflux as well. Under intermediate soil moisture conditions (θ: 0.10 – 0.35) the surface temperature, $T_{surf} = \sqrt[4]{\frac{R_{lo}}{\varepsilon\sigma}}$, appears to give the best result, which suggests that plant respiration and CO_2 efflux from litter decomposition are the most dominant terms in the surface respiration of our grassland area.

For dry (typically, θ < 0.10) or wet (typically, θ > 0.35) soil moisture contents the respiration is reduced, relative to the Arrhenius relation. This respiration reduction can well be calculated by using the simple extended Arrhenius' relation:

$$R_E = (1 - f2(\theta))R_{10}e^{E_o\left(\frac{1}{283.15-T_0} - \frac{1}{T_{surf}-T_o}\right)} =$$

$$= (1 - \frac{a_1}{\theta - b_1} - \frac{a_2}{\theta - b_2} - c_1)R_{10}e^{E_o\left(\frac{1}{283.15-T_0} - \frac{1}{T_{surf}-T_o}\right)}$$

In which the constants are data set dependent variables.

Surface respiration calculated with the above extended Arrhenius relation was compared with nighttime measured respirations with the eddy-covariance technique. The unbiased linear relation was found of $R(exp) = 1.01\ R_E$ with a correlation coefficient of R = 0.86.

Acknowledgements

We gratefully acknowledge the assistance of the technical staff of the Wageningen University Meteorology and Air Quality Group for doing the calibrations and maintenance of the instrumentation during the experimental period. This work was partly support by the Netherlands Foundation for the Advancement of Research (NWO, Instrumentation Support).

References

Arrhenius S. (1889). Über die Reaktionsgeschwiendigkeit bei der Inversion von Rohrzucker durch Säuren. *Zeitschrift für Physik und Chemie.* **4**, 226-248.

Baldocchi, D., Valentini, R., Running, S., Oechel, W. and Dahlman, R. (1995). Strategies for measuring and modelling carbon dioxide and water vapour fluxes over terrestrial ecosystems. *Glob. Change Biol.* **2**, 159-168.

Campbell, G.S. & Norman, J.M. (1997). *An Introduction to Environmental Biophysics.* Springer Verlag, New York, 286 pp.

De Jeu, R.A.M. & Owe, M.. (2004). Derivation of land surface parameters from 3 different vegetation sites with ELBARA. Proc. SPIE Ser., Remote Sensing for Agriculture, Ecosystems and Hydrology V, Vol. 5232, Barcelona, Spain, pp 434 – 444.

De Jong, E., Redmann, R.E. & Ripley E.A. (1978). A comparison of methods to measure soil respiration. *Soil Science* **127**, 5-12.

Dugas, W.A., Reicosky, D.C. & Kiniry, J.R. (1997). Chamber and micrometeorological measurements of CO_2 and H_2O fluxes for three C_4 grasses. *Agric. Forest Meteorol.,* **83**, 113-133.

Ham, J.M. & Knapp, A.K. (1998). Fluxes of CO_2, water vapor, and energy from a prairie ecosystem during the seasonal transition from carbon sink to carbon source. *Agric. Forest Meteorol.,* **89**, 1 – 14.

Hillel, D., 1998. *Environmental Soil Physics.* Academic Press, New York. 757 pp.

Hunt, J.E., Kelliher. F.M., McSeveny, T.M. & Byers, J.N. (2002). Evaporation and carbon dioxide exchange between the atmosphere and a tussock grassland during a summer drought. *Agric. Forest Meteorol.,* **111**, 65-82.

Kim, J., Verma, S.B. & Clement, R.J. (1992). Carbon dioxide budget in temperate grassland ecosystem. *J. Geophysical Res.,* **97**, 6057 – 6063.

Jacobs, A.F.G., Ronda, R.J. & Holtslag, A.A.M. (2003a). Water vapour and carbon dioxide over a bog vegetation. *Agric. Forest Meteorol.,* **116**, 103-112.

Jacobs, A.F.G., Heusinkveld, B.G. & Holtslag, A.A.M. (2003b). Carbon dioxide and water vapour flux densities over a grassland area in the Netherlands. *Int. J. Climatol.* **23**, 1663-1675.

Lloyd, J. & Taylor, J.A. (1994). On the temperature dependence of soil respiration. *Functional Ecology* **8**, 315-323.

Meyers, T. (2001). A comparison of summertime water and CO_2 fluxes over rangeland for well watered and drought conditions. *Agric.Forest Meteorol.* **106**, 205-214.

McMillen, R.T. (1986). A BASIC program for eddy correlation in non-simple terrain, NOAA Tech. Memo. ERLARL-147, NOAA, Air Resources Lab., Oak Ridge MD.

McMillen, R.T. (1988). An eddy-correlation technique with extended applicability to non-simple terrain. *Bound.-Layer Meteorol.* **43**, 231-245.

Moore, C.J. (1986). Frequency response corrections for eddy correlation systems. *Bound.-Layer Meteorol.,* **37**, 17-35.

Nieveen, J.P., Campbell, D.I., Schipper, L.A. & Blair, I.J. (2005). Carbon exchange of grazed pasture on a drained peat soil. *Glob. Change Biol.,* **11**, 607-618.

Norman J.M., Garcia R. & Verma S.B. (1992). Soil surface CO_2 fluxes and carbon budget of a grassland. *Geophys. Res.* **92**, 18845-18853.

Pattey, E., Strachan, I.B., Desjardins, R.L & Massheder, J. (2002). Measuring nighttime CO_2 flux over terrestrial ecosystems using eddy-covariance and nocturnal boundary layer methods. *Agric. Forest Meteorol.,* **113**, 145-158.

Rabbinge, R., Goudriaan, J., Van Keulen, H., Penning de Vries, F.W.T. and Van Laar, H.H. (eds), 1995. Theoretical Production Ecology: Reflections and Prospects. *Simulation Monographs* **34**, Pudoc, Wageningen, 301pp.

Raich, J.W. & Schlesinger, W.H. (1992). The global carbon dioxide flux in soil respiration and its relationship to vegetation and climate. *Tellus,* **44B**, 81-99.

Reichstein, M., Subke, J-A., Angeli, A.C. & Tenhunen, J.D. (2005). Does the temperature sensitivity of decomposition of soil organic matter depend upon water content, soil horizon, or incubation time? *Glob. Change Biol.* **11**, 1754–1767.

Snel, S. (2004). Soil-Atmosphere exchange of CO_2. M.Sc. –thesis, Wageningen University, *Dept. Meteorology and Air Quality*, **54** pp.

Suyker, A.E., Verma, S.B. & Burba, G.G., 2003. Interannual variability in net CO_2 exchange of a native tallgrass prairie. *Glob. Change Biol.* **9**, 255-265.

Van den Hurk, B.J.J.M. (1996). Sparse canopy parameterizations for meteorological models, Ph.D. Thesis, Dept. of Meteorology, Wageningen Agricultural University, Wageningen, 271 pp.

Webb, E.K, Pearman, G.I. & Leuning R. (1980). Corrections of flux measurements for density effects due to heat and water vapour transfer. *Q.J.R. Meteorol. Soc* **106**, 85-100.

Xu, L. & Baldocchi, D.D. (2004). Seasonal variation in carbon dioxide exchange over a Mediterranean annual grassland in California. *Agric.Forest Meteorol.* **1232**, 79-96.

In: Ecosystem Ecology Research Trends
Editors: J. Chen and C. Guo, pp. 327-344

ISBN: 978-1-60456-183-8

Chapter 11

CRAB INFLUENCES ON THE EXPORT OF PLANT DETRITUS FROM SALT MARSHES AND MANGROVES: A REVIEW

Jorge L. Gutiérrez*

Facultad de Ciencias Exactas y Naturales, Universidad Nacional de Mar del Plata, Mar del Plata, Argentina; Consejo Nacional de Investigaciones Científicas y Técnicas (CONICET), Argentina; Grupo de Investigación y Educación en Temas Ambientales (GrIETA), Mar del Plata, Argentina; and Institute of Ecosystem Studies, Millbrook, NY, USA

Abstract

While crabs occur in perhaps most salt-marshes and mangroves worldwide, they are rarely considered in studies dealing with tidal fluxes of plant detritus from these coastal wetlands to adjacent waters. However, crabs can affect the production, availability and tidal transport of plant detritus in a variety of ways. This includes controls via both assimilatory-dissimilatory mechanisms (e.g., herbivory, detritivory, seed and propagule consumption) and non assimilatory, non dissimilatory environmental modification (i.e., *physical ecosystem engineering*; e.g., sediment oxygenation via burrows, detritus excavation and burial, detritus trapping into burrows). Given the high densities and activity rates shown by many crab species, their effects on the export of plant detritus are expected to be important relative to overall marsh production. Therefore, the predictive capacity of models dealing with detritus export by salt marshes and mangroves is likely to be enhanced by considering crab influences.

1. Introduction

Salt marshes and mangroves are highly productive ecosystems that are intimately linked to adjacent estuarine waters by fluxes of particulate and dissolved organic matter (Adam 1990, Mitsch and Gooselink 1993, Lee 1995). A central question regarding the functioning of salt

* E-mail address: gutierrezj@ecostudies.org

marsh and mangrove ecosystems is whether their primary production contributes to fuel trophic webs in adjacent coastal waters. While there is no single answer (see Nixon 1980, Lee 1995, Childers 2000 for reviews), we can predict that the quantity of detritus exported from these coastal wetlands will be a function of detritus availability, its exposure to vectors capable to transport it, and the presence of structures that sequester detritus on their way out of the system. Dominant vegetation, tidal regime and geomorphology are usually postulated as the most important factors causing variation in the export characteristics of tidal marshes and mangroves because of their obvious major influences on the availability and transportability of detritus (Odum et al. 1979, Nixon 1980, Gallagher et al. 1980, Findlay et al 1990, Dame and Allen 1996, Bouillon et al. 2004). Marsh invertebrates may, however, affect detritus production, its exposure to tides and subsequent transport in many ways with potentially significant consequences for marsh export.

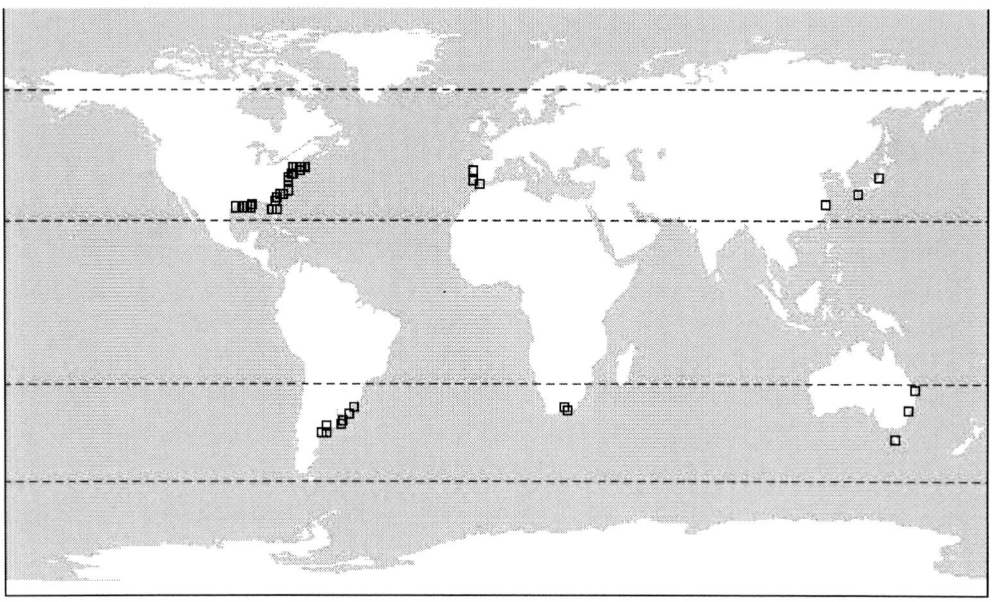

Figure 1. Locations of 26 salt marsh studies (empty squares) and 36 mangrove studies (solid circles) documenting the presence of burrowing crabs. Locations plotted are illustrative rather than exhaustive and were obtained from: Apel and Türkay (1999), Ashtone et al. (2003), Bertness (1985), Bezerra and Matthews-Cascon (2007), Bouillon et al. (2004), Breitfuss et al. (2004), Brosseau et al. (2003), Cammen et al. (1980), Cantera et al. (1999), Castellanos and Rozas (2001) Clarke and Kerrigan (2002), Dahdou-Guebas and Koedam (2001), Delgado et al. (2001), Emmerson (2001), Eshky et al. (1995), Ewa-Oboho (1993), Fagonee (2005), Fell et al. (2003), Giani et al. (1996), Gonçalves et al. (2003), Gribsholt et al. (2003), Griffin (1971), Hanson et al. (2002) Krauss and Allen (2003), Lago (1993), Lee (1989), Lim and Heng (2007), Litulo (2004), Lu et al. (2001), McCraith et al. (2003), McGuiness (1997), Mchenga et al. (2007), McKee (1995), Micheli et al. (1991), Minchinton (2001), Minello et al. (1994), Moulton and Felder (1995), Negreiros-Fransozo et al. (2002), Nordhaus et al. (2006), O'Connor and Judge (1999), Omori et al. (1998), Ólafsson et al. (2002), Pennings et al. (1998), Prieto et al. (2004), Rivera-Monroy and Twilley (1996), Robertson (1986), Rodriguez et al. (1997), Schories et al. (2003), Shih (1995), Smith et al. (1989), Snelling (1959), Sousa and Mitchell (1999), Spivak (1997), Takeda and Kurihara (1987), Taylor and Allanson (1993), Thongtham and Kristensen (2003), Twilley et al. (1997), Valiela et al. (1974), Ward and Busch (1976), Warner (1969), Wolfrath (1992), Zimmerman and Felder (1991).

Table 1. Examples of high crab abundance in salt marshes and mangroves along the five continents.

Location	Crab species	Density	Reference
Salt marshes			
Cape Cod, MA, USA	*Uca pugnax*	42 burrows m^{-2}.	Katz (1980)
Chesapeake Bay, VA, USA	*Uca minax*	Up to 76 burrows m^{-2}. *	Kerwin (1971)
North Inlet, SC, USA	*Uca pugnax*	20 to 300 burrows m^{-2}. *	McCraith et al. (2003)
Sapelo Island, GA, USA	*Uca pugnax*	88 burrows m^{-2}	Montague (1982)
Joseph's Harbor, LA, USA	*Uca longisignalis* and *U. spinicarpa*	26 to 182 and 29 to 43 burrows m^{-2}, respectively. §	Mouton and Felder (1996)
Patos Lagoon, Brazil	*Chasmagnathus granulatus*	23 burrows m^{-2}.	Rosa and Bemvenuti (2005)
Mar Chiquita Coastal Lagoon, Argentina	*Chasmagnathus granulatus*	70 burrows m^{-2}.	Gutiérrez et al. (2006)
Ria Formosa, Portugal	*Uca tangeri*	6 to 16 burrows m^{-2}. §	Wolfrath (1992)
Kariega Estuary, South Africa	*Cleistostoma edwarsii* and *Sesarma catenata*	386 and 47 ind. m^{-2}, respectively.	Taylor and Allanson (1993)
Moreton Bay, Australia	*Helograpsus haswellianus*	More than 80 burrows m^{-2}.	Breitfuss et al. (2004)
Mangroves			
Port Royal, Jamaica	*Aratus pisonii* and *Uca rapax*	Up to 16 ind. m^{-2} each. *	Warner (1969)
Tempisque and Bebedero Rivers, Costa Rica	Primarily *Sesarma sulcatum*	Up to 85 crab holes m^{-2}. *	Delgado et al. (2001)
Bocaripo Lagoon, Venezuela	*Uca rapax*	17.75 ind. m^{-2}.	Prieto et al. (2004)
Pacoti River, Brazil	*Uca thayeri*	8.5 ind. m^{-2}.	Bezerra and Matthews-Cascon (2007)
Gazi Bay, Kenya; Maruhubi and Chwaka Bays, Tanzania	*Neosarmatium meinerti*	3 to 10 ind. m^{-2}. *	Ólaffson et al. (2002)
Mgazana, South Africa	*Neosarmatium meinerti*	5 to 22 burrows m^{-2}. *	Emmerson (2001)
Tanshui, Taiwan	*Mictyris brevidactilus*	62 to 226 ind. m^{-2}. §	Shih (1995)
Pulau Hantu, Singapore	*Uca annulipes*	7.6 to 25.5 ind. m^{-2}. *	Lim and Heng (2007)
Manko, Japan	*Helice formosensis*	26 to 36 holes m^{-2}. *	Mchenga et al. (2007)

* Range of values indicates spatial variations. § Range of values indicates to temporal variations

Table 2. Propagule consumption rates of mangrove crabs.

Location	Crab species	Propagule consumption*	Reference
Everglades National Park, FL, USA	*Aratus pisonii* and *Sesarma curacaoense*	0 to 8% in 4 days	Smith et al. (1989)
Twin Gays, Belize	*Goniopsis cruentata* and *Ucides cordatus*	18 to 60 % in 9 days	McKee (1995)
Tempisque and Bebedero Rivers, Costa Rica	Primarily *Sesarma sulcatum*	5 to 52 % in 25 days	Delgado et al. (2001)
Punta de San Blas, Panama	*Goniopsis cruentata*	2 to 15 %in 4 days	Smith et al. (1989)
Punta Galeta, Panama	*Goniopsis cruentata* and *Ucides cordatus*	3 to 43 % in 4 weeks	Sousa and Mitchell (1999)
Gazi Bay, Kenya	Sesarmids	67 to 100 % in 8 days	Bosire et al. (2005)
Gazi Bay and Mida Creek, Kenya	*Neosarmartium meinerti*	100 % after 2 days	Dahdou-Guebas (1997)
Matang, Malaysia	Sesarmids	6 to 61 % in 4 days	Smith et al. (1989)
Missionary Bay and Cape Ferguson, Australia	Sesarmids	50 to 96 % in 18 days	Smith (1987)
Missionary Bay and Herbert River, Australia	Sesarmids.	20 to 49 % in 4 days	Smith et al. (1989)
Ludmila Creek, Australia	Primarily *Neosarmartium meinerti*	19 to 100 % in 20-22 days	McGuiness (1997)
Bowling Green Bay, Australia	Primarily *Sesarma messa*, *S. smithii* and *S. fourmanoiri*	22 to 100 % in 50 days	Clarke and Kerrigan (2002)
Kosrae, Micronesia	Not reported.	16.7 % in 14 days	Krauss and Allen (2003)
Kosrae, Micronesia	Not reported	22.4 % in 34 days	Allen et al. (2003)

* Within-site variations depend on mangrove species.

Table 3. Increase in the surface area of salt marsh or mangrove sediments due to the presence of crab burrows.

Location	Crab species	Burrow density	Increase in surface area(%)	Reference
Salt marshes				
Cape Cod, MA, USA	*Uca pugnax*	42 burrows m^{-2}	59	Katz (1980)
Sapelo Island, GA, USA	*Uca pugnax*	Not reported	22	Teal and Kanwisher (1961)
Mar Chiquita Coastal Lagoon, Argentina	*Chasmagnathus granulatus*	60-80 burrows m^{-2}	400-600	Fanjul et al. (2007)
Mangroves				
Bangrong, Thailand	*Neoepisesarma versicolor*	0.2 burrows m^{-2}	33.6	Thongtham and Kristensen (2003)
Townsville, Australia	*Sesarma messa*	1 burrow (9 openings) in 0.64 m^{-2}	593.75	Stieglitz et al. (2000)

Table 4. Consumption of leaf litter by mangrove crabs.

Location	Crab species	Litter consumption (% leaf fall)	Reference
Caete estuary, Brazil	*Ucides cordatus*	81.3 §	Nordhaus et al. (2006)
Bragança, Brazil	*Ucides cordatus*	83.4 *	Schories et al. (2003)
Guayas River estuary, Ecuador	*Ucides occidentalis*	100 *	Twilley et al. (1997)
Gazi Bay, Kenya; Maruhubi and Chwaka Bays, Tanzania	*Neosarmatium meinerti*	100 *	Ólafsson et al. (2002)
Gazi Bay, Kenya	*Sesarma guttatum*	40.3 (day) 21.7 (night) *	Slim et al. (1997)
Mgazana estuary, South Africa	*Neosarmatium meinerti*	43.6 *	Emmerson and McGwynne (1992)
Mai Po, Hong Kong	*Chiromantes* spp.	57 *	Lee (1989)
Townsville, Australia	Primarily *Sesarma messa*, *S. smithii* and *S. fourmanoiri*	71 to 79 §	Robertson and Daniel (1989)
Ingham, Australia	Primarily *Cleistostoma wardi*, *Sesarma fourmanoiri* and *S. molluccensis*	33 §	Robertson and Daniel (1989)
Missionary Bay, Australia	*Sesarma messa*	28 §	Robertson (1986)

* Daily leaf fall, § Annual leaf fall

Table 5. Sediment excavation/turnover by crabs in salt marshes and mangroves.

Location	Crab species	Burrow density (burrows m^{-2})	Sediment excavation/turnover	Reference
Salt marshes				
Cape Cod, MA, USA	*Uca pugnax*	42	18 % of the upper 15 cm of sediment per year	Katz (1980)
North Inlet, SC, USA	*Uca pugnax*	40 to 300	12 to 16 cm^3 m^{-2} d^{-1}	McCraith et al. (2003)
Sapelo Island, GA, USA	*Uca pugnax*	88	6.07 g m^{-2} d^{-1}	Montague (1982)
Mar Chiquita Coastal Lagoon, Argentina	*Chasmagnathus granulatus*	40	2.4 kg m^{-2} d^{-1}	Iribarne et al. (1997)
Mar Chiquita Coastal Lagoon, Argentina	*Chasmagnathus granulatus*	70	547.08 g m^{-2} d^{-1}	Gutiérrez et al (2006)
Mangroves				
Mida Creek, Kenya	*Neosarmatium meinerti* and *Cardisoma carnifex*	~ 4	80 to 210 cm^3 m^{-2} d^{-1}	Micheli et al. (1991)
Bangrong, Thailand	*Neoepisesarma versicolor*	0.2	0.3 % of the sediment volume to a depth of 84 cm	Thongtham and Kristensen (2003)

Brachyuran crabs are amongst the most conspicuous invertebrate inhabitants of salt marshes and mangroves worldwide, where they often occur at remarkably high densities (Table 1) and/or constitute an important proportion of the invertebrate biomass and secondary production (e.g., Teal 1962, Cammen et al. 1980). Given their high abundance and activity rates, crabs can be major influence on the fate of marsh primary production. Nevertheless, while there is abundant literature illustrating important effects of crabs on biological, physical and biogeochemical variables at salt marshes and mangroves, the implications of such crab activities for material export (and detritus export in particular) were rarely subject of research attention. In this chapter, I illustrate the multiple ways in which crabs can affect the tidal export of macrophyte-derived detritus from marshes and mangroves. This includes trophic as well as *physical ecosystem engineering* (i.e., the physical modification of biotic and abiotic materials by organisms; *sensu* Jones et al. 1994, 1997, Jones and Gutiérrez 2007) influences of crabs. Trophic influences occur via the consumption of both live and dead plant tissues. Physical ecosystem engineering influences are mediated by burrow excavation – a phenomenon that takes place in many salt marshes and mangroves worldwide (noticeable exceptions are salt marshes at the North Sea and the Pacific coasts of North and South America; Fig. 1). Such influences include effects on detritus production, its accumulation as litter at the sediment surface (and thus, its exposure to tides), and its redeposition after entrainment into tidal transport.

2. Influences on Detritus Production

2.1. Herbivory

The consumption of live plant tissues by crabs can affect the magnitude of detritus export via effects on detritus production. Removal of plant biomass by herbivores often results in decreased detritus production and litter accumulation (e.g., Bazely and Jefferies 1986; Carson and Root 2000). However, in many circumstances, damage and partial consumption of plant tissues by herbivores induces the abscission and renewal of damaged tissues (e.g., Risley and Crossley 1988, Lee 1991), with a consequent increase in the rates detritus production.

A few crab species feed on live tissues of salt marsh plants, including *Chasmagnathus granulatus* in Argentina, Uruguay and Southern Brazil (Bortolus and Iribarne 1999, Bortolus et al. 2004) and *Armases cinereum* in the Eastern coast of the United States (Pennings et al. 1998, Buck et al. 2003). In addition, also a small number of crab species feed on live mangrove tissues. They include tree-climbing species such as *Aratus pisonii* in the Atlantic and Pacific coasts of tropical America (Beever et al. 1979) and *Parasesarma leptosoma* in the Western Indo-Pacific (Dahdouh-Guebas 1999), which reach the mangrove canopy to feed on fresh leaves; and less arboreal species as *Macrophthalmus quadratus* in Indonesia, which feeds on the bark of mangrove pneumatophores (Wada and Wowor 1990).

The magnitude of crab herbivory effects on detritus availability and subsequent tidal export is likely small. Herbivory is usually a minor pathway in the flow of energy and materials in salt marshes and mangroves (i.e., usually less than 20 % of the net primary production is consumed by herbivores); primarily because of the low palatability and nutritional quality of marsh and mangrove tissues (Cebrian 1999). Nevertheless, Bortolus and Iribarne (1999) observed that inclusion of the grapsid crab, *Chasmagnathus granulatus*, during one month in a burned marsh area previously uninhabited by crabs led to 71.5 % clipping and 87.5 % decrease in the aboveground biomass of newly-emerged cordgrass shoots. Such heavy *C. granulatus* grazing on cordgrass during regeneration – which was attribited to the high palatability of newly-emerged shoots – suggest important consequences for detritus production. Therefore, there are particular circumstances (regeneration of plant stands from burning in the above example) when large crab effects on plant detritus production and subsequent export could be expected.

2.2. Seed and Propagule Consumption

Crab feeding on seeds and propagules is common on mangrove forests (though not reported in salt marshes) and can affect detritus production and its availability for tidal export by altering forest regeneration and composition. Either pre- or post-dispersal consumption of mangrove propagules was observed in a variety of crab species (see examples in Table 4), primarily as a supplement of leaf consumption (see *Detritus consumption*). Pre-dispersal propagule consumption by crabs and insects collectively ranges between 0 and 95 % of propagule stocks in mangrove ecosystems worldwide (Farnsworth and Ellison 1997). In addition, post-dispersal propagule consumption by crabs – which was the focus of most studies to date – can occur at rates as high as 100 % consumption in eight days (Table 2), determining tree recruitment and accounting for the dominance and abundance of particular mangrove species (Smith 1987, McKee 1995, Sousa and Mitchell 1999, Delgado et al. 2001, Clarke and Kerrigan 2002, Bosire et al. 2005). Such crab effects on the abundance and dominance of mangrove species are expected to affect forest productivity and, concomitantly, overall detritus production and export.

2.3. Sediment Oxygenation via Burrows

The construction on burrows by many crab species can affect marsh and mangrove production via the oxygenation of otherwise anaerobic sediments. In the absence of burrows, the fine sediments that characterize salt marshes and mangroves are usually anoxic below the top few millimeters (Aller 1988, Kristensen 1988). However, crab burrows can increase the surface area of marsh and mangrove sediments several times (Table 3). In so doing, they extend the sediment-water (or sediment-air) interface; increasing the overall volume of sediments dominated by oxic conditions, and delivering oxygen to subsurface sediments (Aller 1988, Kristensen 1988). Increased oxygen availability in marsh and mangrove sediments is expected to increase primary production via a variety of pathways that are not mutually exclusive. These include: (1) increased mineral nitrogen due to enhancement of aerobic microbial transformations as mineralization and nitrification (Kristensen 1988), (2) increased nitrogen uptake due to increased root respiration (Morris 1984, Morris and Dacey 1984), (3) increased nutrient uptake due to root association with aerobic mycorrhizal symbionts (Miller and Sharitz 2000), and (4) decreased accumulation of toxic reduced compounds in soils (Koch and Mendelssohn 1989, Bradley and Morris 1990).

Several experimental studies demonstrate positive effects of crab burrows on plant production that can be attributed to any of the above-mentioned aerobic pathways. Montague

(1982) observed increased smooth cordgrass, *Spartina alterniflora,* production and ammonium concentration in sediments after introduction of mud fiddler crabs, *Uca pugnax*, in a marsh area not inhabited by them. In the same vein, Bertness (1985) observed decreased *S. alterniflora* production, sediment aeration and oxidation-reduction potential after *U. pugnax* removal. Smith et al. (1990) found decreased cumulative forest growth (as measured by stipule fall) and increased sulphide concentrations in sediments after experimental reduction of burrowing crab densities. Daleo et al. (2007) demonstrated that 35 % of dense-flowered cordgrass *Spartina densiflora* growth depends on root colonization by arbuscular mycorrhizal fungi, which takes place only in the presence of crab *Chasmagnathus granulatus* burrows that oxygenate marsh sediments. In general, the crab burrow environment is characterized by biogeochemical conditions that favor plant growth such as increased oxidation-reduction potentials (Bertness 1985, Fanjul et al. 2007, Mchenga et al. 2007), decreased sulphide concentrations (Smith et al 1991, Nielsen et al. 2003, Thongtham and Kristensen 2003, Fanjul et al. 2007), increased organic matter mineralization (Gribsholt et al. 2003, Nielsen et al. 2003, Fanjul et al. 2007), and increased ammonium and nitrate concentrations (Montague 1982, Takeda and Kurihara 1987, Fanjul 2007) relative to surrounding sediments. All these biogeochemical influences of crab burrows on plant growth and production are expected to directly affect detritus production.

3. Influences on Litter Accumulation

3.1. Detritivory

Many crab species in salt marshes and mangroves ate plant detritus. The degree to which crab detritivory affects the accumulation of detritus on the surface of salt marshes (and, thus, its availability for tidal export) is apparently low. In fact, the fraction of the detritus produced in marshes that is not exported is primarily decomposed by bacteria and fungi (Adam 1990). This suggests that detritus availability in salt marshes would be little affected by crab detritivory.

In contrast, crab detritivory is clearly a major control on the availability of detritus in the floor of mangrove forests. A variety of mangrove crab species feed on the leaf litter (Linton and Greenaway 2007). These crabs can consume between 30 and 100 % of the total leaf fall in mangrove forests (Table 4). By processing an important proportion of the mangrove leaf production, these crabs enhance nutrient re-mineralization at the forest and help to retain nutrients and energy within the mangrove ecosystem (Werry and Lee 2005, Nordhaus et al. 2006). Leaf consumption by these crabs, thus, have an important role in controlling the export of particulate matter from the forests to other nearshore habitats.

The contrasting importance of crab detritivory in salt marshes and mangroves cannot be explained from differences in nutritional value and/or palatability between marsh plant detritus and mangrove leaves. Both marsh plant detritus and mangrove leaves are poor in nitrogen and rich in hardly-digestible structural carbohydrates and tannins (Tenore 1983, Linton and Greenaway 2007). In both cases C-N ratios (> 25) are exceedingly large to meet the crab nutritional demands (Tenore 1983, Linton and Greenaway 2007) and, consequently, crabs usually supplement detritus consumption with foods of higher nitrogen content by means of opportunistic predation, cannibalism, scavenging, algae grazing, or seed and propagule consumption (e.g., Dahdou-Guebas et al. 1999, Buck et al. 2003, Linton and Greenaway 2007).

3.2. Detritus Excavation and Burial

By constructing and maintaining burrows, crabs translocate subsuperficial sediments into the marsh surface. They usually do so at very high rates, resulting in significant sediment overturning (Table 5). This has two counteracting influences on the availability of detritus at the marsh surface. Sediment excavation can result in the transport of belowground organic matter (e.g. live and dead roots and rhizomes) to the marsh surface (De la Cruz and Hackney 1977). However, the deposition of excavated sediments at the marsh surface can cause the burial of surface detritus (Takeda and Kurihara 1987). Gutiérrez et al. (2006) demonstrated that burrowing by *Chasmagnathus granulatus* in an Argentinean salt marsh decreases the organic matter content of surface sediments via the excavation of organic-poor subsurface sediments and its deposition as mounds at the marsh surface. Deposition of excavated sediments at the marsh surface by *C. granulatus* buries plant macrodetritus at rates that approximate those of detritus production at the marsh (i.e. 1 to 9 and 0.5 to 10 g m^{-2} d^{-1} respectively), thus leading to very low litter accumulation at the marsh surface (Gutiérrez et al. unpublished data). Evidence of similar processes was found in a Japanese salt marsh by Takeda and Kurihara (1987), who observed decreasing litter accumulation and increasing amounts of buried macrodetritus with increasing densities of burrowing crabs *Helice tridens*. Although little acknowledged, sediment excavation by crabs is likely a widespread control on detritus accumulation in salt marshes as well as in those mangroves where leaf consumption is low (e.g., Florida mangroves; McIvor and Smith 1995).

4. Influences on Detritus Transport

4.1. Detritus Trapping into Burrows

Crab burrows can function as traps for particles that are transported by the overlying water (DePatra and Levin 1989, Botto and Iribarne 2000). This is specifically true for crab species that maintain their burrows open during tidal inundation (exceptions are several ocypodid species; e.g., Warren 1990, Litulo 2004). Crab burrow openings can cover an important proportion of the sediment surface and intercept bedload-transported detritus in their pathway to the adjacent coastal waters. Estimates obtained from an Argentinean salt marsh indicate that burrows of *Chasmagnathus granulatus* collect 4.15 and 0.3 g m^{-2} d^{-1} of sedimentary organic matter and macrodetritus (i.e., detrital particles larger than 1 cm length), respectively (Gutiérrez et al. 2006, unpublished data). In addition, organic matter collection into burrows is not restricted to salt marshes themselves but also occurs in the tidal flats that characterize the transition between marshes and the adjacent waters. Since these tidal flats are often extensive (i.e., more than 40 % of the area of the estuary; Iribarne et al. 2005) and show high crab densities (up to 80 burrows m^{-2}) they can intercept an important proportion of marsh detritus in their way to the open estuary (Botto et al. 2006). Estimates conducted by Botto et al. (2006) indicate that these burrowed tidal flats can trap an amount of detritus equivalent to the annual production of an equally sized *Spartina* marsh in less than 100 days. While this kind of phenomenon was primarily investigated in Argentinean salt marshes, it is likely that any other burrowing crab species that maintain permanently open burrows in marshes, mangroves, or their fringing tidal flats have similar influences on detritus transport.

5. Conclusion

Crabs can potentially affect the export of plant detritus from salt marshes by affecting its production, its accumulation as litter (and, thus, its exposure to tidal flushing), or its subsequent tidal transport. Six general mechanisms of crab effect on the production, accumulation and transport of plant detritus were identified in this review, namely, herbivory, detritivory, seed or propagule consumption, sediment oxygenation via burrows, detritus excavation and burial, and detritus trapping into burrows. All these mechanisms are related to functional attributes that are pervasive among marsh and mangrove crabs (e.g., herbivory, detritivory, burrowing). In addition, these mechanisms are not necessarily mutually exclusive and can have synergistic influences on detritus export (e.g., Gutiérrez et al. 2006). While the broad global distribution of crabs in marshes and mangroves suggests that the mechanisms identified here are widespread, there is no quantitative measure of the contribution of these crab-mediated mechanisms to detritus export. This review illustrates that crab influences on detrital-related variables determining detritus export (i.e., production, exposure to tides, transport) can be big. Therefore, considering these crab-mediated mechanisms – together with the known influences of vegetation, geomorphology and tidal dynamics – is likely to enhance our capacity to predict detritus export by salt marshes and mangroves.

Acknowledgements

I thank Pablo Ribeiro for access to literature. This is a contribution to the programs of GrIETA and the Institute of Ecosystem Studies.

References

Adam, P. (1990) *Saltmarsh ecology*. Cambridge, UK: Cambridge University Press.

Allen, J. A., Krauss, K. W. & Hauff, R. D. (2003) Factors limiting the intertidal distribution of the mangrove species *Xylocarpus granatum*. *Oecologia,* **135**, 110-121.

Aller, R. C. (1988) Benthic fauna and biogeochemical processes in marine sediments: The role of burrow structures. In Blackburn, T. H. & Sorensen, J. (Eds.) *Nitrogen cycling in coastal marine environments* (pp. 301-338). New York, NY: John Wiley & Sons.

Apel, M. & Türkay, M. (1999) Taxonomic composition, distribution and zoogeographic relationships of the grapsid and ocypodid crab fauna of intertidal soft bottoms in the Arabian Gulf. *Estuarine, Coastal and Shelf Science,* **49**, 131-142.

Ashton, E. C., MacIntosh & D. J., Hogarth, P. J. (2003) A baseline study of the diversity and community ecology of crab and molluscan macrofauna in the Sematan mangrove forest, Sarawak, Malaysia. *Journal of Tropical Ecology,* **19**, 127-142.

Bazely, D. R. & Jefferies, R. L. (1986) Changes in the composition and standing crop of salt-marsh communities in response to the removal of a grazer. *Journal of Ecology,* **74**, 93-706.

Beever III, J. W., Simberloff, D. & King, L. L. (1979). Herbivory and predation by the mangrove tree crab *Aratus pisonii. Oecologia,* **43**, 317–328.

Bertness, M. D. (1985) Fiddler crab regulation of *Spartina alterniflora* production on a New England salt marsh. *Ecology,* **66**, 1042-1055.

Bezerra, L. E. A. & Matthews-Cascon, H. 2007. Population and reproductive biology of the fiddler crab *Uca thayeri* Rathbun, 1900 (Crustacea: Ocypodidae) in a tropical mangrove from Northeast Brazil. *Acta Oecologica,* **31**, 251-258.

Bortolus, A. & Iribarne, O. O. (1999) Effects of the SW Atlantic burrowing crab *Chasmagnathus granulata* on a *Spartina* salt marsh. *Marine Ecology Progress Series,* **178**, 79-88.

Bortolus A., Laterra P. & Iribarne O. O. 2004. Crab-mediated phenotypic changes in *Spartina densiflora* Brong. *Estuarine, Coastal and Shelf Science,* **59**, 97-107.

Bosire, J. O., Kairo, J. G., Kazungu, J., Koedam, N., & Dahdouh-Guebas, F. (2005) Predation on propagules regulates regeneration in a high-density reforested mangrove plantation. *Marine Ecology Progress Series,* **299**, 149-155.

Botto, F. & Iribarne, O. O. (2000) Contrasting effects of two burrowing crabs (*Chasmagnathus granulata* and *Uca uruguayensis*) on sediment composition and transport in estuarine environments. *Estuarine, Coastal and Shelf Science,* **51**, 141-151.

Botto, F., Iribarne, O. O., Gutiérrez, J. L., Bava, J., Gagliardini, A. & Valiela, I. (2006) Ecological importance of passive deposition of organic matter into burrows of the SW Atlantic crab *Chasmagnathus granulatus. Marine Ecology Progress Series,* **312**,201-210.

Bouillon, S., Moens, T., Overmeer, I., Koedam, N. & Dehairs, F. (2004) Resource utilization patterns of epifauna from mangrove forests with contrasting inputs of local versus imported organic matter. *Marine Ecology-Progress Series,* **278**, 77-88.

Bradley, P. M. & Morris, J. T. (1990) Influence of oxygen and sulfide concentration on nitrogen uptake kinetics in *Spartina alterniflora. Ecology,* **71**, 282-287.

Breitfuss, M. J., Connolly, R. M. & Dale, P. E. R. (2004) Densities and aperture sizes of burrows constructed by *Helograpsus haswellianus* (Decapoda: Varunidae) in saltmarshes with and without mosquito-control runnels. *Wetlands,* **24**, 14-22.

Brousseau, D. J., Kriksciun, K. & Baglivo, J. A. 2003. Fiddler crab burrow usage by the Asian crab *Hemigrapsus sanguineus*, in a Long Island Sound salt marsh. *Northeastern Naturalist,* **10**, 415-420.

Buck, T. L., Breed, G. A., Pennings, S. C., Chase, M. E., Zimmer, M. & Carefoot, T. H. (2003) Diet choice in an omnivorous salt-marsh crab: Different food types, body size, and habitat complexity. *Journal of Experimental Marine Biology and Ecology,* **292**, 103-116.

Cammen, L. M., Seneca, E. D. & Stroud, L. M. (1980) Energy flow through the fiddler crabs *Uca pugnax* and *U. minax* and the marsh periwinkle *Littorina irrorata* in a North Carolina salt marsh. *American Midland Naturalist,* **103**, 238-250.

Cantera, K. J. R., Thomassin, B.A. & Arnaud, P.M. (1999) Faunal zonation and assemblages in the Pacific Colombian mangroves. *Hydrobiologia,* **413**, 17-33.

Carson, W. P. & Root R. B. (2000) Herbivory and plant species coexistence: Community regulation by an outbreaking phytophagous insect. *Ecological Monographs,* **70**, 73-99.

Castellanos, D. L. & Rozas, L. P. (2001) Nekton use of submerged aquatic vegetation, marsh, and shallow unvegetated bottom in the Atchafalaya River Delta, a Louisiana tidal freshwater ecosystem. *Estuaries,* **24**, 184-197.

Cebrian, J. (1999) Patterns in the fate of production in plant communities. *American Naturalist,* **154**, 449-468.

Childers, D. L., Day, J. W. & McKellar, H. N. (2000) Twenty more years of marsh and estuarine flux studies: Revisiting Nixon (1980). In Weinstein, M. P. & Kreeger D. A. (Eds.) *Concepts and controversies in tidal marsh ecology* (pp. 391-424). Dordretch, The Netherlands: Kluwer Academic Press.

Clarke, P. J. & Kerrigan, R. A. (2002) The effects of seed predators on the recruitment of mangroves. *Journal of Ecology,* **90**, 728-736.

Dahdouh-Guebas, F., & Koedam N. (2001) Are the northernmost mangroves of West Africa viable? – A case study in Banc d'Arguin National Park, Mauritania. *Hydrobiologia,* **458**, 241-253

Dahdouh-Guebas, F., Giuggioli, M., Oluoch, A., Vannini, M. & Cannicci, S. (1999) Feeding habits of non-ocypodid crabs from two mangrove forests in Kenya. *Bulletin of Marine Science,* **64**, 291–297.

Dahdouh-Guebas, F., Verneirt, M., Tack, J. F., & Koedam, N. (1997) Food preferences of *Neosarmatium meinerti* de Man (Decapoda: Sesarminae) and its possible effect on the regeneration of mangroves. *Hydrobiologia,* **347**, 83-89.

Daleo, P., Fanjul, E., Mendez-Casariego, A., Silliman, B. R., Bertness, M. D. & Iribarne, O. O. (2007) Ecosystem engineers activate mycorrhizal mutualism in salt marshes. *Ecology Letters,* **10**, 902-908.

Dame, R. F. & Allen, D. M. (1996) Between estuaries and the sea. *Journal of Experimental Marine Biology and Ecology,* **200**, 169-185.

De la Cruz. A. A. & Hackney, C. T. (1977) Energy value, elemental composition, and productivity of belowground biomass of a *Juncus* tidal marsh. *Ecology,* **58**, 1165-70.

Delgado, P., Hensel, P. F., Jiménez, J. A. & Day, J. W. (2001) The importance of propagule establishment and physical factors in mangrove distributional patterns in a Costa Rican estuary. *Aquatic Botany,* **71**, 157-178.

DePatra, K. D. & Levin, L. A. (1989) Evidence of the passive deposition of meiofauna into fiddler crab burrows. *Journal of Experimental Marine Biology and Ecology,* **125**, 173-192.

Emmerson, W. D. (2001) Aspects of the population dynamics of *Neosarmatium meinerti* at Mgazana, a warm temperate mangrove swamp in the East Cape, South Africa, investigated using an indirect method. *Hydrobiologia,* **449**, 221-229.

Emmerson, W. D. & McGwynne, L. E. (1992) Feeding and assimilation of mangrove leaves by the crab *Sesarma meinerti* De Man in relation to leaf-litter production in Mgazana, a warm temperate Southern African mangrove swamp. *Journal of Experimental Marine Biology and Ecology,* **157**, 41-53.

Eshky, A. A., Atkinson, R. J. A. & Taylor, A. C. (1995) Physiological ecology of crabs from Saudi Arabian mangrove. *Marine Ecology Progress Series,* **126**, 83-95.

Ewa-Oboho, I. (1993) Substratum preference two estuarine crabs *Uca tangeri* Eydoux (Ocypodidae) and *Ocypode cursor* Linne (Ocypodidae) found in a Nigeria mangrove ecosystem. *Hydrobiologia,* **271**, 119–127.

Fagoonee, I. (2005) Coastal marine ecosystems of Mauritius. *Hydrobiologia,* **208**, 55-62.

Fanjul, M. E, Grela, M. A. & Iribarne, O. O. (2007) Effects of the dominant SW Atlantic intertidal burrowing crab *Chasmagnathus granulatus* on sediment chemistry and nutrient distribution. *Marine Ecology Progress Series,* **341**, 177–190.

Farnsworth, E. J. & Ellison, A. M. (1997) Global patterns of pre-dispersal propagule predation in mangrove forests. *Biotropica,* **29**, 318-330.

Fell, P. E., Warren, R. S., Light, J. K., Rawson, R. L. & Fairley, S. M. (2003) Comparison of fish and macroinvertebrate use of *Typha angustifolia*, *Phragmites australis*, and treated *Phragmites* marshes along the lower Connecticut River. *Estuaries,* **26**, 534-551.

Findlay, S. E. G., Howe, K. & Austin, H. K. (1990) Comparison of detritus dynamics in two tidal freshwater wetlands. *Ecology,* **71**, 288-295.

Gallagher, J. L., Reimold, R. J., Linthurst, R. A. & Pfeiffer, W. J. (1980) Aerial production, mortality, and mineral accumulation: Export dynamics in *Spartina alterniflora* and *Juncus roemerianus* stands. *Ecology,* **61**, 303-312.

Giani, L., Bashan, Y., Holguin, G., & Strangmann, A. (1996) Characteristics and methanogenesis of the Balandra lagoon mangrove soils, Baja California Sur, Mexico. *Geoderma,* **72**, 149-160.

Gonçalves F., Ribeiro R., & Soares A.M.V.M. (2003) Comparison between two lunar situations on emission and larval transport of decapod larvae in the Mondego estuary (Portugal). *Acta Oecologica,* **24**, 183-190.

Gribsholt, B., Kostka, J. E. & Kristensen, E. (2003) Impact of fiddler crabs and plant roots on sediment biogeochemistry in a Georgia saltmarsh. *Marine Ecology Progress Series,* **259**, 237-251.

Griffin, D. (1971) The ecological distribution of grapsid and ocypodid crabs (Crustacea: Brachyura) in Tasmania. *Journal of Animal Ecology,* **40**, 597–621.

Gutiérrez, J. L., Jones, C. G., Groffman, P. M., Findlay, S. E. G., Iribarne, O. O., Ribeiro, P. D. & Bruschetti, C. M. (2006) The contribution of crab burrow excavation to carbon availability in superficial salt-marsh sediments. *Ecosystems,* **9**, 647-658.

Hanson, S. R., Osgood, D. T. & Yozzo, D. J. (2002) Nekton use of a *Phragmites australis* marsh on the Hudson River, New York, USA. *Wetlands,* **22**, 326-337.

Iribarne, O. O., Bortolus, A. & Botto, F. (1997) Between-habitat differences in burrow characteristics and trophic modes in the burrowing crab *Chasmagnathus granulata. Marine Ecology Progress Series,* **155**,137-145.

Iribarne, O. O., Bruschetti, M., Escapa, M., Bava, J., Botto, F., Gutiérrez, J. L., Palomo, G., Delhey, K., Petracci, P. & Gagliardini, A. (2005) Small- and large-scale effect of the SW Atlantic burrowing crab *Chasmagnathus granulatus* on habitat use by migratory shorebirds. *Journal of Experimental Marine Biology and Ecology,* **315**, 87-101.

Jones, C. G., Lawton, J. H. & Shachak, M. (1994) Organisms as ecosystem engineers. *Oikos,* **69**, 373-386.

Jones, C. G., Lawton, J. H. & Shachak, M. (1997) Positive and negative effects of organisms as ecosystem engineers. *Ecology,* **78**, 1946-1957.

Jones, C. G., & Gutiérrez, J. L. (2007) On the meaning, usage and purpose of the ecosystem engineering concept. In Cuddington, K., Byers, J., Hastings, A., & Lambrinos, J. (Eds.). *Ecosystem Engineers: Plants to Protists* (pp. 3-24). New York, NY, USA: Academic Press.

Katz, L. C. (1980) Effects of burrowing by the fiddler crab *Uca pugnax* (Smith). *Estuarine, Coastal and Marine Science,* **11**, 233-237.

Kerwin, J. A. (1971) Distribution of the fiddler crab (*Uca minax*) in relation to marsh plants within a Virginia estuary. *Chesapeake Science* **12**, 180-183.

Koch M. S. & Mendelssohn, I. A. (1989) Sulphide as a soil phytotoxin: Differential responses in two marsh species. *Journal of Ecology,* **77***,* 565-578.

Krauss, K. W, & Allen, J. A. (2003) Factors influencing the regeneration of the mangrove *Bruguiera gymnorrhiza* (L.) Lamk. on a tropical Pacific island. *Forest Ecology and Management,* **176**, 49-60.

Kristensen, E. (1988) Benthic fauna and biogeochemical processes in marine sediments: Microbial activities and fluxes. In Blackburn, T. H. & Sorensen, J. (Eds.) *Nitrogen cycling in coastal marine environments* (pp. 275-299). New York, NY: John Wiley & Sons.

Lago, R. P. (1993) Tidal exchange of larvae of *Sesarma catenata* (Decapoda, Brachyura) in the Swartkops estuary, South Africa. *South African Journal of Zoology,* **28**, 182-191.

Lee, S. Y. (1989) The importance of Sasarminae crabs *Chiromanthes* spp. and inundation frequency on mangrove (*Kandelia candel* (L.) Druce) leaf litter turnover in a Hong Kong tidal shrimp pond. *Journal of Experimental Marine Biology and Ecology,* **131**, 23-43.

Lee, S. Y. (1991) Herbivory as an ecological process in a *Kandelia candel* (Rhizophoraceae) mangal in Hong Kong. *Journal of Tropical Ecology,* **7***,* 337-348.

Lee, S. Y. (1995) Mangrove outwelling: A review. *Hydrobiologia,* **295**, 203-212.

Lim, S. S. L. & Heng, M. M. S. (2007) Mangrove micro-habitat influence on bioturbative activities and burrow morphology of the fiddler crab, *Uca annulipes* (H. Milne Edwards, 1837) (Decapoda, Ocypodidae). *Crustaceana,* **80**, 31-45.

Linton, S. M. & Greenaway, P. (2007) A review of feeding and nutrition of herbivorous land crabs: Adaptations to low quality plant diets. *Journal of Comparative Physiology B – Biochemical, Systemic and Environmental Physiology,* **177***,* 269-286.

Litulo, C. (2004) Fecundity of the pantropical fiddler crab *Uca annulipes* (H. Milne Edwards, 1837) (Brachyura: Ocypodidae) at Costa do Sol Mangrove, Maputo Bay, Southern Mozambique. *Western Indian Ocean Journal of Marine Science,* **3***,* 87–91.

Lu, J. J., He, W. S., Zhou, K. J., Tang, Y. W., Ye, S. F. & Sun, P. Y. (2001) Behavior of Zn, Cu, Pb and Cd in biota of Yangtze Estuary. *Science in China Series B-Chemistry,* **44**, 165-172.

McCraith, B. J., Gardner, L. R., Wethey, D. S. & Moore, W.S. 2003. The effect of fiddler crab burrowing on sediment mixing and radionuclide profiles along a topographic gradient in a southeastern salt marsh. *Journal of Marine Research,* **61**, 359-390.

McGuinness, K. A. (1997) Seed predation in a tropical mangrove forest: A test of the dominance- predation model in Northern Australia. *Journal of Tropical Ecology,* **13**, 293-302.

Mchenga, I. S. S., Mfilinge, P. L. & Tsuchiya. M. (2007) Bioturbation activity by the grapsid crab *Helice formosensis* and its effects on mangrove sedimentary organic matter. *Estuarine, Coastal and Shelf Science,* **73**, 316-324.

McIvor, C. C. & Smith, T. J. (1995) Differences in the crab fauna of mangrove areas at a Southwest Florida and a Northeast Australia location: Implications for leaf litter processing. *Estuaries,* **18**, 591-597.

McKee, K. L. (1995) Mangrove species distribution and propagule predation in Belize: An exception to the dominance predation hypothesis. *Biotropica,* **27**, 334-345.

Micheli, F., Gherardi, F. & Vannini, M. (1991). Feeding and burrowing ecology of two East African mangrove crabs. *Marine Biology,* **111**, 247-254.

Miller, S.P. & Sharitz, R.R. (2000) Manipulation of flooding and arbuscular mycorrhizal formation influences growth and nutrition of two semiaquatic species. *Functional Ecology,* **14**, 738–748.

Minello, T. J., Zimmerman, R. J. & Medina, R. (1994) The importance of edge for natant macrofauna in a created salt-marsh. *Wetlands,* **14**, 184-198.

Minchinton, T. E. (2001) Canopy and substratum heterogeneity influence recruitment of the mangrove *Avicennia marina. Journal of Ecology,* **89**, 888-902.

Mitsch, W. J. & Gosselink, J. G. (1993) *Wetlands*. New York, NY: Van Nostrand Reinhold.

Montague, C. L. (1982) The influence of fiddler crab burrows and burrowing on metabolic processes in salt marsh sediments. In Kennedy, V. S. (Ed.) *Estuarine comparisons* (pp. 283-301). New York, NY: Academic Press.

Morris, J. T. (1984) Effects of oxygen and salinity on ammonia uptake by *Spartina alterniflora* Loisel and *Spartina patens* (Aiton) Muhl. *Journal of Experimental Marine Biology and Ecology*, **78**, 87-98.

Morris, J. T. & Dacey, J. W. H. (1984) Effects of O_2 on ammonium uptake and root respiration by *Spartina alterniflora. American Journal of Botany,* **71**, 979-985.

Negreiros-Fransozo, M. L., Delevati-Colpo, K., & Costa T. M. (2002) Allometric growth in the fiddler crab *Uca thayeri* (Brachyura, Ocypodidae) from a subtropical mangrove. *Journal of Crustacean Biology,* **23**, 273-279.

Nielsen, O. I., Kristensen, E. & MacIntosh, D. J. (2003) Impact of fiddler crabs (*Uca* spp.) on rates and pathways of benthic mineralization in deposited mangrove shrimp pond waste. *Journal of Experimental Marine Biology and Ecology,* **289**, 59-81.

Nixon, S. W. (1980) Between coastal marshes and coastal waters: A review of twenty years of speculation and research on the role of salt marshes in estuarine productivity and water chemistry. In Hamilton, P., MacDonald, K. B. (Eds.) *Estuarine and wetland processes*. (pp. 437-525). New York, NY: Plenum Press.

Nordhaus, I., Wolff, M. & Diele, K. (2006) Litter processing and population food intake of the mangrove crab *Ucides cordatus* in a high intertidal forest in Northern Brazil. *Estuarine, Coastal and Shelf Science,* **67**, 239-250.

O'Connor, N. J. & Judge, M. L. (1999) Cues in salt marshes stimulate molting of fiddler crab *Uca pugnax* megalopae: More evidence from field experiments. *Marine Ecology-Progress Series,* **181**, 131-139.

Odum, W. E., Fisher, J. S. & Pickral, J. C. (1979) Factors controlling the flux of particulate organic carbon from estuarine wetlands. In Livingston, R. C. (Ed.). *Ecological processes in coastal marine systems* (pp. 69-80). New York, NY: Plenum Press.

Ólafsson, E., Buchmayer. S. & Skov, M. W. (2002) The East African decapod crab *Neosarmatium meinerti* (de Man) sweeps mangrove floors clean of leaf litter. *Ambio,* **31**, 569-573.

Omori, K., Irawan, B., & Kikutani, Y. (1998) Studies on the salinity and desiccation tolerances of *Helice tridens* and *Helice japonica* (Decapoda: Grapsidae). *Hydrobiologia,* **386**, 27-36.

Pennings, S. C., Carefoot, T. H., Siska, E. L., Chase, M. E. & Page, T. A. (1998) Feeding preferences of a generalist salt-marsh crab: Relative importance of multiple plant traits. *Ecology,* **79**, 1968-1979.

Prieto, A. S., Ruiz, L. J. & Montes, A. (2004) Abundancia y morfometría de una población de *Uca rapax rapax* (Brachyura: Ocypodidae) en la Laguna de Bocaripo, Estado Sucre, Venezuela. *Boletín del Centro de Investigaciones Biológicas,* **38**, 81-93.

Risley, L. S. & Crossley, D. A. (1988) Herbivore-caused greenfall in the Southern Appalachians. *Ecology,* **69**, 1118-1127.

Rivera-Monroy V, H. & Twilley R. R. (1996) The relative role of denitrification and immobilization in the fate of inorganic nitrogen in mangrove sediments (Terminos Lagoon, Mexico). *Limnology and Oceanography,* **41**, 284-296.

Robertson, A. I. (1986) Leaf-burying crabs: Their influence on energy-flow and export from mixed mangrove forests (*Rhizophora* spp.) in Northeastern Australia. *Journal of Experimental Marine Biology and Ecology,* **102**, 237-248.

Robertson, A. I. & Daniel, P. A. (1989) The influence of crabs on litter processing in high intertidal mangrove forests in tropical Australia. *Oecologia,* **78**, 191-198.

Rodriguez, A., Drake, P. & Arias, A. M. (1997) Reproductive periods and larval abundance patterns of the crabs *Panopeus africanus* and *Uca tangeri* in a shallow inlet (SW Spain). *Marine Ecology-Progress Series,* **149**, 133-142.

Rosa, L. C. & Bemvenuti, C. E. (2005) Effects of the burrowing crab *Chasmagnathus granulata* (Dana) on meiofauna of estuarine intertidal habitats of Patos Lagoon, Southern Brazil. *Brazilian Archives of Biology and Technology,* **48**, 267-274.

Schories D., Barletta-Bergan, A., Barletta, M., Krumme, U., Mehlig, U. & Rademaker, V. (2003) The keystone role of leaf-removing crabs in mangrove forests of North Brazil. *Wetlands Ecology and Management,* **11**, 243-255.

Shih, J. T. (1995) Population densities and annual activities of *Mictyris brevidactylus* (Stimpson, 1858) in the Tanshui mangrove swamp of Northern Taiwan. *Zoological Studies,* **34**, 96-105.

Slim, F. J., Hemminga, M. A., Ochieng, C., Jannink, N. T., de la Moriniere, E. C. & van der Velde, G. (1997) Leaf litter removal by the snail *Terebralia palustris* (Linnaeus) and sesarmid crabs in an East African mangrove forest (Gazi Bay, Kenya). *Journal of Experimental Marine Biology and Ecology,* **215**, 35-48.

Smith, T. J. (1987) Seed predation in relation to tree dominance and distribution in mangrove forests. *Ecology,* **68**, 266-273.

Smith, T. J., Chan, H. T.; McIvor, C. C., & Robblee, M. B. (1989) Comparisons of seed predation in tropical, tidal forests from three continents. *Ecology,* **70**, 146-151.

Smith, T. J., Boto, K. G., Frusher, S. D. & Giddins, R. L. (1991). Keystone species and mangrove forest dynamics: The influence of burrowing crabs on soil nutrient status and forest productivity. *Estuarine, Coastal and Shelf Science,* **33**, 419-432.

Snelling, B. (1959) The distribution of intertidal crabs in the Brisbane River. *Australian Journal of Marine and Freshwater Research,* **10**, 67–83.

Sousa, W. P., & Mitchell, B. J. (1999) The effect of seed predators on plant distributions: Is there a general pattern in mangroves? *Oikos,* **86**, 55-66.

Spivak, E. D. (1997) Cangrejos estuariales del Atlántico sudoccidental (25°-41°S) (Crustacea: Decapoda: Brachyura). *Investigaciones Marinas Valparaíso,* **25**, 105-120.

Stieglitz, T., Ridd, P. & Muller, P. (2000) Passive irrigation and functional morphology of crustacean burrows in a tropical mangrove swamp. *Hydrobiologia,* **421**, 69-76.

Takeda, S. & Kurihara, Y. (1987) The effects of burrowing of *Helice tridens* (De Haan) on the soil of a salt marsh habitat. *Journal of Experimental Marine Biology and Ecology,* **113**, 79-89.

Taylor, D. I. & Allanson, B. R. (1993) Impacts of dense crab populations on carbon exchanges across the surface of a salt marsh. *Marine Ecology Progress Series,* **101**, 119-129.

Teal, J. M. (1962) Energy flow in the salt marsh ecosystem of Georgia. *Ecology* **43**, 614-624.

Teal, J. M. & Kanwisher, J. (1961) Gas exchange in a Georgia salt marsh. *Limnology and Oceanography,* **6,** 388-399.

Tenore, K. R. (1983) What controls the availability to animals of detritus derived from vascular plants: Organic nitrogen enrichment or carbon availability? *Marine Ecology Progress Series,* **10**, 307-309.

Thongtham, N. & Kristensen, E. (2003) Physical and chemical characteristics of mangrove crab (*Neoepisesarma versicolor*) burrows in the Bangrong mangrove forest, Phuket, Thailand with emphasis on behavioural response to changing environmental conditions. *Vie et Milieu,* **53**, 141-151.

Twilley, R. R., Pozo, M., García, V. H., Rivera-Monroy, V. H., Zambrano, R. & Bodero, A. (1997) Litter dynamics in riverine mangrove forests in the Guayas River estuary, Ecuador. *Oecologia,* **111**, 109-122.

Valiela, I., Babiec, D. F., Atherton, W., Seitzinger, S. & Krebs, C. (1974) some consequences of sexual dimorphism - feeding in male and female fiddler crabs, *Uca pugnax* (Smith). *Biological Bulletin,* **147**, 652-660.

Wada, K. & Wowor, D. (1990) Foraging on mangrove pneumatophores by ocypodid crabs. *Journal of Experimental Marine Biology and Ecology,* **134**, 89-100.

Ward, D. V. & Busch D. A. (1976). Effects of temefos, an organophosphorous insecticide, on survival and escape behavior of marsh fiddler crab *Uca pugnax*. *Oikos,* **27**, 331-335.

Warner, G. F. (1969) The occurrence and distribution of crabs in a Jamaican mangrove swamp. *Journal of Animal Ecology,* **38**, 379-389.

Warren, J. H. (1990) Role of burrows as refuges from subtidal predators of temperate mangrove crabs. *Marine Ecology Progress Series,* **67**, 295-299.

Werry, J. & Lee, S. Y. (2005) Grapsid crabs mediate link between mangrove litter production and estuarine planktonic food chains. *Marine Ecology Progress Series,* **293**, 165-176.

Wolfrath, B. (1992) Burrowing of the fiddler crab *Uca tangeri* in the Ria Formosa in Portugal and its influence on sediment structure. *Marine Ecology Progress Series,* **85**, 237-243.

Zimmerman, T. L. & Felder, D. L. (1991) Reproductive ecology of an intertidal brachyuran crab, *Sesarma* sp. (nr. *reticulatum*), from the Gulf of Mexico. *Biological Bulletin,* **181**, 387-401.

INDEX

A

Abies alba, 48
abiotic, vii, 24, 38, 43, 93, 332
abundance, 107, 148
access, 262, 284, 285, 289, 336
accessibility, 27, 197
accounting, 178, 278, 333
accuracy, 11, 18, 314
acid, vii, 3, 5, 6, 7, 25, 32, 43, 46, 47, 48, 52, 53, 55, 60, 61, 141, 146, 193, 303
acidic, 31, 47, 257, 302, 304
acidification, 5, 6, 7, 214, 239
activation, 72, 84, 89, 316
activity rate, xiii, 326
actynomyces, xi, 241, 242, 246, 247, 248, 249, 256, 259, 261, 262, 263, 264, 269, 271, 272, 273, 274
adaptation, vii, xi, 3, 6, 11, 156, 241, 242, 262
adjustment, 225
administration, 73, 75, 104
adsorption, 32, 43, 93, 218, 227, 233
adult organisms, 78, 92
adults, 145, 164, 228
Africa, 217, 218, 238, 311, 328, 330, 338, 340
agar, 252, 263
age, 4, 44, 61, 137, 150, 161, 166, 167, 171, 314
agent, xii, 275, 276, 280, 282, 283, 285, 286, 287, 288, 289, 290, 295
aggregates, 37, 113, 193
aggregation, 38, 60, 61, 138, 192, 204, 311, 315
agonist, 75, 81
agriculture, x, xii, 31, 33, 34, 36, 37, 177, 178, 200, 201, 202, 203, 205, 206, 216, 229, 230, 238, 241, 242, 247, 290, 300, 301, 310
agrochemicals, xi, 30, 209
air pollutants, 297
air quality, 26, 282
Alabama, 204
alanine, 194, 205
Alaska, 310
algae, 69, 70, 71, 101, 160, 334
algorithm, 157, 162
alkaline, 227, 243, 244, 246, 250, 251, 253, 254, 257, 258, 259, 260
allele(s), 161, 162, 164, 165, 171, 172, 173, 174
alternative, xii, 18, 68, 92, 97, 179, 193, 200, 275, 280, 281, 282, 286, 287, 288, 289, 290, 295, 301
alters, 45, 47, 101, 137, 145, 156, 160, 179
ambiguity, 26
amendments, 37, 52, 56, 204, 205
amino acids, 41, 306
ammonia, 25, 302, 303, 305, 341
ammonium, 6, 53, 214, 216, 303, 306, 307, 334, 341
amphibia, xi, 210, 213, 229, 230, 233, 235, 238, 239
amphibians, xi, 77, 156, 209, 210, 213, 223, 229, 230, 232, 233, 235, 238, 239
amphipoda, xi, 210, 232
amplitude, 169, 278
amputation, 79
AMR, 280
Amsterdam, 52
analytical framework, 296
anatomy, 73, 88, 296
androgen(s), 73, 74, 75, 92, 100, 101, 103, 105
androgen receptors, 75, 92, 103
androstenedione, 72, 73, 102, 104
animal health, 27
animal models, viii, ix, 65, 66, 67, 68, 76, 92, 96, 97
animals, vii, viii, 9, 10, 11, 12, 16, 17, 18, 25, 29, 65, 67, 68, 70, 71, 72, 73, 75, 76, 77, 81, 82, 93, 94, 95, 96, 97, 109, 195, 200, 292, 300, 301, 343
ANOVA, 82
antagonism, 41, 273
antagonists, 205, 242, 246, 248, 252, 261, 271
Antedon mediterranea, 80, 81, 98
anthropogenic, 26, 48, 145, 300
anthropology, 166

antibacterial properties, 261, 263, 271
antibiotic(s), 242, 243, 246, 252, 259, 264, 265, 269, 271, 272, 273
antibody, 83
anus, 69, 70
apoptosis, 78
aquatic habitats, 229
AR, 48, 81, 92
Argentina, 112, 209, 234, 235, 236, 237, 238, 326, 328, 330, 331, 332
argument, 33
Arrhenius equation, 316, 317, 322, 323
arthropods, 196
articulation, 154
ash, 178
Asia, 179
Asian countries, 179
assessment, xi, 18, 38, 51, 56, 67, 72, 76, 78, 94, 97, 99, 101, 102, 150, 205, 209, 212, 213, 219, 220, 221, 222, 225, 289, 291, 294
assimilation, 30, 53, 315, 338
assumptions, viii, 9, 10, 11, 18, 137, 164
asymptotic, x, 153, 154, 155, 160, 161, 163
Athens, 3
Atlantic cod, ix, 107, 108, 111, 115, 147, 148, 149, 150, 151
Atlantic Ocean, 109
atmosphere, vii, xiii, 3, 7, 24, 25, 33, 34, 35, 66, 313, 314, 316, 317, 325
atmospheric deposition, 300
atomic force, 49
atrophy, 84, 86
attacks, 79
attention, viii, 41, 49, 65, 66, 67, 137, 154, 159, 242, 243, 332
atypical, 277
Australia, 54, 57, 60, 61, 179, 217, 236, 328, 329, 330, 340, 342
autolysis, 272
autotrophic, 113
availability, xiii, 5, 6, 29, 41, 42, 43, 44, 45, 46, 47, 48, 57, 58, 115, 129, 145, 146, 169, 179, 186, 187, 202, 252, 254, 257, 258, 289, 300, 308, 310, 314, 326, 327, 332, 333, 334, 335, 339, 343
averaging, 16, 182
avoidance, 219
awareness, 26, 178, 282

B

bacteria, 25, 41, 43, 46, 49, 58, 113, 192, 193, 194, 195, 201, 203, 223, 242, 248, 260, 261, 262, 264, 334
barley, 178, 180
barriers, 233
battery, 222
Bayesian, 289, 291, 292, 295
Bayesian analysis, 295
behavior, x, xii, 145, 153, 154, 155, 160, 168, 194, 275, 276, 282, 317, 343
behavioral change, 225
Belarus, 273
Belgium, 99
benefits, x, xii, 50, 66, 177, 179, 180, 182, 199, 275, 276, 277, 278, 280, 282, 293, 294
bias, 11, 19
bicarbonate, 214
binding, 5, 31, 38, 67, 75, 99, 101, 103
bioaccumulation, 68, 71, 76, 230
bioanalytical, 52
bioassay(s), xi, 72, 205, 219, 222, 236, 241
bioavailability, 32
bioconcentration, 93
biodiversity, xi, xii, 26, 37, 38, 40, 41, 49, 50, 55, 149, 178, 200, 207, 241, 242, 243, 270, 275, 277, 301
bioindicators, 69, 78
biological activity, 27, 30, 31, 32, 67, 74, 184, 264
biological control, 197, 234
biological interactions, 49, 138
biological processes, ix, 28, 38, 43, 47, 65, 97
biological systems, 30, 154
biologically active compounds, 269
biomarker, 97
biomass, vii, viii, ix, xiii, 3, 4, 5, 6, 24, 30, 31, 33, 36, 37, 38, 41, 42, 43, 44, 45, 46, 47, 48, 53, 54, 56, 57, 58, 59, 60, 61, 63, 68, 69, 78, 107, 108, 113, 115, 116, 118, 119, 120, 121, 129, 130, 131, 136, 137, 138, 139, 141, 143, 144, 145, 146, 178, 180, 184, 187, 189, 191, 192, 193, 195, 202, 204, 216, 218, 221, 222, 233, 260, 299, 300, 301, 303, 304, 305, 306, 307, 308, 309, 310, 311, 314, 332, 333, 338
biomineralization, 49
biosphere, viii, 23, 26, 27
biosynthesis, 73, 102, 104, 272
biotechnology, 242
biotic, vii, 24, 41, 93, 332
birds, 77, 156, 197
blocks, 179
blood, 74
body fluid, 67
body shape, 70
body size, 337
bonds, 31
bone mass, 101

bootstrap, 61
boreal forest, 52
brain, 67
Brazil, 35, 36, 54, 59, 191, 206, 293, 328, 330, 332, 337, 341, 342
breakdown, 30, 31, 39, 53, 72, 184, 194
breast, 77
breeding, 215, 216, 230
British Columbia, 57, 62, 107, 147, 149
bromodeoxyuridine, 83
brothers, 24
buffer, 46, 82, 216, 233, 237
burn(ing), 33, 34, 35, 55, 178, 278, 280, 283, 312, 333
by-products, 27, 28, 31

C

cadmium, 70, 77, 99, 104
calcium, xii, 214, 239, 241, 246, 253, 261, 270
California, 101, 102, 184, 193, 205, 239, 293, 325, 339
calyx, 69
campaigns, 315
Canada, 105, 107, 108, 109, 143, 144, 147, 148, 150, 151, 179, 196, 201, 222, 235
canals, 68, 89
Canberra, 53, 62
cancer, 77, 167
capacity, 304
capelin, ix, 107, 110, 113, 118, 121, 123, 128, 129, 130, 131, 132, 133, 134, 136, 143, 144, 151
capture domain, x, 153, 154, 166, 167
carapace, 113
carbohydrates, 189, 334
carbon, x, xii, 7, 26, 30, 31, 33, 34, 35, 36, 37, 38, 41, 42, 49, 51, 52, 54, 55, 56, 57, 58, 60, 61, 62, 150, 177, 179, 182, 184, 186, 187, 197, 200, 201, 202, 203, 204, 225, 239, 243, 244, 253, 254, 257, 258, 275, 302, 304, 306, 308, 310, 311, 312, 313, 314, 315, 317, 324, 325, 339, 341, 343
carbon dioxide, 30, 324, 325
case study, 19, 104, 160, 162, 166, 234, 338
casein, 261
cash crops, 178, 179, 188
catabolism, 67
catalysis, 38, 48
catalysts, 38
catchments, 62
cation, 7, 30
cattle, 211, 214, 215, 216, 219, 224
cell, 30, 31, 73, 76, 78, 79, 80, 83, 88, 89, 90, 92, 97, 98, 157, 167, 269
cell cycle, 88
cell death, 30
cellulose, 39, 41, 169, 186, 187, 189, 260, 314
cement, 55
Central Europe, 311
cereals, 180, 187
certainty, 156
certification, 206
CES, 146, 147
changing environment, 155, 343
channels, 109, 114, 115, 216
chemical composition, 31, 184
chemical properties, 27, 29, 30, 31, 45, 60, 63, 93
chemical reactions, 316
chemical reactivity, 32
Chicago, 173, 312
children, 293
China, 179, 340
chloride, 250, 258
chloroform, 43, 53, 303
chlorophyll, xi, 210, 221, 223, 229, 233
chlorpyriphos, xi, 210, 226, 233
cholesterol, 72, 73
chromatography, 72
chromosomes, 164
circulation, xii, 81, 275, 281
classes, ix, 65, 68, 72, 73, 76, 80, 144, 145, 167, 194, 202, 259, 264, 284
classification, 56, 243
clay, viii, xiii, 5, 23, 30, 42, 52, 299, 315, 318
climate change, xii, 206, 275, 276, 278, 279, 280, 281, 282, 287, 288, 289, 292, 295, 296, 300, 313, 314
climatic change, 294
climatic factors, 45
cloning, 79, 99
Clupea harengus, ix, 107, 112, 115
cluster analysis, 17
clusters, viii, 9, 11, 13, 16, 17, 76
C-N, 334
CO2, vii, xiii, 3, 4, 5, 6, 7, 30, 33, 34, 35, 36, 54, 55, 184, 186, 187, 200, 204, 281, 313, 314, 315, 316, 317, 318, 319, 320, 321, 323, 324, 325
coccus, 260
coelom, 73, 76, 83
colloids, 32
colonization, 334
Columbia University, 292
combined effect, 230, 233
combustibility, 281, 283
commercial, ix, x, xii, 107, 108, 109, 112, 115, 122, 144, 149, 211, 214, 215, 275, 278, 283, 286
communication, 321

community, ix, x, xi, xii, 3, 24, 39, 40, 41, 42, 43, 45, 46, 49, 50, 57, 58, 60, 61, 71, 96, 108, 115, 149, 156, 168, 175, 177, 178, 179, 192, 193, 204, 205, 207, 210, 220, 221, 223, 224, 228, 229, 230, 235, 238, 239, 275, 280, 282, 283, 285, 286, 287, 336
compatibility, 182
compensation, 7
competition, 37, 43, 61, 109, 134, 144, 145, 146, 175, 198
competitiveness, 223
competitor, 144
complexity, 27, 31, 32, 60, 148, 154, 278, 289, 290, 337
components, vii, viii, xi, 23, 24, 25, 31, 38, 49, 84, 115, 137, 209, 210, 227, 232, 280, 290, 316
composite, 48
composition, xi, 5, 31, 32, 38, 39, 42, 43, 45, 47, 49, 50, 51, 53, 55, 57, 71, 118, 126, 127, 129, 130, 131, 139, 184, 210, 220, 223, 333, 336, 337, 338
compost, 179, 205
compounds, viii, xii, 4, 5, 27, 29, 30, 31, 32, 39, 41, 43, 47, 48, 49, 52, 53, 65, 66, 67, 71, 72, 74, 77, 80, 81, 82, 83, 85, 86, 87, 89, 94, 96, 97, 102, 103, 179, 194, 220, 233, 237, 241, 242, 243, 261, 262, 264, 269, 270, 272, 307, 311, 333
comprehension, ix, 66, 76, 97
computation, 157
concentration, 3, 4, 7, 31, 32, 34, 78, 82, 86, 88, 92, 94, 95, 96, 186, 191, 198, 211, 212, 214, 215, 216, 217, 218, 219, 223, 225, 227, 233, 236, 247, 263, 271, 286, 303, 306, 307, 309, 314, 334, 337
conception, 155
conceptual model, 41
conductance, 4
conductivity, 227
confidence, 124, 125, 138, 155, 166, 167, 289
confidence interval, 124, 125, 166, 167
Congress, 105, 297
conjugation, 74
Connecticut, 339
connective tissue, 90
consensus, 314, 315
conservation, viii, 11, 20, 23, 26, 33, 34, 35, 36, 43, 46, 48, 52, 54, 160, 174, 178, 182, 191, 200, 229, 285, 294, 297, 311
conspiracy, 295
constraints, x, 27, 32, 116, 117, 138, 139, 153, 154, 156, 157, 158, 159, 160, 163, 164, 165, 166, 168, 169, 170, 173, 286
construction, 333
consumers, 25, 117, 136, 137, 144
consumption, ix, x, xi, xiii, 108, 116, 118, 122, 124, 125, 128, 129, 130, 131, 132, 134, 135, 136, 137, 139, 141, 143, 148, 151, 178, 184, 209, 210, 308, 326, 329, 330, 332, 333, 334, 335, 336
consumption rates, 329
contaminant(s), ix, 48, 65, 66, 69, 70, 71, 76, 80, 97, 98, 99, 190
contaminated soils, 19, 179
contamination, 68, 69, 80, 98, 100, 205, 218, 233, 236
control, viii, 23, 30, 31, 34, 38, 42, 50, 56, 60, 66, 67, 70, 71, 72, 81, 82, 83, 84, 86, 87, 88, 90, 92, 96, 97, 100, 101, 103, 151, 166, 169, 173, 182, 187, 202, 204, 205, 210, 234, 300, 334, 335, 337
control group, 84
convergence, x, xii, 153, 275, 276, 278, 280, 287
conversion, xiii, 30, 33, 57, 73, 299, 300, 304, 305, 306, 308, 309, 310, 311, 314
convex, 36, 156, 157, 161, 168
Copenhagen, 202
copper, 239
coral reefs, 68, 69
corn, 184, 188, 189, 190, 191, 192, 194, 198, 201, 202, 204, 205, 206, 210, 211, 214
correlation, 197, 225, 306, 323, 324, 325
correlation coefficient, 323, 324
Costa Rica, 328, 338
costs, xii, 275, 276, 277, 278, 283, 284, 285, 293
cotton, 57, 190, 191, 200, 204, 236
coupling, 138
covering, viii, 23, 109, 222, 257, 314
creep, 70, 198
crinoid, ix, 65, 69, 76, 77, 78, 80, 81, 82, 84, 86, 92, 98, 101, 103
critical period, 190
critical value, 317
crop production, 181, 198, 199
crop residues, 33, 34, 37, 184, 186, 189, 193, 197
crop rotations, 37
crops, x, 33, 35, 36, 37, 177, 178, 179, 180, 181, 182, 183, 184, 186, 187, 188, 189, 190, 191, 192, 193, 194, 195, 196, 197, 198, 200, 201, 202, 203, 204, 205, 206, 210, 234
crystalline, 38
cultivation, viii, xii, 23, 33, 37, 43, 54, 145, 183, 185, 187, 189, 193, 198, 202, 203, 204, 230, 237, 241, 247, 249, 259, 261, 263, 264, 271, 272, 300, 304, 307, 308, 312
culture, 169, 211, 214, 221, 247, 264
C-viability, x, 153, 154, 155, 157, 158, 159, 160, 161, 162, 163, 165, 167, 168, 169, 170
cyanide, 194
cyanobacteria, 223, 224, 238

cycles, 7, 24, 26, 34, 45, 48, 66, 101, 164, 179, 220, 224, 225, 300
cycling, vii, 3, 4, 6, 7, 27, 31, 37, 38, 41, 42, 43, 44, 45, 46, 47, 52, 53, 56, 60, 62, 160, 169, 172, 179, 182, 184, 195, 200, 205, 300, 309, 336, 340
cypermethrin, xi, 210, 211, 212, 213, 217, 218, 219, 220, 222, 224, 225, 226, 227, 229, 231, 232, 233, 239
cyproterone acetate, ix, 65
cyst, 194
cytochrome, 99

D

dairies, 237
danger, 66, 290
data availability, 115
data set, 16, 50, 324
database, 264, 271
dating, 69
death(s), viii, 23, 30, 66, 116, 167, 171, 283
decay, viii, 23, 39
decision making, 20, 289, 290, 291
decisions, 11, 27, 28, 279, 280, 281, 282, 283, 286, 287, 288, 289, 295
decomposition, vii, xii, 3, 5, 6, 29, 30, 31, 32, 35, 37, 38, 39, 40, 47, 55, 56, 60, 63, 179, 184, 186, 187, 189, 192, 193, 194, 241, 260, 261, 270, 282, 300, 310, 312, 314, 316, 317, 318, 323, 325
decoupling, 46
deficiency, 6, 45, 189, 280
definition, 28, 50, 87, 281
deforestation, 311
degradation, vii, xii, 31, 38, 39, 42, 45, 48, 49, 63, 66, 70, 93, 178, 200, 201, 206, 275, 278, 300, 310
dehydration, 82
delivery, 283
demand, x, 27, 44, 177, 178, 189, 193, 200, 306
denitrification, 56, 190, 216, 237, 306, 314, 342
density, viii, 9, 11, 13, 16, 17, 18, 19, 20, 32, 53, 71, 77, 91, 155, 157, 158, 164, 169, 172, 174, 191, 198, 217, 218, 223, 284, 302, 304, 325, 330, 331, 337
Department of Agriculture, 291, 296
Department of the Interior, 277, 292, 296
dependent variable, 317, 324
deposition, vii, 3, 5, 6, 7, 55, 78, 92, 166, 167, 300, 335, 337, 338
deposits, 245, 257, 258
derivatives, 73, 100
desiccation, 186, 189, 191, 341
desorption, 43
destruction, 46, 160
detection, 9, 10, 11, 14, 15, 16, 18, 19, 32, 211, 218, 220
deterministic, 154, 156, 167
detritivory, xiii, 326, 334, 336
detritus, xiii, 25, 71, 116, 117, 123, 128, 140, 142, 149, 249, 254, 257, 326, 327, 332, 333, 334, 335, 336, 339, 343
developing countries, 178
developmental process, 92
deviation, 134, 136, 228, 231, 317, 319, 320
diet, 71, 117, 118, 126, 127, 129, 130, 131, 132, 133, 134, 137, 141, 143, 145, 146
diet composition, 126, 127, 130
differential equations, 155, 158, 167
differentiated cells, 80, 89
differentiation, 67, 75, 78, 79, 92, 97, 193, 258
diffusion, 290
diffusivity, 314, 318
digestion, 141, 303
dimorphism, 343
direct observation, 315
disaster, 291, 296
discharges, 214, 215
disequilibrium, 154, 173
dispersion, 172, 283
displacement, 86
dissociation, 31
distance sampling, viii, 9, 10, 11, 13, 15, 17, 18, 19
distribution, 13, 15, 49, 50, 51, 76, 93, 104, 141, 154, 161, 164, 167, 171, 181, 183, 205, 212, 242, 246, 252, 256, 292, 300, 302, 336, 339, 340, 342, 343
divergence, 86
diversity, viii, x, 23, 24, 26, 27, 28, 30, 31, 34, 37, 40, 41, 50, 55, 57, 60, 61, 62, 66, 97, 143, 170, 171, 177, 179, 191, 195, 200, 229, 237, 259, 264, 336
division, 72, 109, 114, 243
DNA, 102, 166, 170, 171, 172, 175, 264
dominance, x, 153, 163, 333, 340, 342
drainage, 35
Drosophila, 161, 164, 170, 172
drought, 190, 211, 233, 238, 276, 277, 315, 325
dry matter, 180, 186, 198
drying, 58, 262, 302, 306, 310, 312
duration, 278, 314
dynamic systems, x, 153
dynamical systems, 172

E

E.coli, 264, 265, 266, 269, 271
ears, 33
earth, viii, 23, 27, 40, 53, 200, 313

earthworm(s), 25, 197, 198, 201, 205
East Asia, 179
echinoderms, viii, ix, 65, 68, 69, 70, 73, 75, 76, 77, 78, 79, 80, 90, 92, 96, 97, 98, 99, 101, 103, 105
ecological restoration, 293, 312
ecology, vii, x, 11, 19, 31, 68, 101, 149, 154, 168, 174, 175, 177, 179, 194, 201, 205, 273, 289, 336, 338, 341, 343
economic growth, xii, 275, 276, 280, 281, 283, 287, 288, 289
economic growth rate, xii, 275, 288
economics, 54, 295
ecosystem, vii, viii, ix, x, xi, xii, xiii, 4, 5, 7, 23, 24, 25, 26, 27, 29, 30, 32, 33, 35, 38, 40, 41, 42, 43, 44, 46, 47, 48, 49, 50, 53, 57, 58, 61, 62, 63, 65, 67, 68, 69, 71, 76, 94, 103, 104, 107, 108, 109, 116, 117, 118, 121, 132, 133, 137, 138, 139, 141, 143, 145, 146, 147, 148, 149, 150, 151, 153, 179, 184, 187, 189, 190, 196, 197, 198, 199, 200, 207, 210, 235, 242, 262, 269, 275, 277, 278, 282, 290, 293, 295, 299, 300, 301, 302, 305, 306, 307, 309, 313, 324, 325, 326, 332, 334, 338, 339, 343
ecotoxicological tests, ix, 65, 66, 82, 87, 92, 95, 104
ecotoxicology, ix, 65, 66, 68, 92, 100, 236
Ecuador, 147, 330, 343
eddy-covariance technique, xiii, 313, 314, 315, 323, 324
education, 28, 233
egg(s), 76, 77, 79, 82, 86, 87, 145, 190
elaboration, 96
election, 164, 173, 174
electrons, 242
electrophoresis, 43
elongation, 222
emigration, 25
emission, 35, 36, 72, 187, 200, 252, 339
employment, 68, 76, 96, 278
endocrine, viii, ix, 65, 66, 67, 70, 72, 74, 75, 76, 77, 81, 83, 90, 92, 98, 100, 101, 102, 103, 104, 105, 230
Endocrine Disrupter Compounds (EDCs), viii, ix, 65, 66, 67, 68, 69, 70, 71, 72, 76, 77, 78, 80, 81, 82, 86, 87, 90, 91, 92, 97
endocrine system, 66
endocrinology, ix, 65, 72, 87, 97, 99, 101, 102
endogeneity, 157
endoskeleton, 68, 71
energy, viii, 23, 24, 25, 30, 32, 41, 116, 117, 137, 138, 172, 193, 200, 204, 300, 316, 324, 332, 334, 342
energy transfer, viii, 23
England, 98, 101, 293, 337
environment, ix, 4, 24, 26, 38, 45, 54, 66, 67, 68, 70, 72, 103, 139, 145, 154, 155, 156, 161, 164, 173, 174, 179, 192, 194, 197, 201, 203, 206, 210, 219, 227, 232, 233, 242, 243, 244, 253, 257, 258, 264, 282, 290, 295, 312, 334
environmental advantage, 179
environmental change, 23, 79, 139, 282
environmental chemicals, 66
environmental conditions, 43, 70, 79, 81, 146, 161, 343
environmental contaminants, 66, 97
environmental contamination, 68
environmental factors, 25, 41, 67, 187, 300
environmental impact, 236
environmental protection, 49
Environmental Protection Agency (EPA), 239, 296, 297
enzymatic activity, 55
enzyme(s), 38, 39, 40, 41, 43, 44, 48, 49, 53, 55, 56, 63, 73, 74, 192, 223, 242
ephemeroptera, xi, 210, 232
epidermis, 71
epithelia, 68
epithelium, 73, 76
equilibrium, 25, 71, 80, 154, 155, 160, 161, 164, 171
equipment, 18, 59, 284, 285
equity, 199
erosion, 34, 35, 37, 42, 180, 182, 203, 206, 210, 217, 218, 220, 257, 302
ERS, 199
estimating, viii, 9, 10, 11, 15, 303
estradiol, 72, 86, 88, 101, 103, 104
estrogen, 74, 75, 78, 86, 100, 103
ethanol, 82, 96
Ethiopia, 310
EU, 68, 96, 147
eukaryotes, 223, 242
Europe, 5, 7, 178, 179, 215, 291
European, 55, 63, 80, 99, 271, 311
European Commission, 99
European Parliament, 99
eutrophication, 224
evaporation, 182
Everglades, 293, 329
evidence, 4, 16, 34, 42, 66, 75, 99, 100, 101, 103, 146, 174, 221, 229, 230, 341
evolution, 19, 95, 153, 155, 170, 172, 262
exclusion, 277, 293
excretion, 68, 95
expansions, 16
experimental condition, 93, 94
expertise, 138, 146
exploitation, ix, 11, 108, 138

exposure, ix, 65, 66, 71, 76, 77, 78, 80, 81, 82, 84, 85, 86, 87, 88, 89, 90, 91, 92, 93, 94, 95, 96, 97, 98, 100, 101, 102, 218, 219, 222, 225, 227, 228, 230, 327, 332, 336
external environment, 79
extinction, viii, 23, 160, 163, 164, 169, 174
extraction, 43, 303, 310
exudate, 194
eyes, 69

F

faecal, 141
Fagus sylvatica, 57, 311
failure, 139
family, 75, 137, 163
famine, 310
FAO, 33, 54, 300, 310
farm, 211, 214, 215, 220, 225, 227
farmers, x, 26, 27, 28, 178, 182, 190, 199, 209, 210, 214, 217
farming techniques, 178
farms, 178, 211
fat, 224
fatty acid, 43, 46, 61, 141, 146, 193
fauna, x, xi, 80, 177, 187, 189, 190, 192, 195, 196, 197, 200, 203, 210, 217, 227, 228, 232, 336, 340
FDA, 48
feather star, ix, 65, 68, 69, 80, 98
feces, 113
feedback, 7
feet, 68, 69, 70, 71
females, 73, 74, 75, 83, 84, 85, 86
fenarimol, ix, 65, 78, 81, 84, 86
fertility, vii, 3, 7, 30, 31, 33, 37, 42, 43, 47, 49, 50, 51, 58, 98, 155, 156, 166, 169, 171, 172, 179, 205, 273, 309, 314
fertility rate, 155
fertilization, 34, 190, 191, 193, 215, 216, 232, 300, 311
fertilizer, x, 36, 37, 43, 177, 179, 181, 188, 189, 190, 191, 198, 199, 201, 204, 205, 215, 216, 232
fertilizers, 28, 34, 178, 261, 263, 273
filters, 70
financial support, 234
fire suppression, 276, 277, 278, 280
fires, 277, 278, 292, 293, 295, 296, 297
fish, ix, x, xi, 11, 107, 108, 115, 118, 119, 120, 121, 122, 123, 128, 129, 130, 131, 134, 135, 136, 137, 139, 141, 143, 144, 145, 146, 148, 149, 150, 155, 209, 210, 213, 223, 224, 225, 226, 227, 229, 232, 233, 236, 239, 339
fisheries, ix, x, 107, 108, 109, 115, 117, 118, 131, 132, 134, 135, 136, 137, 139, 141, 142, 143, 144, 146, 147, 149, 150, 151
fishing, ix, x, 70, 71, 107, 108, 109, 116, 137, 139, 141, 142, 143, 145, 146, 148, 149, 150, 151
fission, 70
fitness, 76, 155, 156, 157, 160, 161, 162
fixation, 35, 37, 46, 161, 179, 184, 216
flame, 66, 283, 295
flame retardants, 66
flexibility, 156
floating, 25, 71
flood(ing), 61, 216, 217, 219, 310, 341
flora, xi, 209, 213, 220, 232, 237, 242
fluctuations, 45, 59, 155, 160, 169, 173
fluid, 68, 73, 74, 78, 84, 102
focusing, 41
food, x, 25, 41, 60, 69, 75, 80, 81, 83, 108, 115, 116, 117, 118, 122, 131, 132, 135, 137, 139, 143, 144, 145, 146, 149, 151, 159, 175, 177, 178, 180, 182, 194, 195, 196, 197, 204, 337, 341, 343
food intake, 116, 341
food production, x, 177
forage fishes, x, 108, 110, 119, 120, 121, 122, 129, 130, 131, 135, 136, 137, 139, 143, 144, 145
forest ecosystem, 3, 4, 5, 6, 7, 25, 26, 32, 42, 43, 44, 47, 52, 57, 58, 280, 299, 300, 301, 304, 305, 306, 307, 308, 309, 312
forest management, 45, 51
Forest Service, 275, 276, 277, 278, 282, 290, 291, 292, 293, 296
forests, vii, 3, 5, 6, 25, 33, 43, 44, 47, 48, 51, 57, 59, 62, 71, 253, 255, 256, 264, 271, 278, 283, 293, 300, 301, 303, 307, 308, 333, 334, 337, 338, 339, 342, 343
fossil, 55, 178
fossil fuel, 178
fractionation, 32, 53, 214
fragility, 27
fragmentation, 230
France, 12, 14, 148, 149, 153, 214, 216, 235, 238
freshwater, 234, 235, 239, 338, 339
friction, 317
frost, 257, 260
fruits, 33
fuel, vii, xii, 34, 55, 178, 275, 276, 277, 281, 282, 293, 295, 296, 327
fumigation, xi, 32, 43, 210, 224, 225, 229, 233, 303, 310
fungi, 25, 41, 46, 49, 61, 63, 193, 195, 204, 205, 242, 264, 269, 271, 272, 334
fungus, 192, 260, 272
fur, 66, 67, 98, 101, 102

Fusarium oxysporum, 204
futures, xii, 275, 280, 288, 289, 290

G

Gadus morhua, ix, 107, 111, 147, 148, 150
gamete(s), ix, 65, 70, 76
gametogenesis, 72, 84, 98, 99, 100, 104
gel, 43
gene(s), xi, 67, 75, 101, 166, 241, 242, 264
gene expression, 101
General Accounting Office, 276, 278, 292
generation, 33, 165, 166, 262, 271
genetic drift, 161
genetics, x, 153, 154, 155, 160, 164, 166, 170
Geneva, 294
genome, 75
genotype, 164, 165
geochemical, 215
Geographic Information System (GIS), xii, 18, 82, 275, 281, 289, 290, 291,292, 293
Georgia, 141, 147, 184, 197, 261, 339, 343
germ cells, 75
Germany, 97, 98, 103, 206, 214, 215
germination, 32, 53, 194, 222
gland, 77
glass, 81, 93
global climate change, 313
global warming, vii
glucose, 30, 51
glycosides, 200
glyphosate, x, xi, 209, 210, 212, 213, 219, 220, 222, 223, 224, 225, 226, 228, 229, 231, 234, 238
goals, xi, 210, 233
gonads, 71, 72, 73, 74, 75, 76, 77, 78, 80, 82, 100, 104
government, 206, 280
G-protein, 75
GPS, 18
gracilis, 228, 229, 231
grains, 180
gram-negative bacteria, 260
graph, 84, 163
grass(es), xiii, 51, 61, 194, 204, 211, 214, 230, 243, 244, 250, 253, 254, 257, 258, 281, 283, 313, 315, 316, 317, 324
grasslands, 25, 33, 37, 58, 210, 301, 303, 315
grazing, viii, 23, 25, 28, 203, 283, 312, 315, 333, 334
Greece, 3
greenhouse, 36, 178, 200
greenhouse gas, 200
Greenland, ix, 60, 107, 110, 111, 112, 113, 115, 122, 123, 129, 133, 143, 144, 147
groundwater, 178, 190, 205, 211
grouping, 11, 20
groups, ix, xi, 9, 16, 17, 18, 31, 32, 40, 41, 43, 49, 67, 68, 81, 83, 87, 108, 111, 113, 115, 116, 118, 119, 120, 121, 122, 123, 128, 129, 130, 131, 132, 133, 134, 135, 136, 137, 139, 140, 142, 143, 144, 150, 232, 241, 247, 248, 261, 262, 270, 271
growth, vii, viii, ix, xii, 3, 4, 5, 7, 23, 26, 30, 31, 32, 43, 44, 45, 46, 47, 52, 59, 65, 67, 71, 74, 75, 77, 79, 80, 83, 88, 92, 97, 103, 104, 137, 145, 147, 156, 158, 166, 168, 169, 173, 178, 182, 183, 191, 194, 198, 200, 201, 205, 206, 211, 221, 222, 230, 235, 260, 262, 263, 264, 267, 269, 271, 272, 273, 275, 276, 277, 278, 279, 280, 281, 282, 283, 285, 287, 288, 289, 295, 300, 308, 310, 314, 334, 341
growth factors, 103
growth rate, xii, 4, 156, 169, 182, 262, 275, 281, 288, 295, 308
growth temperature, 262
guessing, 157
guidelines, 234
Gulf of Mexico, 343
Gulf of St. Lawrence, ix, 107, 108, 109, 111, 115, 132, 133, 139, 141, 147, 148, 149, 150, 151

H

habitat, xi, 25, 70, 80, 139, 145, 164, 179, 182, 210, 220, 230, 233, 261, 262, 264, 297, 337, 339, 340, 343
halibut, ix, 107, 110, 111, 112, 113, 115, 121, 122, 123, 124, 125, 126, 127, 129, 133, 143, 144, 147
Hamiltonian, 172
hands, 192
haploid, 160, 171, 172, 174
haplotypes, 161
hardness, 183, 227
harm, 282
harvesting, x, 28, 33, 45, 47, 59, 61, 62, 108, 184, 283, 300
Hawaii, 70
hazards, 291
head, 71
healing, 79
health, viii, xii, 26, 27, 28, 39, 40, 45, 50, 51, 59, 60, 65, 66, 78, 164, 178, 179, 180, 200, 206, 275, 277, 279, 283, 285, 286, 292, 295
health effects, 66, 283, 285, 286
health problems, viii, 65, 66
heart disease, 283
heat, 25, 278, 325
heavy metals, 55, 238, 252, 263, 271
height, 13, 48, 233, 315, 316

hemicellulose, 39, 189
hemp, 194
herbicide(s), xi, 62, 179, 198, 210, 212, 220, 221, 222, 223, 224, 229, 233, 234, 235, 238, 283
herbivory, xiii, 326, 332, 336
herbs, 25
herring, ix, 107, 110, 111, 112, 115, 118, 119, 120, 121, 122, 129, 143, 145
heterogeneity, 32, 60, 155, 341
heterotrophic, 113, 223, 260
heterozygotes, 166
hip, 145
histopathology, 83
Holocene, 232
homeostasis, 74
homeowners, xii, 275, 280, 282, 283, 284, 285, 286, 287
homogeneity, 82
Hong Kong, 330, 340
hormonal control, 90
hormone, 32, 72, 73, 74, 75, 86, 100, 105
horticulture, 205
host, 155, 169, 174
house(ing), 196, 273, 278, 281, 284, 292, 297
hue, 254
human activity, 300
human development, 276
humic substances, 38, 47, 51, 52, 53
humidity, 32, 184, 242, 262, 302, 316
humus, xii, 29, 31, 32, 43, 48, 57, 241, 243, 244, 245, 246, 249, 250, 253, 254, 257, 258, 260, 261, 270, 307
hunting, x, 108
hydrogen bonds, 31
hydrogen cyanide, 194
hydrolases, 38
hydrological, 26, 42, 47
hydrology, 25
hydrolysis, 66
hydrophobic, 31, 80, 217, 220, 233
hydrophobic interactions, 31
hydrophobicity, 96
hyperbolic, 319, 320
hypertrophy, 84, 86
hypothesis, 4, 30, 45, 75, 78, 86, 92, 101, 287, 288, 340

I

id, 139, 149, 150, 214, 218, 223, 296
Idaho, 62, 294
ignitability, 284, 291
images, 216
imbalances, 117
immigration, 25
immobilization, 30, 37, 43, 46, 189, 300, 302, 308, 309, 342
immune response, 174
immune system, 66
immunity, 67
immunocytochemistry, 83
immunotherapy, 167, 173
impact assessment, 289
implementation, 10, 51, 66, 138, 210, 285, 289
in situ, xi, 210, 218, 219, 224, 225, 232, 235, 236, 303
in vitro, 75, 78
in vivo, 75, 78, 83, 86, 101
incidence, 67, 276
inclusion, 165, 332
income, 301
incubation time, 325
independence, 17
India, 43, 57, 58, 299, 301, 310, 312
Indian, xiii, 299, 301, 308, 309, 312, 340
Indian Ocean, 340
indication, 42, 46, 47, 72
indicators, 27, 28, 29, 39, 40, 47, 50, 51, 60, 81, 109, 147, 191, 206, 221
indices, 27, 117, 132, 133, 134, 139, 141, 143, 144, 309
indigenous, viii, 23, 254
Indonesia, 332
inducer, 74
induction, 179
industrial wastes, 242, 252
industry, 264, 269, 271
infancy, 138
infection, 242
inferences, 13
infertility, 77
infestations, 276
infinite, 154, 155, 160, 165
Information System, 278, 291, 292, 293, 296
infrared spectroscopy, 49
inhibition, 53, 78, 80, 194, 221, 222, 235
inhibitor, 41
inhibitory effect, 80
initial state, 158, 159, 160, 165
injury(ies), xii, 79, 179, 221, 275, 278
innovation, 75, 297
insecticide(s), xi, 178, 210, 211, 212, 219, 220, 222, 229, 230, 233, 234, 236, 238, 239, 343, 252
insects, xi, 37, 169, 178, 180, 193, 209, 210, 333
insight, 41, 155
instability, 260

institutions, 66
instruments, 316, 319
insurance, 49, 284
integration, 290
integrity, 38, 49, 296
intelligence, 98, 282
intensity, viii, 23, 42, 219, 263, 277, 283
interaction(s), vii, ix, 24, 28, 30, 31, 32, 38, 39, 46, 48, 49, 53, 55, 63, 66, 86, 92, 108, 137, 138, 139, 141, 145, 151, 154, 155, 156, 160, 164, 167, 168, 169, 173, 194, 196, 233, 273
interface, viii, xii, xiii, 23, 201, 237, 275, 276, 277, 290, 291, 294, 295, 313, 333
interference, 84, 145, 277, 300
international law, 149
interpretation, 149
interrelations, 235
intersex, 67, 77
interval, 135, 155, 166, 167, 218, 253, 280, 288, 289
intrinsic value, 27
invertebrates, viii, ix, x, xi, 37, 65, 67, 68, 72, 74, 77, 99, 100, 101, 102, 105, 108, 112, 121, 122, 131, 134, 143, 144, 210, 232, 239, 327
investment, 282, 285
ions, 6, 214, 300
iron, 214
irrigation, 28, 193, 216, 342
isolation, xi, 230, 241, 243
Italy, 23, 44, 48, 53, 65, 81

J

Jamaica, 328
Japan, 57, 78, 100, 101, 104, 174, 177, 178, 179, 180, 184, 185, 186, 189, 190, 192, 195, 196, 198, 199, 204, 328
jellyfish, 112
jobs, 281
Jung, 194

K

K^+, 303
Kenya, 328, 329, 330, 331, 338, 342
kernel, x, 19, 153, 154, 157, 158, 159, 160, 161, 162, 163, 165, 168, 170, 174
killing, 187, 301
kinetics, 58, 95, 97, 189, 337
King, 337
Korea, 69, 103
Kyoto Protocol, 178, 206
Kyrgyzstan, xi, xii, 241, 242, 243, 244, 245, 247, 249, 251, 253, 254, 255, 257, 259, 261, 263, 264, 269, 271, 273

L

labor, 199
Lagrange multipliers, 169
lakes, 100
land, viii, xi, xii, xiii, 23, 27, 28, 31, 33, 34, 35, 36, 37, 41, 49, 50, 51, 56, 57, 58, 199, 200, 205, 210, 214, 245, 274, 275, 276, 277, 280, 281, 282, 283, 284, 285, 286, 287, 288, 289, 293, 295, 299, 300, 301, 304, 305, 307, 308, 309, 310, 311, 312, 315, 324, 340
land acquisition, 285
land use, xi, xii, xiii, 27, 28, 33, 36, 41, 51, 57, 199, 210, 275, 280, 281, 283, 285, 286, 287, 288, 289, 293, 295, 299, 300, 301, 304, 305, 307, 308, 309, 310, 312, 315
Land Use Policy, 295
landscapes, xii, 28, 60, 229, 242, 249, 252, 277, 282, 284, 290, 291, 292, 293, 296
land-use, xiii, 27, 28, 34, 58, 277, 295, 299, 300
larvae, 75, 76, 79, 99, 145, 164, 228, 339, 340
larval, 78, 158, 173, 239, 339, 342
laser, 300
Latvia, 273
laws, 154, 155
leaching, 5, 6, 47, 178, 179, 190, 191, 203, 206, 215, 216, 306
lead, 4, 5, 11, 15, 25, 39, 47, 70, 79, 94, 96, 97, 145, 146, 154, 160, 294, 300
leadership, 28
leakage, 216, 225
learning, 293
legume(s), 36, 37, 61, 180, 181, 182, 184, 187, 188, 189, 190, 191, 193, 194, 198, 201, 204, 206
lens, 154, 155
LFA, 43
liberation, 35
life span, 25
lignin, 39, 63, 186, 187, 189, 314
likelihood, 278, 296
limitation, 46, 56, 138, 146, 168, 173
linkage, 100, 173, 178
links, 140, 142
lipid, 75, 93
lipopolysaccharides, 60
literature, vii, 3, 4, 11, 16, 29, 96, 138, 153, 163, 230, 238, 311, 332, 336
livestock, 28, 37, 210, 220, 283
localization, 98

location, 10, 27, 69, 180, 244, 279, 340
locus, 160, 161, 164, 171, 172, 173, 174
logging, 47, 61, 278, 284
London, 7, 18, 100, 201, 203
long distance, xi, 210
longleaf pine, 7, 63
long-term impact, 46
Louisiana, 338
lumen, 76, 83, 84, 86
Luxemburg, 99
Lyapunov function, 160
lysis, 38

M

mackerel, ix, 107, 110, 112, 115, 118, 119, 120, 121, 122, 129, 143
macroaggregates, 193
macropores, 60
magnesium, 253
Maine, 237
maintenance, 155, 157, 159, 161, 163, 165, 167, 169, 171, 173, 175, 314
maize, 35, 36, 46, 61
major histocompatibility complex, 173
Malaysia, 329, 336
males, 74, 84, 85
Mallotus villosus, ix, 107, 112, 115
mammal, 18, 109, 110, 117, 118, 131, 132, 143, 144
management, vii, viii, x, xii, xiii, 9, 18, 20, 23, 24, 26, 27, 28, 30, 31, 33, 34, 35, 37, 40, 45, 46, 47, 48, 49, 50, 51, 53, 54, 56, 61, 63, 137, 141, 146, 149, 177, 178, 179, 181, 182, 184, 186, 187, 190, 191, 198, 199, 200, 201, 202, 203, 204, 205, 206, 211, 212, 229, 230, 237, 250, 251, 276, 277, 278, 279, 280, 281, 282, 283, 284, 285, 286, 287, 288, 289, 290, 293, 294, 295, 296, 297, 299, 310, 311, 312
management practices, 45, 47, 184, 200, 203, 230, 282, 283, 285
manganese, 6, 7
mangroves, xiii, 326, 327, 328, 331, 332, 333, 334, 335, 336, 337, 338, 342
manipulation, 235
manure, 179, 202, 205, 215, 261, 301
mapping, 149, 281
marine environment, 67, 68, 70, 336, 340
market, 200, 283, 284, 286
market prices, 283
market value, 284
marsh, xiii, 216, 217, 218, 326, 327, 330, 332, 333, 334, 335, 336, 337, 338, 339, 340, 341, 343
Maryland, 205
mass loss, 39, 55, 56
mass spectrometry, 72
Massachusetts, 191
mathematics, 155, 170
matrix, 163, 168
maturation, ix, 65, 72, 75, 84, 85, 86, 102, 198
Mauritania, 338
Mauritius, 338
MBP, 44
meanings, 26
measles, 174
measurement, 32, 53, 154, 221, 315, 317, 319
measures, 28, 42, 46, 279, 282, 283, 284, 285
mechanical structure, xi, 241, 243, 244, 245, 247, 249, 250, 253, 254, 257, 258, 270
Mediterranean, 5, 39, 59, 71, 80, 215, 325
Mediterranean climate, 59
medusa, 71
membership, x, 153, 156
Merck, 82
mercury, 25
metabolic pathways, 73
metabolism, 31, 32, 43, 45, 53, 67, 70, 74, 77, 78, 93, 99, 100, 101, 104, 105, 242
metabolites, xi, 73, 241, 242, 252
metalloids, 49
metals, 49, 55, 98, 225, 238, 252, 263, 271
metamorphosis, 156, 230
meteorological, 315, 316, 325
methanogenesis, 339
methyl cellulose, 169
methyltestosterone, ix, 65, 78, 81
Mexico, 44, 59, 202, 339, 342, 343
mice, 101
microbes, 193, 272
microbial, viii, xi, xiii, 24, 29, 30, 31, 32, 38, 39, 40, 41, 42, 43, 44, 45, 46, 47, 48, 49, 52, 53, 55, 56, 57, 58, 59, 60, 61, 63, 66, 180, 184, 186, 189, 191, 192, 193, 195, 200, 202, 205, 207, 241, 242, 272, 273, 299, 300, 301, 302, 303, 306, 307, 308, 309, 310, 311, 312, 314, 333, 340
Microbial activity, 300
microbial cells, 311
microbial community(ies), 38, 39, 40, 41, 42, 43, 45, 46, 49, 55, 57, 58, 60, 61, 63, 192, 193, 207
microclimate, 42, 48
microcosms, 55
micrometeorological, 324
micronutrients, vii, 3
microorganism(s), viii, xi, 6, 7, 23, 27, 32, 38, 45, 46, 47, 49, 59, 61, 63, 184, 191, 192, 197, 200, 241, 242, 252, 259, 260, 261, 262, 264, 270, 271, 273, 300

microscope, 82, 195, 196
microscopy, xi, 49, 82, 241
migration, 84, 88, 92, 116, 155, 161, 162, 173, 174
Millennium, 292, 294
mimicry, 172
mineralized, 44, 193, 204
minerals, 27, 38, 49, 71
mining, 179
Minnesota, 281, 294
missions, 55, 312
Missouri, 275, 295
mites, 193, 197
mitochondrial DNA, 170, 171
mitotic, 74
mixing, 71, 115, 214, 227, 340
modeling, xii, 20, 275, 276, 279, 282, 283, 289, 292
models, viii, ix, x, xii, xiii, 16, 41, 65, 66, 67, 68, 76, 80, 86, 92, 94, 96, 97, 109, 113, 116, 129, 130, 131, 132, 133, 134, 137, 138, 141, 143, 144, 145, 147, 148, 149, 150, 151, 153, 154, 155, 160, 167, 169, 170, 172, 198, 199, 275, 281, 282, 289, 290, 292, 325, 326
modules, 54
moisture, xiii, 31, 42, 43, 44, 45, 47, 48, 203, 299, 301, 302, 303, 304, 305, 306, 307, 308, 310, 311, 312, 313, 314, 315, 316, 317, 318, 319, 320, 321, 323, 324
moisture content, 42, 45, 304, 307, 314, 316, 317, 320, 323, 324
mold, 197
molecular mechanisms, 101
molecular weight, 4, 32, 52, 223
molecules, 31, 32, 71, 78
mollusks, 86
monoclonal antibody, 83
Montana, 281, 289, 293, 295
Moon, 101
moratorium, 115
morning, 302
Morocco, 33, 54
morphogenesis, 78, 79, 92
morphology, 42, 264, 340, 342
morphometric, 82
mortality, ix, x, xii, 108, 116, 123, 128, 133, 139, 141, 142, 144, 145, 147, 148, 151, 158, 159, 218, 219, 225, 227, 228, 229, 232, 275, 278, 283, 309, 339
mosaic, 25
Moscow, 272, 273, 274
motion, 149
mountains, 242, 243
movement, 10, 29, 42, 66, 169, 282
Mozambique, 340
mtDNA, 167
multiplication, 222
multiplicity, 40, 41
muscle cells, 76
muscles, 90
mustard oil, 200
mutagen, 263, 271
mutant, 161
mutation(s), 164, 169, 174
mycelium, 262
Myrtus communis, 39, 56

N

nation, 292, 295
native species, 221
natural environment, ix, 45, 49, 66, 161
natural resource management, 289, 291
natural resources, 41, 277
natural selection, 154, 164
Nebraska, 203, 216, 234
needles, 61
negative relation, 145
nematode(s), 37, 50, 112, 177, 193, 194, 195, 196, 202, 204, 205
nerve(s), 72, 73, 76
nervous system, 78
net migration, 116
Netherlands, 98, 99, 100, 101, 104, 105, 292, 313, 315, 323, 324, 325, 338
network, 117, 138, 147, 162, 163, 205
neuroendocrine, 67, 70, 72, 80, 81, 90
New Jersey, 99, 100
new media, 100
New Orleans, 203
New York, 19, 51, 54, 56, 60, 98, 99, 100, 149, 173, 174, 200, 201, 203, 204, 206, 236, 292, 293, 294, 324, 336, 339, 340, 341
New Zealand, 44, 51, 58, 62, 216, 237, 312
Nigeria, 338
NIH, 271
nitrate(s), 6, 25, 189, 190, 205, 206, 214, 215, 216, 232, 233, 235, 302, 303, 305, 306, 307, 310, 334
nitrification, xiii, 6, 45, 299, 301, 302, 305, 306, 307, 309, 310, 311, 314, 333
nitrifying bacteria, 41
nitrogen, vii, x, xii, xiii, 3, 7, 25, 41, 42, 43, 46, 56, 57, 58, 59, 61, 177, 179, 180, 181, 184, 187, 188, 190, 194, 195, 196, 198, 201, 202, 203, 204, 205, 206, 216, 232, 236, 237, 241, 243, 244, 245, 250, 253, 254, 257, 258, 260, 270, 299, 300, 302, 304, 306, 307, 308, 309, 310, 311, 312, 314, 333, 334, 337, 342, 343

nitrogen fixation, 46, 179
N-mineralization, xiii, 299, 300, 301, 302, 303, 305, 306, 307, 309
NMR, 32, 52, 53
nodes, 162, 163
North America, 25, 178, 215
North Atlantic, 147, 149
North Carolina, 190, 216
Norway, 61, 146
nuclear receptors, 75
null hypothesis, 287, 288
nutrient cycling, vii, 3, 6, 31, 37, 38, 42, 43, 44, 45, 46, 53, 60, 179, 182, 200, 309
nutrients, vii, viii, 3, 4, 5, 6, 23, 25, 27, 29, 30, 31, 32, 37, 38, 40, 41, 42, 43, 46, 47, 48, 59, 61, 62, 71, 72, 84, 86, 160, 173, 178, 179, 180, 182, 187, 190, 200, 232, 262, 300, 302, 306, 307, 308, 309, 310, 315, 334
nutrition, vii, 3, 4, 6, 7, 45, 258, 300, 310, 314, 340, 341

O

observations, viii, xi, 7, 9, 10, 73, 75, 84, 85, 96, 116, 138, 145, 159, 169, 210, 224, 226, 229, 230, 232, 233, 315, 323
oceans, 68
odonata, xi, 210, 232
OECD, 100
oil(s), vii, viii, 3, 5, 6, 23, 26, 27, 28, 30, 35, 38, 40, 41, 45, 47, 51, 52, 53, 54, 56, 57, 58, 59, 60, 61, 100, 177, 195, 196, 197, 200, 201, 202, 204, 205, 210, 214, 262, 264, 273, 300, 302, 309, 310, 311, 312, 314, 324
oocyte(s), 72, 74, 75, 82, 83, 84, 85, 86, 87, 102
oogenesis, 74, 84, 85, 86
optimism, 295
optimization, 37
ores, 30, 322
organic chemicals, 230
organic compounds, xii, 5, 27, 29, 30, 31, 41, 48, 241, 261, 270, 272, 307, 311
organic matter, viii, x, 23, 24, 25, 27, 29, 30, 31, 33, 34, 36, 37, 38, 41, 42, 44, 45, 46, 47, 48, 49, 50, 51, 52, 53, 54, 55, 56, 58, 59, 60, 63, 70, 113, 137, 177, 179, 180, 182, 184, 187, 193, 195, 197, 200, 205, 218, 225, 233, 282, 300, 307, 308, 309, 311, 312, 313, 314, 315, 316, 325, 326, 334, 335, 337, 340
organic solvent, 96
organism, viii, x, 23, 38, 46, 66, 67, 70, 74, 92, 93, 155, 156, 177, 200, 202, 219, 227, 311
organization, 25, 76, 89, 242
organotin compounds, 80
orientation, 13
osmium, 82
ossicles, 91
osteoclasts, 78, 92
osteocytes, 78, 92
osteoporosis, 78
Ottawa, 235, 238
ovaries, 74, 86, 98, 99, 100, 102, 104, 105
overproduction, 223
oxidation, 32, 54, 57, 334
oxide(s), 6, 38, 104, 312
oxygen, 31, 45, 79, 242, 333, 337, 341
oxygenation, xiii, 71, 326, 333, 336

P

Pacific, 151, 290, 291, 332, 337, 340
paints, 77, 98
Pakistan, 310
Panama, 329
Pandalus borealis, ix, 107, 112
Paracentrotus lividus, ix, 65, 71, 77, 80, 81, 98, 101, 102
parameter, 19, 38, 83, 86, 87, 88, 90, 116, 124, 125, 137, 138, 139, 150, 193
parameter estimates, 116, 150
parameter estimation, 137
parasite(s), 25, 169, 174
Paris, 153, 172, 200
Parliament, 99
partial differential equations, 167
particles, 71, 113, 214, 217, 218, 257, 264, 273, 335
particulate matter, 66, 70, 225, 283, 297, 334
partition, 211, 233
passive, 217, 218, 337, 338
pasture(s), 36, 57, 58, 59, 198, 215, 216, 300, 311, 325
pathogenic, 204
pathogens, 174, 202
pathways, 41, 73, 138, 145, 333, 341
PCBs, 77, 80, 90, 99
PCR, 264
peat, 58, 257, 325
pectin, 39, 44
penicillin, 264
perceptions, 293
peritoneum, 76
permeability, 257
personal, 73, 75, 85, 281, 321
personal communication, 321
pesticide(s), xi, 28, 38, 67, 81, 178, 194, 209, 210, 211, 212, 213, 217, 218, 219, 220, 221, 222, 223,

224, 225, 226, 227, 228, 229, 230, 231, 232, 233, 234, 236, 237, 238
pests, 178, 179, 182, 197
pH, 5, 33, 42, 55, 214, 239, 243, 244, 245, 246, 249, 250, 253, 254, 255, 257, 258, 259, 302, 304, 314
phagocyte, 84
pharmaceutical(s), 66, 264, 269, 271
pharmacology, xii, 212
phase space, 163
phenol, 49, 78, 97
phenolic acids, 32, 53, 194
phenotype(s), 67, 156, 159
Philippines, 297
phosphatases, 48
phosphate(s), 214, 237, 238, 273
phospholipids, 46, 60
phosphorus, vii, xii, 3, 41, 56, 58, 214, 223, 224, 237, 238, 241, 243, 244, 253, 257, 258, 270, 302, 303
photosynthesis, 4, 25, 34, 184
Phragmites australis, 339
phylum, 68, 69
physical environment, 24
physical properties, 45
physicochemical, 302
physico-chemical characteristics, 41
physico-chemical properties, 93
physics, 154
physiological constraints, 116
physiology, 31, 68, 72, 96, 97, 105
phytoplankton, 113, 116, 223
Pinus laricio, 30, 48, 53, 63
planning, ix, 28, 66, 76, 93, 94, 96, 285, 288, 290, 294
plants, vii, 4, 5, 6, 7, 25, 29, 31, 34, 38, 44, 46, 48, 52, 59, 60, 61, 69, 71, 101, 113, 156, 170, 179, 182, 183, 189, 193, 194, 195, 201, 221, 222, 223, 225, 233, 237, 264, 266, 269, 271, 292, 300, 301, 306, 308, 314, 317, 332, 340, 343
plasticity, 159
pneumonia, 283
Poland, 164, 174, 204
polar body, 72
policy makers, 27
pollen, 197
pollutants, viii, 65, 66, 67, 68, 71, 80, 230, 297
pollution, 66, 69, 70, 98, 100, 103, 180, 190, 219, 238, 252
polyethylene, 303, 310
polymorphism, 153, 156, 157, 160, 161, 162, 164, 165, 166, 171
polyphenols, 53
pools, 31, 32, 43, 44, 46, 47, 57, 58, 182, 183, 201, 211, 300
poor, xi, 6, 16, 28, 43, 138, 146, 148, 164, 173, 210, 259, 260, 261, 271, 286, 334, 335
population, x, 10, 11, 15, 16, 18, 20, 29, 38, 41, 44, 45, 49, 67, 71, 73, 76, 78, 86, 100, 102, 131, 138, 144, 145, 147, 151, 153, 154, 155, 160, 161, 164, 166, 167, 169, 170, 171, 172, 173, 174, 177, 178, 191, 192, 193, 194, 195, 196, 197, 198, 200, 218, 219, 222, 224, 225, 226, 231, 236, 262, 264, 273, 278, 283, 301, 308, 338, 341
population density, 16, 18
population growth, 169, 173, 178, 222
population size, 15, 86, 138, 155, 166
porosity, 31, 45
Portugal, 9, 13, 328, 339, 343
positive relation, 30, 306, 309
positive relationship, 30
potassium, xii, 58, 241, 243, 244, 253, 257, 258, 270
potato, 247
Potomac River, 237
poverty, 178
power, 97, 155, 170
precipitation, viii, 5, 23, 25, 43, 281, 323
predatory fishes, ix, x, 107, 108, 110, 118, 119, 120, 121, 122, 130, 131, 135, 136, 137, 139, 141, 143, 144, 145
prediction, 145, 146, 199
preference, 29, 338
pressure, 139, 301, 316
prevention, 276
prices, 279, 283, 284, 285
prior knowledge, 13
probability, x, 9, 10, 14, 15, 16, 17, 153, 154, 155, 156, 161, 166, 278, 282, 283, 284, 285, 286, 287, 288, 292
probability distribution, 154, 161
probe, 174
producers, 25, 27, 117, 242, 243, 252, 259
production, vii, x, xi, xiii, 3, 5, 6, 7, 31, 35, 36, 37, 46, 50, 54, 56, 63, 72, 73, 86, 116, 143, 147, 151, 177, 179, 181, 184, 193, 198, 199, 200, 201, 203, 204, 205, 206, 210, 222, 223, 224, 233, 309, 310, 313, 314, 317, 326, 327, 332, 333, 334, 335, 336, 337, 338, 339, 343
productive capacity, 108, 139
productivity, vii, x, 26, 27, 28, 29, 31, 33, 37, 38, 40, 45, 46, 49, 50, 51, 52, 57, 59, 60, 139, 146, 147, 150, 177, 178, 200, 201, 203, 272, 282, 300, 333, 338, 341, 342
profits, 182
progesterone, 72, 73, 99, 102, 104
program, 82, 146, 148, 206, 276, 289, 316, 325

programming, 289
prokaryotes, 242
proliferation, 74, 78, 79, 80, 83, 88, 92, 98
promote, 27, 37, 38, 75, 84, 86, 277
promoter, 74, 86
prosperity, xi, 241, 252, 270
prostate, 77
protective mechanisms, 263, 271
protein(s), 25, 32, 41, 67, 73, 75, 86, 92, 99, 100, 103, 271, 314
protein structure, 92
protein synthesis, 32, 100
protocol(s), 16, 82, 83, 144, 167, 225, 227
protozoa, 113, 193, 195, 201
public health, 283
pulse(s), xi, 209, 217, 218, 219, 232, 233, 238, 311
pumps, 227

Q

Quercus ilex, 39, 56

R

radial distance, 10
radiation, 47, 316
radius, 13
rain, 25, 44, 178, 180, 190, 211, 213, 214, 216, 217, 218, 219, 220, 229, 231, 232, 233, 311, 319
rain forest, 311
rainfall, 37, 44, 182, 193, 212, 219, 224, 302, 307
rainforest, 59
rainwater, 211, 214, 216
randomness, 11
range, ix, 10, 11, 14, 28, 29, 31, 32, 38, 46, 47, 48, 65, 66, 67, 138, 156, 173, 179, 180, 189, 190, 200, 211, 212, 230, 282, 304, 311, 312
rangeland, 216, 325
reaction rate, 316, 317
reactivity, 32
reagents, 174
realism, 137
reality, 97
recall, 72
receptors, 75, 92, 99, 103
reclamation, 310
recognition, viii, 23, 178
recombination, 160, 162, 172
reconcile, 169
recovery, x, xi, 41, 62, 71, 83, 84, 108, 109, 141, 144, 145, 146, 147, 150, 154, 170, 174, 187, 189, 190, 191, 193, 201, 204, 206, 210, 221, 239, 309
recreation, 216
recurrence, 161
recycling, viii, 23, 44, 80, 89, 300
redevelopment, 309
redfish, ix, 107, 115, 118, 121, 122, 128, 129, 131, 132, 139, 143, 144, 149
redistribution, 42, 47
reduction, xi, 4, 33, 34, 36, 86, 87, 105, 143, 145, 154, 178, 179, 183, 210, 218, 220, 221, 222, 223, 224, 225, 229, 246, 261, 277, 278, 284, 285, 295, 307, 309, 310, 317, 324, 334
redundancy, 41
refractory, 31, 32
regenerate, 70, 79, 80, 81, 82, 87, 88, 89, 91
regeneration, 32, 48, 63, 72, 76, 78, 79, 80, 81, 82, 87, 90, 91, 98, 103, 104, 333, 337, 338, 340
regional, xi, xii, 26, 210, 224, 232, 275, 279, 280, 281, 282, 283, 285, 286, 287, 289
regression, 309, 320
regression equation, 309
regrowth, 62
regulation(s), 26, 38, 66, 68, 75, 101, 137, 157, 161, 162, 165, 166, 168, 172, 200, 242, 285, 337
regulators, 67, 92
Reinhardtius hippoglossoides, ix, 107, 111, 115
relationship(s), 4, 29, 30, 31, 39, 40, 49, 56, 57, 78, 84, 87, 88, 101, 104, 138, 145, 163, 181, 194, 196, 197, 198, 199, 268, 308, 309, 314, 325, 336
relatives, 20
relevance, viii, ix, 41, 52, 65, 66, 67, 68, 93, 97, 221, 229
reliability, 286, 319
remediation, 49, 179
repair, 79
repetitions, 160
replication, 41
repression, 43
reproduction, ix, 65, 67, 70, 72, 75, 76, 77, 79, 97, 99, 156, 221
reproductive organs, 77, 82
reptile, 239
reserves, 47, 312
residential, xii, 275, 278, 280, 281, 284, 287, 288, 289, 296
residual error, 117
residues, 31, 33, 34, 35, 36, 37, 46, 47, 60, 61, 69, 178, 182, 184, 186, 187, 189, 192, 193, 194, 197, 236, 315
resilience, 26, 49, 149, 174, 179
resin, 193
resistance, 41, 169, 178
resolution, 41, 155, 280, 281, 282, 285, 286
resource availability, 46

resource management, 40, 282, 289, 291
resources, 26, 27, 28, 31, 41, 46, 118, 131, 132, 133, 139, 143, 144, 145, 146, 156, 168, 178, 200, 277, 292, 296
respiration, xiii, 29, 34, 36, 39, 44, 47, 57, 58, 68, 116, 184, 303, 313, 314, 315, 316, 317, 318, 319, 320, 321, 322, 323, 324, 325, 333, 341
respiratory, 71, 283
respiratory problems, 283
responsiveness, ix, 65, 76
restructuring, 47
retardation, 230
retention, 4, 30, 41, 47, 216, 218, 238
returns, 163
Reynolds, 200
Rhizobium, 61
Rhizoctonia solani, 193, 204
rhizosphere, 4, 5, 42, 59
rice, 183, 184, 185, 189, 190, 195, 202, 203, 204, 236, 237, 278, 289
rice field, 236, 237
rings, 169
Rio de Janeiro, 206
Río de la Plata, 238
riparian, xi, 58, 210, 214, 216, 217, 220, 221, 223, 229, 233
risk, viii, xii, 9, 37, 41, 67, 94, 97, 229, 275, 276, 278, 280, 282, 284, 285, 287, 288, 289, 291, 292, 294, 295, 296
risk assessment, 67, 97, 294
risk management, 295
rivers, 66, 211, 247, 295
robustness, 11, 16, 138
Rome, 54, 310
room temperature, 303
rotations, 37, 54, 178
Royal Society, 60
rubber, 93
runoff, xi, 35, 180, 210, 212, 213, 214, 215, 216, 217, 218, 219, 220, 222, 223, 227, 228, 229, 230, 232, 233, 236, 238, 278
rural areas, 210

S

SA, 205, 271, 293, 339
sacrifice, 82
safety, 28
salinity, 79, 93, 314, 341
salinization, viii, 23, 258
salt(s), xiii, 250, 253, 257, 258, 326, 327, 328, 330, 331, 332, 333, 334, 335, 336, 337, 338, 339, 340, 341, 343
sample, 11, 13, 15, 16, 166, 167, 222, 288, 302, 303, 316
sampling, 9, 11, 12, 13, 15, 16, 18, 19, 20, 45, 211, 218, 219, 220, 225, 228, 244, 263, 302
saprophyte, 262
Saudi Arabia, 338
savannah, 59
scatter, 317, 319, 323
scheduling, 193
schema, 181
school, 19, 271
science, vii, 49, 51, 54, 137, 154, 183, 190, 192, 195, 196, 242, 272, 273, 291, 292, 293
scientific community, 3, 96
scientific understanding, 138, 291
SCN, 194
Scomber scombrus, ix, 107, 112, 115
Scots pine, 61
sea level, 301
sea urchin, ix, 65, 68, 71, 72, 73, 74, 75, 78, 80, 81, 82, 83, 84, 85, 86, 97, 100, 101, 102, 104, 112
seals, x, 100, 108, 109, 110, 111, 113, 115, 118, 119, 120, 123, 124, 125, 126, 127, 128, 129, 130, 132, 133, 134, 135, 136, 137, 139, 141, 143, 144, 145, 148, 151
search(ing), x, 71, 153, 160, 206, 271
seasonal pattern, 304, 306, 307
seasonality, 43, 44
Sebastes spp., ix, 107, 115
secondary data, 284
secretion, 38
security, 11, 163, 178, 204
sediment(s), xi, xiii, 66, 68, 69, 70, 71, 72, 80, 81, 169, 209, 210, 211, 212, 215, 217, 218, 219, 220, 222, 225, 227, 232, 233, 235, 236, 237, 238, 239, 330, 333, 334, 335, 336, 339, 340, 341, 342, 246, 247, 257, 326, 331, 332, 333, 334, 335, 336, 337, 339, 340, 343
seed(s), xiii, 32, 33, 34, 53, 62, 198, 199, 206, 222, 223, 233, 326, 333, 334, 336, 338, 342
seedling(s), 32, 59, 63, 182, 191, 193, 194, 198, 202, 210, 221, 237, 310
segregation, 16
self-awareness, 282
self-regulation, 200, 242
sensitivity, ix, 27, 65, 76, 97, 137, 138, 146, 219, 222, 230, 232, 234, 325
separation, 32
sequencing, xi, 241
series, 16, 25, 82, 138, 262, 271, 274, 284
serotonin, 103
severity, 293
sex, 67, 73, 74, 84, 100, 101

sex steroid, 101
sexual development, 99
sexual dimorphism, 343
sexual reproduction, 70
shape, 16, 70, 89, 155, 160
sharing, 118
sheep, 20, 216, 224
shelter, 69
shoot, 35, 303
short supply, 145
shortage, 37
shrimp, ix, 107, 112, 115, 129, 130, 131, 132, 133, 139, 143, 144, 148, 151, 340, 341
shrubland, 39
shrubs, 25
Siberia, 262, 272
side effects, 28
sign(s), 90, 108, 160
signalling, 97, 101, 103
signals, 70
silver, 48, 112, 243, 250
similarity, 42, 92, 118, 284
simulation, xii, 171, 276, 282, 289, 290, 292, 295
Singapore, 328
sinus, 73, 76, 86
sites, xi, 5, 35, 42, 44, 48, 58, 72, 73, 75, 101, 210, 211, 212, 218, 219, 220, 221, 224, 225, 228, 229, 230, 231, 244, 259, 260, 263, 271, 284, 301, 302, 306, 307, 309, 324
skeleton, 250, 253, 254, 257
skin, 27
smoke, 279, 283, 285, 286, 291
snakes, 71
SOC, 182, 184, 185, 192, 200
social behaviour, viii, 9
social network, 162
social sciences, 162
society, 51, 149, 163, 207
soft substrate, 72
software, 15, 16, 17, 18, 117, 294, 316
soil acidification, 7
soil erosion, 180, 182
soil organic C, 33, 37
soil particles, 214, 217, 264, 273
solar, 47
solid phase, 30
solubility, 4, 32, 66
solvent, 81, 82, 84, 86, 87, 88, 96
somatic cell, 75
sorting, 196
South Africa, 217, 218, 238, 311, 328, 330, 338, 340
South America, 179, 311
South Carolina, 191
soybean(s), x, xi, 190, 191, 194, 198, 204, 209, 210, 211, 214, 215, 219, 220, 221, 226, 229, 230, 231, 232, 233, 234
Spain, 9, 13, 14, 20, 55, 324, 342
spastic, 228
species richness, 221, 230
spectroscopy, 49
spectrum, 76, 97, 242, 247, 249, 262
speculation, 341
speed, 181, 182, 314
sperm, 75, 77, 78, 103
spermatogenesis, 74, 75, 84, 86, 97, 100
spermatogonia, 74
spermatozoon, 85
spermiogenesis, 75
spore, 260, 264
SPSS, 303, 312
stability, xi, 25, 37, 38, 40, 42, 46, 54, 66, 154, 163, 172, 175, 184, 200, 241, 270, 300, 309
stabilization, 30, 31, 192, 197, 220
stages, 25, 44, 74, 75, 78, 79, 82, 90, 98, 145, 146, 223, 230
stakeholders, xii, 275, 276, 280, 289
standard deviation, 116, 134, 136, 228, 231, 319
standard error, 317
standards, 51, 272, 283
starch, 261
stars, 68, 69
starvation, 146, 271
statistical analysis, 82
statistics, 284, 289, 294
stem cells, 79, 80, 90
sterile, 264
steroid, 99, 101
steroid hormone(s), 101, 105
steroids, 67, 72, 74, 75, 78, 90, 92, 99, 102, 103, 104
stochastic, 154, 155, 156, 157, 167
stock, 33, 34, 35, 82, 115, 146, 147, 150, 164, 307
stomach, 141
storage, 58, 61, 66, 72, 74, 84, 182
storms, 211
strain, 242
strategies, x, 11, 28, 46, 49, 56, 155, 156, 158, 159, 160, 167, 172, 177, 181, 193, 206, 228, 278, 295, 310
stratification, 13
strength, 5, 164, 204
Streptomyces, xi, xii, 241, 242, 243, 246, 247, 248, 249, 251, 252, 253, 254, 255, 256, 259, 260, 261, 262, 263, 264, 265, 266, 267, 269, 270, 271, 272, 273
stress, xi, 40, 41, 46, 49, 55, 61, 78, 155, 156, 179, 209, 230, 232, 308

stress factors, 46, 230
stroke, 283
strong interaction, 38
structuring, 143
substitution, x, 153, 200
substrates, viii, 23, 30, 72
Sudan, 194
sulphur, 58
summer, x, xiii, 14, 44, 109, 115, 177, 178, 182, 184, 187, 190, 198, 199, 211, 221, 242, 249, 254, 256, 257, 260, 262, 299, 302, 303, 305, 306, 307, 308, 315, 325
Sun, 340
superimposition, 67
supervision, 210
supply, 5, 6, 7, 29, 30, 37, 62, 63, 143, 145, 182, 187, 188, 197, 300, 311
suppression, x, xii, 177, 179, 201, 206, 275, 276, 277, 278, 280, 284, 290
surface area, 30, 330, 333
surface structure, 48
surplus, 178
surprise, x, 153, 154, 155
survey design, 11
survivability, 285
survival, viii, 23, 67, 98, 145, 151, 155, 156, 158, 159, 169, 174, 198, 199, 218, 219, 224, 230, 231, 252, 263, 271, 343
survival rate, 218, 219, 224
survivors, 69
susceptibility, ix, 35, 65, 68
suspensions, 218
sustainability, viii, x, 24, 27, 28, 33, 37, 45, 47, 49, 50, 177
sustainable development, 206
swamps, 315
Sweden, 230
switching, 137
Switzerland, 294
symbiosis, 4
symbiotic, 216
symmetry, 68, 70
symptoms, 228
synthesis, 32, 67, 73, 77, 100, 291, 310, 312
systems, viii, ix, x, xi, 23, 27, 30, 33, 34, 35, 36, 37, 38, 44, 45, 46, 47, 50, 51, 54, 56, 58, 61, 107, 109, 115, 153, 154, 155, 159, 160, 162, 171, 172, 174, 175, 177, 178, 179, 182, 184, 186, 187, 192, 194, 196, 197, 200, 203, 205, 209, 232, 233, 235, 236, 237, 242, 291, 294, 295, 311, 316, 325, 341

T

Taiwan, 57, 328, 342
tanks, 96
tannins, 194, 334
Tanzania, 310, 328, 330
targets, ix, 65, 68, 73, 88, 100, 139
taxonomy, 68, 174
TCC, 221, 222
technology, 41, 179, 290
teeth, 71
temperature, viii, xiii, 23, 25, 31, 36, 42, 43, 44, 45, 48, 58, 59, 79, 83, 93, 179, 212, 248, 249, 254, 257, 258, 260, 262, 281, 302, 303, 311, 313, 314, 315, 316, 317, 318, 319, 323, 325
temperature dependence, 316, 325
territory, 242, 263
test statistic, 287, 288
testosterone, 72, 77, 78, 88, 100, 101, 104
testosterone levels, 77
tetroxide, 82
Thailand, 50, 141, 330, 331, 343
theory, 151, 153, 154, 155, 161, 170, 172, 291
therapy, 167
thinking, 172
Thomson, 164, 173
threat(s), 66, 67, 178, 292
threshold(s), 40, 96, 157, 158, 161, 163, 169, 227
threshold level, 40, 96
tides, 327, 332, 336
timber, xii, 28, 45, 59, 62, 275, 278, 279, 283, 284, 285, 286, 301
time periods, x, 108, 117, 137, 138, 141, 142, 143
time series, 138
timing, 80, 90, 187, 204, 277
tin, 100
tissue, 73, 76, 78, 79, 88, 97, 104, 167
tobacco, 190
Tokyo, 203, 311
total product, 28
toxic effect, 222, 225
toxic substances, 252
toxicity, xi, 32, 49, 53, 97, 201, 209, 217, 218, 219, 220, 222, 224, 225, 227, 230, 232, 233, 234, 236, 239
toxicology, 239
trace elements, 5, 6
tradition, 168
traits, 341
trajectory, 154, 155, 157, 158, 159, 161, 165, 167
transactions, 149
transducer, 75

transformation(s), xiii, 38, 44, 45, 48, 49, 52, 63, 66, 237, 242, 260, 262, 299, 300, 310, 312, 333
transgenic, 211
transition(s), viii, 23, 324, 335
translation, 170
translocation, vii, 3, 4, 6
transmission, 315
transpiration, vii, 3, 4, 6, 47
transport, xiii, 32, 67, 73, 216, 219, 326, 327, 332, 335, 336, 337, 339
trees, 4, 6, 7, 25, 47, 53, 62, 230, 246, 278, 301, 312
trend, 48, 74, 94, 121, 143, 214, 216, 223, 224, 232, 277, 306
triphenyltin, ix, 65, 69, 78, 81, 94, 100, 104
tropical forest(s), xiii, 57, 59, 62, 299, 300, 307, 309, 310, 311, 312
tumor, 167
tundra, 262, 264
turbulence, 317
turbulent, 317
turnover, 30, 32, 37, 38, 45, 47, 51, 52, 53, 58, 308, 311, 331, 340

U

UNCED, 178, 206
uncertainty, 16, 138, 154, 174, 280, 288, 289, 295
UNFCCC, 178, 206
uniform, 16, 164, 174, 212, 249, 302
United Kingdom (UK), 52, 61, 63, 98, 201, 294, 336
United Nations (UN), 178, 206, 291
United States, 51, 179, 190, 230, 290, 295, 296, 332
universal gas constant, 316
urea, 214
urease, 48
Uruguay, 332
USDA, 239, 275, 276, 277, 278, 280, 282, 283, 290, 291, 292, 293, 296, 310
users, 289
USSR, 271, 272, 274
uterus, 74
UV, 102

V

vaccine, 174
vagina, 74
validity, 10, 138
values, xiii, 28, 36, 48, 71, 94, 95, 110, 118, 124, 125, 133, 134, 138, 139, 141, 155, 156, 157, 159, 160, 161, 162, 169, 187, 188, 189, 211, 222, 223, 225, 283, 286, 287, 288, 299, 304, 305, 306, 308, 312, 317, 323, 328
vapor, 324
variability, x, 41, 83, 149, 153, 154, 156, 264, 282, 309, 311, 325
variable(s), 10, 20, 43, 46, 116, 156, 166, 167, 182, 216, 222, 230, 278, 281, 282, 286, 317, 321, 323, 324, 332, 336
variance, 15, 16, 17, 82, 166, 303
variation, 16, 19, 42, 43, 57, 59, 146, 149, 154, 155, 156, 159, 161, 170, 171, 173, 174, 214, 221, 278, 307, 308, 325, 327
vascular system, 68
vegetation, viii, xi, xii, 13, 23, 27, 33, 35, 36, 43, 51, 55, 56, 60, 200, 210, 216, 217, 220, 221, 223, 225, 229, 232, 233, 243, 244, 245, 249, 250, 253, 257, 258, 260, 275, 276, 277, 279, 282, 284, 293, 300, 308, 309, 311, 324, 325, 327, 336, 338
vein, 334
velocity, 220, 317
Venezuela, 328, 342
vertebrates, viii, ix, 65, 67, 68, 72, 73, 74, 75, 77, 92
village, 246, 250, 251, 255
Virginia, 50, 191, 296, 340
viscera, 71
vitamins, 242
vouchers, 228
vulnerability, 230, 283

W

Washington, 61, 239, 290, 291, 292, 293, 294, 295, 296, 297, 310
water evaporation, 182
water permeability, 257
water quality, 178, 180, 217, 219, 224
water resources, 46
water vapor, 324
watershed, 24, 56, 62, 201, 214, 216
web, 25, 41, 69, 115, 117, 139, 143, 146, 149, 151, 172, 174, 195, 196
West Africa, 338
Western Europe, 178
Western Siberia, 272
wetlands, xiii, 33, 216, 230, 238, 326, 327, 339, 341
wheat, 54, 178, 180, 190, 191, 192, 194, 197, 200, 205, 210, 211, 214, 215
wheat germ, 194
wilderness, 297
wildfire, xii, 275, 276, 277, 278, 280, 282, 283, 284, 285, 286, 287, 288, 289, 292, 293, 294, 295, 297

wildland, xii, 275, 276, 277, 278, 279, 280, 281, 282, 283, 284, 285, 286, 287, 290, 291, 292, 294, 295, 296
wildlife, viii, 20, 65, 66, 99, 180, 230
wind, 37, 180, 181, 206, 257, 314, 317
windows, 218, 284
winter, 43, 44, 111, 115, 178, 180, 182, 184, 186, 190, 191, 194, 198, 200, 201, 202, 203, 205, 214, 215, 249, 254, 256, 260, 262, 302, 305, 306
Wisconsin, 189, 203, 296
women, 78
wood, xi, 39, 47, 241, 242, 243, 250, 253, 255, 256, 262, 284
workers, 42, 281
worms, 71, 197
wound healing, 79

X

xenobiotics, viii, 38, 65, 66, 70, 103
X-ray, 49

Y

yield, 7, 38, 139, 186, 189, 194, 202, 204, 205, 206, 236, 259
yolk, 74

Z

zinc, 77, 104
zooplankton, 112, 113, 126, 127, 129, 130, 131, 133, 134, 139, 143, 144